SHAKING THE TREE

Shaking the Tree

Readings from *Nature* in the History of Life

EDITED BY
Henry Gee

The University of Chicago Press * Chicago and London

HENRY GEE is a senior editor at *Nature* magazine.

© 2000 by Nature/Macmillan Magazines, Ltd.
Introductions and afterword © 2000 by Henry Gee
All rights reserved. Published 2000
Printed in the United States of America
09 08 07 06 05 04 03 02 01 00   1 2 3 4 5

ISBN: 0-226-28496-4 (cloth)
ISBN: 0-226-28497-2 (paper)

Library of Congress Cataloging-in-Publication Data

Shaking the tree : readings from *Nature* in the history of life / edited by Henry Gee.
p. cm.
Reprints of articles originally published from 1991 to 1997.
ISBN 0-226-28496-4 (cloth : alk. paper)—ISBN 0-226-28497-2 (pbk : alk. paper)
1. Evolution (Biology). 2. Paleontology. I. Gee, Henry.
QH366.2 .S492 2000
576.8—dc21   99-049796

♾ The paper used in this publication meets the minimum requirements of the American National Standard for Information Sciences—Permanence of Paper for Printed Library Materials, ANSI Z39.48-1992.

CONTENTS

Henry Gee

INTRODUCTION: SHAKEN AND STIRRED

Perhaps nothing is as exciting as coming across a fossil, a relic of past life, so strange when held against the light of what is currently known, that it challenges our understanding of the history of life and our place in it.

This thrill of discovery governs how we at *Nature* have always chosen manuscripts to publish in paleontology and evolutionary biology, as we have done since *Nature*'s very first issue, in November 1869, in which Thomas Henry Huxley wrote about Triassic dinosaurs.

The past few years have seen paleontology grow in vigor and stature. The two primary reasons are the revitalizing contribution of molecular biology to our understanding of ontogeny as well as phylogeny, and the development of cladistics, an objective method of phylogenetic reconstruction that has given comparative biology a new and muscular rigor.

Molecular phylogenies now join those based on anatomy, and there is much argument concerning their relative merit. The two offer different but complementary insights into the history of life.

When used to pinpoint the time when two lineages diverged, anatomical data from stratigraphically dated fossils can, at best, set a minimum estimate of lineage divergence. The earliest known example of a given lineage (let's call it lineage A) can be recognized only when it is *already* a member of lineage A—that is, it has already acquired the characters we use to recognize members of that lineage. It follows that there will be part of a lineage, from that earliest-known member back to the branch point between lineages A and B, whose members we would not be able to recognize for what they were, if we happened to exhume them.

Here's an example: the first members of the lineage that today comprises the elephants did not look like elephants. The first animals we *recognize* as being on the elephant lineage are already rather elephant-like: that is, they have characters that we recognize by comparison with elephants. This leads me inescapably to a playground joke that, to my mind, has deeper implications than are first apparent. Question: why is an elephant large, gray and wrinkled? Because if it were small, white and round, it would be an aspirin.

Molecular evolution has the potential to address this problem, by identifying a suite of character states present at the point of lineage divergence, not some unspecified time afterward. With the amino-acid sequences of proteins, or the nucleotide sequences of genes, one can compare the sequences of living (that is, extant) creatures, and from the pattern of similarities and differences build up a branching pattern—a pattern of diverging lineages—that may reflect the phylogeny of those genes or proteins and, by implication, the creatures that house them.

It has been claimed that molecular sequences change in a regular, clocklike way which could, in principle, reveal the precise point at which lineages diverge. In reality, molecules of different proteins and genes change at different rates; these rates can differ between taxa; and they may change over time. However, the sampling of genes and proteins now extends far beyond the fruit fly *Drosophila* and a few laboratory standards. With complete genome data for various organisms on the horizon, molecular-clock estimates can only become more precise and credible.

Fossil data, then, give one a minimum divergence time between lineages, whereas molecular data give the actual time, within statistical limits. The two are quite different, but their contrast reveals the important difference between lineage splitting (cladogenesis) and the acquisition, in each lineage, of characters we would recognize as lineage-specific (broadly, anagenesis.) Recognizing this difference may resolve persistent quarrels about whether, for example, modern mammal orders diverged after the Cretaceous-Tertiary boundary (as suggested by the fossils) or earlier, in the Cretaceous (as indicated by the molecules). The answer is that the lineages may have diverged a very long time ago, but that the members of each lineage either spent some time looking very similar to animals in other lineages, or had yet to acquire features that present-day zoologists recognize as "distinctively" ungulate, or rodent, or whatever. The danger is that we take the present-day suite of animal forms and use it to read the past retrospectively. If all we seek from the deep past are elephants—because elephants are what we see today—then elephants are all that we shall find. But if the extinct relatives of elephants looked like aspirins, their elephantine essence will not be apparent to us.

More persistent quarrels concern the insight that molecules give into the pattern of phylogenies themselves, especially when these phylogenies have different branching patterns from solutions reached using anatomical data. Which is right? Which is *better*? Do molecular data, being closer to the genotype, subvert the problems of convergence that beset anatomical data? Or is molecular information hobbled by the difficulty of seeing what a string of nucleotides really "means"? Of course, both sorts of data have their problems, but these questions are, to a large extent, distractions. The two sorts of data are complementary, and have the potential to shed light on one another: this trend, this reciprocal illumination, will increase, and the controversies subside, as molecular data continue to accumulate. It should now be more appropriate to ask questions about the possible impact of whole genomes on comparative biology. What problems of data handling and analysis will whole genomes pose—can we use the same methods of comparison, or will we need to shift computational power up a gear? Will the large-scale comparisons of nucleotide sequences and genome structure offered by whole-genome systematics generate a qualitative leap in understanding, or muddy the waters further?

Another trend opened up by our improving technical facility with cells and molecules is the increasingly refined discipline of molecular embryology. This offers new strategies to help solve many of evolutionary biology's most intractable questions, such as the origin of major groups and new body plans. The new discipline of "evolutionary developmental biology," or just "evo-devo" to its devotees, is a 1990s phenomenon.

But perhaps it is really a reinvention: nowadays we'd think of the philosophical musings of "transcendental morphologists" such as Etienne Geoffroy Saint-Hilaire, back in the 1820s and 1830s, as "evo-devo." Long before Darwin, he wondered whether it was possible (if only in some kind of aesthetic sense) to transmute creatures, one into another—creatures otherwise sundered by vast morphological gulfs. By folding a duck in half, would its internal organs not bear the same relations as in a squid? Or, if one wished, could not the same transformation be effected by turning a duck upside-down, so that its major nerve cord became ventral, its major blood vessel dorsal, just like in a lobster?

Since Darwin, people have wondered whether such transformations are possible, or rather, if structures in a vertebrate have some deep homology with structures in the bodies of lobsters, even though their positions have been turned topsy-turvy. By making such comparisons, can we draw the tree of life down as far as these fundamental trunks? Can we defy Georges Cuvier, Geoffroy Saint-Hilaire's opponent, who saw the different *embranchements* of life as fundamentally separate?

We might try, but without the tools to investigate the task, the effort remains something of a parlor game. At the turn of the century, two of the foremost scholars on the origin of vertebrates, schooled in such philosophy, turned their backs on comparative morphology, to plunge instead into the emerging quantitative, experimental biology. There at least, something could be done: concrete results could be achieved. These scholars were William Bateson, who rehabilitated the Mendelian model of the gene and brought it into the Darwinian fold, and Thomas Hunt Morgan, a pioneer worker in the genetics of the fruit fly, *Drosophila.* The fruits of their labors, and those of their colleagues—notably Ernst Mayr, George Gaylord Simpson and, perhaps most of all, Theodosius Dobzhansky—was the "modern synthesis" of genetics and evolutionary biology.

But perhaps their greater legacy was the exploration of the genetic control of embryology in *Drosophila,* taken by their intellectual descendants to an ever wider circle of creatures. Through their flight from comparative biology, Bateson and Morgan have given us the tools to solve the questions that interested Geoffroy Saint-Hilaire, but which they themselves found impossibly daunting. The astonishing demonstrations, repeated time and time again, of close similarities in the genetic organization in, say, *Drosophila* and mice make one wonder not how different these animals look, but how they can possibly look so different given their underlying similarities. The answer seems to be in the miracle of development, whose pinnacles are even now being scaled. Again, we can only guess what insights access to entire genomes will give to understanding developmental processes.

Nevertheless, paleontologists and other comparative biologists have been in the forefront, presenting molecular biologists with a vast panoply of the possible. Within the laboratory, there is but one insect, *Drosophila:* outside, there are millions, each with its own developmental story to tell, its own reasons for being different. And yet each creature is, tantalizingly, a dark continent under the skin. Paleontologists add another dimension still, showing that even the richness of life today pales before the strange creatures that once existed, creatures that may show physical evidence of the kinds of unions that populated Geoffroy Saint-Hilaire's dreams. This is not to suggest that lobsters really did flip over to become vertebrates, but that fossils display combinations of features not seen in the inevitably depauperate extant fauna—combinations which, when handled by cladistics, influence our understanding of the shape of the tree of life.

Let us turn, then, to cladistics, the other reason for the renewed vigor of paleontology. Historically, the modern synthesis of evolutionary biology refers to the reconciliation between evolutionary biology and genetics—disciplines that had followed separate courses since 1900, when Bateson rehabilitated Mendel's work.[1] The result was, essentially, a process-based view of evolution in which organisms were defined not by what they were, but by what they did. This is exemplified by Ernst Mayr's definition of a species as a collection of mutually interfertile individuals that would not interbreed with other such groups—a definition based on habit, not structure. Simpson extended this definition into geological time, defining species as successions of reproductively isolated groups that could be placed in a lineage, defined by genetic continuity. Changes of form in such lineages would be driven by adaptation.

In its process-based view of the world, the modern synthesis—*pace* Simpson—left paleontology out in the cold. In that it represents the remains of a living organism, a fossil may represent, in a general way, the result of adaptive processes and reproductive isolation: beyond that, one can say little about the particular adaptive forces that molded it, and nothing at all about its ancestry. No fossil is ever buried with a dog tag, not even the fossil of a dog. Fossils are mute: the stories they might tell are made up by us, after the fact: stories that rely for their validity on authority rather than experiment.

Cladistics changes that by imposing a radically different outlook on life. Traditional evolutionary paleontology, following Simpson, defined species in terms of their places in lineages. Cladistics, in contrast, defines species in terms of their overall relationships, rather than their ancestry. For even if it is impossible to prove that one fossil species is the ancestor of another, it must be true (if life on Earth has a single origin) that any species, living or extinct, is the "cousin" in some degree of every other. The relative branching orders implied by such relationships—degrees of cousinhood—can be inferred from morphology and presented as "cladograms," branching treelike diagrams, each of which represents a hypothesis of relationship. Because cladograms are testable, they need not rely for their currency on authority. The adoption of cladistics as a rigorous and transparent method of phylogenetic reconstruction has transformed paleontology, and systematic biology as a whole, into a truly scientific endeavor.

Traditional, scenario-based methods of phylogenetic reconstruction are tied to their substrates, are limited by what the investigator imagines to be plausible, and are untestable. For example, a scenario in which birds evolved from animals that leaped ever farther out of trees, inventing flight on the way, is limited only to birds. It says nothing, for example, about how fishes emerged onto land to become amphibians.

Second, such a scenario assumes that birds and flight must have evolved in concert. This makes sense in terms of an adaptive scenario, especially when compared with the large number of extant animals that fly, glide and parachute out of trees: but need this have been true for all time? If some other researcher were to invent a scenario in which birds evolved from animals that ran very fast along the ground, flapping their arms, who is then to judge which is more plausible?

In any case, how can such a scenario ever be shown to be true from the fossil record? Can one take a sequence of fossils of successive ages, spot some transformation, and assert that evolution happened in such-and-such a way, for this or that adaptive reason? Of course, one cannot—or, rather, one should not, because such a scenario makes untestable assumptions that cast doubt on the scientific worth of the scenario as a whole.

Cladistics, in contrast, seeks to reconstruct phylogeny by looking at the evidence as transparently as possible, as a child might: simply reconstructing the Darwinian branching pattern based on shared similarities acquired in evolution. It does this without the clutter of adaptive scenarios, that is, wondering what these similarities are *for,* of what adaptive advantage they might confer on their possessor.

The resulting cladogram is not a real family tree, but an ideal, an expression of relatedness that resolves into nested sets of dichotomously branching groups or "clades." Cladistics can be used to reconstruct the phylogeny of anything whose genealogy can be reconstructed as a dichotomously branching tree, irrespective of the process used to generate it. The motor doesn't have to be natural selection, and the results do not have to be alive: historians reconstruct the history of ancient manuscripts, and linguists trace the origins and diversification of languages, using methods closely similar to cladistics.

Apart from the implicit assumptions of Darwinian-style descent with modification and the relatedness of all creatures, the one assumption in cladistics is that the cladogram one constructs should express the smallest number of evolutionary transformations (acquisitions of characters, or character reversals). This is called the principle of parsimony, and is essentially the same thing as the principle of Occam's razor—that is, in the absence of any special circumstances, the hypothesis that one should choose should be the simplest.

Of course, nobody pretends that evolution is parsimonious: nevertheless, it is a solution, admittedly imperfect, to the problem of convergence, in which two unrelated creatures come to look like each other by chance, or because they are responding to similar adaptive pressures.

Critics of cladistics claim that it is useless in the face of convergence. This criticism is misplaced, because the principle of parsimony aims to address exactly that problem. Convergence afflicts all methods of phylogenetic reconstruction, the scenario-based approach included. The advantage of cladistics is that it reflects the scientific approach of the modern evolutionary biologist, who seeks to create testable hypotheses rather than untestable assertions.

Cladistic analyses conclude, not infrequently, with not one maximally parsimonious tree, but with a range—several, a dozen, or hundreds—of possible tree topologies of equal parsimony. Critics suggest that this indecision reflects a failing of cladistics. Far from it: such situations often yield valuable information about the quality of the data, the strengths of the character states used to recreate the phylogeny, and the occurrence of convergence.

Other critics accuse cladistics of being anti-Darwinian, even antievolutionary. This criticism is likewise misplaced: cladistics depends utterly on the production, by evolution, of a dichotomously branching tree of life of the kind envisaged by Darwin. Having said that, modern evolutionary biologists recognize that in order to test ideas about how a process such as natural selection shapes a phylogeny, one should have some way of producing that phylogeny beforehand that is free from any prior assumptions. Otherwise the process is circular and self-defeating.

Many would disagree with this thesis, including some of the authors of the papers collected in this volume. But if cladistics is about one thing, it is about choice. You, the reader, must read what follows and decide for yourself. There are no answers at the back of the book. And most of all, you should never, *ever,* take anyone's word for it.

We at *Nature* are often accused of being dedicated followers of fashion. We embrace that accusation: we cherish it and uphold it. Being the devotees that we are of anything that is new and exciting, we have been following the renaissance in comparative biology with enthusiasm. Over the past few years we have published a number of review articles in which leaders of the field discuss and evaluate the latest developments, in their various aspects. What should be immediately evident is the vitality and excitement of a fast-moving and intellectually challenging field.

Nineteen such reviews are collected here. They vary greatly in style and focus, which is to be expected: unlike longer articles in review journals, or book chapters, reviews in *Nature* are brief, shoot-from-the-hip accounts, which will—hopefully—be much more digestible than conventional reviews. We encourage authors of reviews to be opinionated, speculative and timely rather than comprehensive, encyclopedic and archival. The last three qualities, admirable though they might be, are not for us. When commissioning review articles for *Nature,* we advise prospective authors that what we expect from them, above all, is something that will change the way that readers look at the world—as well as being a rattling good read. This aim is timeless, and transcends the particulars of when the articles were written, and even who wrote them.

Given these constraints, writing a review for *Nature* is a tough assignment. Not everyone is equipped for the task, and most review commissions never come to fruition. To plan a comprehensive collection of works, in book form, that would meet the standards we set for review articles would be a work of monumental folly, like the Key to All Mythologies that drove the scholar Casaubon to his death in *Middlemarch.* This book, then, was never planned. Each review grew out of its own substrate: each has its own story to tell. The genesis of many of these reviews lies in chance meetings, snatches of conversation and suggestions for kite-flying. I shall reveal some of these circumstances in the introductions to each section of the book.

There are a few gaps. There is little to be found in these pages, for example, about dinosaurs. A review on dinosaurs was planned, but never made it. Another review, on anatomically modern humans, ran into the shoals of implacably hostile peer review: an occupational hazard, sadly, of paleoanthropology. It would have been wonderful to have a survey on arthropod evolution. That, too, never got off the ground. A review on the contribution of vertebrate ichnology—the study of tracks and traces—came too late to be included in this collection. Reviews on many other topics of interest remain, at the time of writing, up in the air, promises unfulfilled. There is clearly much more to say.

The book you see is, therefore, a makeshift, a communiqué, a progress report on unfinished business: just like science, and, indeed, like evolution. Given such piebald circumstances, perhaps the best way is to select reviews that shake the whole tree of life, above and beyond the delineation of one or two of its branches. In that way, a review about a part of the tree says something of value about the entire tree.

I have grouped this collection thematically, into five loosely ordered parts. The first part, "Shaping the Tree," looks at general problems presented by the pattern of the tree of life, but from three widely different perspectives. While Stephen Jay Gould and Niles Eldredge look at punctuated equilibrium, their own "rumbustious child," as it reached its majority, Eörs Szathmáry and John Maynard Smith wonder how life could have increased in complexity through a number of seemingly unlikely and maladaptive transitions.

These reviews are polar opposites: one about pattern, the other about process, one from authors who set much store by adaptation, another from authors who question its universal authority in shaping the tree of life.

Neither group of authors is associated with cladistics, or the principle of parsimony: orthogonal to these two, then, is a discussion by Caro-Beth Stewart of the power and pitfalls of parsimony, particularly as applied to molecular data. What are you, the reader, to make of these three contrasting views? That is very much up to you. Science is not about received wisdom, of authority, but about participation, which means thinking things through for oneself and making up one's own mind.

Part 2 looks at some of the remarkable advances in molecular embryology, in particular as regards the light it sheds on the problems that so perplexed Geoffroy Saint-Hilaire and his colleagues. Sean Carroll looks at the common heritage of multicellular animals in the specification of their longitudinal body axes, while E. M. De Robertis and Yoshiki Sasai do the same thing for the dorsoventral axis.

Paleontology is playing an increasingly interactive part in these discussions in certain areas accessible to the fossil record. One of these has been the evolution of the tetrapod limb from the fin of a fishy ancestor. Fossils have also shed light on the diversity of arthropod-limb design, and through that diversity has come an increasing understanding of the common ground plan of the arthropod limb. Neil Shubin, Cliff Tabin and Sean Carroll look at how genetics helps us understand these important evolutionary transitions. In so doing, they raise difficult questions about what we mean by terms such as "homology."

The diversity and strangeness of fossils, especially from earlier episodes in metazoan evolution, may also cast a new and different light on the tree of life. When doing what Geoffroy Saint-Hilaire did, that is, to imagine leaps between creatures with very different body plans, embryology and development come to the fore. Yet there is still space for leaping in the dark. A typically bold piece from Simon Conway Morris places a whole host of problematic fossils in the context of metazoan phylogeny. The interpolation of a gaggle of unfamiliar names into the tree of life overwhelms the things with which we are familiar. For every clam, centipede or chordate there is a host of chancelloriids, conulariids and carinachitiids, emphasizing the weakness of the commonly held assumption that life in the past need have been anything like it is in the present day.

Part 3 takes a look at the forest in which the tree of life germinated, the metaphorical soil in which it is rooted and the conditions that affected it as it grew. The three reviews in this section each look at a major theme in what one might call life's "context." Andrew Knoll and Malcolm Walter look at the Proterozoic, the latest part of the Precambrian just before the apparent "explosion" in metazoan life that characterizes the Cambrian fossil record. What, if anything, in the Proterozoic environment triggered this sudden flowering? William Shear looks at another transition, from water to land. How did terrestrial ecosystems become established? Finally, Douglas Erwin looks at that great leveller—mass extinction. The disappearance of the majority of living forms at the end of the Permian period, 251 million years ago, represents the largest known such event. But can we identify causes reliably?

Part 4—"Shaking the Tree"—is the core of the book. Each of the six reviews in this section looks at the radiation of a major group of multicellular organisms from a cladistic perspective. In contrast to comfortable but untestable stories based on adaptation, these treatments make no assumptions about how the creatures concerned lived: yet the perspectives they offer are often surprising, raising our understanding of the history of life, and questioning, at every turn, our assumptions about what we humans imagine is plausible or even possible.

The topics covered include early land plants, flowering plants, agnathans and jawed vertebrates, tetrapods, birds and mammals. The range is not comprehensive—it is not intended to be, nor could it ever be—but it should establish the point, which is that we now have at our disposal, in cladistics, a way of reconstructing phylogenies that is testable and transparent, and that allows the fossils and molecules to speak for themselves, clearly, without our presuming to speak on their behalf. Because cladistics makes no assumptions about what we, as humans, think of as possible or plausible when applied to a scenario, cladistic solutions occasionally come up with startling, even uncomfortable conclusions. Who would have thought, for example, that the flightless, dinosaurlike *Mononykus* is more closely related to modern birds than is *Archaeopteryx,* as described by Luis Chiappe in his article on bird evolution?

Part 5 is all about the primates, the mammalian order that includes ourselves. Taken together, the three reviews should incite a feeling of unease in any human reader. The evidence concerning human origins has always been meager and ambiguous: here we see that it fails us at every turn. In his review, Robert Martin shows that the primate fossil record is so sparse that one can probably find some justification for any evolutionary trajectory one cares to drive through the evidence. (This fact alone seems ample justification for the use of cladistics.) Peter Andrews looks at the apelike creatures of the Miocene epoch, between about 20 and 10 million years ago. Many of these creatures disappeared toward the end of the Miocene: but which, if any, were the antecedents of humanity? The final punch is delivered by Bernard Wood, who looks at the persistent problems of understanding the definition of our own genus, *Homo.* Was *Homo habilis,* "handy man," a real species, or a heterogeneous collection of fragments from this and that? And how much of this collection deserves inclusion in the genus *Homo*?

This final, brutal confrontation with the fragmentary nature of the real data exposes many (if not most) speculations about the evolution of behavior, language and so on, as so much hot air. It may seem unpalatable to some, but no such speculations are possible without a phylogeny, objectively built; and creating such a phylogeny is impossible if we unsure whether the fossils we are dealing with belong to one species, or many. This conclusion may seem depressing, but science does not exist to provide comfort or solace. If anything, science exists to expose the reality behind our preconceptions. As such, it is bound to be uncomfortable. The task of the scientist is to stand before the evidence and not flinch.

All the reviews included here are accompanied by their own extensive and detailed references. Nevertheless, each of the five sections has a short introduction that offers some ideas for further reading, published since the reviews appeared. These additional bibliographies are not meant to be comprehensive. If they were, this book would be too great a burden on the shelf and on the pocketbook, and would suffocate under the weight of its own citations. Such an encyclopedic treatment would also defeat the object of the exercise, which is to present a snapshot portrait of fast-moving disciplines—disciplines that do not hang around long enough to sit for an oil painting. The title of this book, after all, is *Shaking the Tree*—what we are meant to appreciate is the tree as a whole, rather than the beauty of any particular leaf or twig, enticing though these might be. Some of the additional citations represent my own personal selection. Many, however, have been suggested by the authors of the following articles, who draw on many of them as teaching aids. As such, I hope that the reviews offered herein, with the additional references, will serve as a general grounding for further studies on particular issues in evolutionary biology.

To conclude, the whole package is offered as a provocation, not as a panacea. In it you will find questions and ideas that will make you think about the history of life, and how we have striven—and continue to strive—to understand it. The intention is to provide added spice to nourishing-but-bland textbook fare, for the student who likes to look a little farther, and aim a little higher. As always with cutting-edge science, you will find more questions than answers. That's where *you* come in.

Reference

1. D. L. Hull. *Science as a process: An evolutionary account of the social and conceptual development of science.* Chicago: University of Chicago Press, 1988.

PART 1

SHAPING THE TREE: HOW TO CREATE A TREE OF LIFE

Stephen Jay Gould and Niles Eldredge first mooted their idea of "punctuated equilibrium" in 1972. We thought it would be fun for the same authors to revisit their controversial concept in 1993, a full twenty-one years later. This they did, and although the article was (forgivably) self-indulgent, *Nature*'s postbag subsequent to publication showed that the idea still had potential to get pulses racing, especially among those who had no qualms about extending Darwinian natural selection, as a matter of course, over geological time.

Why is this idea controversial? Gould and Eldredge simply advocate looking at the incomplete stratigraphic record at face value, and admitting that the pattern it reveals says something about evolution. The pattern is striking. Species appear rather suddenly, and spend a long time looking just the same. Gould and Eldredge came up with a model to explain this, of allopatric speciation in small, peripheral groups (unlikely to get fossilized), and of a prevalent "stasis." What enraged people, I think, was the idea of natural selection as primarily a force for conservatism, not change: natural selection enforces the norm in times of stasis. In speciation, relying on small groups, stochastic processes such as genetic drift might assume greater importance.

In parallel with this, Gould has promoted a view that embraces the idea of contingency as a force in evolutionary change,[1] and that refuses to take at face value any explanation for patterns in the fossil record that relies on adaptationist scenarios. This is because one can, arguably, come up with an adaptive value for anything, if one tries hard enough—and who will then say it's wrong? Gould's healthy skepticism has got him into trouble with self-imposed keepers of the Darwinian flame. Criticisms of Gould as somehow anti-Darwinian are unjustified (see ref. 2 for a philosophical critique of Gould's ideas). Gould is as Darwinian as anybody, yet there is still a current of thought that sees phyletic gradualism and punctuated equilibrium as opposites that must somehow be reconciled.[3]

Gradual or jagged, even the smoothest Darwinian would admit that several distinct leaps in complexity have occurred in evolution, such as the transition from prokaryotic to eukaryotic organization, the evolution of sex, the evolution of multicellularity, and the origin of language. Unless one is to adopt the view that these things just "happened," one is forced to seek adaptive explanations for transitions that imply qualitative changes in complexity, which at first sight seem maladaptive.

Thinking about such transitions, one gets a picture of a kind of evolution that, if not explicitly "progressive," tends to an increasing facility with the storage and transmission of ever larger packets of information. Eörs Szathmáry and John Maynard Smith explore these ideas in the second item in this section, ideas that can be explored further in a deliberately eclectic range of further reading.[4–14]

Users of cladistics will find much to criticize in the first two selections. Punctuated equilibrium comes under suspicion, as to derive evolutionary patterns from stratigraphy is, somehow, to make presumptions about specific lines of ancestry and descent, which may not be testable: as such, they are less scientific hypotheses than "just-so stories."[15]

At the same time, explanations for large-scale evolutionary change, such as those tackled by Maynard Smith and Szathmáry, make, perhaps (in the cladistic mind), too great an appeal to adaptation—or, rather, to what human minds perceive as being possible or plausible. How do we know that evolution followed the seemly courses suggested by Maynard Smith and Szathmáry, no matter how ingenious their solutions?

To take this idea to an extreme, one need only recall the discovery, by Misia Landau, that supposedly scientific accounts of human ancestry followed the conventions of narrative structure, not just in some general way, but in the forms of the "hero-myths" of folklore (see ref. 16 for a full account).

The review by Caro-Beth Stewart, the third in this section, is thus orthogonal to the two preceding it. It is about the power of parsimony—the central assumption of cladistics—and the problems one encounters in constructing phylogenetic trees in which parsimony is a guiding principle. Her review concentrates on molecular-sequence data, though the lessons she draws could be applied to anatomical data, or indeed any data generated by Darwinian descent with modification, in which convergence may confuse a method that seeks, as data, genuinely shared, derived features.

The controversy that cladistics has generated within the field of comparative biology[17] is matched only by the incomprehension outside. The reasons for this are simple: many cladistic papers are couched in a furious language that accretes unwieldy neologisms. Reference 18, for example, is a classic paper of the genre but of a stridency and a density that would deter all but the most patient. Yet Stewart's review here gives a full account for the novice that can be used as a guide to (and a defense against) the more rigorous treatments in the literature.

Parsimony is not the only criterion used by contemporary researchers in phylogenetic reconstruction. Other workers, particularly those who look at molecules rather than morphology, use a variety of criteria and techniques, such as "maximum likelihood," in addition to parsimony. Historically, however—and particularly where morphology is concerned—parsimony has attracted the most attention, and this should explain its emphasis in this book. That, and the fact that cladistics (which uses parsimony) is not so much a technique—a recipe for reconstructing phylogenies, as it were—but a distinctive way of looking at the history of life, which contrasts with the traditional, process-based approach.

References

1. S. J. Gould. *Wonderful life.* New York: Norton, 1991.
2. D. C. Dennett. *Darwin's dangerous idea.* New York: Simon and Schuster, 1995.
3. J. C. von Vaupel Klein. Punctuated equilibria and phyletic gradualism: Even partners can be good friends. *Acta Biotheoretica* 42 (1994): 15–48.
4. J. Maynard Smith and E. Szathmáry. *The major transitions in evolution.* Oxford: Freeman, 1995.
5. E. G. Leigh and T. E. Rowell. The evolution of mutualism and forms of harmony at various levels of biological organization. *Ecologie* 26 (1995): 131–158.
6. E. Szathmáry and J. Maynard Smith. From replicators to reproducers: The first major transitions leading to life. *Journal of Theoretical Biology* 187, 4 (1997): 555–571.
7. T. W. Deacon. *The symbolic species.* New York: Norton, 1997.
8. R. A. Raff. *The shape of life.* Chicago: University of Chicago Press, 1996.
9. D. Bickerton. *Language and human behaviour.* London: UCL Press, 1996.
10. C. De Duve. *Vital dust.* New York: Basic Books, 1997.
11. W. G. Runciman, J. Maynard Smith and R. I. M. Dunbar, eds. *Evolution of social behavioural patterns in primates and man.* Oxford: Oxford University Press, 1996.
12. J. Diamond. *The third chimpanzee.* London: Harperperennial, 1993.

13. S. Pinker. *The language instinct.* London: Penguin Books, 1995.
14. I. Stewart and J. Cohen. *Figments of reality.* Cambridge: Cambridge University Press, 1997.
15. P. L. Forey. Neontological analysis versus paleontological stories. In K. A. Joysey and A. E. Friday, eds., *Problems of phylogenetic reconstruction,* pp. 119–157. London: Academic Press, 1982.
16. R. Lewin. *Bones of contention.* London: Penguin, 1987.
17. D. L. Hull. *Science as a process: An evolutionary account of the social and conceptual development of science.* Chicago: University of Chicago Press, 1988.
18. D. E. Rosen, P. L. Forey, B. G. Gardiner and C. Patterson. Lungfishes, tetrapods, paleontology and plesiomorphy. *Bulletin of the American Museum of Natural History* 167 (1981): 163–275.

Stephen Jay Gould and Niles Eldredge

PUNCTUATED EQUILIBRIUM COMES OF AGE

The intense controversies that surrounded the youth of punctuated equilibrium have helped it mature to a useful extension of evolutionary theory. As a complement to phyletic gradualism, its most important implications remain the recognition of stasis as a meaningful and predominant pattern within the history of species, and in the recasting of macroevolution as the differential success of certain species (and their descendants) within clades.

Punctuated equilibrium has finally obtained an unambiguous and incontrovertible majority—that is, our theory is now 21 years old.[1] We also, with parental pride (and, therefore, potential bias), believe that primary controversy has ceded to general comprehension, and that punctuated equilibrium has been accepted by most of our colleagues (a more conventional form of majority) as a valuable addition to evolutionary theory.

Kellogg[1] began the best book written to celebrate Darwinism's fiftieth birthday by noting how often (and how continually) various critics had proclaimed the *Sterbelager* (death bed) of natural selection. Punctuated equilibrium[2] has also prospered from announcements of its death or triviality[3–5] and has been featured in much recent discussion.[6–8]

As a neonate in 1972, punctuated equilibrium entered the world in unusual guise. We claimed no new discovery, but only a novel interpretation for the oldest and most robust of paleontological observations: the geologically instantaneous origination and subsequent stability (often for millions of years) of paleontological "morphospecies." This observation had long been ascribed, by Darwin and others, to the notorious imperfection of the fossil record, and was therefore read in a negative light—as missing information about evolution (defined in standard paleontological textbooks of the time[9] as continuous anagenetic transformation of populations, or phyletic gradualism).

In a strictly logical sense, this negative explanation worked and preserved gradualism, then falsely equated with evolution itself, amidst an astonishing lack of evidence for this putative main signal of Darwinism. But think of the practical or heuristic dilemma for working paleontologists: if evolution meant gradualism, and imperfection precluded the observation of such steady change, then scientists could not access the very phenomenon that both motivated their interest and built life's history. As young, committed and ambitious parents, we therefore proposed punctuated equilibrium, hoping to validate our profession's primary data as signal rather than void. We realized that a standard biological account, Mayr's[10] peripatric theory of speciation in small populations peripherally isolated from a parental stock, would yield stasis and punctuation when properly scaled into the vastness of geological time—for small populations speciating away from a central mass in tens or hundreds of thousands of years, will translate in almost every geological circumstance as a punctuation on a bedding plane, not gradual change up a hill of sediment, whereas stasis should characterize the long and recoverable history of successful central populations.

Punctuated equilibrium then grew during its childhood and adolescence, in ways both unruly and orderly. Unruly accidents of history included the misunderstandings of colleagues (who, for example, failed to grasp the key claim about geological scaling, misread geological abruptness as true suddenness, and then interpreted punctuated equilibrium as a saltational theory), and the purposeful misuses of creationist foes as this political issue heated up in the United States during the late 1970s (although we took pride in joining with so many colleagues for a successful fight against this philistine scourge, as one of us testified in the Arkansas "monkey" trial in 1981[11] and the other wrote a book on creationist distortions[12]).

But orderly extensions, implicit in the undeveloped logic of our original argument, fuelled the useful growth of punctuated equilibrium to fruitful adulthood. (We now realize how poorly we initially grasped the implications of our original argument; we thank our colleagues, especially S. M. Stanley[13] and E. S. Vrba,[14] for developing several extensions.) We originally focused on tempo, but more important theoretical arguments flowed from implications concerning evolution's mode[15–17]—particularly the causes surrounding our two major claims for equilibrium, or stasis of established species, and the need to reformulate macroevolution, notably the key phenomenon of trends, as an accumulation of discrete speciation events treated as entities rather than undefinable segments of continua—a subject encompassed by debate about species selection[13] or species sorting.[18]

Punctuated Equilibrium and Macroevolution

Stasis and Its Meaning

We opened our original paper with a section on what philosopher N. R. Hanson called "the cloven hoofprint of theory,"[19] or the structuring of all supposedly objective observation by expectations of prevailing general views. Stasis, as palpable and observable in virtually all cases (whereas rapid punctuations are usually, but not always, elusive), becomes the major empirical ground for studying punctuated equilibrium. Putting together the philosophical insight of ineluctable theoretical bias, with the empirical theme of the tractability of stasis, we devised a motto: "stasis is data." For no bias can be more constricting than invisibility—and stasis, inevitably read as absence of evolution, had always been treated as a non-subject. How odd, though, to define the most common of all paleontological phenomena as beyond interest or notice! Yet paleontologists never wrote papers on the absence of change in lineages before punctuated equilibrium granted the subject some theoretical space. And, even worse, as paleontologists didn't discuss stasis, most evolutionary biologists assumed continual change as a norm, and didn't even know that stability dominates the fossil record. Mayr has written:[20] "Of all the claims made in the punctuationalist theory of Eldredge and Gould, the one that encountered the greatest opposition was that of 'pronounced stasis as the usual fate of most species,' after having completed the phase of origination . . . I agree with Gould that the frequency of stasis in fossil species revealed by the recent analysis was unexpected by most evolutionary biologists."

As the most important change in research practice provoked by punctuated equilibrium, stasis has now exited from its closet of non-definition to become a subject of quantitative investigation in all major fossil groups—from microfossils[21,22] (27,000 measured specimens from 400 closely spaced samples spanning 8 million years in the latter study), to molluscs,[23-27] to mammals.[28-30] Although punctuated equilibrium deals directly only with stability of species through time, the higher-level analogue of non-trending in larger clades has also graduated from an undefined non-subject to a phenomenon worth documenting.[31] Moreover, because species often maintain stability through such intense climatic change as glacial cycling,[32] stasis must be viewed as an active phenomenon, not a passive response to unaltered environments. Many leading evolutionary theorists, while not accepting our preference for viewing stasis in the context of habitat tracking[17] or developmental constraint,[33,34] have been persuaded by punctuated equilibrium that maintenance of stability within species must be considered as a major evolutionary problem.

Macroevolution as a Problem in Species Sorting

If punctuated equilibrium has provoked a shift in paradigms for macroevolutionary theory (see ref. 35 for a defense of this view), the main insight for revision holds that all substantial evolutionary change must be reconceived as higher-level sorting based on differential success of certain kinds of stable species, rather than as progressive transformation within lineages (see Eldredge[36] on taxic versus transformational views of evolution; Simpson,[37] however, in the canonical paleontological statement of the generation before punctuated equilibrium, had attributed 90% of macroevolution to the transformational mode, and only 10% to speciation). Figure 1, our original diagram of punctuated equilibrium, shows how a trend may be produced by differential success of certain species without directional change in any species following its origin.

Darwin's theory of natural selection locates the causality of evolutionary change at one domain on one level: natural selection operating by struggle among individual organisms for reproductive success. Given Darwin's crucial reliance upon Lyellian uniformity for extrapolating this mode of change to encompass all magnitudes through all times, the interposition of a level for sorting among stable species breaks this causal reduction and truly, in Stanley's felicitous term,[38] "decouples" macro- from microevolution. Decoupling is not a claim for the falseness or irrelevancy of microevolutionary mechanisms, especially natural selection, but a recognition that Darwinian extrapolation cannot fully explain large-scale change in the history of life.

The main point may be summarized as follows. Most macroevolution must be rendered by asking what kinds of species within a clade did better than others (speciated more frequently, survived longer), or what biases in direction of speciation prevailed among species within a clade. Such questions enjoin a very different program of research from the traditional "how did natural selection within a lineage build substantial adaptation during long stretches of time?" The new questions require a direct study of species and their differential success; older queries focused downward upon processes within populations and their extrapolation through time. Darwin's location of causality in organisms must be superseded by a hierarchical model of selection, with simultaneous and important action at genic, organismal and taxic levels.[39,40] Williams,[34] who so stoutly defended classical Darwinism against older, invalid, and very different forms of

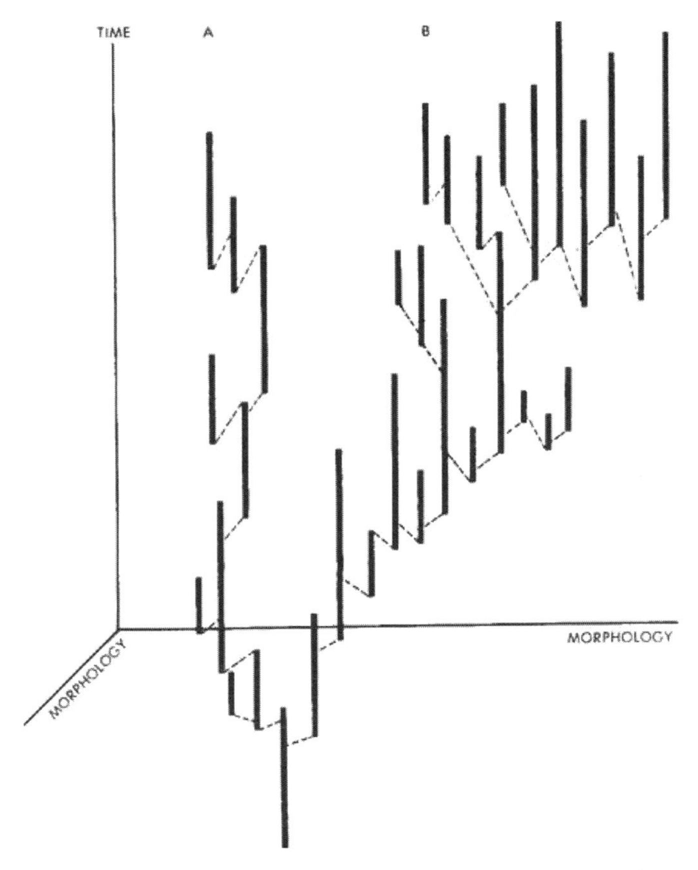

Fig. 1 Three-dimensional sketch contrasting a pattern of relative stability (*A*) with a trend (*B*), where speciation (*dashed lines*) is occurring in both major lineages. Morphological change is depicted here along the horizontal axes, while the vertical axis is time. Though a retrospective pattern of directional selection might be fitted as a straight line in *B*, the actual pattern is stasis within species, and differential success of species exhibiting morphological change in a particular direction. For further explanation, see ref. 1.

group selection,[41] now acknowledges the importance of such clade selection in macroevolution. Punctuated equilibrium has been used as a central concept in the development of hierarchy theory in evolutionary biology.

Implications

Any theory with a claim to novelty in broad perspective must enlighten old problems and suggest extensions. The speciational view of macroevolution, which does not strictly require punctuated equilibrium, but which was nurtured and has thrived in its context, requires a reformulation of nearly all macroevolutionary questions. For example, so-called living fossils, once treated as lineages rendered static by optimal adaptation, unusually stable environment, or lack of genetic variation, should be reconceptualized as members of groups with unusually low speciation rates, and therefore little opportunity to accumulate change.[42] (We have no evidence that the species of "living fossil" groups are particularly old. For example, the western Atlantic horseshoe crab, *Limulus polyphemus*—the "type example" of the phenomenon—has no fossil record at all, whereas the genus can only be traced to the Miocene.)

Going further, the entire tradition of expressing evolutionary change in darwin units (where 1 darwin equals character change by a factor of *e* in 1 million years)[43] makes no sense in a speciational context. (If a lineage goes from species A to D in 10 million years through three episodes of rapid change with intervening stasis, a cited rate of so many millidarwins becomes a meaningless average.) We learn as a received truth of evolution, for example, that human brain size increased at an extraordinary (many say unprecedented) rate during later stages of our lineage. But this entrenched belief may be a chimera born of an error in averaging rates over both punctuations and subsequent periods of stasis. *Homo sapiens* is a young species, perhaps no more than 200,000 years old. If most of our increment accrued quickly at our origin, but we then express this entirety from our origin to the present time as a darwin rate, we calculate a high value because our subsequent time of stasis has been so short. But if the same speciation event, with the same increment in the same time, had occurred 2 million years ago (with subsequent stasis), the darwin rate for the identical event would be much lower.

Cope's rule, the tendency for phyletic increase in body size, had generally been attributed to selective value of large size within anagenetic lineages, but is probably better interpreted[44,45] as greater propensity for speciation in smaller species, for whom increasing size is the only "open" pathway (see Martin[46] on the negative correlation of generic species richness and body size). Raup and Sepkoski[47] proposed a conventional explanation for decreasing rate of background extinction through geological time: generally better adaptation of later species. But Valentine[48] and Gilinsky (personal communication) offer an interesting speciational alternative: if extinction intensities were constant through time, groups with equally high speciation and extinction rates would fare just as well as groups with equally lower rates. But extinction intensities vary greatly, and the geological record features episodes of high dying, during which extinction-prone groups are more likely to disappear, leaving extinction-resistant groups as life's legacy. Valentine concludes that "these clade-characteristic rates are of course not adaptations *per se,* but effects flowing from clade properties that were established probably during the early radiations that founded the clades."

The most exciting direct extensions of punctuated equilibrium now involve the study of correlated punctuational events across taxa, and the ecological and environmental sources of such cohesion. Eminently testable are Vrba's[49] "turnover-pulse hypothesis" of evolution concentrated in punctuational bursts at times of worldwide climatic pulsing, one of which, about 2.5 million years ago, may have stimulated the origin of the genus *Homo;* and Brett's[50] hypothesis of "coordinated stasis" for the structuring of major paleontological faunas. What might be the ecological source of such striking coherence across disparate taxa through such long times?[51]

The Empirics of Punctuated Equilibrium

Like all major theories in the sciences of natural history, including natural selection itself, punctuated equilibrium is a claim about relative frequency, not exclusivity.[52] Phyletic gradualism has been well documented, again across all taxa from microfossils[53] to mammals.[54,55] Punctuated equilibrium surely exists in abundance, but validation of the general hypothesis requires a relative frequency sufficiently high to impart the predominant motif and signal to life's history. The issue remains unsettled, but we consider (in our biased way) that four classes of evidence establish a strong putative case for punctuated equilibrium in this general sense.

Individual Cases

Examples of stasis alone (cited earlier) and simple abrupt replacement, although conforming to expectations of punctuated equilibrium, are not direct evidence for our mechanism: for stasis might just be a lull in anagenetic gradualism (though pervasive stasis for long periods in all species of a fauna [a common finding] would require special pleading from gradualists), and replacement might represent rapid transformation without branching, or migration of a distant (phyletic or geographic) relative rather than evolution *in situ.* A good test of punctuated equilibrium requires (in addition to the obvious need for documented rapidity in an interval known to be sufficiently short) both a phyletic hypothesis to assert sister-group relationship of the taxa involved, and survival of putative ancestors to affirm an event of true branching rather than rapid phyletic transformation.

Given these stringent requirements, and in the light of such an imperfect fossil record, we are delighted that so many cases have been well documented, particularly in the crucial requirement of ancestral survival after punctuated branching.[32,56–61] Williamson's discovery[62] of multiple molluscan speciation events in isolated African lakes, with return of ancestral lineages upon reconnection with parental water bodies, has been most widely discussed[63] and disputed[64] (although all accept the punctuational pattern). Cheetham's elegant and meticulously documented[65,66] story of evolution in the bryozoan *Metrarabdotos* from Tertiary strata of the Caribbean is particularly gratifying (Fig. 2) in the number of purely punctuational events, the full coverage of the lineage, and the unusual completeness of documentation, especially as Cheetham began his study expecting

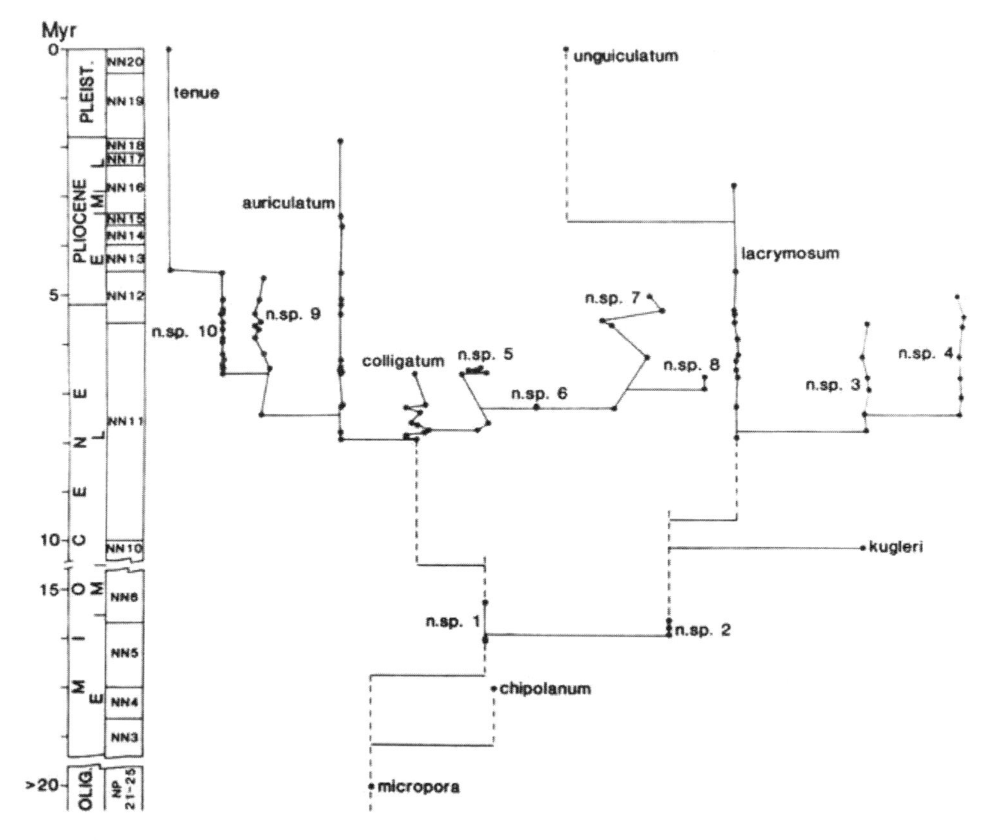

Fig. 2 Phylogenetic (stratophenetic) tree of Caribbean species of *Metrarabdotos.* Relationships were constructed by calculating euclidean distances between sampled populations using all canonical scores from discriminant analysis, and connecting nearest morphological neighbors in stratigraphic sequence. Morphological distances (horizontal axis) between inferred ancestor and descendant species are to scale; those between species of different pairs are not necessarily so (for example, *M. kugleri* is morphologically more distant from *M.* n. sp. 3 than from its ancestor, *M.* n. sp. 2, even though the diagram makes it appear otherwise). Within-species fluctuation in species-distinguishing morphology is based on pairwise discriminant scores scaled to the distances between species pairs. Reproduced with permission from ref. 65.

to reconfirm a gradualistic interpretation (writing to McKinney: "The chronocline I thought was represented . . . is perhaps the most conspicuous casualty of the restudy, which shows that the supposed cline members largely overlap each other in time. Eldredge and Gould were certainly right about the danger of stringing a series of chronologically isolated populations together with a gradualist's expectations").

On the subject of punctuational corrections for received gradualistic wisdom, Prothero and Shubin[67] have shown that the most "firmly" gradualistic part of the horse lineage (the general, and false, exemplar of gradualism in its totality), the Oligocene transition from *Mesohippus* to *Miohippus,*[68] conforms to punctuated equilibrium, with stasis in all species of both lines, transition by rapid branching rather than phyletic transformation, and stratigraphic overlap of both genera (one set of beds in Wyoming has yielded three species of *Mesohippus* and two of *Miohippus,* all contemporaries). Prothero and Shubin concluded: "This is contrary to the widely-held myth about horse species as gradualistically-varying parts of a continuum, with no real distinctions between species. Throughout the history of horses, the species are well-marked and static over millions of years. At high resolution, the gradualistic picture of horse evolution becomes a complex bush of overlapping, closely related species."

Relative Frequencies

Elegant cases don't make punctuated equilibrium any more than a swallow makes a summer, but there are a growing number of reports documenting an overwhelming relative frequency (often an exclusivity) for punctuated equilibrium in entire groups or faunas. Consider the lifetime testimonies of taxonomic experts on microfossils,[69] on brachiopods[70,71] and on beetles.[72] Fortey[73] has concluded for trilobites and graptolites "that the gradualistic mode does occur especially in pelagic or planktic forms, but accounts for 10% or less of observations of phyletic change, and is relatively slow."

Other studies access all available lineages in entire faunas and assert the dominance of punctuated equilibrium. Stanley and Yang[26] found no gradualism at all in the classic Tertiary molluscan sequences of the Gulf and Atlantic Coasts. With the exception of *Gryphaea,* Hallam[23] detected no phyletic change in shape (but only for body size) in any Jurassic bivalve in Europe. Kelley[24,25] documented the prevalence of punctuation for molluscs in the famous Maryland Miocene sequence, and Vrba[74] has done the same for African bovids. Even compilations from the literature, so strongly biased by previous traditions for ignoring stasis as non-data and only documenting putative gradualism, grant a majority to punctuated equilibrium, as in Barnovsky's[75] compendium for Quaternary mammals, with punctuated equilibrium "supported twice as often as phyletic gradualism . . . the majority of species considered exhibit most of their morphological change near a speciation event, and most species seem to be discrete entities." When controlled studies are done by one team in the field, punctuated equilibrium almost always seems to predominate. Prothero[76] "examined all the mammals with a reasonably complete record from the Eocene-Oligocene beds of the Big Badlands of South Dakota and related areas in Wyoming and Nebraska . . . With one exception (gradual dwarfing in the oreodont *Miniochoerus*), we found that all of the Badlands mammals were static through millions of years, or speciated abruptly (if they changed at all)."

Inductive Patterns

Even more general inductive patterns should be explored as criteria. Stanley[38,77] has proposed a series of tests, all carried out to punctuated equilibrium's advantage. Others suggest that certain environments and ecologies should be conducive to one preferred mode along the continuum of possibilities. Johnson[78,79] suggests that punctuated equilibrium should dominate in the benthic environments that yield most of our fossil record, while gradualism might prevail in pelagic realms. Sheldon[80] proposes the counter-intuitive but not unreasonable idea that punctuated equilibrium may prevail in unstable environments, gradualism in stable regimes.

Tests from Living Organisms

Distinct evolutionary modes yield disparate patterns as results; punctuated equilibrium might therefore be tested by studying the morphological and taxonomic distributions of organisms, including living faunas. (Several of Stanley's tests[38] use modern organisms, and other criteria from fossils should be explored—especially the biometric discordance or orthogonality, favorable to punctuated equilibrium and actually found where investigated,[25,81] of within and between species trends.)

Cladistic patterns should provide a good proving ground. Avise[82] performed an interesting and much quoted test, favorable to gradualism, by comparing genetic and morphological differences in two fish clades of apparently equal age and markedly different speciation frequencies. But as Mayden[83] argued, this test was wrong in its particular case, and nonoptimal as a general procedure; a better method would compare cladistic sister groups, guaranteed by this status to be equal in age. Mindell *et al.*[84,85]

have now performed such a test on the reptilian genus *Sceloporus* and on allozymic data in general, and have validated punctuated equilibrium's key claim for positive correlation of evolutionary distance and speciation frequency. Lemen and Freeman's[86] interesting proposal for additional cladistic tests cannot be sustained because they must assume that unbranched arms of their cladograms truly feature no speciation events along their routes, whereas numerous transient and extinct species must populate most of these pathways. Wagner[87] has developed a way of estimating rapidly branching speciation versus gradual speciation or transformation from cladograms, and his initial results favor predominant rapid branching in Paleozoic gastropods.

Difficulties and Prospects

Many semantic and terminological muddles that once impeded resolution of this debate have been clarified. Opponents now accept that punctuated equilibrium was never meant as a saltational theory, and that stasis does not signify rock-hard immobility, but fluctuation of little or no accumulated consequence, and temporal spread within the range of geographic variability among contemporary populations—by Stanley's proper criterion, so strikingly validated in his classic study.[26] We trust that everyone now grasps the centrality of relative frequency as a key criterion (and will allow, we may hope, that enough evidence has now accumulated to make a case, if not fully prove the point).

Evolutionary biologists have also raised a number of theoretical issues from their domain of microevolution. Some, like the frequency of sibling speciation, seem to us either irrelevant or untroublesome as a bias against, rather than for, our view (as we then underestimate the amount of true speciation from paleontologically defined morphospecies, and such an underestimate works against punctuated equilibrium). Others, like the potential lack of correspondence between biospecies and paleontological morphospecies, might be worrisome, but available studies, done to assess the problem in the light of punctuated equilibrium, affirm the identity of paleontological taxa with true biospecies (see Jackson and Cheetham[88] on bryozoan species, and Michaux[27] on paleontological stasis in gastropod morphospecies that persist as good genetic biospecies).

But continuing unhappiness, justified this time, focuses upon claims that speciation causes significant morphological change, for no validation of such a position has emerged (while the frequency and efficacy of our original supporting notion, Mayr's "genetic revolution" in peripheral isolates, has been questioned). Moreover, reasonable arguments for potential change throughout the history of lineages have been advanced,[6,34] although the empirics of stasis throws the efficacy of such processes into doubt. The pattern of punctuated equilibrium exists (at predominant relative frequency, we would argue) and is robust. *Eppur non si muove;* but why then? For the association of morphological change with speciation remains as a major pattern in the fossil record.

We believe that the solution to this dilemma may be provided in a brilliant but neglected suggestion of Futuyma.[89] He holds that morphological change may accumulate anywhere along the geological trajectory of a species. But unless that change be "locked up" by acquisition of reproductive isolation (that is, speciation), it cannot persist or accumulate and must be washed out during the complexity of interdigitation through time among varying populations of a species. Thus, species are not special because their origin permits a unique moment for instigating change, but because they provide the only mechanism for protecting change. Futuyma writes: "In the absence of reproductive isolation, differentiation is broken down by recombination. Given reproductive isolation, however, a species can retain its distinctive complex of characters as its spatial distribution changes along with that of its habitat or niche . . . Although speciation does not accelerate evolution within populations, it provides morphological changes with enough permanence to be registered in the fossil record. Thus, it is plausible to expect many evolutionary changes in the fossil record to be associated with speciation." By an extension of the same argument, sequences of speciation are then required for trends: "Each step has had a more than ephemeral existence only because reproductive isolation prevented the slippage consequent on interbreeding with other populations . . . Speciation may facilitate anagenesis by retaining, stepwise, the advances made in any one direction." Futuyma's simple yet profound insight may help to heal the remaining rifts and integrate punctuated equilibrium into an evolutionary theory hierarchically enriched in its light.[17,18]

In summarizing the impact of recent theories upon human concepts of nature's order, we cannot yet know whether we have witnessed a mighty gain in insight about the natural world (against anthropocentric hopes and biases that always hold us down), or just another transient blip in the history of correspondence between misperceptions of nature and prevailing social realities of war and uncertainty. Nonetheless, contemporary science has massively substituted notions of indeterminacy, historical contingency, chaos and punctuation for previous convictions about gradual, progressive, predictable determinism. These transitions have occurred in field after field; Kuhn's[90] celebrated notion of scientific revolutions is, for example, a punctuational theory for the history of scientific ideas. Punctuated equilibrium, in this light, is only paleontology's contribution to a *Zeitgeist,* and *Zeitgeists,* as (literally) transient ghosts of time, should never be trusted. Thus, in developing punctuated equilibrium, we have either been toadies and panderers to fashion, and therefore destined for history's ash heap, or we had a spark of insight about nature's constitution. Only the punctuational and unpredictable future can tell.

STEPHEN JAY GOULD is at the Museum of Comparative Zoology, Harvard University, Cambridge, Massachusetts 02138, USA; and Niles Eldredge is at the Department of Invertebrates, American Museum of Natural History, Central Park West at 79th Street, New York, New York 10024, USA.

References

1. Kellogg, V. L. *Darwinism Today* (Bell, London, 1907).
2. Eldredge, N. & Gould, S. J. in *Models in Paleobiology* (ed. Schopf, T. J. M.) 82–11 (Freeman, Cooper, San Francisco, 1972).
3. Dawkins, R. *The Blind Watchmaker* (Norton, New York, 1986).
4. Levinton, J. *Genetics, Paleontology and Macroevolution* (Cambridge Univ. Press, New York, 1988).
5. Hoffman, A. *Arguments on Evolution* (Oxford Univ. Press, New York, 1989).
6. Ridley, M. *Evolution* (Blackwell Scientific, Boston, 1993).
7. Erwin, D. *Speciation in the Fossil Record* 138–140 (Geol. Soc. Am. Ann. Meeting, Abstr., Boulder, CO, 1992).
8. Mayr, E., Boulding, K. E. & Masters, R. D. in *The Dynamics of Evolution: The Punctuated Equilibrium Debate in the Natural and Social Sciences* (eds. Somit, A. & Peterson, S. A.) (Cornell Univ. Press, Ithaca, NY, 1992).
9. Moore, R. C., Lalicker, C. G. & Fischer, A. G. *Invertebrate Fossils* (McGraw-Hill, New York, 1952).
10. Mayr, E. *Animal Species and Evolution* (Harvard Univ. Press, Cambridge, MA, 1963).
11. Gould, S. J. *Hen's Teeth and Horse's Toes* (Norton, New York, 1983).
12. Eldredge, N. *The Monkey Business: A Scientist Looks at Creationism* (Pocket Books, New York, 1982).
13. Stanley, S. M. *Proc. natn. Acad. Sci. USA* **72,** 646–650 (1975).
14. Vrba, E. S. *S. Afr. J. Sci.* **76,** 61–84 (1980).
15. Gould, S. J. & Eldredge, N. *Paleobiology* **3,** 115–151 (1977).
16. Gould, S. J. in *Perspectives on Evolution* (ed. Milkman, R.) 83–104 (Sinauer, Sunderland, MA, 1982).
17. Eldredge, N. *Macroevolutionary Dynamics* (McGraw-Hill, New York, 1989).
18. Vrba, E. & Gould, S. J. *Paleobiology* **12,** 217–228 (1986).
19. Hanson, N. R. *Perception and Discovery* (Freeman, Cooper, San Francisco, 1969).
20. Mayr, E. in *The Dynamics of Evolution* (eds. Somit, A. & Peterson, S. A.) 21–53 (Cornell Univ. Press, Ithaca, NY, 1992).
21. Sorhannus, U. *Hist. Biol.* **31,** 241–247 (1990).
22. Thomas, E. *Utrecht Micropal. Bull.* **23,** 1–167 (1980).

23. Hallam, A. *Paleobiology* **4,** 16–25 (1978).
24. Kelley, P. H. *J. Paleontol.* **57,** 581–598 (1983).
25. Kelley, P. H. *J. Paleontol.* **58,** 1235–1250 (1984).
26. Stanley, S. M. & Yang, X. *Paleobiology* **13,** 113–139 (1987).
27. Michaux, B. *Biol. J. Linn. Soc.* **38,** 239–255 (1989).
28. West, R. M. *Paleobiology* **5,** 252–260 (1979).
29. Schankler, D. M. *Nature* **293,** 135–138 (1981).
30. Lich, D. K. *Paleobiology* **16,** 384–395 (1990).
31. Budd, A. F. & Coates, A. G. *Paleobiology* **18,** 425–446 (1992).
32. Cronin, T. M. *Science* **227,** 60–63 (1985).
33. Maynard Smith, J. A. *Rev. Genet.* **17,** 11–25 (1984).
34. Williams, G. C. *Natural Selection: Domains, Levels and Challenges* (Oxford Univ. Press, New York, 1992).
35. Ruse, M. in *The Dynamics of Evolution* (eds. Somit, A. & Peterson, S. A.) 139–168 (Cornell Univ. Press, Ithaca, NY, 1992).
36. Eldredge, N. *Bull. Carnegie Mus. Nat. Hist.* **13,** 7–19 (1979).
37. Simpson, G. G. *The Major Features of Evolution* (Columbia Univ. Press, New York, 1953).
38. Stanley, S. M. *Macroevolution* (Freeman, San Francisco, 1979).
39. Lloyd, E. A. *The Structure and Confirmation of Evolutionary Theory* (Greenwood, New York, 1988).
40. Brandon, R. N. *Adaptation and Environment* (Princeton Univ. Press, Princeton, NJ, 1990).
41. Williams, G. C. *Adaptation and Natural Selection* (Princeton Univ. Press, Princeton, NJ, 1966).
42. Eldredge, N. & Stanley, S. M. *Living Fossils* (Springer, New York, 1984).
43. Haldane, J. B. S. *Evolution* **3,** 51–56 (1949).
44. Stanley, S. M. *Evolution* **27,** 1–26 (1973).
45. Gould, S. J. *J. Paleont.* **62,** 319–329 (1988).
46. Martin, R. A. *Hist. Biol.* **6,** 73–90 (1992).
47. Raup, D. M. & Sepkoski, J. J. Jr. *Science* **215,** 1501–1503 (1982).
48. Valentine, J. W. in *Molds, Molecules and Metazoa* (eds. Grant, P. R. & Horn, H. S.) 17–32 (Princeton Univ. Press, Princeton, NJ, 1992).
49. Vrba, E. S. *Suid-Afrikaanse Tydskvif Wetens* **81,** 229–236 (1985).
50. Brett, C. E. *Coordinated Stasis and Evolutionary Ecology of Silurian-Devonian Marine Biotas in the Appalachian Basin* 139 (Geol. Soc. Am. Ann. Meeting, Abstr., Boulder, CO, 1992).
51. Morris, P. J., Ivany, L. & Schopf, K. M. *Paleocological Stasis in Evolutionary Theory* 313 (Geol. Soc. Am. Ann. Meeting, Boulder, CO, 1992).
52. Gould, S. J. *Am. Sci.* **74,** 60–69 (1986).
53. MacLeod, N. *Paleobiology* **17,** 167–188 (199).
54. Gingerich, P. D. *Am. J. Sci.* **276,** 1–28 (1976).
55. Chaline, J. & Laurin, B. *Paleobiology* **12,** 203–216 (1986).
56. Bergström, J. & Levi-Setti, R. *Geol. Palaeontol.* **12,** 1–40 (1978).
57. Smith, A. B. & Paul, C. R. C. *Sp. Pap. Palaeont.* **33,** 29–37 (1985).
58. Flynn, L. J. *Contr. Geol. Univ. Wyoming* **3,** 273–285 (1986).
59. Finney, S. C. *Geol. Soc. Sp. Pub.* **20,** 103–113 (1986).
60. Wei, K. Y. & Kennett, J. P. *Paleobiology* **14,** 345–363 (1988).

61. Nehm, R. H. & Geary, D. H. *Paleontol. Soc. Sp. Pub.* **6,** 222 (1992).
62. Williamson, P. G. *Nature* **293,** 437–443 (1981).
63. Williamson, P. G. *Biol. J. Linn. Soc.* **26,** 307–324 (1985).
64. Fryer, G., Greenwood, P. H. & Peake, J. F. *Biol. J. Linn. Soc.* **20,** 195–205 (1983).
65. Cheetham, A. H. *Paleobiology* **12,** 190–202 (1986).
66. Cheetham, A. H. *Paleobiology* **13,** 286–296 (1987).
67. Prothero, D. R. & Shubin, N. in *The Evolution of Perissodactyls* (eds. Prothero, D. R. & Schoch, R. M.) 142–175 (Oxford Univ. Press, Oxford, 1989).
68. Simpson, G. G. *Horses* (Oxford Univ. Press, Oxford, 1951).
69. MacGillavry, H. J. *Bijdragen tot de Dierkunde* **38,** 69–74 (1968).
70. Ager, D. V. *Proc. Geologists' Ass.* **87,** 131–160 (1976).
71. Ager, D. V. *Palaeontology* **26,** 555–565 (1983).
72. Coope, G. R. A. *Rev. Ecol. Syst.* **10,** 247–267 (1979).
73. Fortey, R. A. *Sp. Pap. Palaeontol.* **33,** 17–28 (1985).
74. Vrba, E. S. in *Living Fossils* (eds. Eldredge, N. & Stanley, S. M.) 62–79 (Springer, New York, 1984).
75. Barnovsky, A. D. *Curr. Mammal.* **1,** 109–147 (1987).
76. Prothero, D. R. *Skeptic* **1,** 38–47 (1992).
77. Stanley, S. M. *Paleobiology* **4,** 26–40 (1978).
78. Johnson, J. G. *Paleontol.* **49,** 646–661 (1975).
79. Johnson, J. G. *Paleontol.* **56,** 1329–1331 (1982).
80. Sheldon, P. R. *Nature* **345,** 772 (1990).
81. Shapiro, E. A. *Natural Selection in a Miocene Pectinid: A Test of the Punctuated Equilibria Model* 490 (Geol. Soc. Am. Ann. Meeting, Abstr., Boulder, CO, 1978).
82. Avise, J. C. *Proc. natn. Acad. Sci. USA* **74,** 5083–5087 (1977).
83. Mayden, R. L. *Syst. Zool.* **35,** 591–602 (1986).
84. Mindell, D. P., Sites, J. R. Jr. & Graur, D. *Cladistics* **5,** 49–61 (1989).
85. Mindell, D. P., Sites, J. R. Jr. & Graur, D. *J. evol. Biol.* **3,** 125–131 (1990).
86. Lemen, C. A. & Freeman, P. W. *Evolution* **43,** 1538–1554 (1989).
87. Wagner, P. J. *Cladograms as Tests of Speciation Patterns* 139 (Geol. Soc. Am. Ann. Meeting, Abstr., Boulder, CO, 1992).
88. Jackson, S. B. C. & Cheetham, A. H. *Science* **248,** 579–583 (1990).
89. Futuyma, D. J. *Am. Nat.* **130,** 465–473 (1987).
90. Kuhn, T. S. *The Structure of Scientific Revolutions* (Univ. Chicago Press, Chicago, 1962).

Eörs Szathmáry and John Maynard Smith

THE MAJOR EVOLUTIONARY TRANSITIONS

There is no theoretical reason to expect evolutionary lineages to increase in complexity with time, and no empirical evidence that they do so. Nevertheless, eukaryotic cells are more complex than prokaryotic ones, animals and plants are more complex than protists, and so on. This increase in complexity may have been achieved as a result of a series of major evolutionary transitions. These involved changes in the way information is stored and transmitted.

The major evolutionary transitions[1] are listed in Table 1. There are common features that recur in many of the transitions: (1) Entities that were capable of independent replication before the transition can only replicate as parts of a larger unit after it. For example, free-living bacteria evolved into organelles.[2] (2) The division of labor: as Smith[3] pointed out, increased efficiency can result from task specialization (for a comprehensive review of this subject in the classical literature, see ref. 4). For example, in ribo-organisms nucleic acids played two roles, as genetic material and enzymes, whereas today most enzymes are proteins. (3) There have been changes in language, information storage and transmission. Examples include the origin of the genetic code, of sexual reproduction, of epigenetic inheritance and of human language.

Complexity

There is no generally accepted measure of biological complexity. Two possible candidates are the number of protein-coding genes, and the richness and variety of morphology and behavior. Table 2 shows the sizes of the coding regions of various organisms.[5] The trend is fairly robust: eukaryotes have a larger coding genome than prokaryotes, higher plants and invertebrates have a larger genome than protists, and vertebrates a larger genome than invertebrates. The last observation is puzzling: perhaps the nervous system of vertebrates requires the extra genetic information. Unfortunately, the data do not tell us much about structural or func-

Table 1 The major transitions[1]

Replicating molecules to populations of molecules in compartments
Unlinked replicators to chromosomes
RNA as gene and enzyme to DNA and protein (genetic code)
Prokaryotes to eukaryotes
Asexual clones to sexual populations
Protists to animals, plants and fungi (cell differentiation)
Solitary individuals to colonies (non-reproductive castes)
Primate societies to human societies (language)

Table 2 Genome size and DNA content[5]

	Genome Size (base pairs $\times 10^9$)	Coding DNA (%)
Bacterium (*E. coli*)	0.004	100
Yeast (*Saccharomyces*)	0.009	70
Nematode (*Caenorhabditis*)	0.09	25
Fruitfly (*Drosophila*)	0.18	33
Newt (*Triturus*)	19.0	1.5–4.5
Human	3.5	9–27
Lungfish (*Protopterus*)	140.0	0
Flowering plant (*Arabidopsis*)	0.2	31
Flowering plant (*Fritillaria*)	130.0	0.02

tional complexity, because we do not know the mapping between genotype and phenotype.

Bonner[6] measures complexity in terms of the variety of behavior. For example, the emergence of humans depended on a greater behavioral variety. The point need not be confined to ethology: complexity increases with the diversity of actions an organism can carry out. For example, phagocytosis is a complex behavior that depends on the eukaryotic cytoskeleton: prokaryotes cannot do it. The number of cell types in an organism can be taken as a measure of its complexity. Unfortunately, it is hard to quantify this aspect of complexity, or to get beyond the common-sense, but rather boring, conclusion that complexity has indeed increased in some lineages.

It is more interesting to list the mechanisms whereby the quantity of genetic information can increase. The three main possibilities—duplication and divergence, symbiosis and epigenesis—are shown in Fig. 1.

Transition from Independent Replicators

In many of the transitions listed in Table 1 we find the common phenomenon that entities capable of independent replication before the transition

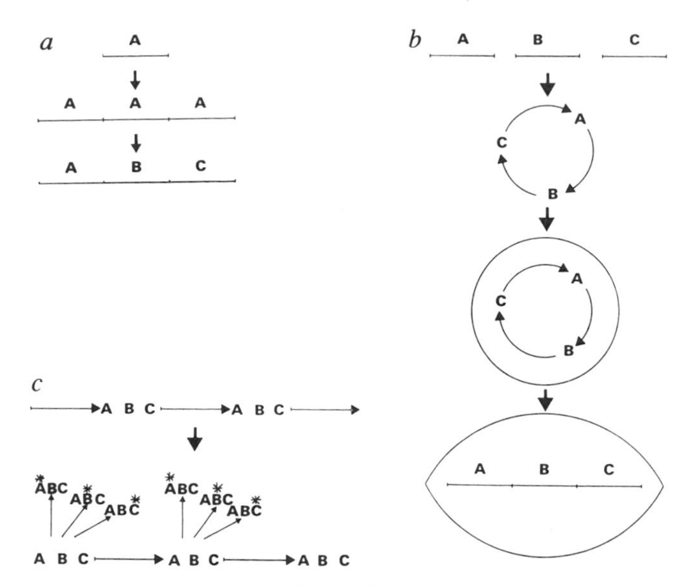

Fig. 1 Processes whereby the quantity of information can increase during evolution. *a,* Duplication followed by divergence: this is the main process whereby information increases between the major transitions. *b,* Symbiosis: the figure illustrates first a set of independent replicators, then a hypercycle,[60] in which the replicators interact to form a stable ecological cycle, then the enclosure of the hypercycle in a compartment, and finally the physical linkage of replicators, so that when one replicates, all do. *c,* Epigenesis: A, B and C are different genes; asterisks indicate states of gene activity transmitted through cell division.

can only replicate as parts of a larger whole afterwards. Examples include the origin of chromosomes; the origin of eukaryotes with symbiotically derived organelles; the origin of sex; the origin of multicellular organisms (the cells of animals, plants and fungi are descended from unicellular protists, each of which could survive on its own: today, they exist only as parts of larger organisms); and the origin of social groups. Note that the last two examples differ from the previous ones: the cells of multicellular animals did not form the organism through a symbiosis of independent entities, but they consist of entities (the cells), the analogues of which do exist as independent forms. Thus, units of evolution at the higher level may either be analogous (multicellular organisms) or homologous (eukaryotes) to an "ecosystem" of lower-level units.

Given this common feature of the major transitions, there is a common question we can ask of them. Why did natural selection, acting on entities at the lower level (replicating molecules, free-living prokaryotes, asexual protists, single cells, individual organisms), not disrupt integration at the higher level (chromosomes, eukaryotic cells, sexual species, multicellular organisms, societies)? The problem is not an imaginary one: there is a real danger that selection at the lower level will disrupt integration at the higher. Some examples are:[1] (1) If Mendel's laws are rigorously obeyed, a gene can only increase its representation in future generations by ensuring the success of the cell in which it finds itself, and of the other genes in the cell. Hence Mendel's laws ensure the evolution of cooperative, or "coadapted," genes. But the laws are broken, in meiotic drive,[7] and by transposable elements.[8] These are examples of the more general phenomenon of intragenomic conflict.[9] (2) A sexual population has an advantage, in rate of evolution, and in the elimination of harmful mutations, over an asexual one. But a parthenogenetic female has, in the short run, a two-fold advantage over a sexual one, and parthenogens are not uncommon.[10] (3) A gene in a somatic cell of a plant might best ensure the transmission of replicas of itself by giving rise to a flower bud, even if this reduced the success of the whole plant. (4) A bee colony produces more reproductives if the workers raise the queen's offspring. But workers do lay eggs (which are unfertilized, and hence male).[11]

We cannot explain these transitions in terms of the ultimate benefits they conferred. For example, it may be that, in the long run, the most important difference between prokaryotes and eukaryotes is that the latter evolved a mechanism for chromosome segregation at cell division that permits DNA replication to start simultaneously at many origins, whereas prokaryotes have only a single origin of replication.[12] At the very least, this was a necessary precondition for the subsequent increase in DNA content, without which complexity could not increase. But this is not the reason why the change occurred in the first place: the new segregation mechanism was forced on the early eukaryotes by the loss of a rigid cell wall, which plays a crucial role in the segregation of eubacterial chromosomes. Or, to take a second example, meiotic sex was an important preadaptation for the subsequent evolutionary radiation of the eukaryotes, but it could not have originated for that reason.

The transitions must be explained in terms of immediate selective advantage to individual replicators. We are committed to the gene-centered approach outlined by Williams[13] and made still more explicit by Dawkins.[14] There is, in fact, one feature of the transitions listed in Table 1 that leads to this conclusion. At some point in the life cycle, there is only one copy, or very few copies, of the genetic material: consequently, there is a high degree of genetic relatedness between the units that combine in the higher organism. The importance of this general principle was first emphasized by Hamilton[15] in his explanation of the evolution of social behavior, but we believe it to be quite general. To give two other examples: multicellular organisms develop from a single fertilized egg, so that their cells are genetically identical, except for somatic mutation; most eukaryotes inherit their organelles from one parent only, so that the organelles in a single individual are almost always genetically identical.[16,17] We think that a similar principle operated in the origin of the earliest cells:[18–20] this example is discussed further in Box 1.

In several of the listed transitions, one is effectively dealing with a group of replicators: when does such a group qualify as an organism, or—viewed from the level of the component replicators—a "superorganism"? Wilson and Sober[21] define a superorganism as a "collection of single creatures that together possess the functional organization implicit in the formal definition of organism." They suggest that groups are superorganisms

BOX 1 From naked genes to compartments to chromosomes

The idea that the primordial genome must have consisted of unlinked genes comes from a paradox of Eigen. He demonstrated that the tolerable mutational load of a population sets an upper limit to the length of the genome: it is proportional to the reciprocal of the mutation rate per base per generation (the error threshold).[60] Early genomes could not have been much longer than a contemporary transfer RNA owing to the low fidelity of replication. A population in mutation-selection balance of such molecules (a so-called quasispecies[61]) cannot harbor a set of sufficiently dissimilar genes (to encode different functions in appreciable concentration). Therefore, a collection of unrelated and unlinked genes is needed, but they will compete with each other until only one gene (with the highest replication rate) survives, and in turn the genome is doomed to extinction; hence the paradox that long chromosomes are unstable because of excessive mutational load, and a set of small genes is unstable because of internal competition. As Eigen recognized, some functional coupling among the genes is necessary.[60]

A possible resolution of Eigen's paradox is the "stochastic corrector" model[18–20] (figure). The assumptions are as follows: (1) Unlinked genes replicate in compartments. There are two types of gene, which replicate at different expected rates, so that there is between-gene selection within compartments. (2) Compartments reproduce when the number of genes they contain has doubled. The rate of compartment growth depends on the kinds of genes it contains, and is fastest when different kinds are present. Thus different genes contribute non-additively to compartment fitness, as emphasized in our discussion on the division of labor. (3) Replication is a stochastic process and assortment of genes into offspring compartments is random.

Given these conditions, there is efficient group selection at the compartment level. Despite internal competition, natural selection maintains a stable compartment distribution, and neither type of gene is lost. This is due to the stochastic processes generating variation between compartments, on which selection—between the protocells—can act.

The spread of chromosomes has also been analyzed in the context of the stochastic corrector model. Chromosomes (linking complementing genes), when introduced in small numbers into some protocells of a simulated population, are established in the population, despite a twofold within-cell replicative disadvantage relative to individual genes.[62] The reasons are first, that linkage is a safeguard against internal competition of genes, as one cannot replicate without the other, and second, that it pays for a gene to sit on a chromosome, because it does not then run the risk of finding itself, after cell division, in a cell with low fitness due to the absence of its complementing partner. The chemical feasibility of this transition has been worked out,[63] resting in part on the "genomic tag" model,[64] which argues for an ancestral role of tRNA-like structures as signals for RNA replication.

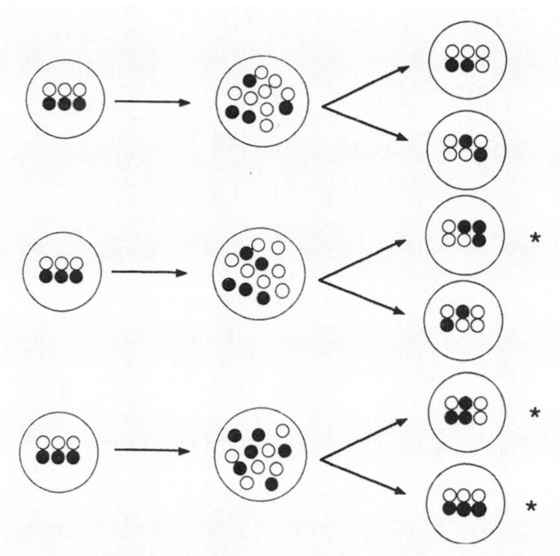

The stochastic corrector model.[18–20] Empty and filled circles represent two kinds of gene; the former have an average within-cell replicative advantage, but all genes replicate faster in cells with equal numbers of the two kinds. Stochastic processes (in replication and cell division) ensure the reappearance of cells, marked with asterisks, with the optimal gene composition, despite contrary within-cell selection.

if they satisfy the following criteria:[21] the population consists of several groups; there is a difference between the groups in their contribution of progeny to the next generation (differential group fitness); variation of group fitness is due to heritable genetic variation; individuals within the group have the same fitness.

Obviously, the last criterion cannot hold at the time of origin itself; it is precisely this absence of between-individual, within-group selection that has to be explained. The real question is whether a mechanism, suppressing internal competition, will invade when rare. The answer is known to be yes for certain cases (Box 1). For such a mechanism to spread, selection between groups must be effective. It is most efficient if the following criteria are met:[22] the number of groups must be much larger than that of the units within each group; there is no migration (horizontal transfer) between groups; each group has no more than one parental group.

The effect of these conditions is that there will be genetic differences between groups, but individuals in a single group will be similar. If so, the groups will be units of evolution[23] and will evolve by natural selection.

The principle of the small number of founders is important at the time of the transition. Two other processes—contingent irreversibility and central control—help to explain the maintenance of higher-level entities once they have arisen, although they are less relevant to the origin of such entities.[1]

Contingent Irreversibility

If an entity has replicated as part of a larger whole for a long time, it may have lost the capacity for independent replication that it once had, for accidental reasons that have little to do with the selective forces that led to the evolution of the higher-level entity in the first place. For example, mitochondria cannot resume independent existence, if only because most of their genes have been transferred to the nucleus; cancer cells may escape growth control, but have no independent future as protists; worker bees may lay male eggs, but cannot establish a new colony on their own.

The contingent nature of irreversibility is perhaps best illustrated by the reversion from sex parthenogenesis. Mammals are never parthenogens, probably because, at some loci in some tissues, only the allele inherited from the father is active: hence, in an embryo with no father, some essential gene activities are missing. Gymnosperms are also never parthenogens, perhaps for a different reason: chloroplasts are transmitted in the pollen. Anamniote vertebrates, although they may be parthenogens, always require sperm from the males of other species to initiate development, perhaps because the sperm provides a centriole. The relevance of these sexual hangups—and there are many others—is to show how various and accidental are the reasons why reversal is difficult or impossible.

Central Control

If a "selfish" mutation occurs in a chromosomal gene, a suppressor mutation at any other locus in the genome would be favored by selection. Hence the rest of the genome may win the contest, not because of any analogue of majority voting, but because of the large number of loci, and hence of possible suppressor mutations, that are available for each selfish mutation. It may be relevant that attempts to use driving chromosomes in biological control have so far failed because of the rapid evolution of suppression. It is in this sense that Leigh's idea[24] of a "parliament of genes" should be understood.

The Division of Labor

The most familiar examples of the advantages arising from a division of labor concern caste differentiation in the social insects. Bell[25] applies

Adam Smith's ideas to a less familiar example, cell differentiation in the Volvocales.[26] The specific name *Volvox weismannia* is a reminder that these algae are an excellent example of the segregation of germ line and soma. Most members of the order possess only a single cell type, which fulfills all vegetative and reproductive functions. In *Pledorina,* there is partial division of labor: some cells start with vegetative functions, but later differentiate into gonidia (asexual propagules). The genus *Volvox* has a *bona fide* segregation of germ line and soma: germ cells are immotile in the center of the spheroid, and somatic cells bear cilia but cannot divide. The benefit of differentiation has been demonstrated: colonies produce a larger bulk of smaller offspring than do single cells of a similar size.

One precondition for the division of labor in the Volvocales is that motile cells cannot divide, and mitosing cells cannot move, because the same organelles are used either as basal bodies or as centrioles.[27] A similar argument was used by Buss[28] for the flagellated blastulae of the lower Metazoa.

Some other cases in which a division of labor is evident are:[1,29] (1) The evolution of many specific enzymes from a set of multifunctional low-efficiency enzymes. If the first protocells were equipped with a few multi-functional enzymes, more efficient enzymes could evolve only by duplication and divergence.[30] (2) In the RNA world, RNA served both as genetic material and catalyst: today, DNA is the genetic material, and most enzymes are proteins.[31] (3) In prokaryotes there is a single cell compartment, whereas in eukaryotes the genetic nucleus and metabolic cytoplasm are separated, and additional organelles have evolved, some recruited symbiotically.[2] (4) In sexual populations, isogamy has repeatedly evolved to anisogamy, with differentiated sperm and ova.[10] (5) Hermaphrodites are replaced by separate sexes: the most convincing explanation for why some organisms are hermaphrodite and some dioecious is in terms of the advantages of a division of labor.[32]

If cooperation is to evolve, non-additive, or synergistic, fitness interactions are needed. If two or more cooperating individuals can achieve something that a similar number of isolated individuals cannot, the preconditions exist: the image to bear in mind is that two men, each with one oar, can propel a boat, but one man with one oar will row in circles.[1] But the dangers of intragenomic conflict remain: both relatedness and synergistic fitness interactions are likely to be needed.

The Evolution of Heredity

Heredity means that like begets like: it requires some means whereby information can be transmitted. A crucial distinction is between systems of "limited heredity," in which only a few distinct states can be transmitted, and systems of "unlimited heredity," capable of transmitting an indefinitely large number of messages. We suggest the following stages.[1]

1. The origin of simple autocatalytic systems with limited heredity.[33–35] Autocatalysis, whereby a single molecule gives rise to two molecules of the same kind, is essential for growth, but does not by itself imply heredity, which requires that, if the nature of the initial molecule is changed, two molecules of the new kind are produced. Several authors have suggested that autocatalytic networks could have this property, but at best they could display limited heredity, with only a few molecular types able to reproduce themselves.

2. The origin of polynucleotide-like molecules, providing unlimited heredity.[31] This transition has proven surprisingly difficult to explain. Even if one assumes the presence of all the necessary chemical constituents, there are severe obstacles to continued replication, such as enantiomeric cross-inhibition (mirror-image building blocks pair but do not form a covalent link with the growing chain[36]) and non-separation of template and replica due to the many hydrogen bonds formed between them. Oligonucleotide replication is a possible intermediate stage leading to that of polynucleotides. Most such experiments use chemical analogues of oligonucleotides (such as the first successful attempt by von Kiedrowski[37]). The short length allows for the spontaneous dissociation ("melting") of template and copy, so ongoing replication is possible, although an increased concentration is unfavorable for the latter because two complete strands find each other more readily. This results in a parabolic (subexponential) growth of such replicators,[37] which in a competitive situation leads to a stable "survival of everybody"[38,39] (see also ref. 40 for review). "Survival of the fittest" needs an exponential growth tendency,[39] and therefore more efficient strand separation. The latter could have been accomplished by RNAs with replicase function.

3. The origin of the genetic code in the context of the RNA world, before translation. The essence of the code is that specific amino acids should be attached to specific oligonucleotides: today, it depends on the attachment of amino acids to transfer RNA molecules. Several workers in the field have realized that translation and coding are difficult to evolve simultaneously. One way to avoid such an evolutionary trap is preadaptation: rudiments of a complex adaptation may have evolved by selection for some other function. Concrete versions of this idea suggest that aminoacylation helped the replication of RNA, or that peptide-specific ribosomes with an internal message antedate general protein synthesis using external templates (messenger RNA). The idea that we favor is that amino

acids were used as coenzymes of ribozymes, and were equipped with unambiguous trinucleotide handles, enabling them to bind to a ribozyme by base pairing.[41] Such handles would enable the same amino acid to be used by several ribozymes. Each new amino acid that acquired a specific handle would increase the enzymatic versatility of the organism, so that the difficulty of a complete adaptation being acquired in a single step largely disappears.

4. The origin of translation and encoded protein synthesis. The details of this transition are discussed in ref. 1.

5. The replacement of RNA by DNA as the genetic material could well have happened before the origin of genetic code, because it is chemically a much less complicated transition.[42] The primary selective force for this may have been the increased chemical stability of thymine (as opposed to uracil) and deoxyribose (as opposed to ribose).[43] The usual argument that there is no mismatch repair or repair of damage in RNA misses the point that, chemically, all these processes would be feasible in double-stranded RNA.

6. The emergence of hereditary regulative states in prokaryotes and simple eukaryotes. Already in prokaryotes, patterns of methylation are transmitted through cell division and can be responsible for states of phenotypic differentiation. Thus there is a dual inheritance system, in which heredity depends either on differences in DNA sequence or on transmissible states of gene activation.[44,45] Such a system is crucial for the development of animals and plants, but what selective forces were responsible for its evolution in single-celled organisms? Jablonka suggests that it was the need for protists to adapt to regular changes in the environment, the timescale of which was too large in comparison with generation times, and too small relative to the time required for typical evolutionary changes.[44] In this case, a heritable mark M1 on some gene could have been beneficial in environment E1, and an alternative mark M2 (maybe simply the lack of M1) could have been beneficial in environment E2. An alternative suggestion is that morphological and physiological adaptations of sexual protists could have been preadaptations for simple forms of multicellularity, as alternative phenotypes, specific cell adhesion, cell-to-cell signalling and cell division arrest play a crucial role in both.[46]

7. The evolution of epigenetic inheritance with unlimited heredity: the emergence of animals, plants and fungi. The transition to multicellular organisms with many kinds of differentiated cells occurred on three occasions, suggesting that it may not have been particularly difficult. This would be explained if the main cellular novelty required was an epigenetic inheritance system, as this existed already in protists. If so, the emergence

and radiation of the Metazoa had to wait only for suitable environmental conditions.[47]

8. The emergence of protolanguage in *Homo erectus*—a cultural inheritance system with limited potential in which, because of the absence of grammar, only certain types of statement can be made.[48]

9. The emergence of human language with a universal grammar[49] and unlimited semantic representation.[50] Grammar enables a speaker with a finite vocabulary to convey an indefinitely large number of meanings, just as the genetic code enables DNA to specify an indefinitely large number of proteins. We accept Chomsky's argument that grammatical competence is unique, both in the sense of being peculiar to humans, and of being special to language, and not merely an aspect of general learning ability. But we are puzzled by the reluctance of many linguists, including Chomsky himself, to think about the evolution of this competence. The objection takes the form of asserting not only that human language is different in kind from animal communication, but that no intermediate is possible between the two.

It is argued that any rudimentary form of grammar would not allow one to generate some types of sentence. This is true but irrelevant: by analogy, it is better to have some light-sensitive cells than none at all; a perfect eye is not the only useful solution to the problem.[51]

It is in fact rather easy to think of intermediates between protolanguage and true language. There remains the question of the evolutionary origin of grammatical novelties. It is reasonable to assume that this happened by genetic assimilation,[51] new rules being made up by individuals as nongenetic innovations, then learned by members of the community, then hard-wired into the "language organ" subsequently. It has been demonstrated that learning and selection can lead to such an assimilation in extreme cases when the latter alone could not get anywhere:[52] learning can transform an initially flat fitness landscape with a needle-like peak into a well-behaved Fujiyama-like one.

Perhaps the most convincing evidence for the belief that grammatical competence is to some degree independent of general learning ability, and for the possibility of functional intermediates between no grammar and perfect grammar, comes from studies of hereditary variation in linguistic competence. One remarkable case involves a family in which a so-called feature-blind dysphasia seems to be inherited in a Mendelian fashion, a single dominant gene being responsible.[53] Members cannot automatically generate plurals and past tense. Although they understand the meaning of plural and past perfectly well, they have to learn each new case anew: *paint* and *painted, book* and *books* must be learned separately (in the case of exceptions such as *go* and *went,* we must do the same). To be sure, this is not a genetical violation of one of Chomsky's rules, but it demonstrates that there can be something useful between perfect grammar and protolanguage: it also holds out the hope that we will in future be able to dissect language genetically, as we are today dissecting development.

Constructive Evolution

Although some of the key intermediate stages of evolution seem to have vanished, their experimental re-creation could teach us a lot. Examples include:

The *de novo* synthesis of a living chemical system, such as the chemoton[54] (see Box 2 for a summary of this idea, and how several of our discussed points integrate into a unified picture at a certain level of organization).

In vitro construction of a truly self-replicating RNA.

In vitro generation of ribozymes, using amplification and selection by affinity chromatography.[19,55] The first such example has been given.[56] In a similar vein, the generation of RNA molecules of importance for primordial coding and translation, essentially by the same protocol, should provide us with useful information about feasible scenarios of the code's origin.[19,55] Recently Famulok reported the *in vitro* selection of an RNA binding ornithine and citrulline.[57] Similar tests using proteinogenic amino acids would be welcome.

The establishment of artificial symbioses should help to clarify several aspects of some of the transitions. A first example is Jeon's bacteria, originally parasitizing an amoeba, which became obligatorily dependent on these bacteria later.[58]

Finally, re-creation of extant species or forms may be in certain cases possible. The re-creation of a fossil fern species from genomes of extant polyploids is a remarkable example.[59]

Conclusions

A central idea in contemporary biology is that of information. Developmental biology can be seen as the study of how information in the genome is translated into adult structure, and evolutionary biology of how the information came to be there in the first place. Our excuse for writing an article concerning topics as diverse as the origins of genes, of cells and of language is that all are concerned with the storage and transmission of information. The article is more an agenda for future research than a summary of what is known. But there is sufficient formal similarity between the various transitions to hold out the hope that progress in understanding any one of them will help to illuminate others.

BOX 2 The chemoton as a sensible protocell model and its importance in explaining the first major transitions

The basic model of the chemoton[34,54,65] (figure) consists of three subsystems: the metabolic "engine," which is an autocatalytic cycle; a self-replicating template macromolecule; and a bilayer membrane. The autocatalytic cycle produces the building blocks of the two other subsystems as well, at the expense of the energy and material difference between X and Y. The condensation by-product R serves as a stringent stoichiometric coupling between template polycondensation and membrane growth. It is easy to see that the whole system grows in synchrony. Division is a more tricky problem. There are calculations (which need to be verified experimentally) showing that a chemical system with the described couplings would indeed undergo spontaneous fission into two offspring compartments, owing to the interplay between growth, osmotic relations and surface tension of the membrane.[66,67] The two questions we will discuss in turn concern the realistic feasibility of the subsystems and how our central themes manifest themselves in the origin and evolution of such a system.

A remarkable example of an autocatalytic network is Butlerow's formose "reaction," synthesizing sugars out of formaldehyde with the catalytic aid of pre-existing sugars (see, for example, ref. 68). Other, still non-enzymatic, cycles were suggested as variations on the theme of the reductive citric acid cycle,[69] as yet without experimental evidence. The origin of RNA-like self-replicating molecules is still a problem, as discussed in the main text (compare ref. 31). It is worth emphasizing the special kinetic effects within the chemoton, however: replication occurs only upon reaching a certain threshold concentration of V within the compartment, and this happens only once during the protocell cycle.[34] The autocatalytic formation of membranes without enzymes is now proven.[70] Experiments should now concentrate on the division mechanism.

As to our common themes, the following considerations are worth noting.

Complexity. Nobody thinks that a simple cycle in its drawn form could be realistic. Inevitably, one should have a network. The increase in complexity of such a network is an open problem. Chemical symbiosis,[71] or the grafting of novel extensions onto the pre-existing network,[35] could have resulted in heritable, "macroevolutionary" changes in the system. Apart from this, it is a family of templates, arising by mutation, duplication and divergence that can lead to a complex set of templates, which make use of digital information, first in the form of ribozymes catalyzing steps of the metabolic network, and later as protein-coding genes.

Division of labor. It is common that symbiotic partners provide complementary metabolic "tool-kits" for the new unit.[2] The chemoton's subsystems serve exactly such complementary roles, and their unit can be regarded as a very special case of chemical symbiosis. It is obvious that the membrane itself would be an inferior metabolic subsystem, and that templates could provide only leaky boundaries, so indeed there is an advantage in the union of specialized subsystems: molecular "jacks-of-all-trades" are replaced by "masters."

Competition of replicators. The organization of the basic chemoton model is such that the complementary subsystems cannot get "out of phase."[34,54] This is not so within subsystems, most notably for the digital information carriers.[19,20] Through microevolution, selfish mutants can arise. It is the stochastic corrector principle (Box 1) that can prevent the system as a whole from deteriorating.

Heredity. Two of the subsystems (the membrane and the metabolic cycle) carry only analogue information[72] and are at the most limited hereditary replicators. It is the realistic versions of the pV_n that can provide the system with unlimited heredity, because of their digital information.[72] First this is used for ribozymic activity, then in translation.

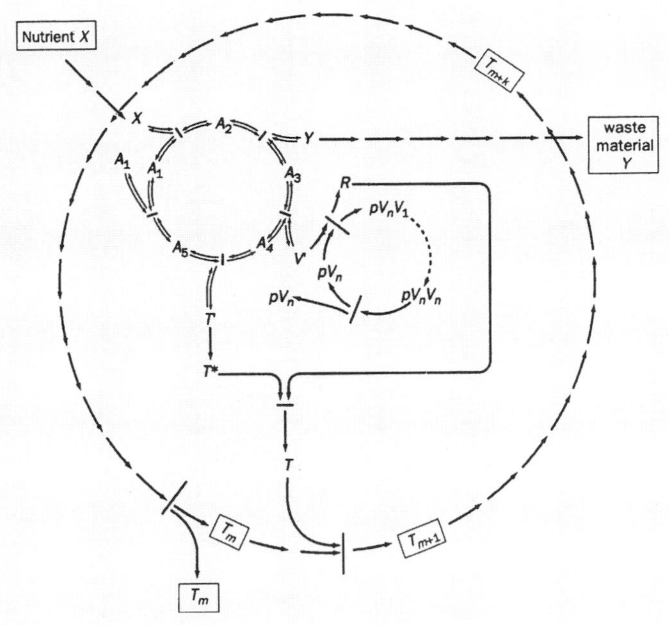

The chemoton.[54] The metabolic subsystem, with intermediates A_i, is an autocatalytic chemical cycle, consuming X as nutrient and producing Y as waste material; pV_n is a polymer of n molecules of V, which undergoes template replication; R is a condensation by-product of this replication, needed to turn T' into T, the membranogenic molecule; the symbol T_m represents a bilayer membrane composed of m units made of T molecules. It can be shown that such a system can grow and divide spontaneously.

Eörs Szathmáry is in the Collegium Budapest (Institute for Advanced Study), Szentháromság u. 2., H-1014 Budapest and Ecological Modelling Research Group, Department of Plant Taxonomy and Ecology, Eötvös University, Ludovika tér 2, H-1083 Budapest, Hungary; John Maynard Smith is at the School of Biological Sciences, the University of Sussex, Brighton BN1 9QG, UK.

References

1. Maynard Smith, J. & Szathmáry, E. *The Major Transitions in Evolution* (Freeman, Oxford, 1995).
2. Margulis, L. *Symbiosis in Cell Evolution* (Freeman, San Francisco, 1981).
3. Smith, A. *The Wealth of Nations* (1777).
4. Rensch, B. *Evolution above the Species Level* (Wiley, New York, 1966).
5. Cavalier-Smith, T. (ed.) *The Evolution of Genome Size* (Wiley, Chichester, 1985).
6. Bonner, J. T. *The Evolution of Complexity by Means of Natural Selection* (Princeton Univ. Press, Princeton, NJ, 1988).

7. Crow, J. F. *BioEssays* **13,** 305–312 (1991).
8. Charlesworth, B., Sniegowski, P. & Stephan, W. *Nature* **371,** 215–220 (1994).
9. Hurst, L. *Proc. R. Soc. Lond.* B **248,** 135–140 (1992).
10. Maynard Smith, J. *The Evolution of Sex* (Cambridge Univ. Press, Cambridge, 1978).
11. Wilson, E. O. *Sociobiology: The New Synthesis* (Belknap, Harvard Univ. Press, Cambridge, MA, 1975).
12. Cavalier-Smith, T. *Ann. N.Y. Acad. Sci.* **503,** 17–54 (1987).
13. Williams, G. C. *Adaptation and Natural Selection* (Princeton Univ. Press, Princeton, NJ, 1966).
14. Dawkins, R. *The Selfish Gene* (Oxford Univ. Press, Oxford, 1976).
15. Hamilton, W. D. *J. theor. Biol.* **7,** 1–52 (1964).
16. Hurst, L. D. & Hamilton, W. D. *Proc. R. Soc. Lond.* B **247,** 189–194 (1992).
17. Hutson, V. & Law, R. *Proc. R. Soc. Lond.* B **263,** 43–51 (1993).
18. Szathmáry, E. & Demeter, L. *J. theor. Biol.* **128,** 463–486 (1987).
19. Szathmáry, E. *Oxf. Surv. Evol. Biol.* **6,** 169–205 (1989).
20. Szathmáry, E. *Trends Ecol. Evol.* **4,** 200–204 (1989).
21. Wilson, D. S. & Sober, E. *J. theor. Biol.* **136,** 337–356 (1989).
22. Leigh, E. G. *Trends Ecol. Evol.* **6,** 257–262 (1991).
23. Maynard Smith, J. *Proc. R. Soc. Lond.* B **219,** 315–325 (1983).
24. Leigh, E. G. *Adaptation and Diversity* (Freeman, Cooper and Co., San Francisco, 1971).
25. Bell, G. in *The Origin and Early Evolution of Sex* (eds. Halvorson, H. O. & Mornoy, A.) 221–256 (Liss, New York, 1985).
26. Kirk, D. L. *Trends Genet.* **4,** 32–36 (1994).
27. Koufopanou, V. *Am. Nat.* **143,** 907–931 (1994).
28. Buss, L. *The Evolution of Individuality* (Princeton Univ. Press, Princeton, NJ, 1987).
29. Molnár, I. *Anstr. botanica (Budapest)* **17,** 207–224 (1993).
30. Kacser, H. & Beeby, R. *J. molec. Evol.* **20,** 38–51 (1984).
31. Joyce, G. *Nature* **338,** 217–224 (1989).
32. Charnov, E., Maynard Smith, J. & Bull, J. J. *Nature* **263,** 125–126 (1976).
33. Ycas, M. *Proc. natn. Acad. Sci. USA* **41,** 714–716 (1955).
34. Gánti, T. *A Theory of Biochemical Supersystems* (Akadémiai Kiadó, Budapest and Univ. Park Press, Baltimore, 1979).
35. Wächtershäuser, G. *Microbiol. Rev.* **52,** 452–484 (1988).
36. Joyce, G. F., Schwartz, A. W., Orgei, L. E. & Miller, L. S. *Proc. natn. Acad. Sci. USA* **84,** 4398–4402 (1987).
37. von Kiedrowski, G. *Angew. Chem. Inter. Ed.* **25,** 923–935 (1986).
38. Szathmáry, E & Gladkih, I. *J. theor. Biol.* **138,** 55–58 (1989).
39. Szathmáry, E. *Trends Ecol. Evol.* **6,** 366–370 (1991).
40. von Kiedrowski, G. *Bioorg. Chem. Frontiers* **3,** 113–146 (1993).
41. Szathmáry, E. *Proc. natn. Acad. Sci. USA* **90,** 9916–9920 (1993).
42. Benner, S. A. *et al. Cold Spring Harbor Symp. quant. Biol.* **52,** 56–63 (1987).
43. Lazcano, A., Guerrero, R., Margulis, L. & Oró, J. *J. molec. Evol.* **27,** 283–290 (1988).
44. Jablonka, E. *J. theor. Biol.* **170,** 301–309 (1994).
45. Jablonka, E. & Lamb, M. J. *Epigenetic Inheritance and Evolution: The Lamarckian Dimension* (Oxford Univ. Press, Oxford, 1995).

46. Szathmáry, E. *J. theor. Biol.* **169,** 125–132 (1994).
47. Wolpert, L. *Biol. J. Linn. Soc.* **39,** 109–124 (1990).
48. Bickerton, D. *Language and Species* (Univ. Chicago Press, Chicago, 1990).
49. Chomsky, N. *Aspects of the Theory of Syntax* (MIT Press, Cambridge, MA, 1965).
50. Pinker, S. *The Language Instinct: The New Science of Language and Mind* (Penguin, London, 1994).
51. Pinker, S. & Bloom, P. *Behav. Brain. Sci.* **13,** 707–784 (1990).
52. Hinton, G. E. & Nowlan, S. J. *Complex Syst.* **1,** 495–502 (1987).
53. Gopnik, M. *Nature* **344,** 715 (1990).
54. Gánti, T. *The Principle of Life* (OMIKK, Budapest, 1987).
55. Szathmáry, E. *Nature* **344,** 115 (1990).
56. Prudent, J. R., Uro, T. & Schultz, P. G. *Science* **264,** 1924–1927 (1994).
57. Famulok, M. *J. Am. chem. Soc.* (in the press).
58. Jeon, K. W. in *Symbiosis as a Source of Evolutionary Innovation* (eds. Margulis, L. & Fester, R.) 118–131 (MIT Press, Cambridge, MA, 1991).
59. Rasbach, H., Rasbach, K., Reichstein, J. J. & Vida, G. *Ber. Bayer, Bot. Ges.* **50,** 23–27 (1979).
60. Eigen, M. *Naturwissenschaften* **58,** 465–523 (1971).
61. Eigen, M. & Schuster, P. *Naturwissenschaften* **64,** 541–565 (1977).
62. Maynard Smith, J. & Szathmáry, E. *J. theor. Biol.* **164,** 437–446 (1993).
63. Szathmáry, E. & Maynard Smith, J. *J. theor. Biol.* **164,** 447–454 (1993).
64. Weiner, M. & Maizels, N. *Proc. natn. Acad. Sci. USA* **91,** 6729-6734 (1994).
65. Gánti, T. *BioSystems* **7,** 189–195 (1975).
66. Koch, A. L. *J. molec. Evol.* **21,** 270–277 (1985).
67. Tarumi, K. & Schwegler, H. *Bull. Math. Biol.* **47,** 307–320 (1987).
68. Cairns-Smith, A. G. & Walker, G. L. *BioSystems* **5,** 173–186 (1974).
69. Wächtershäuser, G. *Progr. Biophys. molec. Biol.* **58,** 85–201 (1992).
70. Bachmann, P. A., Luigi, P. L & Lang, J. *Nature* **357,** 57–59 (1992).
71. King, G. A. M. *BioSystems* **13,** 23–45 (1980).
72. Wächtershäuser, G. *Proc. natn. Acad. Sci. USA* **91,** 4283–4287 (1994).

Acknowledgments

This work has partly been supported by the Hungarian National Scientific Research Fund (OTKA).

CARO-BETH STEWART

THE POWERS AND PITFALLS OF PARSIMONY

Parsimony analysis is a powerful tool for the study of biological evolution. It is used to construct phylogenetic trees, to evaluate alternative hypotheses objectively, and to study evolutionary pattern and process. Yet, as comparative data sets expand, the pitfalls of parsimony analysis are catching experts and novices alike.

The principle of parsimony states that the simplest explanation consistent with a data set should be chosen over more complex explanations, and is a guiding tenet in scientific study (for critical review, see ref. 1). Parsimonious reasoning is a fundamental way of "knowing" in comparative evolutionary biology, whether the raw data are molecular, physiological, morphological or behavioral.[1–3] In particular, the principle of parsimony is the foundation of a powerful method, called parsimony analysis, that allows reconstruction of the past to be undertaken with logical and statistical rigor.[4–6] Here I outline the powers and pitfalls of parsimony analysis, with emphasis on its application to molecular-sequence data.[7–11] The following represents my opinion as regards a reasonable consensus in the field; it is written with the novice in mind.

Darwin[12] referred to evolution as "descent with modification," and this simple phrase still embodies our best current understanding of the process.[1,4,13] The goal of building phylogenetic trees is to discover the genealogical relationships between "taxa" (biological entities such as genes, proteins, individuals, populations, species, or higher taxonomic units). A phylogeny is a hierarchy of nested sets of taxa indicating relative recentness of common ancestry.[4–6,12–15]

A widespread misconception, especially concerning molecular sequences,[16] is that the building of evolutionary trees simply requires the grouping of taxa according to overall similarity.[1,4,13] Several methods that embody this concept have been invented: they are referred to as distance-matrix methods.[9,17–20] These methods ignore the possibility that apparent overall similarity and true evolutionary relationship are not necessarily the same thing.[1,4–6,21] Two taxa can appear quite similar to each other yet be related relatively distantly. (After all, who among us would assume that an Elvis Presley look-alike is actually his twin brother?) Conversely, two closely related taxa may appear quite different from each other.

This distinction can be illustrated by the genealogical relationship of tetrapods (the four-legged vertebrates) to their fish-like ancestors. In Fig. 1, we consider the possible evolutionary relationships between three vertebrate lineages, one leading to sharks, another to lungfishes, a third to primates. If these species were grouped according to overall morphological similarity, then the tree that unites sharks and lungfishes (Fig. 1, tree 1) would be chosen. Yet, careful morphological comparisons of extant and fossil vertebrates clearly indicates that lungfishes and tetrapods are more closely related to one another than either are to sharks[22] (that is, tree 3 is correct). The apparent similarity between sharks and lungfishes is due to the retention of ancestral characteristics; these two lineages have evolved more slowly than has the lineage leading to primates.

Although molecular sequences generally evolve at a more regular "clock-like" rate than morphological characters,[23] there are many known examples of the same molecule evolving at different rates in different species.[24] For example, baboon α-globin differs from rhesus monkey α-globin by 9 amino acids and from human α-globin by 11 amino acids, whereas the human and rhesus proteins differ by only 5 amino acids.[24,25] Of these three species, the two monkeys are most closely related. Assuming that the three α-globins are the products of the "same" globin gene duplicate in the three primates which diverged when the species diverged (that is, the genes are "orthologous"[11]), then the two monkey α-globins are most closely related; the large difference in their amino-acid sequences is most probably due to very rapid evolution of the baboon protein.[24,25] Yet, if these three primate molecules were grouped by overall similarity, the human and rhesus α-globins would cluster together. The above dis-

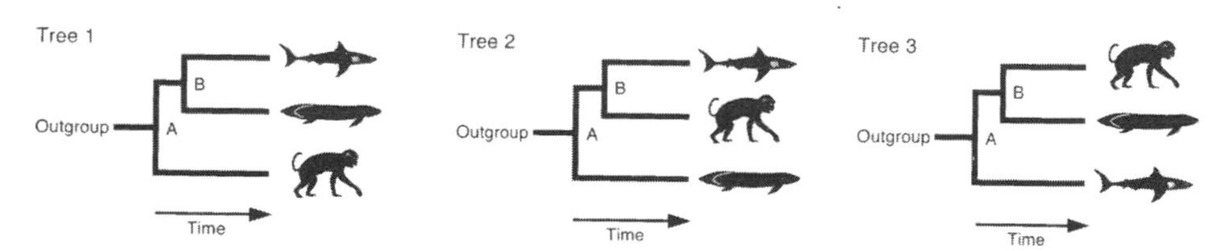

Fig. 1 Three possible trees relating sharks, lungfishes, and tetrapods. Shown here are the three possible bifurcating, rooted (node A) evolutionary trees relating sharks, lungfishes and tetrapods. In tree 1, sharks and lungfishes are depicted as most closely related; that is, they share a more recent common ancestor (node B) than either does with tetrapods (represented here by a monkey). In tree 2, sharks and monkeys are depicted as most closely related. In tree 3, lungfishes and monkeys are depicted as most closely related. Tree 3 is thought to be correct.[22]

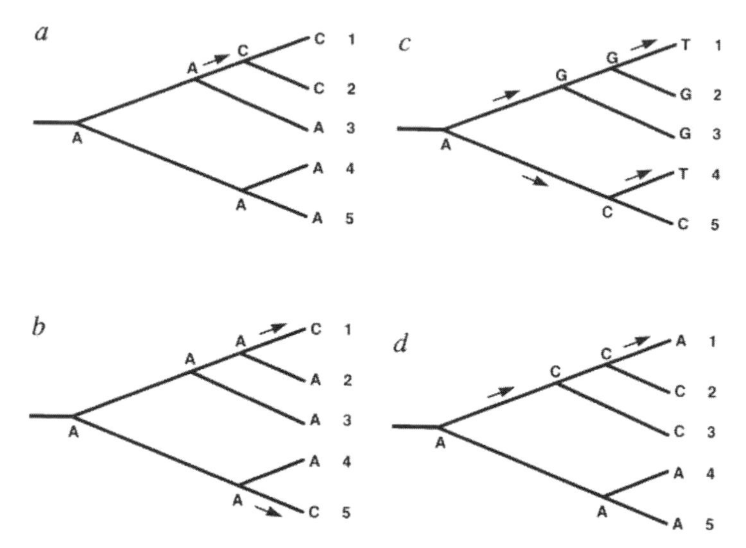

Fig. 2 The different types of similarity. The same phylogenetic tree uniting five taxa (labelled 1–5) is shown in each of the figures. The various reasons why homologous nucleotides from different taxa can be the same[56] are illustrated. These principles also apply to other unpolarized characters (that is, characters such as molecular sequences for which the ancestral state is not known *a priori*). For purposes of illustration, the ancestral states of the characters are assumed to be known. *G,* guanine; *A,* adenine; *T,* thymine; *C,* cytosine. *a,* Shared-ancestral and shared-derived characters. The Cs observed in taxa 1 and 2 are examples of shared-derived characters, whereas the As in taxa 3, 4 and 5 are shared-ancestral characters. Shared-derived characters indicate relative recentness of common ancestry, whereas shared-ancestral characters do not. Characters that are shared-derived for one level in the hierarchy may be shared-ancestral for more closely related taxa. *b,* Parallel substitutions. The Cs observed in taxa 1 and 5 are the result of nucleotide substitutions that occurred in parallel along those lineages, as indicated by the arrows. If the phylogeny were not known, this site would appear to support a tree that links taxa 1 and 5. By parsimonious reasoning on this known phylogeny, the ancestral states at the internal nodes (that is, A at each node) and the direction of substitution (a change from an ancestral A to a derived C) can be inferred from the observed sequences in the five taxa. *c,* Convergent substitutions. The Ts observed in taxa 1 and 4 are the result of convergent nucleotide substitutions because they arose from different ancestral nucleotides (G and C), and caused the sequences of taxa 1 and 4 to regain sequence similarity over time.[26,56] (In this case, the direction of the substitutions could not be inferred from the observed sequences without further information.) The term convergence is used to mean different things in different contexts.[26,27] For example, unrelated molecules are said to be convergent if they have gained similar functions, regardless of how this was accomplished. In common usage, the term *convergence* is often used in place of the term *homoplasy*. *d,* Reversal. The A observed in taxon 1 is the result of a change back to the ancestral state, A, which was retained in taxa 4 and 5. This reversal makes taxon 1 appear more similar to taxa 4 and 5 than it is to its closer relatives, taxa 2 and 3. (If the ancestral states were not known in this case, there would be other equally parsimonious solutions.)

tinction between overall similarity and evolutionary relationship clearly applies with equal force to molecules; therefore, the task of building trees that accurately reflect evolutionary history is often more complicated than simply clustering taxa by overall similarity.

It has long been recognized[14,26] that overall similarity can be broken into three components (Fig. 2): (1) shared-ancestral characters, which are due to retention of traits found in the common ancestor; (2) shared-derived characters, which are due to new traits or modifications that arose along more recent lines of common descent; and (3) homoplasies (convergences, parallelisms and reversals), which are due to the same new trait or modification having been derived independently along different lineages.[27] The concept of overall similarity consciously groups "true" evolutionary resemblance (shared-ancestral characters and shared-derived characters) together with "false" resemblance (homoplasy).[4] Willi Hennig[14] formalized the concept that relative closeness of evolutionary relationship should be inferred solely on the basis of shared-derived characters. Hennig's ideas spawned the school of thought called cladistics[4–6,15,28] from which modern parsimony analysis developed.[7–11,13]

Powers of Parsimony

The fundamental powers of parsimony analysis are that it uses derived characters to infer phylogenetic trees (see Box 1), and that it can be used to tweeze apart overall similarity into its various components[14] (Fig. 2). Parsimony analysis may well be the most versatile and powerful tool in evolutionary biology; but, like any power tool, its proper use requires training, practice and attentiveness.

Building Phylogenetic Trees

The essence[18,29] of building parsimony trees from molecular sequences[7–11] is illustrated in Box 1. At its simplest, the method is a purely logical one that partitions similarities on a character-by-character basis. Alternative trees are evaluated, one character at a time, to determine how many evolutionary events they each require. By the criterion of parsimony, the "best" or most parsimonious tree is the one that requires the fewest total events.[30,31] At the operational level, parsimony analysis does not distinguish between shared-derived similarities and homoplasies.[32] The most parsimonious tree will be the correct phylogeny only if the number of shared-derived characters is high enough and the number of homoplasies low enough. Recent studies using known phylogenies and actual molecular data have shown that parsimony analysis is generally reliable for the

BOX 1 Building parsimony trees using nucleotide sequences

The simplest case[18] for undirected parsimony analysis involves a minimum of four taxa. There are three possible unrooted bifurcating trees for four taxa, as illustrated in *a,* using shark (S), lungfish (L), monkey (M) and a non-vertebrate outgroup (O). Parsimony analysis of nucleotide sequences requires homologous (evolutionarily related) and properly aligned sequences; each nucleotide position in the sequence is considered a "character" which can have five states (G, A, T, C or gap). Ten nucleotides from a hypothetical gene for the four taxa are aligned in *b.* Characters 1, 3 and 7 are invariant (shared-ancestral for all four taxa); invariant and highly conserved characters allow alignments and inference of homology (evolutionary relationship) between sequences. Only "phylogenetically informative" sites (those that can be used to choose between alternative trees, in this case because two taxa have one nucleotide and two other taxa have another nucleotide) are useful in building parsimony trees.[18] The phylogenetically informative characters are marked with an asterisk (*). The other variable sites (characters 4 and 10) are not phylogenetically informative because all trees require the same number of substitutions.

Choosing between alternative trees. To find the most parsimonious tree, each character is evaluated on the unrooted trees to determine how few substitutions are needed to explain its observed distribution. For example, character 2 is analyzed on the three unrooted trees in *a.* As indicated by the arrows, tree 1 requires a minimum of two nucleotide substitutions (one of two equally parsimonious solutions is shown, with A assumed at the nodes, and changes to G occurring along the lineages leading to M and L), tree 2 requires a minimum of two substitutions (again, only one of the parsimonious solutions is shown), and tree 3 requires only one substitution. If each character in the sequence were analyzed in this manner (and the reader should do this), the events per tree would be as shown in *b.*

The optimal or "best" tree by the parsimony criterion is the one that requires the fewest total evolutionary events[30,31] (nucleotide substitutions, in this case); that tree is said to be the "most parsimonious" or "shortest" tree. The shortest tree in this hypothetical example is tree 3, because it requires 10 substitutions, whereas trees 1 and 2 each require 12. The sites that are compatible with each tree are underlined; the number of compatible sites per tree can be used in statistical tests in hopes of ruling out alternatives.[29,39]

Rooting the tree. The tree produced by parsimony programs such as PAUP[32] is "unrooted," meaning that the ancestral node has not been identified.

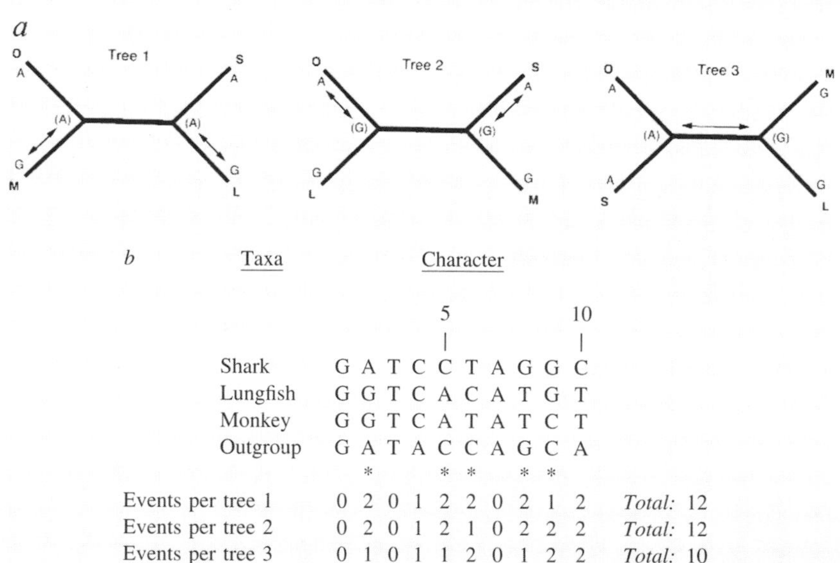

Taxa	1	2	3	4	5	6	7	8	9	10	
Shark	G	A	T	C	C	T	A	G	G	C	
Lungfish	G	G	T	C	A	C	A	T	G	T	
Monkey	G	G	T	C	A	T	A	T	C	T	
Outgroup	G	A	T	A	C	C	A	G	C	A	
		*			*	*		*	*		
Events per tree 1	0	2	0	1	2	2	0	2	1	2	*Total:* 12
Events per tree 2	0	2	0	1	2	1	0	2	2	2	*Total:* 12
Events per tree 3	0	1	0	1	1	2	0	1	2	2	*Total:* 10

Rooting any tree requires knowledge or assumptions about the data set; decisions regarding rooting cannot be abdicated to the computer. Three common ways of rooting molecular trees are as follows. (1) Outgroup. One or more lineages can be included that are known to have diverged before the divergence of the taxa of interest. Ideally, the outgroup taxon should be a member of a closely related sister group to the taxa of interest, the ingroup. (2) Gene duplication. When dealing with multigene families, a known earlier gene duplicate can be used as an outgroup to root the tree.[10] If gene duplicates are used, a final step is needed to complete the rooted tree: gene duplications must be assigned at internal nodes in sufficient number to explain the known distribution of the duplicates within extant species. (3) Midpoint. If no information is available regarding outgroups, the root can be placed at the half-way mark, or midpoint, along the longest reconstructed lineage between two taxa.[32] Midpoint rooting rests on the shaky assumptions of relatively equal rates of evolution along different lineages and appropriate sampling of the lineages.

To visualize turning an unrooted tree into a hierarchical phylogeny, think of the unrooted tree as though it were made of string, and mentally tug on the ancestral node until the other lineages move into place (or make a string tree and do this manually). Tugging on the outgroup of each of the unrooted trees shown here will produce the respective rooted trees in Fig. 1.

inference of the correct tree.[33,34] Parsimony and other methods of phylogenetic tree building have been reviewed recently.[9,17,19]

It is simple to examine all possible trees for four or five taxa manually. Indeed, building trees by hand is the best way to understand the method, and is a highly recommended exercise (see Box 1). But, as the number of taxa increases, the number of possible trees increases in a greater than exponential manner,[30] and building trees quickly becomes a task best suited for a computer.

A versatile and sophisticated computer package for inferring parsimony trees, Phylogenetic Analysis Using Parsimony (PAUP),[32] has been developed by David Swofford. Although other good parsimony programs are available (listed in refs. 9 and 20), PAUP is the most generally useful package, and will be discussed here. PAUP can guarantee to find the most parsimonious trees for relatively large data sets, and has many invaluable features and options. Using various user-defined assumptions, PAUP can analyze any type of character data (such as nucleic-acid sequences, protein sequences, restriction site polymorphisms, morphological characters, behavioral characters, and so on). The ability to analyze diverse data sets with the same basic method and computer program is helping to unite the subdisciplines of evolutionary biology.

In theory, a tree built from molecular sequences (the "gene tree") should be an accurate reflection of the evolutionary history of those sequences. An accurate gene tree will mirror the phylogeny of the species from which the molecules were obtained if, and only if, the molecules being compared are orthologous, are not different alleles that have been retained across speciation events, and have not been the victims of gene conversion or other disruptive events after the speciation.[18,29,35] The inference of species phylogenies from gene trees is in widespread practice today, and has yielded many important and unexpected results.[36,37] Gene trees also inform us about the evolution of multigene families.[10,18,38]

Objective Evaluation of Trees

Although the statistical properties of phylogenetic trees are not completely understood, several statistical methods have been developed to evaluate tree hypotheses.[19,39,40] One widely used approach is the "bootstrap" resampling method,[19] which is used to evaluate the support for internal branches of a tree; this method is available as an option in PAUP. Another class of methods counts the number of sites compatible with different (usually four-taxa) phylogenies, and applies a statistical test to see if alternative trees can be ruled out.[29,39] The ability to test alternative hypotheses objectively makes phylogenetic reconstruction a "falsifiable" discipline.[40]

Character and Process Analysis

Parsimony analysis attains its greatest power for the study of evolutionary pattern and process when done on rooted trees (Fig. 2), here referred to as cladistic or character analysis. Cladistic analysis can be used to reconstruct character states at the internal and ancestral nodes, and can suggest whether a given character is shared-ancestral, shared-derived, uniquely derived, or a result of homoplasy.[27,38,41–46] This information can be used to infer modes of evolution (what happened), to calculate rates of evolution (how quickly it happened) and to detect adaptation (why it happened).[24,45,46] Parsimonious reasoning on phylogenies is a rigorous method in comparative biology, as is elegantly explained in two recent books.[2,3] Although this approach is not widely used by molecular biologists, character analysis is a powerful tool for studying the evolution of the molecules themselves.[10,11,38,44–46] For example, the most parsimonious explanation concerning introns in protein-coding genes is that they were inserted during early eukaryotic evolution.[47]

Although character analysis can be done manually, MacClade,[43] a visual and interactive Macintosh computer program, greatly aids in these calculations. The manual to this program provides a helpful review of parsimony and character analysis.

Pitfalls of Parsimony

Parsimony analysis is simple and straightforward: therefore, its problems are relatively well understood. The problems encountered while building parsimony trees generally fall into two broad categories: (1) failure to find the shortest tree, and (2) the shortest tree not being the correct phylogeny. These pitfalls are discussed below, along with possible ways to detect and avoid them.

Many of the pitfalls may be avoided entirely through judicious choices regarding the number and type of taxa and characters collected. Before beginning a study, careful thought should be given to what scientific questions are to be addressed. Regardless of the major scientific questions, certain practical and methodological questions should be answered (see Box 2).

Failure to Find the Shortest Tree

The goal of building parsimony trees is to find the shortest tree (or trees) that exists for a given data set under the chosen assumptions.[31,32] This pursuit can fail for two practical reasons: too many taxa or too few phylogenetically informative sites, or a combination of both.

Too Many Taxa

Whenever possible, evolutionary tree programs should be used in a mode that guarantees to find all optimal trees. PAUP[32] has two exact algorithms that guarantee to find all most-parsimonious trees for a limited number of taxa. The task of finding the shortest tree becomes prohibitively time-consuming as the number of taxa increases, even for the most sophisti-

BOX 2 Practical and methodological questions concerning evolutionary trees

Below are examples of some basic questions that should be answered regarding phylogenetic analyses; generally this information should be presented with published phylogenies. If space permits, the complete sequence alignment should also be presented, as should any derived distance matrices used in the analyses. The primary data ought to be easily available to reviewers and readers who wish to verify results or try additional analyses; submitting the data on diskettes with the manuscript is recommended.

What computer program and algorithm was used to build the tree(s)?

Was the program used in a manner that guarantees to find the "best" tree?

If not, what measures were taken to find the best tree? (For example, how many times was the data set reordered and the program rerun?)

What criterion was used to select the tree or trees presented?

Were there other trees that tied for best? (If so, how many? What do they look like; that is, what are their evolutionary implications?)

If many equally parsimonious trees were found, does it make more sense to present a consensus tree?

How many evolutionary events (that is, nucleotide substitutions, amino-acid replacements and so on) does the best tree require?

Are there other trees that require only a few more events?

Can these alternative trees be ruled out statistically?

How much support (that is, bootstrap value) is there for any given branch?

Is the gene tree a reasonable species tree? If not, what are the possible explanations?

cated and rapid computers and programs. For up to 10 taxa, PAUP can do an exhaustive search of all possible trees, and produce a frequency distribution showing the tree lengths. Using a "branch and bound" algorithm, PAUP can guarantee to find the shortest tree for a maximum of about 12 to 25 taxa, depending on the data, but will not give the distribution.[32] The practical upper limit for the number of taxa depends on the length and complexity of the character data, the assumptions and options chosen, the speed of the computer, and the patience of the investigator. To analyze a large number of taxa, a computer run may take hours, days or even weeks. Because it often takes longer to analyze data properly than to generate it, phylogenetic analysis should not be tacked on to the end of a study as an afterthought. A few suggestions concerning the analysis of large data sets follow.

If the data set has too many taxa for exact algorithms, question whether all taxa are necessary. Rarely is the exact placement of every available taxon relevant to the question of interest. Sometimes taxa can be selectively omitted, provided that important conclusions are not altered significantly by the choice of omitted data.[48]

Alternatively, a large set of taxa can be broken into smaller groups.[49] For example, one may be interested in the branching order of certain major lineages (such as plants, animals and fungi), for which numerous sequences are available. Rather than omitting taxa, one might find the most parsimonious tree for the species within each major lineage, then define the topologies of these sub-trees to PAUP.[32] PAUP will treat each defined sub-tree as a single taxon while building trees that relate the lineages, thereby greatly reducing the number of alternative trees that must be considered. Moreover, PAUP will reconstruct ancestral sequences for the sub-trees which should better represent the lineages than would any of the individual sequences. Thus, the power of phylogenetic reconstruction[10,11,42,43] can be combined with the hypothesis testing abilities of the four-way test[29,39] to focus on the question of interest.

For some questions, neither of the above taxa reduction approaches may be appropriate. If so, parsimony programs can be run using faster "heuristic" algorithms that attempt, but do not guarantee, to find the shortest trees. Heuristic parsimony algorithms rearrange starting trees to look for shorter ones, and can get trapped in local optima where no simple rearrangement will yield the globally most parsimonious tree. A way to escape local optima is to use many different starting trees;[50] this can be done automatically in PAUP through its random addition option.[32]

Distance trees also can be used as starting trees in heuristic parsimony searches. Currently, most distance tree algorithms are heuristic and will analyze a large number of taxa quickly, yet will produce only one "optimal" tree per run. With heuristic algorithms, the input order of the data can influence the branching order of the tree that is found. Therefore, the input matrix should be reordered and the program run repeatedly to search for optimal trees.[20] Ideally, optimal distance and parsimony trees should be found and compared; congruence of the topologies is reassuring, but does not guarantee that the correct tree for that data has been inferred. The computer software package PHYLIP[20] contains programs for most distance-matrix methods, as well as some molecular parsimony programs. Taken together, PAUP and PHYLIP cover most methods for phylogenetic inference.

Too Few Phylogenetically Informative Sites

If a data set does not contain enough shared-derived, phylogenetically informative characters to resolve the branching order of all of the taxa, many equally parsimonious trees may be found. What constitutes "enough" such characters depends on the type and quality of data; as a minimum, the data set generally should contain more phylogenetically informative characters than it does taxa. Poorly supported branches in the trees will have low bootstrap scores. Complete lack of resolution of lineages can be detected by branches of zero length. Distance methods sometimes can perform better than parsimony for data sets having few phylogenetically informative sites, because all variable sites are used in the distance calculations.[51]

The problem of too few phylogenetically informative characters can arise if too short a segment of DNA is sequenced, or if a region is chosen for study that is not evolving rapidly enough to answer the question at hand. It is a good idea to make analysis an ongoing part of the research strategy, because it can provide valuable information concerning adequacy of the data. The obvious solution to this problem is to collect more appropriate characters per taxa, but this is not always feasible. It may be difficult or impossible to collect enough phylogenetically informative characters to fully resolve a gene phylogeny for very closely related taxa, such as individuals within populations or species.

The data set from the well-publicized human mitochondrial DNA study[52] illustrates this point. In this extensive study, 610 nucleotides from the rapidly evolving mitochondrial control region were sequenced from 189 individuals; of the 610 characters, 201 were variable and 119 were phylogenetically informative.[52] This data set contains too many taxa for an exact search, and too few phylogenetically informative sites to allow complete resolution of the phylogeny by parsimony; indeed, numerous equally parsimonious trees exist for these data.[50,52] This example should be a cautionary tale for those who wish to study population genetics phylogenetically: the combined problems of too many taxa and too few phylogenetically informative sites are likely to plague all such studies. Indeed, the literature contains numerous examples that have not been so well publicized.

The Shortest Tree Not Being the Correct Phylogeny

Even if the data set appears to have plenty of phylogenetically informative sites and an exact search finds one most parsimonious tree, that tree may not accurately reflect the true evolutionary history of the taxa. There are conditions under which parsimony analysis can fail to find the correct phylogenetic tree.

Homoplasy, in its various disguises, is the ultimate trickster of parsimony. Identities or similarities due to convergence, parallel evolution, and reversals can cause historically incorrect trees to be most parsimonious. Convergence and parallel evolution are often considered indications of adaptation or positive selection, but a certain amount of homoplasy happens by chance.[27,53]

Chance Homoplasy

The more distantly related the taxa, the more likely that multiple substitutions have occurred at variable sites in the sequences, especially at synonymous sites in codons. Multiple hits obscure the phylogenetic "signal" (the informative sites that support the true phylogeny) with "noise" (homoplasy). A noisy data set cannot produce the correct phylogeny with any certainty;[54] yet low levels of random homoplasy are unlikely to produce an incorrect tree having significant support because the events are usually scattered over the tree and cancel each other.[53]

A quick way to check a data set for phylogenetic signal is to examine the distribution of possible trees for skewness.[7,54] PAUP automatically produces tree distributions during exhaustive searches, and a random sample of possible trees can be produced under other search modes.[32,54] A data set with high phylogenetic signal should have a positive skew, with the shortest tree(s) on a tail of the distribution.[7,53]

How might phylogenetic signal be extracted from noisy DNA data? One approach is to weight transversions more heavily than transitions,[55] which can be done easily in PAUP.[32] For protein-coding genes, third positions in codons can be omitted, and only the more conservative first and second positions analyzed.[29] For distantly related proteins, it may be more productive to analyze the amino-acid sequences[16] with protein parsimony programs such as PROTPARS.[20,32]

Directed Homoplasy

More interesting cases of homoplasy are those due to adaptation rather than chance.[27] In unusual cases, "convergence" of amino-acid sequences can cause distantly related taxa to be pulled together on amino-acid parsimony trees.[29,45,46] A warning sign for such taxonomically localized homoplasy is the discovery of two or more short trees that dramatically rearrange a lineage, combined with an unusually long reconstructed branch length for the "misplaced" lineage.[45] Cladistic analysis of the sequences on the "correct" rooted phylogeny will pinpoint the homoplastic characters.[45,46] Molecular "convergence" or "parallelism" generally presents itself at the protein, not DNA, sequence level.[24,27,45,46] This highlights the need for cladistic analysis of protein sequences to study adaptive evolution of protein structure and function.[38,45]

Although homoplasy may be a pitfall of parsimony tree building, the detection of convergence and parallel evolution is one of the major powers of parsimony analysis. Character analysis on a rooted phylogeny, regardless of how the phylogeny was constructed, is the only method by which homoplasy can be detected.[56]

Conclusion

Genome evolution is a rich and complex tapestry interwoven with chromosome and gene duplications, gene conversions, mobile genetic elements, allelic diversity, hybridization and, in rare cases, sequence convergence and horizontal gene transfer. Evolutionary trees built from molecular sequences reflect these complex processes. Thus, the failure of a molecular phylogeny to reflect the phylogeny of the species perfectly should not be taken as a failure of parsimony analysis. If a gene tree conflicts with an accepted species tree, one should stop and ponder why. If the same tree is found using other genes from the same species, then the molecules are probably correct about the species phylogeny. If not, the different genes may have different evolutionary histories, which can be reconstructed through careful comparative studies. Comparative analysis of genes and proteins in a phylogenetic framework informs us about molecular evolutionary processes, and sheds light on the evolution of genomes. Until recently, the primary use of parsimony analysis has been largely limited to the study of organismal evolution: its potential to resolve questions about molecular evolution is only now being realized.

Caro-Beth Stewart is at the Department of Biological Sciences, State University of New York at Albany, Albany, New York 12222, USA.

References

1. Sober, E. *Reconstructing the Past: Parsimony, Evolution, and Inference* (MIT Press, Cambridge, MA, 1988).
2. Harvey, P. H. & Pagel, M. D. *The Comparative Method in Evolutionary Biology* (Oxford Univ. Press, New York, 1991).
3. Brooks, D. R. & McLennan, D. A. *Phylogeny, Ecology, and Behavior* (Univ. Chicago Press, Chicago, 1991).
4. Eldredge, N. & Cracraft, J. *Phylogenetic Patterns and the Evolutionary Process* (Columbia Univ. Press, New York, 1980).
5. Wiley, E. O. *Phylogenetics: The Theory and Practice of Phylogenetic Systematics* (Wiley, New York, 1981).
6. Hull, D. L. *Science as a Process* (Univ. Chicago Press, Chicago, 1988).
7. Fitch, W. M. in *Cladistics* (eds. Duncan, T. & Stuessy, T. F.) 221–252 (Columbia Univ. Press, New York, 1984).
8. Fitch, W. M. *Am. Nat.* **111,** 223–257 (1977).
9. Swofford, D. L. & Olsen, G. J. in *Molecular Systematics* (eds. Hillis, D. M. & Moritz, C.) 411–501 (Sinauer, Sunderland, MA, 1990).
10. Dayhoff, M. O. & Eck, R. V. *Atlas of Protein Sequence and Structure* Vol. 2 (National Biomedical Research Foundation, Silver Spring, MD, 1966).
11. Fitch, W. M. *Syst. Zool.* **19,** 99–113 (1970).
12. Darwin, C. *On the Origin of Species by Means of Natural Selection* (Murray, London, 1859).
13. Hull, D. L. in *The Hierarchy of Life: Molecules and Morphology in Phylogenetic Analysis* (eds. Fernholm, B., Bremer, K. & Jörnvall, H.) 3–15 (Elsevier, Amsterdam, 1989).
14. Hennig, W. *Phylogenetic Systematics* (translated by Davis, D. D. & Zangerl, R.) (Univ. Illinois Press, Urbana, 1966).
15. Wiley, E. O., Siegel-Causey, D., Brooks, D. R. & Funk, V. A. *The Compleat Cladist: A Primer of Phylogenetic Procedures* (Univ. Kansas Museum of Natural History, Special Publication No. 19, 1991).
16. Doolittle, R. *Of URFs and ORFS: A Primer on How to Analyze Derived Amino Acid Sequences* (University Science Books, Mill Valley, CA, 1986).
17. Nei, M. in *Phylogenetic Analysis of DNA Sequences* (eds. Miyamoto, M. M. & Cracraft, J.) 90–128 (Oxford Univ. Press, New York, 1991).
18. Li, W.-H. & Graur, D. *Fundamentals of Molecular Evolution* (Sinauer, Sunderland, MA, 1991).
19. Felsenstein, J. A. *Rev. Genet.* **22,** 521–565 (1988).
20. Felsenstein, J. *PHYLIP (Phylogeny Inference Package).* Version 3.5 (Computer software package and manual distributed by the author, Dept. Genetics, University of Washington, Seattle, WA, 1993).

21. Gould, S. J. *Nat. Hist.* **92,** 14–21 (1992).
22. Colbert, E. H. & Morales, M. *Evolution of the Vertebrates* 4th edn. (Wiley, New York, 1977).
23. Wilson, A. C., Carlson, S. S. & White, T. J. A. *Rev. Biochem.* **46,** 573–639 (1977).
24. Gillespie, J. *The Causes of Molecular Evolution* (Oxford Univ. Press, New York, 1991).
25. Shaw, J.-P., Marks, J., Shen, C. C. & Shen, C.-K. J. *Proc. natn. Acad. Sci. USA* **86,** 1312–1316 (1989).
26. Haas, O. & Simpson, G. G. *Proc. Am. Phil. Soc.* **90,** 319–349 (1946).
27. Patterson, C. *Molec. Biol. Evol.* **5,** 603–625 (1988).
28. Nelson, G. & Platnick, N. *Systematics and Biogeography: Cladistics and Vicariance* (Columbia Univ. Press, New York, 1981).
29. Irwin, D. M. & Wilson, A. C. in *Mammalian Phylogeny* (eds. Szalay, F. S., Novacek, M. J. & McKenna, M. C.) 257–276 (Springer, New York, 1992).
30. Cavalli-Sforza, L. L. & Edwards, A. W. F. *Evolution* **32,** 550–570 (1967).
31. Farris, J. S. *Syst. Zool.* **19,** 83–92 (1970).
32. Swofford, D. L. *PAUP: Phylogenetic Analysis Using Parsimony.* Version 3.0 s (Computer program and manual distributed by the Center for Biodiversity, Illinois Natural History Survey, Champaign, IL 61820, 1992).
33. Atchley, W. R. & Fitch, W. M. *Science* **254,** 554 (1991).
34. Hillis, D. M., Bull, J. J., White, M. E., Badgett, M. R. & Molineux, I. J. *Science* **255,** 589–592 (1992).
35. Doyle, J. *Syst. Bot.* **17,** 144–163 (1992).
36. Wilson, A. C., Zimmer, E. A., Prager, E. & Kocher, T. D. in *The Hierarchy of Life* (eds. Fernholm, B., Bremer, K. & Jörnvall, H.) 407–419 (Elsevier, Amsterdam, 1989).
37. Graur, D., Hide, W. A. & Li, W.-H. *Nature* **351,** 649–652 (1991).
38. Stewart, C.-B. *Meth. Enzym.* (in the press).
39. Li, W.-H. & Gouy, M. *Meth. Enzym.* **183,** 645–659 (1990).
40. Penny, D., Hendy, M. D. & Steel, M. A. in *Phylogenetic Analysis of DNA Sequences* (eds. Miyamoto, M. M. & Cracraft, J.) 155–183 (Oxford Univ. Press, New York, 1991).
41. Swofford, D. L. & Maddison, W. P. in *Systematics, Historical Ecology, and North American Freshwater Fishes* (ed. Mayden, R. L.) 186–223 (Stanford Univ. Press, Stanford, 1992).
42. Maddison, W. P. & Maddison, D. R. *Folia Primatol.* **53,** 190–202 (1989).
43. Maddison, W. P. & Maddison, D. R. *MacClade: Analysis of Phylogeny and Character Evolution* (Sinauer, Sunderland, MA, 1992).
44. Stackhouse, J., Presnell, S. R., McGeehan, G. M., Nambiar, K. P. & Benner, S. A. *FEBS Lett.* **262,** 104–106 (1990).
45. Stewart, C.-B., Schilling, J. W. & Wilson, A. C. *Nature* **330,** 401–404 (1987).
46. Swanson, K. W., Irwin, D. M. & Wilson, A. C. *J. molec. Evol.* **33,** 418–425 (1991).
47. Palmer, J. D. & Logsdon, J. M. Jr. *Curr. Opin. Genet. Dev.* **1,** 470–477 (1991).
48. Patterson, C. in *The Hierarchy of Life* (eds. Fernholm, B., Bremer, K. & Jörnvall, H.) 471–488 (Elsevier, Amsterdam, 1989).
49. Sankoff, D., Cedergren, R. J. & McKay, W. *Nucleic Acids Res.* **10,** 421–431 (1982).
50. Maddison, D. R., Ruvolo, M. & Swofford, D. L. *Syst. Biol.* **41,** 111–124 (1992).
51. Cornish-Bowden, A. *J. theor. Biol.* **101,** 317–319 (1983).

52. Vigilant, L., Stoneking, M., Harpending, H., Hawkes, K. & Wilson, A. C. *Science* **253,** 1503–1507 (1991).
53. Peacock, D. & Boulter, D. *J. molec. Biol.* **95,** 513–527 (1975).
54. Hillis, D. M. in *Phylogenetic Analysis of DNA Sequences* (eds. Miyamoto, M. M. & Cracraft, J.) 278–294 (Oxford Univ. Press, New York, 1991).
55. Fitch, W. M. & Ye, J. in *Phylogenetic Analysis of DNA Sequences* (eds. Miyamoto, M. M. & Cracraft, J.) 147–154 (Oxford Univ. Press, New York, 1991).
56. Sneath, P. H. A. & Sokal, R. R. *Numerical Taxonomy: The Principles and Practice of Numerical Classification* (Freeman, San Francisco, 1973).

Acknowledgments

I thank R. Collura, K. Helm-Bychowski, D. Irwin, W. Messier, L. Taylor and D. Swofford for commenting on various versions of the manuscript; R. Collura for figures; D. Hillis for preprints and discussions; D. Swofford for a test version of PAUP; D. Maddison and W. Maddison for a test version of MacClade; and J. Felsenstein for PHYLIP. This paper is dedicated to the memory of Allan C. Wilson.

PART 2

GEOFFROY'S LEGACY: DEVELOPMENT, EVOLUTION AND PALEONTOLOGY

How do body plans and body parts evolve? asks Sean Carroll in the first of four review articles collected in this section. Remarkable insights into this question have come from the realization that most (if not all) multicellular animals share a cluster of genes, the so-called *Hox* cluster, that seems to direct the regional specification of the anteroposterior axis. In a way that is both deep and true, the front and back ends of humans, worms and flies are homologous—in Geoffroy's transcendental world, they are the "same."

Indeed, as I said in the introduction to this volume, the astonishing demonstrations of close similarities in the genetic organization in, say *Drosophila* and other animals such as mice, or the amphioxus (*Branchiostoma*), a primitive chordate,[1] make one wonder how these two animals can manage to look so different. Sean Carroll reviews the natural history and behavior of the *Hox* genes in arthropods and chordates in an attempt to answer this question. He argues that diversity comes from changes in how *Hox* genes are regulated, and in how they regulate other genes in their turn.

The anatomical complexity of the vertebrates may stem from large-scale duplication of *Hox* and other genes in evolution, with the duplicates being co-opted for other uses.[2–6] In general, the diversity of arthropods such as insects and vertebrates has arisen mainly through changes in how genes are regulated.[7–11]

For a general, highly readable account of the evolution of body plans, from genetic as well as embryological perspectives, I recommend a recent and very clear exposition by Wallace Arthur.[12]

Geoffroy's ghost has been stirred by a particularly intriguing recent theoretical offshoot: the idea, from gene homologies, that arthropods and chordates share the same dorsoventral axis, except that the genes concerned specify inverse functions. That is, the arthropod homologue of a vertebrate gene that specifies a dorsalizing factor in vertebrates, specifies a ventralizing factor in arthropods—and vice versa. This has echoes of Geoffroy's work on inverting a lobster to produce a vertebrate body plan (see ref. 13 for more discussion on this point).

These similarities were highlighted by a series of papers from Nübler-Jung and Arendt, and short commentaries, many in reply to these papers.[14–18] Much of the molecular work on the dorsoventral axis comes from the laboratory of E. M. De Robertis at the University of California, Los Angeles. This resurgence of interest in the topic, largely prompted by the speculations of Nübler-Jung and Arendt, prompted De Robertis and his colleague Yoshiki Sasai to write the short review in *Nature* that forms the second item in this section, in which they use the teasing homologies and inversions in an attempt to understand the organization of the common ancestor of arthropods and chordates. They consider, in particular, the suggestion that the common ancestor of arthropods and chordates might have been segmented. Some of their ideas seemed very speculative at the time, but correspondences that have appeared since—such as the similarities of expression of the gene *engrailed* in the chordate *Branchiostoma* with respect to that in the arthropod *Drosophila*[1,19]—have added conviction. For further reading, see refs. 20–22.

An increasingly well studied "model system" in developmental biology is that of the vertebrate limb. Interest in this topic goes back to the late nineteenth century, but the "modern era" tends to take its cue from a 1986 paper from Neil Shubin and the late Pere Alberch[23] in which a close study of the formation of the tetrapod limb was used to make evolutionary speculations. Shubin is primarily a paleontologist, working on vertebrates: in the third review in this section, he teams up with Sean Carroll (primarily interested in arthropods) and Cliff Tabin (a leading vertebrate embryologist) to produce a comprehensive treatment of the origin and evolution of limbs in animals generally. As with the formation of the dorsoventral and anterior-posterior axis, there are many parallels. The main effect of this review is to raise, once again, the specter of "homology." The delineation of two structures as homologous rather depends on how you look at them—from perspectives that may be anatomical, phylogenetic, embryological or genetic.

The paleontological perspectives of this review lead conveniently to the fourth and last review in this section, a characteristically lively piece from Simon Conway Morris that was originally entitled "Let's Play Metazoan Phylogeny." Conway Morris has one of the most fearless imaginations in paleontology, honed by his long study of some of the more unusual and problematic fossils from the Middle Cambrian Burgess Shales of British Columbia and other Middle Cambrian assemblages.[24]

In the review here (see also refs. 25–27), Conway Morris (and others: see ref. 28) argues that the progress of molecular biology toward understanding the phylogeny, origin and early evolution of multicellular animals—the Metazoa—can only be assisted by the study of fossils, many of them representatives of groups now long extinct. References 29–41 provide further reading on specific instances of how fossils contribute to debates on phylogeny, and on other problematic fossils.

One of these concerns the enigmatic fossil *Yunnanozoon* from the Burgess-like Chengjiang fauna from the Cambrian of China. This creature is variously interpreted as a cephalochordate, the same group that includes the extant *Branchiostoma;*[36] a hemichordate;[37] or a stem-group deuterostome.[38] This debate provides a graphic illustration of the problems paleontologists face when trying to interpret a fossil whose morphology defies interpretation in terms of the necessarily limited range of extant forms.

When discussing the origin of the Metazoa, the subject invariably turns to the vexed question of the so-called Cambrian explosion. The seemingly sudden appearance of a wide range of metazoan forms in the Cambrian period has led to a great deal of thought and investigation. Some molecular, developmental and paleontological work has cast doubt on the suddenness of the event, though paleontologists often (though not always) argue strongly in its favor.

For example, Wray and colleagues[42] reported molecular evidence for the existence of separate metazoan lineages well back into the Precambrian. This finding has been questioned purely from methodological[43] and from paleontological[44] perspectives. Yet there still seems to be evidence that lineage splitting—cladogenesis—should be decoupled from the physical appearance of fossils in the stratigraphy, even using purely paleontological data sets.[45]

Whatever the outcome, the continuing debate[46] is an instance of the disjunction between molecular and morphological evidence I discussed above: that is, molecular evidence will reveal when lineages split, but fossils will give only a minimum estimate of the timing of the split, based on the recognition of lineage-specific features. Perhaps metazoan lineages became distinct far back in the Precambrian, but animals remained small, undistinguished and planktonic until the appearance of developmental mechanisms that allowed them to assume large sizes as adults (see ref. 47, and refs. 48–51—useful reviews by Conway Morris—for counterarguments and further perspectives on the Cambrian explosion).

References

1. L. Z. Holland, M. Kene, N. A. Williams and N. D. Holland. Sequence and embryonic expression of the amphioxus *engrailed* gene *(AmphiEn):* The metameric pattern of transcription resembles that of its segment-polarity homolog in *Drosophila. Development* 124 (1997): 1723–1732.
2. P. W. H. Holland, J. Garcia-Fernàndez, N. A. Williams and A. Sidow. Gene duplications and the origins of vertebrate development. *Development* suppl. (1994): 125–133.
3. A. P. Bird. Gene number, noise reduction and biological complexity. *Trends in Genetics* 11 (1995): 94–100.
4. P. W. H. Holland and J. Garcia-Fernàndez. *Hox* genes and chordate evolution. *Developmental Biology* 173 (1996): 382–395.
5. J. Garcia-Fernàndez and P. W. H. Holland. Archetypal organization of the amphioxus *Hox* gene cluster. *Nature* 370 (1994): 563–566.
6. H. Gee. *Before the backbone: Views on the origin of the vertebrates.* London: Chapman and Hall, 1996.
7. J. K. Grenier, T. L. Garber, R. Warren, P. M. Whitington and S. Carroll. Evolution of the entire arthropod Hox gene set predated the origin of the onychophoran/arthropod clade. *Current Biology* 7 (1997): 547–553.
8. M. Averof and N. Patel. Crustacean appendage evolution associated with changes on Hox gene expression. *Nature* 388 (1997): 682–686.
9. M. Averof, R. Dawes and D. Ferrier. Diversification of arthropod Hox genes as a paradigm for the evolution of gene functions. *Cellular and Developmental Biology* 7 (1996): 539–551.
10. R. S. Mann. The specificity of homeotic gene function. *BioEssays* 17 (1995): 855–863.
11. J. Zhang and M. Nei. Evolution of Antennapedia-class homeobox genes. *Genetics* 142 (1996): 295–303.
12. W. Arthur. *The origin of animal body plans.* Cambridge: Cambridge University Press, 1997.
13. B. K. Hall. *Evolutionary developmental biology.* London: Chapman and Hall, 1992.
14. D. Arendt and K. Nübler-Jung. Inversion of dorsoventral axis? *Nature* 371 (1994): 26.
15. K. Nübler-Jung and D. Arendt. Is ventral in insects dorsal in vertebrates? *Roux's Arch. Dev. Biol.* 203 (1994): 357–366.
16. K. Nübler-Jung and D. Arendt. Enteropneusts and chordate evolution. *Current Biology* 6 (1996): 352–353.

17. T. C. Lacalli. Dorso-ventral axis inversion. *Nature* 373 (1995): 110–111.
18. K. J. Peterson. Dorso-ventral axis inversion. *Nature* 373 (1995): 111–112.
19. E. M. De Robertis. The ancestry of segmentation. *Nature* 387 (1997): 25–26.
20. S. Piccolo, E. Agius, B. Lu, S. Goodman, L. Dale and E. M. De Robertis. Cleavage of Chordin by the Xolloid metalloprotease suggests a role for proteolytic processing in the regulation of Spemann organizer activity. *Cell* 91 (1997): 407–4161997.
21. C. B. Kimmel. Was *Urbilateria* segmented? *Trends in Genetics* 12 (1996): 320–331.
22. J. Gerhart and M. Kirschner. *Cells, embryos and evolution.* Oxford: Blackwell Science, 1997.
23. N. H. Shubin and P. Alberch. A morphogenetic approach to the origin and basic organization of the tetrapod limb. *Evolutionary Biology* 20 (1986): 319–387.
24. S. Conway Morris. *The crucible of creation: The Burgess shale and the rise of animals.* Oxford: Oxford University Press, 1998.
25. S. Conway Morris. Early metazoan evolution: First steps to an integration of molecular and morphological data. In S. Bengtson, ed., *Early life on Earth,* 450–459, Nobel Symposium no. 84. New York: Columbia University Press, 1994.
26. S. Conway Morris. Why molecular biology needs paleontology. *Development,* suppl. (1994): 1–13.
27. S. Conway Morris. A palaeontological perspective. *Current Opinion in Genetics and Development* 4 (1994): 802–809.
28. J. W. Valentine, D. H. Erwin and D. Jablonski. Developmental evolution of metazoan bodyplans: The fossil evidence. *Developmental Biology* 173 (1996): 373–381.
29. G. E. Budd. The morphology of *Opabinia regalis* and the reconstruction of the arthropod stem-group. *Lethaia* 29 (1996): 1–14.
30. N. J. Butterfield. Plankton ecology and the Proterozoic-Phanerozoic transition. *Paleobiology* 23 (1997): 247–262.
31. D. Collins. The "evolution" of *Anomalocaris* and its classification in the arthropod class Dinocarida (Nov.) and order Radiodonta (Nov.). *Journal of Paleontology* 70 (1996): 280–293.
32. S. Conway Morris and J. S. Peel. Articulated halkieriids from the Lower Cambrian of North Greenland and their role in early protostome evolution. *Philosophical Transactions of the Royal Society* B 347 (1995): 305–358.
33. J. G. Gehling and J. K. Rigby. Long expected sponges from the Neoproterozoic Ediacara fauna of South Australia. *Journal of Paleontology* 70 (1996): 185–195.
34. G. A. Logan, J. M. Hayes, G. B. Hieshma and R. E. Summons. Terminal Proterozoic reorganization of biogeochemical cycles. *Nature* 376 (1995): 53–56.
35. D.-G. Shu, S. Conway Morris and X.-L. Zhang. A *Pikaia*-like chordate from the Lower Cambrian of China. *Nature* 384 (1996): 157–158.
36. J. Chen, J. Dzik, G. D. Edgecombe, L. Ramsköld and G. D. Zhou. A possible early Cambrian cephalochordate. *Nature* 377 (1995): 720–722.
37. D. Shu, X. Zhang and L. Chen. Reinterpretation of *Yunnanozoon* as the earliest known hemichordate. *Nature* 380 (1996): 428–430.
38. J. Dzik. *Yunnanozoon* and the ancestry of chordates. *Acta Palaeontologica Polonica* 40 (1995): 341–360.
39. B. M. Waggoner. Phylogenetic hypotheses of the relationships of arthropods to Precambrian and Cambrian problematic taxa. *Systematic Biology* 45 (1996): 190–222.

40. M. A. Fedonkin and B. M. Waggoner. The late Precambrian fossil Kimberella is a mollusc-like bilaterian organism. *Nature* 388 (1997): 868–871.
41. J.-Y. Chen, L. Ramsköld and G.-Q. Zhou. Evidence for monophyly and arthropod affinity of Cambrian giant predators. *Science* 264 (1994): 1304–1308.
42. G. A. Wray, J. S. Levinton and L. S. Shapiro. Molecular evidence for deep Precambrian divergence among metazoan phyla. *Science* 274 (1996): 568–573.
43. F. J. Ayala, A. Rzhetsky and F. J. Ayala. Origin of the metazoan phyla: Molecular clocks confirm paleontological estimates. *Proceedings of the National Academy of Sciences USA* 95 (1998): 606–611.
44. S. Conway Morris. Defusing the Cambrian "explosion"? *Current Biology* 7 (1997): R71–R74.
45. R. A. Fortey, D. E. G. Briggs and M. A. Wills. The Cambrian evolutionary "explosion": Decoupling cladogenesis from morphological disparity. *Biological Journal of the Linnean Society* 57 (1996): 13–33.
46. R. A. Raff, C. R. Marshall and J. M. Turbeville. Using DNA sequences to unravel the Cambrian radiation of the animal phyla. *Annual Review of Ecology and Systematics* 25 (1994): 351–375.
47. E. H. Davidson, K. J. Peterson and R. A. Cameron. Origin of bilaterian body plans: Evolution of developmental regulatory mechanisms. *Science* 270 (1995): 1319–1375.
48. S. Conway Morris. Eggs and embryos from the Cambrian. *BioEssays* 20 (1998): 676–682.
49. S. Conway Morris. Metazoan phylogenies: Falling into place or falling to pieces? A palaeontological perspective. *Current Opinions in Genetics and Development* 8 (1998): 662–667.
50. S. Conway Morris. The evolution of diversity in ancient ecosystems: A review. *Philosophical Transactions of the Royal Society* B 353 (1998): 327–345.
51. S. Conway Morris. Palaeontology: Grasping the opportunities in the science of the twenty-first century. *Géobios* 30 (7) (1998): 895–904.

Sean B. Carroll

HOMEOTIC GENES AND THE EVOLUTION OF ARTHROPODS AND CHORDATES

Clusters of homeotic genes sculpt the morphology of animal body plans and body parts. Different body patterns may evolve through changes in homeotic gene number, regulation or function. Recent evidence suggests that homeotic gene clusters were duplicated early in vertebrate evolution, but the generation of arthropod and tetrapod diversity has largely involved regulatory changes in the expression of conserved arrays of homeotic genes and the evolution of interactions between homeotic proteins and the genes they regulate.

Arthropods, in terms of their diversity and sheer numbers, and chordates, because of their anatomical and behavioral complexity, are two of the most successful animal phyla. The evolution of both groups has been marked by numerous innovations and modifications to their respective body plans, the archetypes of which arose more than 500 million years (Myr) ago.[1] Although arthropod and vertebrate phylogeny have long been approached through systematics and paleontology, the genetic basis for the morphological diversity of these or any other animals has, until recently, been beyond the reach of biology.

How do body parts evolve? Differences in morphology are, of course, the developmental products of genetic differences between animals, but it has not been clear how many or what kind of differences underlie changes in body patterns. Do gross changes in morphology, such as the evolution of fish, snakes and mammals, or the invention of the insect wing, involve distinct genetic mechanisms from the evolution of different kinds of butterflies?

Early investigations into the nature of genetic evolution identified two potential mechanisms underlying the origin of new features. Ohno,[2] for example, stressed the role of gene duplication and divergence in evolution, whereas Wilson[3] and Jacob[4] emphasized the power of regulatory changes in gene expression. But in order to determine how new genes or regulatory innovations might effect morphological evolution, it is essential to know which genes control morphology.[3,5,6]

One of the most important biological discoveries of the past decade is that arthropods and chordates, and indeed most or all other animals, share a special family of genes, the homeotic (or *Hox*) genes, which are important for determining body pattern. The diversity of *Hox*-regulated features in arthropods (segment morphology, appendage number and pattern) and vertebrates (vertebral morphology, limb and central nervous system pattern) suggest that the *Hox* genes are implicated at some level in the morphological evolution of these animals.

After a brief review of *Hox* gene organization and function in development, I will describe several examples of morphological differences among arthropods and chordates involving evolution at the *Hox* gene level that have recently been discovered. These include changes in the regulation of *Hox* genes and the evolution of interactions between *Hox* gene products and the genes they regulate. The evidence indicates that the duplication of *Hox* clusters and other developmental genes between primitive chordates and early vertebrates enabled the evolution of the anatomical complexity of vertebrates. However, the diversity of arthropods, insects and vertebrates has arisen primarily through regulatory evolution.

Homeotic Genes in Development

Early studies of homeotic genes were almost entirely restricted to *Drosophila melanogaster,* where these genes were known to control the identity (that is, the unique appearance) of different body segments along the anteroposterior (AP) axis of the embryo and adult fly. Subsequently, *Hox* genes have been found in all sorts of animals, including hydra,[7] nematodes,[8] and all arthropods[9] and chordates.[10] Three remarkable conserved features unite the *Hox* genes of higher animals: (1) their organization in gene complexes; (2) their expression in discrete regions in the same relative order along the main (AP) body axis;[11] and (3) their possession of a sequence of 180 base pairs (the homeobox) encoding a DNA-binding motif (the homeodomain).

Hox Gene Organization and Expression

In *Drosophila* there are eight *Hox* genes among two complexes, the Antennapedia complex (ANT-C)[12] and Bithorax complex (BX-C),[13] which act in different regions of the animal (other homeobox genes of the ANT-C that are not homeotic genes will not be considered here; see ref. 9 for discussion). There is a striking correlation between the order of these genes on the chromosome and the position of their expression in the developing animal (Fig. 1, top). In beetles, one gene complex contains the same set of genes, and they control the unique appearance of different beetle segments in an anterior to posterior order.[14] Vertebrate *Hox* genes are also organized in clusters which are deployed in an anterior to posterior order ac-

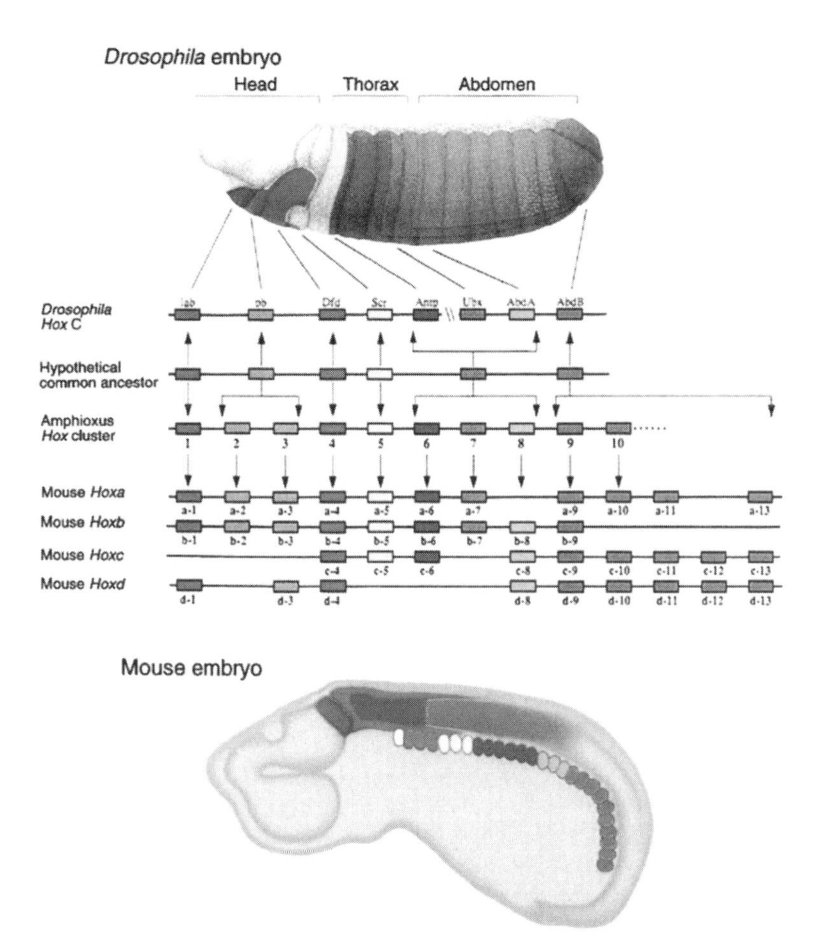

Fig. 1 *Hox* gene organization and expression. *Top,* the AP domains of *Drosophila Hox* gene expression correspond to the order of the genes within the Hox complex. *Middle,* the evolutionary relationship between the *Drosophila,* amphioxus and mouse *Hox* clusters, and the deduced complement of *Hox* genes in the presumed common ancestor of arthropods and chordates. *Bottom,* the AP domains of mouse *Hox* genes within the developing mouse also correspond to gene order in the Hox complexes. Adapted from refs. 50, 52 and 75. (A color version of this figure is available on-line at <http://www.press.uchicago.edu/books/gee>.)

cording to their position in the complex.[15] Vertebrates have four clusters of 9 to 11 *Hox* genes which can be aligned into 13 sets (called paralogous groups) based upon their sequence, organization and thus homology to the *Drosophila* relatives (Fig. 1). Not all genes are represented in each vertebrate cluster, and some genes within clusters appear to have been duplicated since the invertebrate/vertebrate divergence. For example, the *Hox* gene most similar to *Abdominal-B* has been duplicated several times and exists in multiple copies (*Hox-9–13*) in three of the four clusters (A, C and D), suggesting that its representation was expanded before the clusters were duplicated.

The conservation of *Hox* genes may at first seem paradoxical. How can animals that have the same array of *Hox* genes appear so different? For example, flies and beetles have the same complement of *Hox* genes. Moreover, mammals and teleost fish[16] (R. Krumlauf, personal communication; D. Duboule, personal communication) each have four *Hox* clusters (and at least the *Hox B* clusters contain the same set of genes (R. Krumlauf, personal communication), and these genes are expressed in a conserved relative order along the main body axis.[11] For insight into this puzzle, we must appreciate that *Hox* genes act only to demarcate relative positions in animals rather than to specify any particular structure.[11] Within one species, different *Hox* genes control the morphology of different body regions (without *Hox* genes all insect segments appear alike[13,17]). Between species, the same *Hox* gene can regulate the homologous segment or body region in different ways. The key to understanding how *Hox* genes control morphology and diversity is based on their action as regulatory proteins and the wide range of target genes regulated by different *Hox* genes in one animal, and by the same *Hox* gene in different animals.

Hox Proteins and the Genes They Regulate

The large effects of *Hox* genes on morphology suggest that they regulate, directly and indirectly, large numbers of genes. The promiscuity of Hox protein function may stem from the relatively simple sequences recognized by this class of DNA-binding proteins. The Hox proteins bind to a four-base core sequence,[18,19] and there is a lot of flexibility concerning the sequence surrounding this core. This makes it easy to imagine how *Hox* binding sites can be gained or lost by target genes. Indeed, the regulation of several *in vivo* targets of the yeast α2 homeodomain protein has been examined,[20] and a notable tolerance for different bases with α2 recognition sites was found. Most mutations in target sequences produced only small effects on homeodomain binding *in vitro* and target expression *in vivo,* suggesting the "relaxed" DNA-binding specificity of homeodomains may allow new regulatory interactions to evolve readily among Hox proteins and potential target genes.

The identity of genes regulated by the *Hox* proteins is a crucial issue because these genes must mediate the cellular processes involved in morphogenesis. In *Drosophila,* proven *Hox* targets include genes encoding other transcriptional regulatory proteins, secreted signalling proteins, and structural proteins, but these few examples may be just the tip of the iceberg.[21] It has recently been estimated that the *Drosophila* genome contains 85 to 170 genes that are regulated by the product of the *Ultrabithorax* gene.[22] This estimation underscores the monumental challenge ahead in deciphering how *Ubx* regulation of these genes in *Drosophila* distinguishes the morphology of body segments and appendages. It also helps to explain why the homeodomain protein sequences are so constrained in evolution (because mutations in the Ubx protein would alter the regulation, directly and indirectly, of potentially more than 100 genes). Finally, it suggests that divergence among these large sets of target genes may distinguish animal patterns.

Homeotic Genes in Evolution

There are six potential genetic mechanisms through which *Hox* genes could influence morphological evolution: (1) an expansion in the structural diversity of *Hox* genes within a *Hox* complex (for example, when did the different *Hox* genes such as *lab, pb, Dfd, Antp* and *abd-B* arise?); (2) an expansion in the number of *Hox* genes of a given class (for example, the multiple *abd-B*–like genes (paralogues 9–13) found in vertebrates); (3) an expansion in the number of *Hox* complexes; (4) the loss of one or more *Hox* genes; (5) a change in the position, timing or level of *Hox* gene expression; and (6) changes in the regulatory interactions between Hox proteins and their targets. Several of these mechanisms are now correlated with the diversification of arthropods and chordates.

The Lewis Hypothesis: New Genes for New Animals?

In his landmark 1978 review, years before any *Hox* genes had been cloned, Lewis proposed one of the first explicit hypotheses concerning the genetic basis of morphological evolution, in this case the evolution of insects and flies.[13] Founded upon his pioneering genetic studies of homeotic genes in *Drosophila,* this "Lewis hypothesis" consisted of three central ideas. First, during the evolution of insects, "leg-suppressing genes" evolved which removed legs from abdominal segments of millipede-like ancestors. Second, "haltere-promoting genes" evolved to suppress the second pair of wings of a four-winged ancestor. Third, a tandem array of redundant genes presumably diversified by mutation to produce the BX-C. It was this third idea that prompted the initial structural comparisons of homeotic genes a decade ago, and led directly to the identification of the homeobox.[23-25] The discovery of the homeobox-encoded DNA-binding motif in each *Drosophila* homeotic gene tied together both the history and the biochemical function of homeotic genes, and provided the crucial tool for the study of *Hox* genes in the development and evolution of animals.

The first two ideas suggest that genes of the BX-C, which control the morphology of the posterior thorax and abdomen of *Drosophila,* arose in

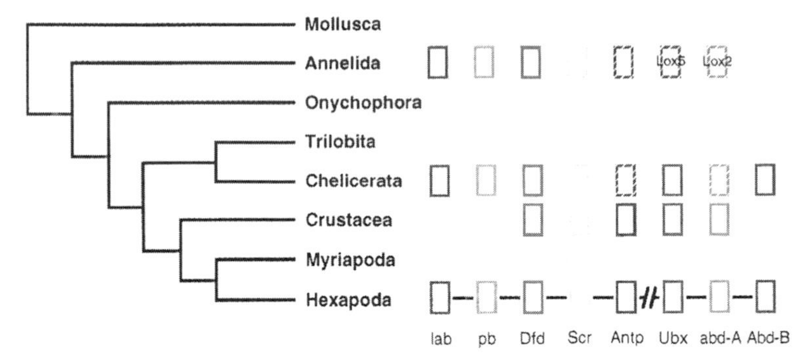

Fig. 2 *Hox* genes and arthropod phylogeny. The array of *Hox* genes detected in arthropods and other invertebrates is displayed in comparison to the *Drosophila Hox* genes. By comparing annelids with arthropods and insects, it appears that most *Hox* genes pre-date the annelid/arthropod divergence, and all insect *Hox* genes pre-date the insect/crustacean divergence. Solid boxes represent confirmed *Hox* genes; stippled boxes are *Hox* genes that are too similar to another to be distinguished on the basis of available data; missing boxes represent the absence of information. Data are from surveys of the *Hox* genes of annelids,[29] chelicerates[59] and crustacea;[28] the arthropod phylogeny on the left is from ref. 76. Other arthropod phylogenies are possible but do not affect the conclusions regarding *Hox* gene origins. (A color version of this figure is available on-line at <http://www.press.uchicago.edu/books/gee>.)

the course of insect and fly evolution. A survey of annelid[26,27] and arthropod[9,28,29] *Hox* genes demonstrates that this is not the case. All of the arthropod *Hox* genes evolved before the origin of insects, as demonstrated by the occurrence of these genes in crustacea and chelicerates (Fig. 2), two arthropod classes that pre-date the insects. In addition, the unexpectedly diverse array of *Hox* genes that exists in annelids (Fig. 2) suggests that the diversification of the *Hox* genes must have occurred before the annelid/arthropod divergence. Similarly, the array of vertebrate *Hox* genes demonstrates that six or seven *Hox* genes were present in the common ancestor of all three groups[30,31] (Fig. 1, middle), although it is not known what that creature would have looked like.[1]

Evolution of *Hox* Gene Regulation and Arthropod Body Plans

If the array of *Hox* genes is conserved among insects, crustacea and chelicerates, how do we explain the different body plans characteristic of these classes? The major differences between arthropods involves the number, type and organization of body appendages, such as antennae, claws, mouthparts and legs, all of which evolved from a common, ancestral arthropod limb.[32] In insects the homeotic genes regulate where body appendages may form and the type of appendage found on a particular segment. For example, in *Drosophila,* the products of the BX-C suppress limb formation on the abdomen by repressing the *Distal-less* gene which is required for proximodistal axis formation,[33] whereas the Antp protein promotes leg formation and suppresses antennal development in the second thoracic segment.[34] The homeotic genes are not actually required to make a limb,[17,35] as that potential appears to exist in all segments; rather, the homeotic genes either suppress limb development or modify it to create unique appendage morphologies.

The insect body plan is divided into three main body regions (or tagma), with a head bearing numerous appendages, a thorax bearing three pairs of walking legs, and a limbless abdomen. This pattern is sculpted by the homeotic genes. By contrast, chelicerates have just two distinct tagma, and the number and type of appendages vary considerably within the three crustacean tagma.[32] To determine the extent to which the deployment of homeotic genes reflects the tagmatization of the arthropods, the expression of the three Hox proteins, Antp, Ubx and abd-A, has been examined in a crustacean.[36] Surprisingly, *Ubx* and *abd-A* are expressed in nearly all of the limb-bearing thoracic segments, far more anteriorly than in insects (Fig. 3). This demonstrates that the expression domains of *Hox* genes have shifted considerably between crustacea and insects, and that the interac-

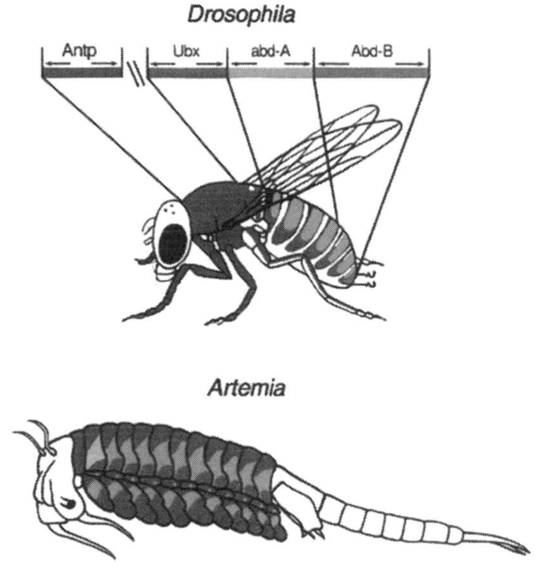

Fig. 3 Evolution of *Hox* gene regulation and the crustacean and insect body plans. The domains of BX-C gene expression have shifted with respect to each other in the thorax and abdomen of insects and crustacea. *Top,* the *Antp* (purple), *Ubx* (brown), *abd-A* (orange) and *Abd-B* (green) genes sculpt the morphology of the insect trunk. *Bottom, Ubx* and *abd-A* are expressed in relatively more anterior positions in crustacea.[36] (A color version of this figure is available on-line at <http://www.press.uchicago.edu/books/gee>.)

tions between Hox proteins and some key developmental programs, such as limb formation, have changed as well.

The scope of regulatory evolution among the *Hox* genes of arthropods is not yet known. The analysis of different crustacea (which exhibit a wide range of body plans[32]), chelicerates, myriapods and the putative sister group of the arthropods, the Onychophora, will be crucial. It is of special interest to know where the same *Hox* genes are deployed in these last two groups, in which all trunk segments are alike, as this represents the probable ancestral arthropod condition. The *Hox* genes are likely to be present, but they might not regulate limb formation or even ectodermal patterns in these animals.

Modifications of the Insect Body Plan

It now seems likely that all insect diversity has evolved from a body plan sculpted by the same set of homeotic genes. Yet the insect body plan has not been static; there have been numerous changes in the number and type of appendages, which are reflected by extant and extinct orders. For example, differences in larval limb and adult wing number, as well as appendage morphology, are regulated by homeotic genes. The study of these characters in modern insects may allow us to gain fundamental insights into their origin and diversification.

Prolegs: A *Hox*-Regulated Atavism?

Abdominal limbs known as prolegs are found on the larvae of various insect species belonging to several orders, but are ubiquitous in the Lepidoptera (moths and butterflies). These limbs were probably present in insect ancestors, so it may be that prolegs have reappeared through the derepression of an ancestral limb developmental program (that is, they may be atavistic). To investigate how homeotic genes influence proleg formation, the expression of several homeotic genes during butterfly development was examined. Proleg formation involves an obvious change in the regulation of *abd-A* expression during embryogenesis.[37] The initial expression of this gene is in the anterior abdomen, as in all insects examined thus far (where it represses limb formation), but it is selectively turned off in small patches of cells within four abdominal segments which then activate limb-patterning genes and begin to form a proleg. It is not known whether switching off *abd-A* underlies proleg formation in other insects, although the phylogenetic and segmental distribution of prolegs[38] suggests that they may arise by a regulatory switch operating on homeotic genes.

The Origin and Evolution of Winged Insects

The most important insect innovation was flight, which catalyzed the radiation of the most successful subclass of insects, the pterygotes. What was the role of *Hox* genes in this crucial development? The first pterygote orders are long extinct, so we can gain insight only by studying modern insects. The answer is surprising: *Hox* genes were not involved in the development of wings.[39] This conclusion relies mainly on the developmental genetics of wing formation in *Drosophila,*[34] and partly on the available fossil record.[40,41] In *Drosophila* and all pterygotes, wings form on the second thoracic segment. This is principally the domain of the *Antp* gene, which suggests that *Antp* would be part of a regulatory code for making a wing. However, this is not true because the wing primordia, the imaginal wing disc and the adult wing all appear normal when the *Antp* gene is removed.[39] Indeed, *Antp* is barely expressed at all in the developing wing.[39,42] The lack of homeotic input into wing formation makes even more sense when we consider the fossil record. When wings first evolved they appeared on all thoracic and abdominal segments;[40,41] this lack of segmental restriction suggests that no homeotic gene positively or negatively regulated wing formation.

In the subsequent course of pterygote evolution, however, homeotic genes did indeed become important, but as repressors and modifiers of wing formation. Orders appeared that lacked abdominal wings, and wings were also lost from the first thoracic segment. This suggests homeotic regulation and, in *Drosophila,* the BX-C and *Scr* genes do repress the wing primordia in these body segments.[39] In flies, which bear one pair of wings on the second thoracic segment and a miniature flight appendage (the haltere) on the third thoracic segment, the *Ubx* gene also represses the size of the flight appendage primordia[38] and modifies the morphology of the developing haltere.[43]

The Morphological Evolution of Homologous Structures between Species

Different *Hox* genes regulate the morphology of serially homologous structures within a species, but what about the homologous structures of different species? In insect terms, for example, can we explain how the hindwings of flies (the haltere), beetles and butterflies are distinct not only from their respective forewings but also from each other? We know in each of these orders that the *Ubx* gene is expressed in and differentiates the hindwing from the forewing.[14,37,44,45] Because *Antp* is not involved in the forewing patterning,[39] the differences between insect forewings are all independent of homeotic genes. However, the considerable differences in the relative size and morphology of the hindwings must also be due to differences in the wide range of target genes regulated by *Ubx* (Fig. 4). If the set of potential target genes in *Drosophila* is as large as recent studies suggest,[22] it is easy to imagine that many differences in *Ubx*-regulated target genes exist between species, and that such differences may evolve readily. For example, butterfly hindwing patterns are usually distinct from forewings and are species specific.[46] This diversity could be generated at least in part through the gain or loss of *Ubx* binding sites in the *cis*-regulatory sequences of a myriad of potential target genes.

Rephrasing the Lewis Hypothesis

The general theme emerging from developmental studies of *Hox* genes,[10,11,47] and echoed by the above examples, is that homeotic genes do

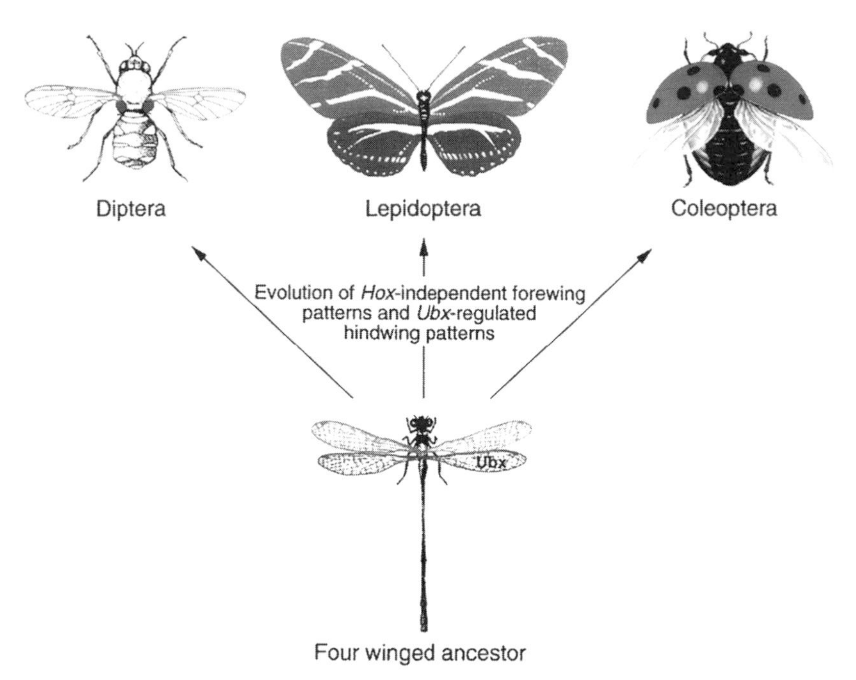

Fig. 4 *Hox* genes and the morphological divergence of homologous structures. The forewings and hindwings differ between insects, and usually from each other. Based on the developmental genetics of wing formation in *Drosophila* and comparisons of homeotic gene expression and function between insects, it appears that forewing patterns evolve independently of homeotic genes, whereas hindwing patterns are *Ubx*-regulated modifications of the respective forewing patterns. In primitive four-winged insects the forewings and hindwings appear similar but have evolved quite different morphologies in various insect orders. The different hindwing morphologies are probably due in part to the divergence of the potentially large sets of wing-patterning genes regulated by *Ubx* in different orders. This divergence is brought about by changes in *cis*-regulatory sequences in target genes that affect their regulation by *Ubx*. (A color version of this figure is available on-line at <http://www.press.uchicago.edu/books/gee>.)

not constitute an instructional code that says "make a wing" or "make a leg." Rather, they modify developmental programs such that what would otherwise be a wing becomes a haltere, or what would develop as an antenna becomes a leg. In terms of homeotic genes and Lewis's original hypothesis, what has evolved in the course of insect and fly evolution are not new genes but new regulatory interactions between BX-C proteins and genes involved in limb formation and wing morphogenesis.

Hox Genes and Chordate Evolution

The course of vertebrate evolution has been marked by many changes in the chordate body plan. The evolution of the head with paired eyes, jaws and teeth, the origin of the tetrapods and subsequent evolution of limbless forms, and the diverse axial morphologies of modern vertebrates are dramatic examples of large-scale morphological evolution. Developmental analyses of *Hox* gene expression and function implicate the homeotic genes in craniofacial development, hindbrain organization, axial morphology, limb patterning and many other facets of vertebrate development,[15,47] which begs the question of their potential roles in the evolution of the chordates. The issues parallel those of the arthropods, discussed above: did the appearance of new genes enable the origin of new structures? How have these new structures been modified in the course of evolution?

The examination of these issues in the chordates is facilitated by the relative richness of vertebrate paleontology, but made more difficult by the genetic and anatomical complexity of the vertebrates. The duplicated *Hox* complexes of modern vertebrates may have allowed the diversification of various homeotic genes but also increase the possibility of genetic redundancy, which conceals individual *Hox* functions in gene knockout experiments.[48] Furthermore, during the formation of a given structure such as a limb, the number of *Hox* genes involved, the spatial and temporal dynamics of gene expression, and the increasing cellular complexity of a growing three-dimensional structure make individual mutant phenotypes difficult to interpret.[49] Nevertheless, the comparison of *Hox* gene organization and expression in primitive chordates and representative vertebrates has provided several examples of *Hox* gene involvement in chordate evolution.

The Origin of the Vertebrate Body Plan

Could the origin of the new *Hox* genes underlie the origin of vertebrate characters? To address this question, the organization of *Hox* genes was examined[50] in amphioxus, a cephalochordate possessing a notochord, dorsal nerve cord and paired somites, but lacking most vertebrate head structures, appendages and various other features. Remarkably, this primitive creature possesses a single *Hox* complex that appears to be the archetype of *Hox* clusters of modern vertebrates. Two features of the amphioxus cluster stand out. First, it contains homologues of at least the first ten groups of vertebrate *Hox* genes in a collinear array. Second, it contains at least two *Abd-B*–like genes, which indicates that the tandem duplication of these genes and their selective retention in the cephalochordates proceeded, or was at least underway before, the evolution of features unique to vertebrates (such as the limbs in which these genes are important[49]). Thus the origins of new genes via the expansion and diversification of the *Hox* clusters may well have been important in enabling the evolution of more complex chordate body plans.

The number and organization of *Hox* genes in other primitive chordates and vertebrates are now being examined to determine whether there were intermediate stages of *Hox* cluster duplication in the evolution of the vertebrates. In jawless fish (such as lamprey and hagfish) there appear to be fewer *Hox* genes and perhaps as few as two *Hox* clusters (although there could be more).[51] Based upon the available *Hox* data and several other gene families, it has been suggested[30] that there were two phases of cluster duplication, one close to the origin of jawed vertebrates.

Hox Gene Regulation and the Evolution of Vertebrate Axial Morphology

The mesodermal segments of vertebrates, called somites, are serially homologous and give rise to the vertebrae. The number and morphology of vertebrae that contribute to each distinct region of the vertebral column (the cervical, thoracic, lumbar, sacral and caudal vertebrae) often differ between vertebrates (Fig. 5). For example, mammals (whether mice or giraffes) usually have seven cervical vertebrae, birds have between 13 and 25 and snakes have up to 454 pre-caudal vertebrae. Because *Hox* genes are expressed at distinct levels along the AP axis of vertebrate embryos, and loss or ectopic expression of individual *Hox* genes can transform vertebrae from one type to another, it has been proposed that the combination of *Hox* genes expressed in each somite specifies the different vertebral morphologies.[52] This would imply that different axial morphologies could be created by shifting *Hox* gene expression domains up or down the AP axis or by changing the response of *Hox*-regulated genes to a fixed pattern of *Hox* gene expression (for example, at a given somite number).

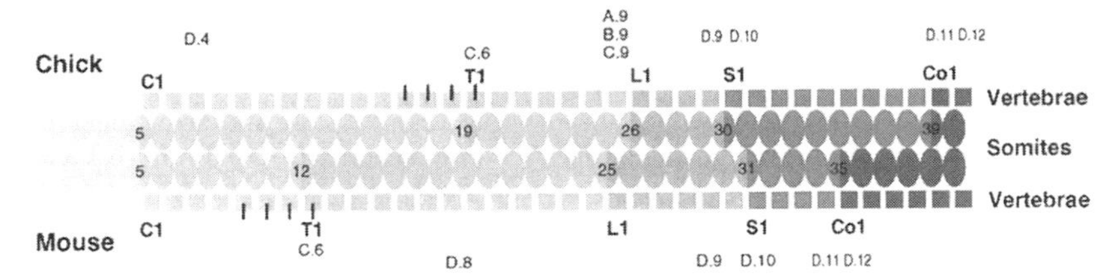

Fig. 5 *Hox* genes and the evolution of vertebrate axial morphology. The expression of various *Hox* genes along the chick and mouse body axes are depicted. The different regions of the respective vertebral columns, that is, cervical *(C)*, thoracic *(T)*, lumbar *(L)*, sacral *(S)* and caudal *(Co)*, contain different numbers of vertebrae. The anterior expression boundaries of several members of the Hox D cluster mark morphological transitions such as the lumbar/sacral boundary *(Hoxd-9–10)* and the sacral/caudal transition *(Hoxd-11–12); Hoxc-6* expression marks the cervical/thoracic transition in both species, even though it arises at different somite positions. Note that in the chick all members of the ninth paralogous group mark the posterior of the thoracic region *(Hoxa-9, b-9, c-9)*, except for *Hoxd-9* which is regulated differently and marks the posterior lumbar region. Adapted from ref. 53. (A color version of this figure is available on-line at <http://www.press.uchicago.edu/books/gee>.)

Comparing the expression of a large set of *Hox* genes between mice and birds has demonstrated[53] that *Hox* gene domains shift in parallel with vertebral anatomy and are independent of the number of vertebrae within a given body region (summarized in Fig. 5). For example, *Hox* genes of the fifth paralogue group (for example, *Hoxc-5*) are expressed at the level of the forelimb, which arises at somite level 17–18 in the chick and 10–11 in the mouse. Similarly, the anterior expression boundary of *Hoxc-6* falls at the boundary of the neck (cervical) and thorax in mice, chickens, geese (which have 17 cervical vertebrae, 3 more than chickens) and frogs (with a highly derived vertebral morphology and only 3 or 4 cervical vertebrae). The thoracic-lumbar transition appears to be associated with expression of the *Hoxa*-9, *Hoxb*-9 and *Hoxc*-9 genes, but the fourth member of this paralogue group, *Hoxd*-9, is expressed out of register. This may be significant because the thoracic-lumbar distinction is not general among tetrapods. It may be that shifts within the *Hox-9* group were important in the evolution of this transition from a more uniform trunk, perhaps even in the evolution of the tetrapods from fish. For *Hox* gene expression domains to shift according to anatomy and not somite level, either the upstream regulators of *Hox* gene expression are deployed differently in different vertebrates, and/or the response of *Hox* genes to these regulators has changed. The fact that individual members of paralogous groups have different sites of expression within a species demonstrates that the response to the same upstream regulators can differ, and suggests that the evolution of regulatory diversity within paralogue groups may have enabled major changes in axial morphology (as well as in limb patterns; see below).

Origin of the Tetrapods

The earliest tetrapod-like fossils suggest that the vertebrate hindlimb evolved first from the pelvic fin of fish,[54] with the forelimb evolving subsequently from the pectoral fin. It is difficult to explain how the fore- and hindlimbs of tetrapods are so similar to each other when they evolved from different fish fins. It has been suggested that a major change in *Hox* gene expression brought about the serial homology of the forelimb and hindlimb.[55]

Hox genes are expressed both in the mesoderm adjacent to the limbs and in the developing limb buds of modern tetrapods and are required for AP and proximodistal patterning. Different *Hox C* genes are expressed in the fore- and hindlimb with their respective expression extending from the adjacent axial mesoderm. However, the *Hox A* and *Hox D* cluster genes expressed in the two limbs are those that are normally expressed in the posterior of the main vertebrate body axis, and they are deployed in very similar spatiotemporal patterns in each limb. Because the *Hox A* and *Hox D* genes are being expressed far anterior to their axial site of expression, it has been suggested[55] that, in the course of tetrapod evolution, the *Hox A* and *Hox D* cluster genes were ectopically activated in the pectoral fin which led to its transformation from fin to limb and accounts for the serial homology between fore- and hindlimbs of modern terrestrial vertebrates. The ectopic activation of *Hox* genes could most readily be accomplished by the expression of a new signalling center in the nascent forelimb. The inspection of *Hox* gene expression patterns in living relics, such as the lungfish, coelacanths and sharks, could test the validity of this intriguing model for *Hox* gene function in tetrapod origins.

Hox Genes and Constraint: Could Snakes Learn to Walk (Again)?

Much of this review has focused on the evolutionary inferences drawn from studies of *Hox* gene organization, expression and function in a few animals. It may also be worthwhile to consider what might be evolutionarily possible but has not been observed (yet). For example, there is no clear case of *Hox* gene loss being implicated in the modification of the vertebrate body plan. Is there such constraint on *Hox* organization and function that even single members of paralogue groups are indispensable? Where might we test such an idea? Perhaps in snakes, which forfeited their limbs, much of their axial morphology, and elements of craniofacial architecture (for example, ears) in the course of their evolution as burrowing reptiles approximately 150 Myr ago.[56] (Their recent radiation as surface-dwelling mammal-hunters is linked to the abundance of prey and the evolution of pre-immobilizing toxins, not the acquisition of new structures.) Given that individual *Hox* genes were lost after the expansion to four *Hox* clusters in vertebrate ancestors,[57] and that mice can survive without certain individual genes,[15] it seems possible that snakes could possess fewer *Hox* genes than other vertebrates.

Snakes could reveal what sort of constraints exist on the *Hox* clusters. For example, if the *Hox* genes have all been preserved then regulatory changes can probably explain their unique axial morphology. This could also suggest that the potential for specifying limb position and morphology is still preserved (a possibility supported by the presence of hindlimb rudiments in pythons and boas). If so, is it conceivable that snakes or other limbless tetrapods might regain their limbs? Probably not, as the evolution of limblessness is strongly correlated with body elongation and increased vertebral number, and the constraints against a return toward the primitive tetrapod form may be prohibitive.[58] If *Hox* genes have been lost, however, then we might learn some valuable lessons about genetic redundancy from the identity of the lost genes.

The other side of the constraint issue is illustrated by the horseshoe crab, *Limulus polyphemus.* This chelicerate arthropod may possess up to four *Hox* clusters which include the genes found in the *Drosophila* cluster as well as additional genes present in vertebrate *Hox* clusters.[59] Although the multiple *Hox* clusters could be due to a genome-wide polyploidization event, this must have occurred at least 75 Myr ago. The static body pattern of horseshoe crabs suggests that *Hox* complexity and anatomical complexity are not strictly correlated. If *Hox* genes create possibilities, *Limulus* has apparently not found any creative uses for a lot of *Hox* genes (or other genes), even though selection has retained them. Might this reflect some constraint imposed by the chelicerate or arthropod body plan that prevents its modification?[60] After 600 Myr to expand their *Hox* clusters, it seems surprising that arthropods don't have more *Hox* genes; is this because they couldn't find ways to use them?

The Primacy of Regulatory Evolution

The available evidence suggests that primitive arthropods and chordates each possessed a single *Hox* complex containing the diverse array of *Hox* genes found in their modern descendants. Although this cluster was duplicated in the chordates, presumably in one or two early phases in their evolution, it appears that the subsequent course of vertebrate evolution from primitive bony fishes to mammals, and the entire course of arthropod evolution, was founded upon conserved sets of *Hox* genes. The phylogeny of *Hox* genes and the many examples cited above of large-scale morphological changes associated with diversity in *Hox* gene regulation and target regulation suggest that the primary genetic mechanism enabling morphological diversity among arthropods and vertebrates is regulatory evolution.

The anatomical complexity of vertebrates, reflected by a larger relative number of different cell types,[61] may be a consequence of a greater number of developmental genes. For example, important developmental gene families such as the *Hox, Wnt, TGF*-β and *Dll/Dlx* genes are generally severalfold larger in vertebrates than primitive chordates, arthropods or annelids (Table 1 and references cited therein). The structural and regulatory diversity among these genes (especially, for example, the *Wnt* and *TGF*-β families) demonstrates that gene duplication and diversification have probably been an important force in the evolution of the cellular and anatomical complexity of vertebrates. However, the complement of *Hox,*[62,16] *Wnt*[63] and *Dlx*[64] genes is comparable in fish and mammals. This suggests that morphological diversity within vertebrates is the product of regulatory evolution within larger, but essentially fixed, families of developmental genes.

How can regulatory evolution be sufficient to explain the differences between trilobites and butterflies, or dinosaurs and sparrows? The creative potential of regulatory evolution lies in the hierarchical and combinatorial nature of the regulatory networks that guide the organization of body plans and the morphogenesis of body parts. We now know that *Hox* genes are regulated by many upstream factors, and that *Hox* proteins act as sculptors that modify the basic arthropod or chordate metamere by modulating the expression of potentially dozens of interacting genes, the products of which determine the cellular events of morphogenesis. Variation in the morphogenetic output of such a multigenic network can arise at many

Table 1 Different trends in chordates and arthropods

	Principal *Hox* Genes	*Wnt* Genes	*TGF*-β–Related Genes	*Dlx* Genes
Annelids	≥7	≥2		
Insects	8	4	3	1
Cephalochordates	≥10	≥2		
Bony fish	~38	14		≥5
Mammals	38	14	>25	6

Gene family data are based on the following references: for *Hox,* refs. 15, 27, 50, and D. Duboule and G. Wagner, personal communication; for *Wnt,* refs. 30, 63, 70; for *TGF*-β, refs. 71, 72; for *Dlx,* refs. 64, 73, 74 and, in the zebrafish, M. Westerfield, personal communication.

levels simply by tinkering with the relative timing of developmental gene expression (heterochrony) or the interactions between members of the regulatory network. Such regulatory changes are presumably the consequence of alterations in *cis*-regulatory sequences through simple mutations as well as duplications or deletions of regulatory elements. In this manner, one aspect of gene function can evolve without altering others. Single genes are often regulated by arrays of discrete regulatory elements which control the pattern, position, timing and level of gene expression, and these features can differ between species.[65] Key body-patterning genes such as the homeotic[66,67] or *Distal-less*[33] genes contain many such elements, which suggests that the evolution of regulatory elements is much more common, and therefore a more continuous source of variation, than the duplication of entire genes.

Conclusion

Our understanding of the role of *Hox* genes in evolution depends on both developmental and comparative studies. One of the most challenging problems is the elucidation of Hox protein function in regulating morphology in model experimental species. When the regulatory targets of individual Hox proteins are better known then the differential regulation of these genes by different Hox proteins in a single species, or by the same protein in different species, can be assessed. The evolutionary value of this information will rely upon the judicious choice of species selected for comparative study.

At a broader level, it is not known how *Hox* genes are organized or expressed in several major arthropod and vertebrate taxa. At the very least, we need to investigate the *Hox* genes of primitive arthropods and possible sister groups as well as all protochordate and vertebrate classes. Of course, arthropods and chordates are only 2 of the 35 or so metazoan phyla and, although studying them provides an understanding of general ways of modifying body plans, it does not explain the origins of *Hox* gene diversity or the evolution of other body plans. The exploration of other phyla, such as the cnidaria, echinoderms, molluscs and flatworms, might lead us to discover when this ancient regulatory "toolbox" came into existence, and to understand its role in the genesis and modification of these diverse animal body plans.

Finally, it must be appreciated that the comparisons of differences between higher taxa reveal only what has changed, not how it changed. We do not know the rate at which these changes arose or the extent of variation in *Hox*-regulated characters in populations. The idea of macroevolution in a single step, the "hopeful monster" so often insinuated in the discussion of homeotic genes,[68] is widely discredited.[69] The new perspective emerging from the study of *Hox* genes in phylogeny and their regulatory roles in development needs to be integrated within the evolutionary frameworks of paleontology and population biology.

Note added in proof: Sardino *et al.* (*Nature* **375**, 678–681, 1995) have recently shown that the tetrapod foot may be a new structure formed under the control of a distinct phase of *Hox* gene expression in the limb bud.

SEAN B. CARROLL is at the Howard Hughes Medical Institute, University of Wisconsin at Madison, R. M. Bock Laboratories, 1525 Linden Drive, Madison, Wisconsin 53706, USA.

References

1. Conway Morris, S. *Nature* **361,** 219–225 (1993).
2. Ohno, S. *Evolution by Gene Duplication* (Springer, New York, 1970).
3. Wilson, A. C. *Scient. Am.* **253,** 164–173 (1985).
4. Jacob, F. *Science* **196,** 1161–1166 (1977).
5. Gould, S. *Ontogeny and Phylogeny* (Belknap, Cambridge, MA, 1977).
6. Raft, R. & Kaufman, T. *Embryos, Genes and Evolution* (Indiana Univ. Press, Bloomington, 1983).
7. Schummer, M., Scheurlen, I., Schafer, C. & Galliot, B. *EMBO J.* **11,** 1815–1823 (1992).
8. Wang, B. *et al. Cell* **74,** 29–42 (1993).
9. Akam, M. *et al. Development* (suppl.) 209–215 (1994).
10. Kenyon, C. *Cell* **78,** 175–180 (1994).
11. Slack, J., Holland, P. & Graham, C. *Nature* **361,** 490–492 (1993).
12. Kaufman, T., Seeger, M. & Olsen, G. in *Genetic Regulatory Hierarchies in Development* (ed. Wright, T.) (Academic, San Diego, 1990).
13. Lewis, E. B. *Nature* **276,** 565–570 (1978).
14. Beeman, R., Stuart, J., Brown, S. & Denell, R. *BioEssays* **15,** 439–444 (1993).
15. Krumlauf, R. *Cell* **78,** 191–201 (1994).
16. Misof, B. & Wagner, G. *CCE Tech. Report 24, Yale Univ.* (1995).
17. Stuart, J., Brown, S., Beeman, R. & Denell, R. *Nature* **350,** 72–74 (1991).
18. Ekker, S. C. *et al. EMBO J.* **13,** 3551–3560 (1994).
19. Gehring, W. *et al. Cell* **78,** 211–233 (1994).
20. Smith, D. L. & Johnson, A. D. *EMBO J.* **13,** 2378–2387 (1994).
21. Botas, J. *Curr. Biol.* **5,** 1015–1022 (1993).
22. Mastick, G., McKay, R., Oligino, T., Donovan, K. & Lopez, A. *Genetics* **139,** 349–363 (1995).
23. McGinnis, W., Levine, M., Hafen, E., Kuroiwa, A. & Gehring, W. *Nature* **308,** 428–433 (1984).
24. Scott, M. P. & Weiner, A. J. *Proc. natn. Acad. Sci. USA* **81,** 4115–4119 (1984).
25. McGinnis, B. *Genetics* **137,** 607–611 (1994).

26. Wysocka-Diller, J. W., Aisemberg, G. O., Baumgarten, M., Levine, M. & Macagno, E. R. *Nature* **341,** 760–763 (1989).
27. Shankland, M., Martindale, M., Nardelli-Haefliger, D., Baxter, E. & Price, D. *Development* (suppl.) 29–38 (1991).
28. Averof, M. & Akam, M. *Curr. Biol.* **3,** 73–78 (1993).
29. Dick, M. & Buss, L. *Molec. Phylog. Evol.* **3,** 146–158 (1994).
30. Holland, P. W. H., Garcia-Fernández, J., Williams, N. A. & Sidow, A. *Development* (suppl.) 125–133 (1994).
31. Schubert, F., Nieselt-Struwe, K. & Gruss, P. *Proc. natn. Acad. Sci. USA* **90,** 143–147 (1993).
32. Brusca, R. & Brusca, G. *Invertebrates* (Sinauer, Sunderland, MA, 1990).
33. Vachon, G. *et al. Cell* **71,** 437–450 (1992).
34. Struhl, G. *Proc. natn. Acad. Sci. USA* **79,** 7380–7384 (1982).
35. Mann, R. *Development* **120,** 3205–3212 (1994).
36. Averof, M. & Akam, M. *Nature* **376,** 420–423 (1995).
37. Warren, R., Nagy, L., Selegue, J., Gates, J. & Carroll, S. *Nature* **372,** 458–461 (1994).
38. Birket-Smith, S. J. R. *Prolegs, Legs and Wings of Insects* (Scandinavian Science, Copenhagen, 1984).
39. Carroll, S., Weatherbee, S. & Langeland, J. *Nature* **375,** 58–61 (1995).
40. Kukalová-Peck, J. *J. Morph.* **156,** 53–126 (1978).
41. Kukalová-Peck, J. *Can. J. Zool.* **61,** 1618–1669 (1983).
42. Condie, J., Mustard, J. & Brower, D. *Drosoph. Inf. Serv.* **70,** 52–54 (1991).
43. Morata, G. & Garcia-Bellido, A. *Wilhelm Roux Arch. Entw. Mech. Org.* **179,** 125–143 (1976).
44. Beachy, P. A, Helfand, S. L. & Hogness, D. S. *Nature* **313,** 545–551 (1985).
45. White, R. A. H. & Wilcox, M. *Cell* **39,** 163–171 (1984).
46. Nijhout, H. F. *The Development and Evolution of Butterfly Wing Patterns* (Smithsonian Institution, Washington, DC, 1991).
47. McGinnis, W. & Krumlauf, R. *Cell* **68,** 283–302 (1992).
48. Rancourt, D., Teruhisa, T. & Capecchi, M. *Genes Dev.* **9,** 108–122 (1995).
49. Graham, A. *Curr. Biol.* **4,** 1135–1137 (1994).
50. Garcia-Fernández, J. & Holland, P. *Nature* **370,** 563–566 (1994).
51. Pendleton, J., Nagai, B., Murtha, M. & Ruddle, F. *Proc. natn. Acad. Sci. USA* **90,** 6300–6304 (1993).
52. Kessel, M. & Gruss, P. *Science* **249,** 374–379 (1990).
53. Burke, A., Nelson, C., Morgan, B. & Tabin, C. *Development* **121,** 333–346 (1995).
54. Ahlberg, P. *Nature* **354,** 298–301 (1991).
55. Tabin, C. & Laufer, E. *Nature* **361,** 692–693 (1993).
56. Carroll, R. *Vertebrate Paleontology and Evolution* (Freeman, New York, 1988).
57. Ruddle, F., Bentley, K., Murtha, M. & Risch, N. *Development* (suppl.) 155–161 (1994).
58. Gans, C. *Am. Zool* **15,** 455–467 (1975).
59. Cartwright, P., Dick, M. & Buss, L. *Molec. Phylog. Evol.* **2,** 185–192 (1993).
60. Jacobs, D. *Proc. natn. Acad. Sci. USA* **87,** 4406–4410 (1990).
61. Bonner, J. *The Evolution of Complexity* (Princeton Univ. Press, Princeton, NJ, 1988).
62. Duboule, D. *BioEssays* **14,** 375–384 (1992).

63. Sidow, A. *Proc. natn. Acad. Sci. USA* **89,** 5098–5102 (1992).
64. Simeone, A. *et al. Proc. natn. Acad. Sci. USA* **91,** 2250–2254 (1994).
65. Dickinson, W. J. in *Evolutionary Biology* (eds. Dobzhansky, T., Hecht, M. & Steere, W. C.) 127–173 (Appleton-Century-Crofts, New York, 1991).
66. Galloni, M., Gyurkovics, P., Schedl, P. & Karch, F. *EMBO J.* **12,** 1087–1097 (1993).
67. Gindhart, J. Jr., King, A. & Kaufman, T. *Genetics* **139,** 781–795 (1995).
68. Goldschmidt, R. *The Material Basis of Evolution* (Yale Univ. Press, New Haven, CT, 1940).
69. Wallace, B. *Q. Rev. Biol.* **60,** 31–42 (1985).
70. Kostriken, R. & Weisblat, D. *Devl. Biol.* **151,** 225–241 (1992).
71. Kingsley, D. *Genes Dev.* **8,** 133–146 (1994).
72. Hogan, B., Blessing, M., Winnier, G. & Suzuki, N. *Development* (suppl.) 53–60 (1994).
73. Cohen, S., Bronner, F., Kuttner, G., Jurgens, G. & Jäckle, H. *Nature* **338,** 432–434 (1989).
74. Panganiban, G., Nagy, L. & Carroll, S. *Curr. Biol.* **4,** 671–675 (1994).
75. McGinnis, W. & Kuziora, M. *Scient. Am.* **270,** 58–66 (1994).
76. Wheeler, W., Cartwright, P. & Hayashi, C. *Cladistics* **9,** 1–39 (1993).

Acknowledgments

We thank members of our laboratory and J. Langeland, A. Burke, P. Carroll and the referees for constructive comments; N. Patel, M. Averof, M. Akam, D. Duboule, G. Wagner, M. Westerfield, R. Krumlauf and P. Holland for communication of unpublished data; J. Wilson for preparation of the manuscript; and L. Olds for artwork. This work was supported by the NSF, the Shaw Scientists Program of the Milwaukee Foundation and the HHMI.

E. M. De Robertis and Yoshiki Sasai

A COMMON PLAN FOR DORSOVENTRAL PATTERNING IN BILATERIA

Functional studies seem now to confirm, as first suggested by E. Geoffroy Saint-Hilaire in 1822, that there was an inversion of the dorsoventral axis during animal evolution. A conserved system of extracellular signals provides positional information for the allocation of embryonic cells to specific tissue types both in *Drosophila* and vertebrates; the ventral region of *Drosophila* is homologous to the dorsal side of the vertebrate. Developmental studies are now revealing some of the characteristics of the ancestral animal that gave rise to the arthropod and mammalian lineages, for which we propose the name *Urbilateria.*

One of the challenges in evolutionary biology is to ascertain to what extent animals have homologous structures that have been derived by descent from a common ancestor (as in the case of the pectoral fin of fishes, forelimbs of tetrapods, and wings of birds and bats) or whether any similarities are due to convergent evolution resulting from the need to perform similar functions (as in the case of wings of flies and birds). Fifty years before Darwin, French naturalist E. Geoffroy Saint-Hilaire proposed that the ventral side of the arthropods was homologous to the dorsal side of the vertebrates.[1] He dissected a lobster (Fig. 1), but instead of placing it in its usual orientation with respect to the ground, he placed it upside down. In this orientation the lobster's central nervous system (CNS) was located above the digestive tract, which in turn was located above the heart. In his own words (our translation): "What was my surprise, and I add, my admiration, in perceiving an ordering that placed under my eyes all the organic systems of this lobster in the order in which they are arranged in mammals?" This idea of a dorsoventral inversion between arthropods and mammals led to a dispute with Georges Cuvier[2] in the French Academy, and has been revived[3–6] and contested[3,7–9] several times. Here we re-examine the issue of homology versus convergence in the development of animal species from the perspective of recent results in molecular embryology.

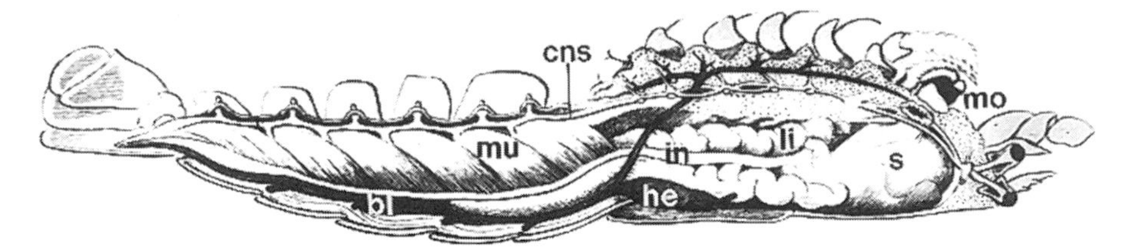

Fig. 1 Geoffroy Saint-Hilaire's famous lobster. In this dissection the animal is presented in the orientation opposite to that it would normally have with respect to the ground. The central nervous system (*cns* or nerve cord) is above, and is traversed by the mouth *(mo)*. Underneath is the digestive tract, with the stomach *(s)*, liver *(li)* and intestine *(in)*. Below the gut are the heart *(he)* and main blood vessels *(bl)*. Muscles *(mu)* flank the CNS. In this orientation the body plan of the arthropods resembles that of the vertebrate. From ref. 1, by courtesy of the History and Special Collection Division, Louise M. Darling Biomedical Library, UCLA.

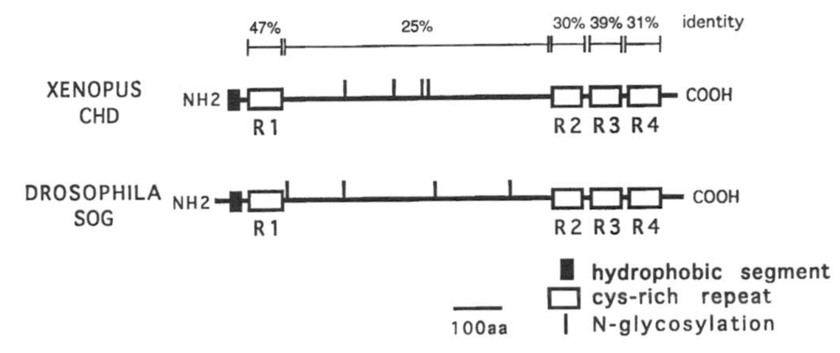

Fig. 2 The amino-acid sequences of *Xenopus chd* and *Drosophila sog* share similarities. Both proteins have a secretory signal sequence, or hydrophobic segment, at the amino *(NH2)* end *(dark box)*, several putative N-glycosylation sites *(vertical lines)* and four cysteine-rich repeats. The spacing of the nine cysteines contained in the cys-rich repeats is conserved in other secreted proteins such as thrombospondin, von Willebrand factor and procollagen. The first repeat (R1) of *chd* is more similar to R1 of *sog* than to any of the other repeats in *chd*. The same is true for R4, indicating that both genes are derived from a common ancestor. The percentage of amino-acid identity between the two proteins is indicated. The amino-acid identity is only 28% along the entire protein, so it was especially important to show that *sog* and *chd* were functionally homologous. Because of its homology of *sog*, and for other reasons,[49] we propose the designation *s-chordin* to refer to the vertebrate homologue *(chd)* in future.

Dorsoventral Patterning by *chd* and *sog*

In vertebrates the formation of the dorsal mesoderm and of the CNS is induced by a region of the embryo called the organizer. *Xenopus* chordin (chd) is an organizer-specific secreted protein that can mimic the activity of the organizer, resulting in a twinned body axis.[10] A *Drosophila* complementary DNA of related structure (Fig. 2) was reported[11,12] and, interestingly, it encoded *short gastrulation (sog)*, a gene long known to play an important role in *Drosophila* dorsoventral patterning.[13,14] As shown in Fig. 3, a number of zygotic genes are required for dorsoventral patterning of the *Drosophila* embryo. The regions in which these genes are expressed in the embryo are controlled by the activity of the maternal gene *dorsal,* which can repress or activate their transcription. Most are expressed in the dorsal 40% of the embryo, and are required to produce active *decapentaplegic (dpp)* gene product, which is a growth factor of the transforming growth factor-β (TGF-β) family. Dpp diffuses, generating a morphogen gradient that patterns the ectoderm.[15,16] Conversely, *sog* is expressed ventrally and its protein diffuses dorsally, where it antagonizes the activity of *dpp* (Fig. 3).

When the embryonic expression patterns of *chd* in the vertebrate[10] and *sog* in the fly[11] are compared, it can be observed that they are inverted with respect to each other (Fig. 4). *sog* is initially expressed in the ventral 60% of the blastoderm, and by the end of gastrulation it becomes localized to two rows of cells, called the mesectodermal cells, along the ventral midline. *chd* is expressed initially in the dorsal side of the blastopore (that is, in the organizer region), and by the end of gastrulation it is restricted to the dorsal midline (Fig. 4).

The case of *sog* and *chd* allows us to test whether *ventral* in the fly corresponds to *dorsal* in the vertebrate, because the gene products (in the form of messenger RNAs) have biological activity when injected into embryos. Injected *sog* mRNA ventralizes fly embryos, leading to formation of ventral denticle belts and ectopic patches of CNS.[17] In *Xenopus, sog* mRNA causes dorsal development (for example, formation of dorsal

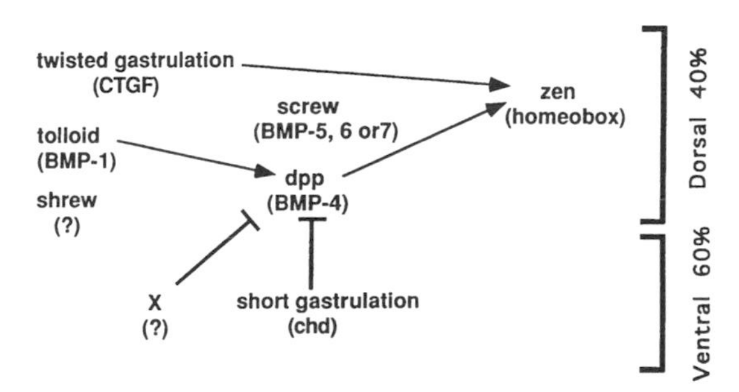

Fig. 3 Zygotic genes involved in dorsoventral patterning of the ectoderm in *Drosophila* embryos, and their vertebrate relatives. Several genes are expressed in the dorsal 40% of the embryo; *tolloid* encodes a metalloprotease required for *dpp* activity, and is similar in sequence to vertebrate *Bmp-1; twisted-gastrulation* encodes a protein related to vertebrate connective tissue growth factor; *screw* is a secreted growth factor related to either *Bmp-5, 6* or *7* that might form heterodimers with *dpp; zen* encodes a homeobox gene for which a vertebrate counterpart has not been isolated. Injection of *short-gastrulation (sog)* product in wild-type *Drosophila* embryos leads to formation of ectopic ventral tissues and CNS, but in embryos mutant for *dorsal* it can only produce a partial ventralization;[17] this led to the proposal that a second ventralizing gene (indicated as X), that cooperates with *sog* and is activated by *dorsal,* should also exist.[17]

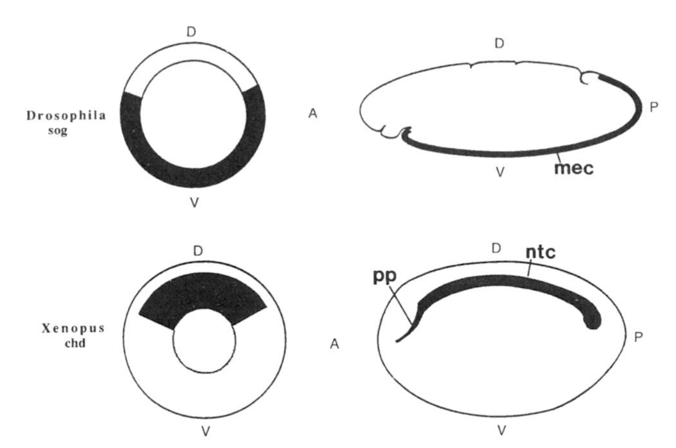

Fig. 4 The expression of *Drosophila sog (top)* and *Xenopus chd (bottom)* is reversed with respect to the dorsoventral axis. In *Drosophila sog* is first activated in the ventral 60% of the blastoderm, is then switched off in the mesoderm and persists in the neurogenic ectoderm, and by late gastrulation the expression resolves into two rows of cells located in the ventral midline, the mesectodermal cells *(mec)*.[11] In *Xenopus, chd* is expressed initially in the dorsal side of the blastopore, whereas by the end of gastrulation it is resolved to the mesoderm of the dorsal midline, occupying the head mesoderm or prechordal plate *(pp)* and the notochord *(ntc)*.[10] *sog* is ventral in *Drosophila* whereas *chd* is dorsal in *Xenopus.*

tissues such as notochord and CNS).[17] The vertebrate homologue, *chd,* behaves functionally much in the same way as does *sog:* injection of *chd* mRNA, a dorsalizing factor in *Xenopus,*[10] promotes ventralization of cell fates in *Drosophila.*[17] Thus, the function of *sog* and *chd* is reversed in insects and vertebrates; in both cases injection of the gene product promotes the development of the side of the embryo that contains the CNS.

dpp and *Bmp-4* Antagonize *sog/chd*

In *Drosophila, dpp* is expressed dorsally and promotes dorsal development;[15,16] that is, it is the opposite of *sog.* Microinjection of increasing amounts of *dpp* mRNA leads to threshold responses in the pattern of tissue differentiation in *Drosophila* ectoderm, shifting the balance in the dorsal direction until at high doses ventral development is prevented.[15,16] In the vertebrate, Bmp-4 (bone morphogenetic protein 4) and Bmp-2 share extensive sequence similarity with Dpp. In the *Xenopus* gastrula *Bmp-4* is expressed in the ventral and lateral marginal zone (which gives rise to ventral mesoderm) as well as in the animal cap (which gives rise to skin ectoderm when cultured in explants).[18] In *Xenopus* embryos *Bmp-4* has potent activity, changing the fate of dorsal mesoderm into ventral cell fates (blood and mesenchyme).[19,20] *Drosophila dpp* can mimic the ventralizing activity of *Bmp-4* in *Xenopus* and this activity can be antagonized by injected *sog* mRNA.[17] The gradient of *dpp* activity observed in *Drosophila* may result, at least in part, from the antagonism between the Sog and Dpp diffusible factors.[21] It is not yet known whether Sog antagonizes by binding directly to Dpp or its receptor, or whether it acts through its own receptor. A similar antagonism has been reported for *chd* and *Bmp-4* (ref. 22).

The suggestion that *Bmp-4* is the vertebrate homologue of *dpp* sparked the recent interest in the inversion of the dorsoventral axis.[4] The functional experiments discussed above indicate that dorsoventral patterning in both *Drosophila* and *Xenopus* is dependent upon a system of antagonistic extracellular signals provided by *dpp/Bmp-4* and *sog/chd.* Despite the morphological differences between embryos of the two species, the *sog/chd* gene is expressed on the side from which the CNS arises while the *dpp/Bmp-4* gene is expressed on the opposite side of the embryo. The functional conservation of the *sog/chd* and the *dpp/Bmp-4* secreted proteins suggests a homologous mechanism of dorsoventral patterning that must have existed in the common ancestor from which insects and vertebrates diverged. The results support the view that a reversal of the dorsoventral axis occurred during the course of evolution.

The Inversion Challenged

The recent revival of Geoffroy Saint-Hilaire's inversion hypothesis[4] was opposed[7–9] and this merits some discussion in light of recent findings. Three main objections have been raised.

First, it has been argued that the use of marker genes, such as those of the *achaete-scute* complex, that are conserved in vertebrates and insects[4] is inappropriate because they mark the presence of similar tissues, in this case nerve cells, and not homologous topology.[7,8] This objection does not apply to the case of *sog/chd* and *dpp/Bmp-4* because these molecules are upstream regulators providing patterning information that determines tissue differentiation according to specific dorsoventral fates. The importance of identifying the upstream regulatory switches is highlighted by the recent discovery that eye development is controlled by a conserved gene, *Pax-6/eyeless,* in both vertebrates and *Drosophila.*[23,24] This is a good example because previously the formation of eyes in insects and mammals was one of the classic cases of convergent evolution, despite the presence of common molecular elements in the light-detection system, such as conserved opsin proteins.[25] Because eye development is controlled by the same genetic switch in flies and mice, it is now considered that eyes are homologous structures related by descent from an ancestral photoreceptor.[26]

Second, the homology between *dpp* and *Bmp-4* has been challenged on the basis of the similarities between *Bmp-4* and its close relative *Bmp-2* (refs. 7, 8). Indeed, *Bmp-2* and *Bmp-4* are very similar in sequence, and both can functionally substitute for *dpp* in *Drosophila.*[27] However, recent loss-of-function studies using antisense RNA in *Xenopus*[28] and gene disruption in mice[5,29] indicate that Bmp-4, but not Bmp-2, plays an essential role in early dorsoventral patterning of the vertebrates, supporting the view 4 that *Bmp-4,* rather than *Bmp-2,* is the homologue of *dpp* that functions in early embryos.

Third, it has been argued that the displacement of the CNS from ventral to dorsal could have resulted from the appearance of a novel region of ectoderm around the mouth region.[7] In this view, first proposed by Garstang, the region corresponding to the ventral midline of the ventral CNS would become displaced, forming the lateral edges of the dorsal neural plate in vertebrates.[7] Recent molecular findings strongly argue that this is not the case, because the midline of the CNS in *Drosophila* and in the vertebrate share a conserved signalling molecule called netrin. The vertebrate *netrin-1* gene encodes a secreted protein expressed in the midline of the CNS (the floor plate) that guides commissural axons to the opposite side of the developing spinal cord.[30] A netrin homologue is also expressed in the midline of the *Drosophila* CNS (C. Goodman, personal communication) and in the *Caenorhabditis elegans* ventral nerve cord.[31] Similarly, in the case of enteropneusts[8,9] and other vermiform animals with multiple nerve cords, the region homologous to the CNS will ultimately have to be defined molecularly by the expression of *chd* and *netrin* homologues.

Thus, although arguments opposing the dorsoventral inversion have been raised, the recent results from molecular biology tend to vindicate the hypothesis of Geoffroy Saint-Hilaire.

Common Signals

An unexpected conclusion from recent studies on dorsoventral patterning is that the ectoderm and the mesoderm appear to be patterned by a common set of signalling molecules involving *dpp/Bmp-4* and *sog/chd.*[22,32] In *Drosophila,* there is strong genetic evidence that the amount of ectoderm allocated to the neurogenic region is controlled by positional information provided by this system. Both loss-of-function and gain-of-function studies show that *dpp* has an antineurogenic activity,[15,16,21] promoting the formation of dorsal epidermis, whereas *sog* promotes neurogenesis in the ectoderm.[11,17,21] Dpp is widely recognized as the main morphogen patterning the ectoderm, but recent evidence indicates that *dpp* patterns the *Drosophila* mesoderm as well. The mesoderm expresses a Dpp receptor, *thickveins,*[33] and Dpp produced by the ectoderm activates the expression of dorsal mesodermal genes (*bagpipe* and *tinman*) and represses the expression of a ventral mesodermal marker *(pox meso).*[34,35]

In *Xenopus,* CNS formation is thought to be different from *Drosophila.* The vertebrate neural plate is considered to be induced by signals secreted by the organizer. The organizer is also considered to be responsible for the release of signals that promote dorsal differentiation of mesoderm.[36] It was generally assumed that these signals would be different, but it now appears that the same molecules that pattern mesoderm also regulate CNS formation, as depicted in Fig. 5. Microinjection of *chd* (or *sog*) mRNA changes the fate of gastrula animal cap explants from ventral ectoderm (epidermis) to dorsal ectoderm (neural tissue).[22] In ventral marginal zone explants (which form ventral mesoderm such as blood and mesenchyme) *chd* induces dorsal mesoderm (notochord and muscle).[10] A similar situation has been observed for two other organizer-specific secreted factors, noggin and follistatin, which are neural inducers as well as dorsalizing

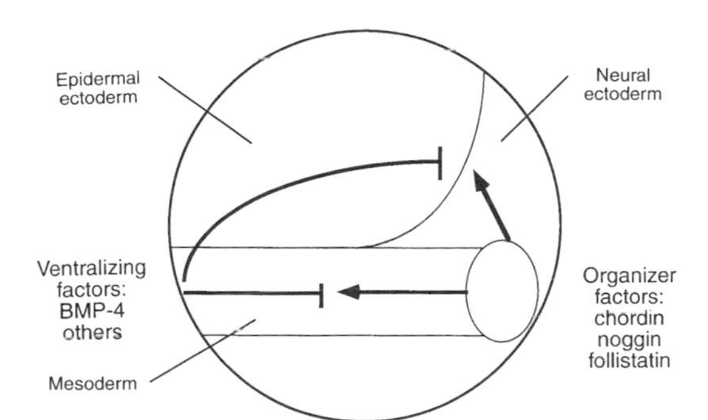

Fig. 5 Model indicating that the same set of regulatory signals may provide the positional information that patterns both ectoderm and mesoderm in *Xenopus.* On the dorsal side (right), the organizer *(oval)* provides dorsal positional values to ectodermal (animal cap, *top*) and mesodermal (marginal zone, *middle*) tissues by secreting organizer factors such as chd, noggin and follistatin. On the opposite side (left), ventralizing factors such as Bmp-4 and presumably other signals give ventral positional values to the tissues, antagonizing organizer signals. High dorsal values promote neural differentiation in the ectoderm and formation of notochord and muscle in mesoderm, whereas high ventral values lead to epidermogenesis in ectoderm and the differentiation of blood island and mesenchyme in the mesoderm. In this model, a default status of tissues does not exist and the balance between the dorsal and ventral signals decides dorsoventral fates.

agents.[22,37–39] This indicates that the signalling molecules used for dorsal differentiation of both ectoderm and mesoderm are the same, and that the differences must reside in the responding tissues (Fig. 5).

Studies on animal cap explants suggest that Bmp-4 has potent antineurogenic effects in *Xenopus.* Animal cap cells contain endogenous *Bmp-4* mRNA[18] which is required to prevent neural differentiation.[22] In addition, a recent experiment suggests that the neuralizing effect of cell dissociation on animal caps[40,41] may be due to the dilution of the Bmp-4 signal by diffusion into the culture medium. Indeed, when Bmp-4 was added to dissociated cells, they differentiated into epidermis (ventral ectoderm), and neural differentiation was prevented.[32] Overexpression of *Bmp-4* can suppress neural induction caused by *chd* and, intriguingly, also neuralization by *noggin* and *follistatin.*[22] The antagonism between *Bmp-4* and *chd* in neural induction is consistent with the roles of their homologues, *dpp* and *sog,* in *Drosophila* ectodermal patterning. It is not known whether *Drosophila* has homologues for *noggin* and *follistatin,* but this is an attractive area for future research because it has been proposed that additional factors should cooperate with *sog* in *Drosophila* neurogenesis (see X in Fig. 3).[17]

Although *Drosophila* neurogenesis and *Xenopus* neural induction have traditionally been considered to be different, the molecular mechanisms underlying the patterning of two distinct germ layers, ectoderm and mesoderm, seem to be regulated by the conserved *sog/chd* and *dpp/Bmp-4* system. The concept of dorsoventral patterning by positional information, which has been so useful in *Drosophila,* might also be applicable to vertebrates. In this view, the organizer would be the source of dorsal positional values which are counteracted by ventral values provided by Bmp-4 and presumably other ventralizing factors (Fig. 5). Thus, the unity of plan revealed by the dorsoventral axial reversal issue can be extended to detailed tissue patterns within individual germ layers.

The *Urbilateria* or Common Ancestor

In 1874 Ernst Haeckel proposed a sweeping homology: that the ectoderm and endoderm in all metazoans was related by descent from a hypothetical animal called the *Gastrea.*[42] The *Gastrea* consisted of two cell layers, the ectoderm and the endoderm, forming a primitive gut cavity opening to the outside. Although simplistic, the *Gastrea* theory historically was very useful because it proposed that all multicellular animals were monophyletic.

A large amount of data have now become available on genes controlling development from *Drosophila* and the vertebrates. The hypothetical ancestral animal, for which we propose the name *Urbilateria* (primitive bilateral animal), from which the arthropod and the chordate lineages diverged 600 million years ago may have presented the following characteristics. (1) It had anteroposterior polarity determined by the *Hox* gene complexes.[43,44] Additional genes present in *Drosophila* and vertebrates, for example homologues of *orthodenticle, empty spiracles* and *caudal,* formed a network that cooperated with *Hox* genes in the generation of anteroposterior pattern.[45] (2) It had a conserved system of dorsoventral patterning provided by the antagonistic *sog/chd* and *dpp/Bmp-4* extracellular signals, as described above. (3) It presumably had a subepidermal longitudinal CNS that had well-defined midline structures secreting axon guidance molecules such as netrin-1. (4) The function of *Pax-6/ eyeless,*[23,24] as well as the conservation of opsins,[25] suggests that the common ancestor had primitive photoreceptor cells from which the eyes of *Drosophila* and vertebrates evolved. (5) It presumably also had a circulatory system with a contractile blood vessel. *Drosophila* has a contractile dorsal vessel whose differentiation is controlled by the homeobox gene *tinman* as well as by the transcription factor DMEF2, both of which have vertebrate homologues expressed in the developing heart tissue.[46] In addition, other characteristics for which present evidence suggests independent evolutionary origins and convergent evolutionary solutions in insects and vertebrates, such as segmentation (metamerism) and the formation of appendages, may require revisions if new upstream regulatory systems are identified in future.

An important question is whether all deuterostomes evolved from the same ancestral animal. Evidence from 18S ribosomal RNA sequence comparisons is consistent with a common grouping for all deuterostomes.[47] Thus it is possible that the inversion event may have occurred concomitantly with a great innovation in gastrulation mechanisms that led to the two main groups of metazoans: the protostomes and the deuterostomes.[4,48] Most of the evidence on dorsoventral inversion of the axis reviewed here comes from comparisons of developmental genes in two very distantly related groups of organisms, the arthropods and the vertebrates. In future it will be useful to extend the comparisons of the *sog/chd* and *dpp/Bmp-4* signalling systems to other phyla in order to fill in the gaps, as has been so fruitful in the case of the *Hox* genes.[44]

Note added in proof: Since this manuscript was first submitted, several papers have appeared that further substantiate the idea of a conserved ventral specification pathway in vertebrates. These include: (1) the expression patterns of *Bmp-4* and *Bmp-2* in *Xenopus;*[50–52] (2) additional support for the requirement of Bmp signalling to repress neutralization of animal cap explants;[53–55] (3) a role for Bmp-7, in collaboration with Bmp-4, as a ventralizing factor;[56] (4) evidence for an inhibitory interaction between the ventralizing factor *Bmp-7* and *follistatin;* and (5) further evidence for the dorsalizing activity of *sog* in *Xenopus* embryos.[57]

E. M. De Robertis and Yoshiki Sasai are at the Howard Hughes Medical Institute, Department of Biological Chemistry, University of California, Los Angeles, California 90095-1737, USA.

References

1. Geoffroy Saint-Hilaire, E. *Mém. du Mus. Hist. Nat.* **9,** 89–119 (1822).
2. Appel, T. A. *The Cuvier-Geoffroy Debate* (Oxford Univ. Press, Oxford, 1987).
3. Nübler-Jung, K. & Arendt, D. *Wilhelm Roux Arch. dev. Biol.* **203,** 357–366 (1994).
4. Arendt, D. & Nübler-Jung, K. *Nature* **371,** 26 (1994).
5. Hogen, B. L. M. *Nature* **376,** 210–211 (1995).
6. Jones, C. M. & Smith, J. C. *Current Biol.* **5,** 574–576 (1995).
7. Lacalli, T. C. *Nature* **373,** 110–111 (1995).
8. Peterson, K. J. *Nature* **373,** 111–112 (1995).
9. Jefferies, R. P. S. & Brown, N. A. *Nature* **374,** 22 (1995).
10. Sasai, Y. *et al. Cell* **79,** 779–790 (1994).
11. François, V., Solloway, M., O'Neill, J. W., Emery, J. & Bier, E. *Genes Dev.* **8,** 2602–2616 (1994).
12. François, V. & Bier, E. *Cell* **80,** 19–20 (1995).
13. Wieschaus, E., Nüsslein-Volhard, C. & Jürgens, G. *Wilhelm Roux Arch. dev. Biol.* **193,** 296–307 (1984).
14. Zusman, S. B., Sweeton, D. & Wieschaus, E. F. *Devl. Biol.* **129,** 417–427 (1988).
15. Ferguson, E. L. & Anderson, K. V. *Cell* **71,** 451–461 (1992).
16. Wharton, K. A., Ray, R. P. & Gelbart, W. M. *Development* **117,** 807–822 (1993).
17. Holley, S. A. *et al. Nature* **376,** 249–253 (1995).
18. Fainsod, A., Steinbeisser, H. & De Robertis, E. M. *EMBO J.* **13,** 5015–5025 (1994).
19. Dale, L., Howes, G., Price, B. M. & Smith, J. C. *Development* **115,** 573–585 (1992).
20. Jones, C. M., Lyons, K. M., Lapan, P. M., Wright, C. V. & Hogan, B. L. *Development* **115,** 639–647 (1992).
21. Ferguson, E. L. & Anderson, K. V. *Development* **114,** 583–597 (1992).
22. Sasai, Y., Lu, B., Steinbeisser, H. & De Robertis, E. M. *Nature* **376,** 333–336 (1995).
23. Quiring, R., Walldorf, U., Kloter, U. & Gehring, W. J. *Science* **265,** 785–789 (1994).
24. Halder, G., Callaerts, P. & Gehring, W. J. *Science* **267,** 1788–1792 (1995).
25. Zuker, C. S., Cowman, A. F. & Rubin, G. M. *Cell* **40,** 851–858 (1985).
26. Zuker, C. S. *Science* **265,** 742–743 (1994).
27. Padgett, R. W., St. Johnson, R. D. & Gilbart, W. M. *Proc. natn. Acad. Sci. USA* **90,** 2905–2909 (1993).

28. Steinbeisser, H., Fainsod, A., Niehrs, C., Sasai, Y. & De Robertis, E. M. *EMBO J.* **14,** 5230–5243 (1995).
29. Winnier, G., Blessing, M., Labowsky, P. A. & Hogan, B. L. M. *Genes Dev.* **9,** 2105–2116 (1995).
30. Serafini, T. *et al. Cell* **78,** 409–424 (1994).
31. Colamarino, S. A. & Tessier-Lavigne, M. *Cell* **81,** 621–629 (1995).
32. Wilson, P. A. & Hemmati-Brivanlou, A. *Nature* **376,** 331–333 (1995).
33. Nellen, D., Affolter, M. & Basler, K. *Cell* **78,** 225–237 (1994).
34. Staehling-Hampton, K., Hoffmann, F. M., Baylies, M. K., Rushton, E. & Bate, M. *Nature* **372,** 783–786 (1994).
35. Frasch, M. *Nature* **374,** 464–467 (1995).
36. De Robertis, E. M. *Nature* **374,** 407–408 (1995).
37. Smith, W. C., Knecht, A. K., Wu, M. & Harland, R. M. *Nature* **361,** 547–549 (1993).
38. Lamb, T. M. *et al. Science* **262,** 713–718 (1993).
39. Hemmati-Brivanlou, A., Kelly, O. G. & Melton, D. A. *Cell* **77,** 283–295 (1994).
40. Grunz, H. & Tacke, L. *Cell Differ. Dev.* **28,** 211–217 (1989).
41. Sato, S. M. & Sargent, R. D. *Devl. Biol.* **134,** 263–266 (1989).
42. Haeckel, E. Q. *J. microsc. Sci.* **14,** 142–247 (1874).
43. Slack, J. M. W., Holland, P. W. H. & Graham, C. F. *Nature* **361,** 490–494 (1993).
44. Carroll, S. B. *Nature* **376,** 479–485 (1995).
45. De Robertis, E. M. in *Guidebook of Homeobox Genes* (ed. Duboule, D.) 11–23 (IRL, Oxford, 1994).
46. Scott, M. P. *Cell* **79,** 1121–1124 (1994).
47. Phillipe, H., Chenuil, A. & Adoutte, A. *Development* (suppl.) 15–25 (1994).
48. Willmer, C. H. *Invertebrate Relationships* (Cambridge Univ. Press, Cambridge, 1990).
49. Preobrazhenskii, A. A. & Glinka, A. V. *Dokl. Akad. Nauk. SSSR* **284,** 1489–1491 (1985).
50. Schmidt, J. E., Suzuki, A., Ueno, N. & Kimelman, D. *Devl. Biol.* **169,** 37–50 (1995).
51. Hemmati-Brivanlou, A. & Thornsen, G. H. *Devl. Genet.* **17,** 78–89 (1995).
52. Clement, J. H., Fettes, P., Knöchel, S., Lef, J. & Knöchel, W. *Mech. Dev.* **52,** 357–370 (1995).
53. Xu, R.-H. *et al. Biochem. biophys. Res. Comm.* **212,** 212–219 (1995).
54. Suzuki, A., Shioda, N. & Ueno, N. *Dev. Growth & Different.* **35,** 581–588 (1995).
55. Hawley, S. H. B. *et al. Genes Dev.* **9,** 2913–2935 (1995).
56. Yamada *et al. J. Cell Biol.* **130,** 217–266 (1995).
57. Schmidt, J., François, V., Bier, E. & Kimelman, D. *Development* **121,** 4319–4328 (1995).

Acknowledgments

The references cited in this Progress article are not comprehensive owing to space constraints. We thank L. Gont and L. Leyns for comments on the manuscript. Our work is supported by a grant from the NIH. E.M.D.R. is an HHMI Investigator. Y.S. was an HFSPO postdoctoral fellow and is an HHMI Research Associate.

Neil Shubin, Cliff Tabin and Sean Carroll

FOSSILS, GENES AND THE EVOLUTION OF ANIMAL LIMBS

The morphological and functional evolution of appendages has played a crucial role in the adaptive radiation of tetrapods, arthropods and winged insects. The origin and diversification of fins, wings and other structures, long a focus of paleontology, can now be approached through developmental genetics. Modifications of appendage number and architecture in each phylum are correlated with regulatory changes in specific patterning genes. Although their respective evolutionary histories are unique, vertebrate, insect and other animal appendages are organized by a similar genetic regulatory system that may have been established in a common ancestor.

The origin of evolutionary novelties raises some of the most fundamental questions of biology. How do new structures arise? Can they evolve *de novo* or are they generally derived from pre-existing structures? And what is the developmental and genetic basis for their origin and modification?[1]

The adaptive evolution of vertebrates and arthropods to aquatic, terrestrial and aerial environments was accomplished by the invention of many novel features, especially new types of appendages. Enormous progress has been made in the past few years in understanding appendage development in both phyla. These genetic discoveries can be integrated with paleontological data to address some of the principal events in the history of animal designs.

We will first examine the origin and evolution of vertebrate limbs and digits and of arthropod legs and insect wings. In both phyla we are confronted with a similar issue, namely the origin and adaptive modification of serially homologous organs. We will integrate paleontological and developmental evidence that suggests that major innovations are largely derived from pre-existing developmental systems and will illustrate the potential genetic regulatory changes that enabled appendage evolution. Then we will explore the significance of newly discovered genetic similarities between arthropod and vertebrate appendages—similarities that have been retained despite more than 500 million years (Myr) of independent evolution. We will develop the hypothesis that the evolution of successively derived limb types, from lobopods to insect wings, and from agnathan fins to tetrapod limbs, appears to be due, in part, to the successive co-option and redeployment of signals established in primitive metazoans. These examples illustrate how comparative developmental genetics can provide a mechanistic explanation of the origin and evolution of structures when paleontological data are robust and important new hypotheses about evolutionary history when the fossil record is silent.

Origin and Diversification of Tetrapod Limbs

Vertebrate limb diversity was produced by changes in the number, position and shape of structures that can be traced to Ordovician[2,3] (463–439 Myr) through late Devonian[2–4] (409–362 Myr) fossils. The demands of feeding and locomotion in Ordovician and Silurian seas led to a surprising variability of the earliest known appendages: some forms possessed a continuous anterior fin that ran the length of the body, others had paired fins that projected immediately behind a head shield, and still other primitive vertebrates had no paired fins at all (Fig. 1). A body plan with two sets of paired appendages, pectoral and pelvic, is a derived feature that first appears in later jawed vertebrates (gnathostomes).[2–4] The number of paired appendages has been highly conserved ever since their origin: the evolution of new gnathostome body plans primarily involved a modification of existing paired appendages rather than the invention of whole new sets (acanthodians are the only exception to this generalization). Therefore, the origin of more recent novelties, such as digits, involved the modification of genetic systems first established in more primitive vertebrates.

Serial Homology and Adaptive Diversification

Primitive genetic systems must have provided a framework for the evolutionary integration of pectoral and pelvic appendages. Digits, for example, arose at the same time in the hand and foot: there is no Devonian tetrapod that has fingers and no toes.[2–5] Even in the post-Devonian world many unique designs appeared simultaneously in forelimbs and hindlimbs, as witnessed by chameleons, ungulates and ichthyosaurs. Obviously, serially homologous appendages can also evolve independently, an extreme case being the modification of pectoral appendages into wings in bats, birds and pterosaurs. Even in these extremely modified forelimbs, however, numerous similarities are retained between wing and leg.

The linkage between forelimbs and hindlimbs appears to be an ancient feature that resulted from patterns of gene co-option during the evolution

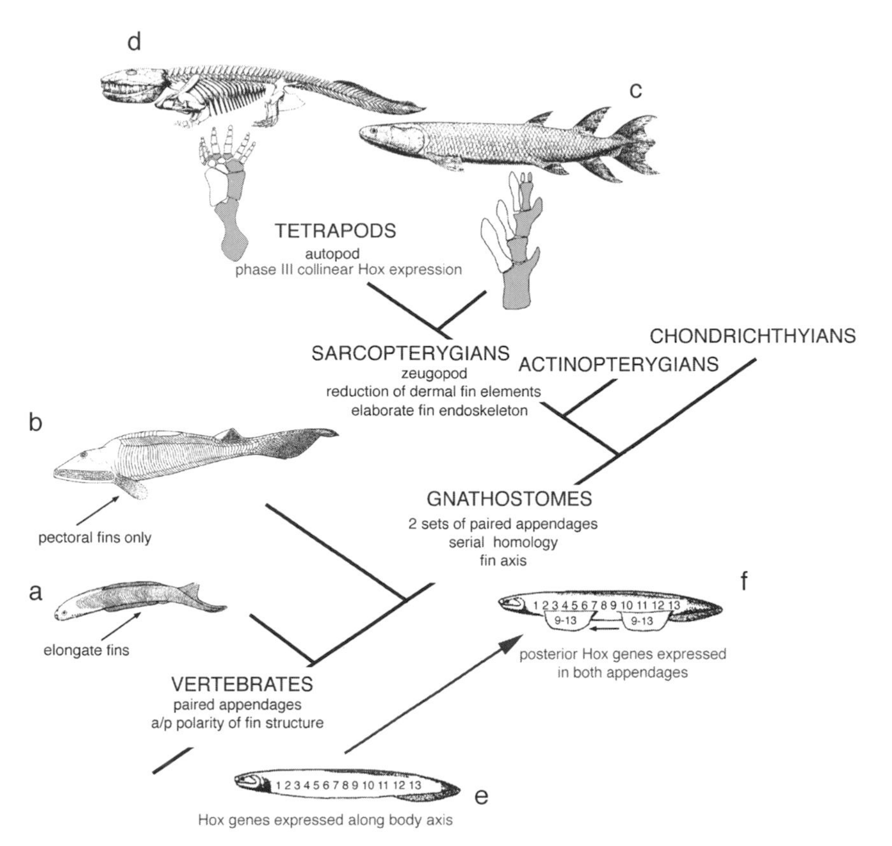

Fig. 1 Major innovations of vertebrate paired appendages. Basal chordates, such as *Amphioxus* (not shown), do not possess appendages homologous to those of vertebrates. Unpaired, median fins are the earliest known vertebrate appendages.[2] Paired appendages are first encountered in Ordovician and Silurian jawless fish as elongate fins that extend laterally along the body wall (for example, *Jamoytius; a*) or as paired pectoral fins (osteostracans; *b*). Other basal vertebrates (not shown) do not possess any paired appendages. Multiple sets of paired appendages are a derived characteristic of jawed fish (gnathostomes). In many gnathostomes, pectoral and pelvic fins have often evolved in parallel. This pattern of concerted evolution suggests that pectoral and pelvic appendages shared similar regulatory genes in early stages of gnathostome evolution. A fin axis *(shaded)* is seen in the fins of many gnathostomes, and it is a primitive characteristic for sarcopterygian fish (for example, *Eusthenopteron; c*) and tetrapods (for example, *Ichthyostega; d*). Sarcopterygian fins are derived in having a zeugopod and an elaborate endoskeletal fin skeleton. Digits develop within the distal portion of this extensive endoskeleton. The establishment of serially homologous appendages is proposed to result from gene co-option during the evolution of Paleozoic vertebrates. *HoxD* genes were probably not involved in the origins of body wall outgrowths in basal vertebrates because unpaired fins do not express these genes.[9] These *Hox* genes were initially involved in specifying regional identities along the primary body axis, particularly in caudal segments *(e)*. One key step in the origin of jawed fish was the co-option of similar nested patterns of expression of *HoxD* genes in the development of both sets of paired appendages *(f)*. This co-option may have happened in both appendages simultaneously, or *Hox* expression could have been initially present in a pelvic appendage and been co-opted in the development of an existing pectoral outgrowth.[8] The reconstructions in *a* and *b* are modified from those in ref. 106, that in *c* from ref. 107, and that in *d* from ref. 108. The hindlimb of *Ichthyostega (d)* is modified from ref. 5. (A color version of this figure is available on-line at <http://www.press.uchicago.edu/books/gee>.)

of Paleozoic fish. Serially homologous paired appendages are seen in Paleozoic placoderms, acanthodians, chondrichthyans and osteichthyans.[2–4] In addition, pectoral and pelvic fins have evolved in parallel in almost all major gnathostome clades. There are numerous genetic parallels in pectoral and pelvic development that could account for these patterns of concerted evolution.[6,7] *Hox* genes, in particular, are likely to have been involved in the evolution of serial homology.[8] The earliest vertebrate appendages (unpaired fins) presumably did not utilize *Hox* genes; *HoxA* and *HoxD* genes are not expressed during the outgrowth of zebrafish unpaired fins.[9] Although these *Hox* genes were probably not involved in the origin of outgrowths in basal vertebrates, their later recruitment in the development of paired appendages was a key step in establishing serially homologous designs. The *HoxD* genes that came to play a role in appendage development are a subset of those involved in specifying regional identities along the caudal body axis (caudal neural tube, gut, somitic and lateral plate mesoderm).[6–10] This suggests one of two situations: either nested patterns of *Hox* expression were originally present in a caudal set of paired outgrowths, and were later recruited in the development of a cranial set of outgrowths,[8] or similar *Hox* genes were recruited in pectoral and pelvic outgrowths at the same time in the evolutionary history of vertebrates (Fig. 1). In either case, pectoral and pelvic appendages were genetically linked early in their history and could have evolved together, presumably because the development of these appendages had already been brought under similar regulatory controls.

Superimposed on these ancient genetic parallels are secondary differences in gene expression and interaction that may have served as the basis for the independent evolution of pectoral and pelvic appendages. *Hox* gene expression in extant tetrapod limbs is dynamic and encompasses at least three distinct phases initiated successively in the primordia of the stylopod, zeugopod and autopod[7] (Fig. 2). The presence of three distinct phases of *Hox* expression in limbs may reflect the observation that all tetrapods maintain a standard pattern of organization (Fig. 2), whereas specific differences in expression (or gene interaction) in each phase could result in the independent modification of pectoral and pelvic appendages.[7] Phase II *Hox* expression is practically identical in the forelimb and hindlimb buds of mice, which possess generally similar skeletal patterns. In contrast, the wings and legs of chicks have very different skeletal patterns and different patterns of phase II *Hox* expression as well.[7,11] Surprisingly, phase III expression is very similar in chick wing and leg buds, indicating that aspects of the derived structure of chick wings are established by some other genetic means. Candidates include *Hox* genes of other clus-

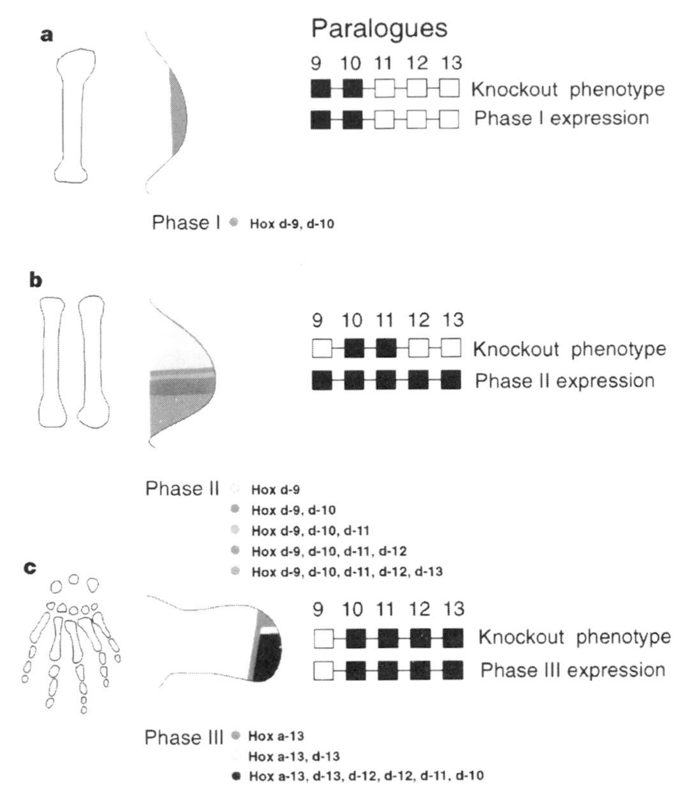

Fig. 2 Stylopod, zeugopod and autopod: patterning the limb. The tetrapod limb consists of three distinct compartments: *a,* the stylopod (upper arm and thigh); *b,* zeugopod (lower arm and calf); and *c,* autopod (hand and foot). This subdivision of the limb is supported by phylogenetic comparison,[4] analysis of gene expression,[7] and experimental manipulation.[14,36–41,109–111] There is a broad correlation between the position of a compartment and its evolutionary history.[4] The stylopod *(a)* is the most ancient (possibly of late Silurian origins) whereas the zeugopod and autopod are the most recent (being first encountered in Devonian sarcopterygians). This same order of appearance of the three limb segments is recapitulated during development. Early removal of the apical ectodermal ridge (AER) results in a limb with only a stylopod; the zeugopod and autopod are produced after successively later surgeries.[112,113] The *Abd-B*–related genes of the *HoxD* cluster are expressed in a complex, dynamic pattern encompassing at least three distinct,[7] independently regulated phases.[24,26] In the first phase (*a;* phase I), two of these genes (*HoxD-9* and *HoxD-10*) are expressed across the entire limb bud.[7] This expression correlates with the time that the stylopod is specified.[7,112] Subsequently, a second phase of expression (*b;* phase II) is initiated in response to the secreted factor, Sonic hedgehog. Here, *Hox* genes are expressed in a nested set centered around the Sonic-expressing cells, with *HoxD-13* being expressed in the most restricted domain, and *HoxD-12* and *HoxD-11* each encompassing a broader domain.[7] This pattern of expression coincides with the time of specification of the zeugopod and takes place in cells fated to form this segment. Finally, a third phase of expression (*c;* phase III) is initiated later during limb development, when these *Hox* genes are all expressed across the majority of the distal portion of the limb bud.[7] During this phase, the expression of *Hox* genes still appears to be a consequence of the Sonic hedgehog signal, but the relative responsiveness of the different genes has changed so that *HoxD-13* now has the broadest expression domain and *HoxD-12* and *HoxD-11* are nested

ters that are differentially expressed in the wing and leg buds[12] and the *T-box* genes, another family of putative transcription factors differentially expressed in the forelimb and hindlimb buds.[13] Combinatorial action between genes may explain different functional requirements for the *HoxA* and *HoxD* genes in the forelimb and hindlimb. For example, homozygous deletion of both *HoxA-11* and *HoxD-11* results in almost complete loss of the zeugopod in the forelimb but not in the hindlimb,[14] despite the fact that these genes have equivalent patterns of phase II expression in both limbs. A possible explanation for the observed differences between the appendages may be that there is expression of the paralogous gene *HoxC-11* during phase II in the hindlimb but not the forelimb, where it may act redundantly with *HoxD-11*.

Learning to Crawl: The Fin-to-Limb Transition

Some regions of vertebrate appendages are more variable than others.[4,15,16] The invention of flippers, wings and other specialized limbs often involved significant changes in the pattern of distal structures rather than proximal ones.[15] Two broad notions of the homology of distal structures have emerged over the past 130 years: one that sees digits as being unique to tetrapods[17,18] and another that sees antecedents of digital structure in the fins of sarcopterygian fish.[19,20] Both genetic and fossil data support the hypothesis that digits are evolutionary novelties[21,22] (Fig. 3).

The origin of digits is associated with the evolution of new temporal and spatial patterns of gene expression and regulation.[7,9,21] In extant tetrapods, the development of digits correlates with a reversal in the anteroposterior order of expression of *Hox* genes in phase II and phase III[7,23] (Figs.

---

within it.[7] The phase III expression *(c)* patterns occur in the presumptive autopod at the time that segment is specified. The combination of the change in relative size of *Hox* expression domains with a phenomenon known as "posterior prevalence" (the general rule that more 5′ genes in the *Hox* cluster are phenotypically dominant) results in different *Hox* genes playing pre-eminent roles in different limb segments: for example, *HoxD-9* during phase I in the stylopod *(a)*, *HoxD-11* during phase II in the zeugopod *(b)*, and *HoxD-13* during phase III in the autopod *(c)*. The expression of the dominant *Hox* genes in each phase is essential for the formation of the bones in each segment, as seen in their knockout phenotypes. The knockout phenotype of a gene consists of alterations in the pattern and shape of skeletal elements. The location of these modifications depends on the position of the gene within the cluster. For example, mice engineered to be deficient in both *HoxD-11* and *HoxA-11* (the paralogous gene of the *HoxA* cluster) form limbs that are essentially missing the zeugopod[14] *(b)*. Phenotypes of *Hox D-9*–deficient mice, in contrast, are specific to the stylopod[109] *(a)*, whereas *Hox D-13* -deficient mice primarily have defects in the autopod[39] (as do mice engineered to be deficient of *Hox A-13* paralogues;[110,111] *c*). Knockout data are derived from refs. 14, 36–41 and 109–111. (A color version of this figure is available online at <http://www.press.uchicago.edu/books/gee>.)

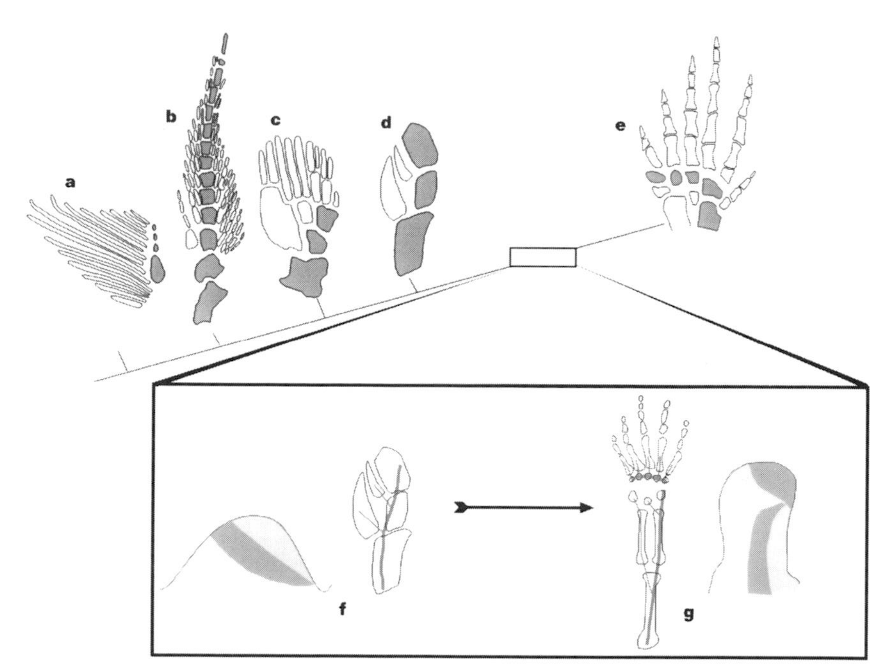

Fig. 3 The origin of digits. A fin axis *(shaded)* is present in the fins of chondrichthyans *(a, Cladoselache)*, basal actinopterygians (not shown), and sarcopterygians *(b–e)*. In sarcopterygians such *as Neoceratodus (b), Sauripteris (c), Panderichthys (d)* and *Tulerpeton (e)*, a single element (stylopod) articulates with the pectoral or pelvic girdle; all other proximal bones of the fin have been lost. The autopod is considered to be a synapomorphy of tetrapods because their nearest relatives do not have any apparent homologues of digits. *Box:* we propose that the origin of digits correlates with a novel pattern of *Hox* expression. *f,* The axis *(thick line)* of the fin of *Panderichthys* is short and radials *(thin lines)* branch preaxially. Patterns of phase II *Hox* expression *(f; Hox D-11* in *light shading, Hox D-13* in *darker shading)* were most likely already in place in sarcopterygian fins. *g,* The proximal portion of the axis *(thick lines)* of tetrapods compares with the entire axis of *Panderichthys.* Phase III *Hox* expression reflects a reversal of the nested domains of expression of *Hox D-11 (g, light shading)* and *Hox D-13 (g, dark shading)* during the time when the autopod is specified. This reversal in the polarity of *Hox* expression is considered to be correlated with the origin of the autopod; digits differ from more proximal structures in that they lie on the postaxial side of the axis. Because this hypothesis relies on genetic comparisons between phylogenetically disparate taxa (for example, teleosts and tetrapods), the shift from phase II to phase III collinear expression may have evolved in more basal sarcopterygians. Analogues of digits are seen in the fins of other sarcopterygian fish *(c, Sauripteris);* the expanded endoskeletons of rhizodontids include as many as eight branched preaxial radials. Different lineages of sarcopterygians appear to be inventing similar solutions to life in shallow freshwater ecosystems. The reconstruction of *Tulerpeton (e)* is modified from ref. 114. (A color version of this figure is available on-line at <http://www.press.uchicago.edu/books/gee>.)

2, 3). Recent studies of teleosts (zebrafish) have revealed patterns of *Hox* expression that are similar to patterns seen in proximal regions of tetrapod limbs.[9,21] Phase III *Hox* expression is not seen in the zebrafish and appears to be unique to the digital region of tetrapod limbs[7,9] (the expression of other sarcopterygians or more basal actinopterygians is not known). In addition, the different phases of *Hox* expression are not only discrete in tetrapod limbs, but are regulated by separate *cis*-regulatory enhancer elements in each phase.[24–26] During phase II *Hox* expression in tetrapods, a complex set of enhancer elements is used within the regulatory region of each *Hox* gene of the cluster (as in the regulation of the *Hox* genes along the main body axis).[25,26] However, regulation of all the *HoxD* genes in phase III depends upon a single enhancer upstream of the entire cluster.[24,26] The utilization of this distinct enhancer is consistent with the hypothesis that digits are evolutionary novelties because the development of the autopod is regulated differently from that of the rest of the limb.

The presence of phase III *Hox* expression in tetrapod limbs, and its absence in teleost fins, suggests that this pattern may be an apomorphy for tetrapods or a more inclusive group. In addition, the presence of a uniquely tetrapod enhancer for phase III *Hox* expression implies that this regulatory element is also more derived relative to conserved phase I and II enhancers. The shift from phase II to phase III collinear expression involves multiple genes expressed at different times and in different regions of the limb. If these changes were genetically independent, then they would have required the joint evolution of numerous regulatory elements. If only a single enhancer was involved, then this shift could have produced a change in the expression of multiple genes in a small number of evolutionary steps. Furthermore, the utilization of the same enhancer in forelimbs and hindlimbs provides a developmental explanation of the observation that fingers and toes arose simultaneously in the fossil record.

We propose that the temporal and spatial shift in the expression of *Hox* genes during limb development correlates with transformations inferred from the fossil record. Devonian fossils provide morphological links between structures in fins and limbs. Sarcopterygian fins are dominated by an axis of segmented endoskeletal elements that extends from proximal to distal[2–4,27] (Figs. 1, 3). This axis is most similar to tetrapod limbs proximally, where the humerus, radius and ulna (femur, tibia and fibula) can readily be compared between taxa.[2–5,15,22] Embryological and paleontological data suggest that the axis of fins was developmentally bent during the origin of tetrapod limbs.[27] This scheme holds that there is a dramatic difference between the autopod and the zeugopod because the branching of the axis shifts from the anterior (preaxial) to the posterior (postaxial) compartment of the limb.[7,27] Proximal elements, such as the radius, project anteriorly from the axis, whereas distal elements, such as the digits, project from the posterior side of the axis. We propose that the reversal of morphological polarities in the appendages of Devonian vertebrates correlates with the reversal of *Hox* gene expression seen in phase III (Fig. 3f, g). As *Hox* expression in phase III is driven by a novel enhancer element, the axis was not bent *per se;* rather, a novel extension (with reversed morphological polarities) is considered to be added to it in the late Devonian. This hypothesis is supported by a comparison of panderichthyid fins and tetrapod limbs. The fins of this sister group of tetrapods (Fig. 3d, f) are highly reduced in comparison to other sarcopterygians (Figs. 1e, 3b, c, e) and this reduction is most prominent distally. No potential homologues of digits, wrist or ankle bones are preserved in these fish.[22]

Are digits, or their functional equivalents, unique to tetrapods? Fins of rhizodontid fish have stunning similarities to tetrapod limbs[20] (Fig. 3). The fins of these Devonian fish contain up to eight endoskeletal radials that project distally; in *Sauripteris* (Fig. 3c), six of these rods terminate at the same proximodistal level. Either these radials are directly homologous to the six to eight digits of Devonian tetrapods (a hypothesis not supported by phylogenetic inference) or they are functional analogues. In either case, these rods reflect a site at which morphological polarities are reversed (for example, the radials of *Sauripteris* branch postaxially[20]), suggesting that phase III *Hox* expression may have arisen in this clade. Several different lineages of Devonian sarcopterygians appear to have evolved the same morphological solution to life in shallow freshwater environments. The tetrapod clade evolved true digits, whereas rhizodontids developed functional analogues. In both cases, the genetic shifts may well have been similar.

Adaptive Diversification: How Many Fingers?

Digit reduction is a dominant theme of tetrapod limb evolution; deviations from a pentadactyl pattern virtually always involve the loss of fingers or toes.[5,28] Polydactylous hands and feet have almost never been fixed in phylogeny, despite the presence of polydactylous variants within populations (individuals of many species including cats, dogs, mice, chickens and humans carry mutations that cause the formation of extra digits). This paradox can be explained in terms of evolutionary constraints by postulating a genetic limitation to digital evolution. One approach holds that the genetic mechanisms that determine the number of digits are distinct from those that regulate morphology, and that there are currently only five discrete genetic programs for specifying unique digit morphology.[28] The primary genetic limitation is on the number of kinds of digits, not their absolute number. A specific prediction of this hypothesis is that polydactyly can arise, but at least two of the digits will have the same identity (that is, morphology). In this regard, it is easier to modify other carpal or tarsal bones to new functions than to create a new digit. Supporting this notion is the observation that the additional "digits" in extant polydactylous taxa are typically modified carpal or tarsal bones (as in frogs and panda bears, for example). Unfortunately, we cannot as yet evaluate the genetic basis of this constraint because we do not understand the genetic mechanisms that regulate the differences between digits.

Many classical morphologists were interested in defining "laws of form"—common trends that appear in widely different groups. Comparative analysis of diverse taxa now offers the promise of fundamental insights into these long-dormant questions. The 360-Myr history of tetrapod limbs is witness to dramatic regularities in digital reduction.[4,16,29] One notion, "Morse's law of digital reduction," contrasts the stability of the inside digits (III, IV) with the lability of outside ones (V, II, I).[30,31] In virtually every known example of digital reduction, digits V, II and I are among the first to be lost and digits III and/or IV are typically retained in tetrapods that have the most extreme patterns of digital reduction. This pattern is widespread and has evolved independently in lizards,[32,33] dinosaur and bird feet,[34] and mammals[35] (for example, ungulates). The major exception to this trend lies in the hands of theropod dinosaurs (that lose postaxial digits); theropod feet conform to Morse's law, as do the hands and feet of other dinosaurs.[34]

Are regularities of digital evolution the product of developmental constraints upon variation? Knockouts of different *Hox* genes *(HoxD-11, HoxD-12, HoxD-13, HoxA-13)* lead to changes in the shape and number of bones in affected mice and these different genes often have overlapping effects.[14,36–41] One common result is the stability of the internal digits (III, IV) in knockouts of single genes or combinations of genes.[41] The parallels between the expectations of Morse's law and the results of experimental manipulation suggest that trends of digital evolution may have a developmental basis. The morphological effects of different gene knockouts may reflect the sequence of digital formation: the first digits to be affected are typically the last to form in development.[37] Although evolutionary patterns of digital reduction are unlikely to involve coding mutations of *Hox* genes, limb reduction may involve changes in the regulation of *Hox* genes or the genes that they control. Comparative analysis of gene expression and function in representative taxa could elucidate the mechanisms behind these general evolutionary trends. The lizard genus *Lerista,* for example, has species with five, four, three, two, and no toes—the range of states of digital reduction in this genus parallels that seen in virtually all other tetrapods.[32]

Origin and Diversification of Arthropod Limbs

> Exites and endites of the proximal limb segment of arthropods, from trilobites to insects, have an extraordinary history. They have furnished most of the remarkable tools of the phylum: mandibles and other mouth parts; gills of trilobites and Crustacea; swimming and grasping appendages; gill-plates of Ephemeroptera . . . Their evolutionary potential is comparable with that of the vertebrate limb . . . (Wigglesworth 1973).[42]

The adaptive radiation of arthropods started much earlier than that of tetrapods. The Cambrian fossil record abounds with trilobites, arachnomorphs and crustacean-like forms, and contains many bizarre animals with spectacular appendages and body armor (Fig. 4). The "arms race" of the Cambrian explosion may well have been a "limbs race" among arthropods to evolve better sensory, locomotory, feeding, grasping and defensive appendages.

The most obvious feature of arthropod diversity is the number, morphology and function of their appendages (Figs. 4, 5). Antennae, mouthparts, walking legs, grasping and swimming appendages are all modifications of a basic jointed limb structure that defines the phylum. It is generally thought that jointed legs evolved from simple unjointed appendages such as the lobopodia found in the probable sister group of the arthropods, the Onychophora (Fig. 5a). The diversity of Cambrian arthropods and lobopodans (Fig. 4) indicates that the transition from lobopods to jointed appendages occurred before the Cambrian (570–510 Myr). Fossils from this period are scarce and the reconstruction of these transitions mostly relies on the comparative analyses of later, Cambrian fossils.[43,44]

Morphological studies of Cambrian lobopodans have investigated the series of innovations that led to the basal arthropod design. These novelties include: the evolution of external segmentation; sclerotization; and, most important to our discussion, the origin of jointed, biramous appendages. Whereas the taxonomic relationships between Cambrian forms and extant arthropods are uncertain, Cambrian taxa with different types of "arthropodization" and limb morphology can be identified.[43] For example, the Cambrian *Aysheaia* and the Recent *Peripatus* possess simple, unjointed uniramous limbs, but lack the annulation and armor typical of more derived lobopodans such as *Hallucigenia*[43] (Fig. 4). Other lobopodans, such as *Opabinia,* are externally segmented and possess lateral lobes above the ventral lobopods[43] (Figs. 4, 5b). Fusion between the gill-like

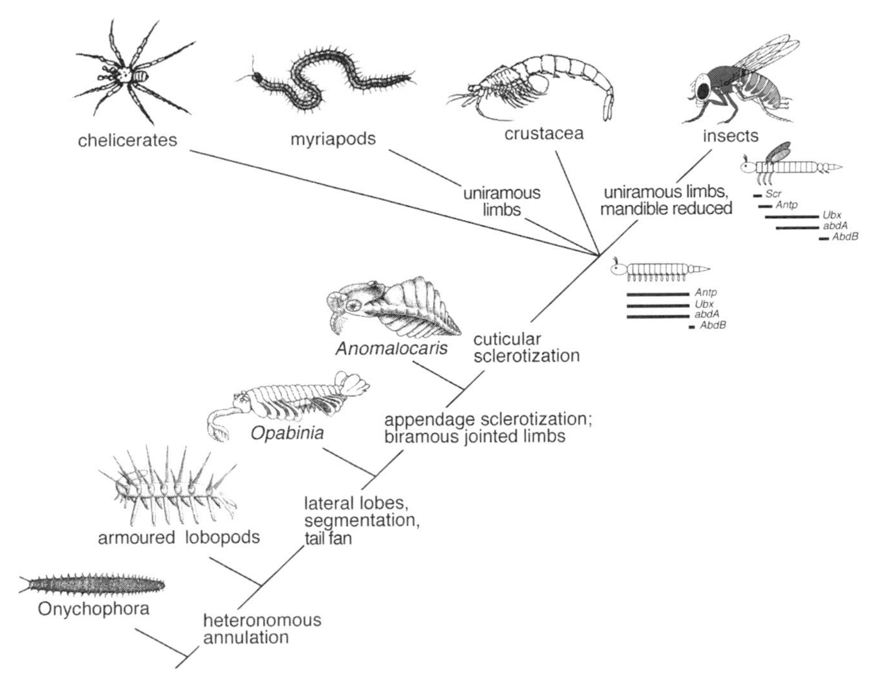

Fig. 4 The lobopod-arthropod transition and the diversification of arthropod limb patterns. Several innovations occurred in the lobopod-arthropod transition.[43] Various lobopods that may represent different degrees of "arthropodization" are depicted. *Opabinia* is shown in partial cutaway view[43] to reveal the lateral lobes and ventral lobopodia. The relationships among the major arthropod groups is an unresolved polychotomy. The most basal euarthropod was probably fully sclerotized with jointed, biramous limbs,[115] and a homonomous trunk. Uniramous limbs evolved in the terrestrial arthropods. The subdivision of the trunk and differentiation of individual limbs in modern insects involved regulatory changes in *Hox* gene domains along the main body axis from an ancestor in which trunk appendages and *Hox* gene domains were mostly identical. *Hox* scheme is adapted from ref. 53.

lateral lobes and ventral lobopodia may have given rise to the biramous limb[43,45–47] (Fig. 5c). In forms such as *Anomalocaris,* this fusion was accompanied by limb segmentation and sclerotization (Figs. 4, 5). Full cuticular sclerotization, then, arose in primitive arthropods.

Serial Homology and Adaptive Diversification

The adaptive radiation of primitive arthropod limbs entailed considerable changes in their number, pattern and function. Two trends are evident. First, in arthropods in general and certain lineages in particular, there have been increases in the number of different limb types. For example, advanced lobopods had perhaps four types of appendages (frontal, jaw, trunk and tail fan) and only one type of trunk appendage, whereas insects pos-

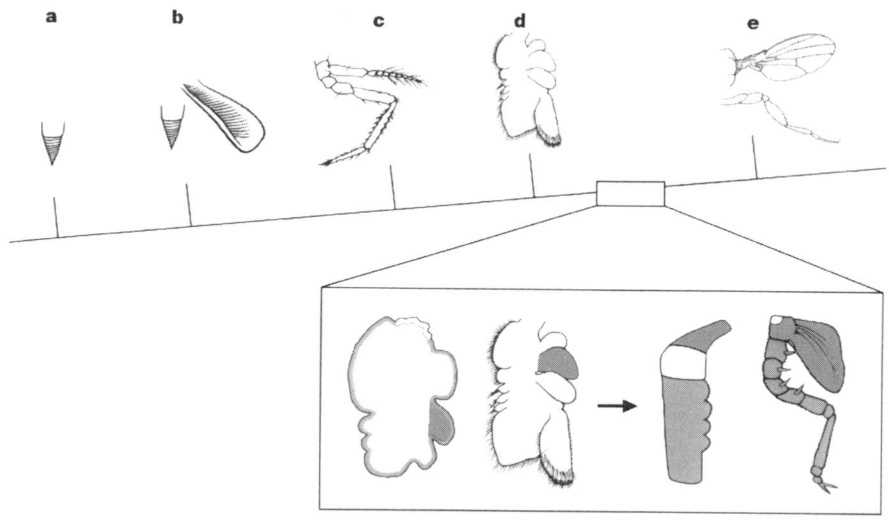

Fig. 5 The evolution of the arthropod limb and the origin of the insect wing. *a–e,* Some of the major transitions in limb architecture. *a,* A simple unjointed, annulated lobopodium; *b,* separate lateral lobes may have served a gill-like function; *c,* a jointed biramous limb in which the two limb branches are joined at the base, an upper branch is often derived from lateral lobes; *d,* a multibranched limb found in the branchiopod crustacean *Artemia; e,* the insect wing and uniramous leg appear to derive from a polyramous limb in an aquatic ancestor. *Box: left,* a developing polyramous limb and corresponding differentiated structure. The *apterous* gene is expressed *(solid block)* in a dorsal respiratory lobe, whereas the *Distal-less* gene is expressed *(edges)* in the other limb branches.[54] An ancestor-descendent relationship between the limbs in the box is not implied. *Right,* separation of the dorsal respiratory lobe from the ventral limb primordium in a primitive pterygote such as a Palaeodictyoptera nymph. The proto-wing at this stage was probably a gill-like structure on all trunk segments and still attached to the base of the limb. The *apterous* and *Distal-less* genes play critical roles in wing and leg formation in *Drosophila.* (A color version of this figure is available on-line at <http://www.press.uchicago.edu/books/gee>.)

sess up to ten distinguishable limb-derived appendages and four or five different trunk appendages (Fig. 4). A second trend has been the dramatic diversification of homologous appendages—not just in size or morphological detail. Major changes in limb organization have evolved, such as the evolution of unbranched walking legs in insects and changes in mandible architecture in myriapods, crustacea and insects (Fig. 4).

The developmental genetics of limb formation and identity in Recent arthropods may help to shed light on the diversification of arthropod limbs. *Hox* genes have played a different role in arthropod evolution from that seen in tetrapods. Studies of *Drosophila melanogaster* have revealed that each type of appendage is typically specified by a single or a pair of *Hox* genes acting in the individual body segment that gives rise to a particular appendage.[48] For example, the *Antennapedia* gene acts in all three pairs of walking legs, but the distinct morphology of the first, second and third pair of walking legs is determined by the *Sex combs reduced, Antennapedia,* and *Ultrabithorax Hox* genes,[49] respectively. In the antenna, no *Hox* gene is active. If *Hox* gene function is lost or ectopically activated in individual segments, the identity of the corresponding appendage is transformed. Thus, loss of *Antennapedia* transforms second leg to antennal structures[50] and expression of *Antennapedia* in the antenna transforms it to a leg.[51] Importantly, loss of all *Hox* gene functions in insects results in a dead embryo bearing antennae on all segments.[52] This demonstrates that the potential to form a limb exists in all segments, but the type of limb formed is determined by individual *Hox* genes.

The specification of different limb types by different *Hox* genes or combinations of *Hox* genes differs from the nested pattern of *Hox* genes expressed in the limbs of vertebrates and has important implications for the pattern of morphological evolution in the arthropods. Different, but serially homologous, arthropod limbs are distinguished by the action of different *Hox* genes that modify the interpretation of a common set of positional signals. Thus, the increase in appendage diversity between lobopods, primitive crustacea and insects must have involved the diversification of *Hox* gene regulation and function in the arthropod trunk. Comparative studies of *Hox* gene expression between crustacea (for example, branchiopods) and insects support this notion. In the branchiopod thorax, the expression of the *Antp, Ubx* and *abd-A Hox* genes is coincident and the morphology of thoracic limbs is uniform.[53] In insects, these same *Hox* genes differentiate the middle thorax, posterior thorax and anterior abdomen (Fig. 5). The adult hexapod abdomen is legless and this is due to the direct repression of limb formation by products of the *Ubx* and *abd-A Hox* genes.[54] These gene products do not repress limb formation in branchiopods.[55]

The diversity of the architecture of putatively homologous appendages has fuelled many debates about arthropod relationships. For example, walking legs and mandibles can differ so much between taxa that it has been suggested that the arthropods had multiple ancestors (that is, they are polyphyletic),[56] but phylogenetic[57,58] studies have refuted this. In addition, developmental studies suggest that different limb architectures arise through modifications of a common genetic program. For example, the *Distal-less (Dll)* gene controls the development of the distal portion of *Drosophila* limbs[59] and is expressed in the distal domains of limbs in all arthropods studied so far.[50,60,61] However, *Dll* is not expressed in insect or developing adult crustacean mandibles[55] but is expressed in myriapod mandibles.[62] These data agree with fossil evidence suggesting that crustacean and insect mandibles were reduced from the primitive whole-limb mandible by truncation of the mandibular proximodistal axis.

The branching of arthropod limbs has been very important to their functional evolution. Different limb branches can be specialized for respiration, locomotion and a variety of other functions. Chelicerates, trilobites and aquatic crustacea have biramous or polyramous limbs and first appear in the Cambrian, whereas the terrestrial myriapods, insects and crustacea have unbranched (uniramous) limbs and appear much later in the Silurian[63] and Devonian,[64] respectively. It has been argued that the ancestral arthropod was biramous.[43,45–47] Comparisons of *Dll* expression and limb outgrowth in various types of crustaceans and insects reveal that all limb types arise from the same relative anteroposterior position within the body segment but differ in their dorsoventral branch points.[55] This suggests that uniramy, biramy and polyramy are the products of shifts in signals along the dorsoventral axis of the body wall (or appendage) and that additions or reductions in branch number may evolve readily. This flexibility was crucial to the later evolution of perhaps the most significant invention by any arthropod—wings.

Learning to Fly: The Leg-to-Wing Transition

Early in the Devonian, before tetrapods arose, one major animal group had already invaded land—the insects. The subsequent evolution of flight presaged the enormous radiation of insects: these taxa now comprise more than two-thirds of all known animal species. The evolution of insects from an as-yet uncertain arthropod ancestor and the emergence of winged (pterygote) forms involved major transitions in limb architecture and function.

Few evolutionary mysteries have inspired more theories than the origin of insect wings. It is not certain when wings first arose because the early Devonian insect fossil record is scanty.[64] All of the major pterygote groups appear by the Carboniferous (362–290 Myr) and are assumed to have arisen earlier. Fundamentally, discussion on the origin of wings has focused on whether they are novelties or whether they are modified versions of ancestral structures. If wings were derived from existing structures, what were their anatomical origins and initial functions?

One of the longest-held models is the "paranotal" theory, which holds that wings are novelties derived from hypothetical rigid extensions of the body wall of a terrestrial ancestor.[65] A second hypothesis, the "limb-exite" model, proposes that insect wings evolved in a series of transitions beginning with the polyramous exite-bearing legs of an aquatic pterygote ancestor.[42,66] According to this model, proximal limb elements were modified to flap-like structures and adapted to spiracular or movable gill covers to facilitate respiration. These sac-like pro-wings were found on all thoracic and abdominal segments and became stronger as ancestral pterygote nymphs (similar to Recent mayfly larvae) used them presumably for propulsion. As insect lifestyles became more amphibiotic, some insects might have used a rudimentary proto-wing for surface skimming, as seen in extant stoneflies.[67] Finally, wings acquired the mechanical strength and flexibility (with corrugation and veins) and the supporting musculature to support active flight.

Recent studies of the developmental regulatory mechanisms controlling wing formation and number in *Drosophila* suggest a close ontogenetic and evolutionary relationship between legs and wings. For example, in *Drosophila*[68] and other Diptera, the wing arises in close association with the leg. Cells that give rise to the wing field in *Drosophila* actually migrate out of the developing ventral limb field.[68] There are important regulatory differences that distinguish the development of the sheet-like wing from the tubular structure of the leg. One key difference is that adult and developing wings are divided into discrete dorsal and ventral compartments whereas legs are not. In *Drosophila,* the definition of dorsal versus ventral cell fates is orchestrated by the *apterous* gene which is expressed only in dorsal cells and is necessary for wing but not leg formation.[69] Clearly, *apterous,* which regulates several crucial downstream signalling components, and is involved in a conserved dorsal pattern of expression in insect wings,[70] was co-opted into a distinct role in dorsoventral patterning at some stage of wing evolution. *apterous* primitively could have specified a dorsal compartment (or branch) of an ancestral polyramous appendage or an evolving proto-wing. Remarkably, this is exactly what has been found in a recent study.[71] The dorsal branch of a branchiopod crustacean respiratory epipodite specifically expresses the *apterous* gene and one other developmental marker of the insect wing field (Fig. 5d). This suggests that Recent wings evolved from the respiratory lobe of an ancestral polyramous limb, probably first appearing in the immature aquatic stages as gill-like structures, such as those found on all trunk segments of extinct Paleodictyoptera or extant mayfly larvae (Fig. 5d). Wings subsequently emerged as adult appendages and acquired greater strength and flexibility for sustained flight (Fig. 5e).

Interestingly, if the respiratory epipodite origin of insect wings is correct, then wings may have an even deeper origin, not just in the aquatic ancestor of pterygotes, but in lobopodans. The origin of the biramous limb has been postulated to involve the fusion of the gill-like structure of lateral lobes (such as those in *Opabinia*) with the ventral lobopod[43,45–47] (Fig. 5c). If this is true, then the wing may indeed be derived from lateral lobes, not of a terrestrial insect as once thought, but of a much more distant lobopodan ancestor.

Deep Homology and Origin of Appendages

It is clear from the fossil record that chordates and arthropods diverged at least by the Cambrian. The appendages of these two groups are not homologous because phylogenetically intermediate taxa (particularly basal chordates) do not possess comparable structures. The most surprising discovery of recent molecular studies, however, is that much of the genetic machinery that patterns the appendages of arthropods, vertebrates and other phyla is similar. These findings suggest that the common ancestor of many animal phyla could have had body-wall outgrowths that were organized by elements of the regulatory systems found in extant appendages. We now describe these similarities and use them to consider the origin of animal limbs.

In *Drosophila,* the anteroposterior (AP) axis of the leg or wing imaginal disk (the larval precursor to the adult appendages) is divided into two compartments (reviewed in ref. 49). The posterior half of the disk expresses the gene *hedgehog,*[72,73] which encodes the key signal that initiates AP patterning (Fig. 6). In response to *hedgehog,* a thin layer of cells running along the border of the anterior and posterior compartments is induced to produce another secreted protein encoded by the gene *decapentaplegic (dpp).*[74] *dpp,* in turn, is a long-range signal providing positional information, and hence differential AP fates, to cells in both compartments.[75–79] Misexpression of either *hedgehog* or *dpp* in the anterior of the disk results in AP mirror-image duplications of limb structures.[74]

The AP axis of the vertebrate limb is set up in a very similar manner (Fig. 6). A key organizing signal is *Sonic hedgehog (Shh),* one of the three direct homologues of the *Drosophila* gene *hedgehog.*[80–83] Like *hedgehog, Shh* is localized posteriorly in the limb bud. Misexpression of *Shh* anteriorly causes AP mirror-image duplications analogous to those caused by *hedgehog* misexpression in the fly imaginal disk.[74,83] In addition, Bmp-2, one of the two vertebrate homologues of the arthropod dpp signalling protein, is expressed in the limb bud in response to Shh.[23] Unlike *Drosophila* dpp, Bmp-2 does not have the ability to cause full limb duplications. However, it clearly functions as a secondary signal in the Shh pathway, polarizing the overlying ectoderm.[84]

Signals organizing proximodistal (PD) outgrowth in vertebrate and insect appendages also operate similarly. In the insect imaginal wing disk, PD outgrowth is organized by a specialized set of cells running the length of the dorsoventral (DV) border, the "wing margin" (Fig. 6). The dorsal compartment of the wing is characterized by the expression of the transcription factor *apterous. apterous* specifies dorsal-specific cell fate[69] and controls the expression of a secreted protein called fringe.[85] The interface between cells expressing and cells not expressing *fringe* becomes the wing edge or margin.[85] A key downstream effector of *fringe* activity is encoded by *Serrate.*[86] In response to *fringe, Serrate* is induced, leading to the activation of several downstream effector genes[87–90] and the production of a signal at the margin which organizes the growth of the wing blade.[87–89]

Unexpectedly, outgrowth of vertebrate limbs appears to be established by a very similar genetic cascade. Outgrowth of the limb bud is driven by signals from a specialized ectodermal structure, the apical ectodermal ridge[91] (AER), which, like the wing margin, runs along the DV border of the limb. Remarkably, a vertebrate homologue of *fringe,* called *Radical-fringe,* is expressed in the dorsal half of the limb ectoderm prior to formation of the AER.[92] At the border between cells expressing *Radical-fringe* and cells not expressing *Radical-fringe,* a homologue of *Serrate, Ser-2,* is induced and the AER forms. A *Radical-fringe* boundary is required to form the AER and ectopic *Radical fringe* can induce an additional AER on the ventral surface.[93]

There are also parallels between the regulation of the DV axis in vertebrate and arthropod appendages (Fig. 6). Genes specifying DV polarity in both groups have been identified. In *Drosophila,* the early ventral expression of the gene *wingless,* a member of the *Wnt* family of secreted factors, is necessary for the proper DV patterning of the wing.[94,95] Subsequently, the expression of the transcription factor *apterous* defines the dorsal compartment and specifies dorsal cell fates.[69] In the vertebrate limb, the early expression of a different *Wnt* family member is also required for DV patterning. *Wnt7a* is specifically expressed throughout the dorsal ectoderm and is necessary and sufficient for many aspects of dorsal patterning.[96–98] *Wnt7a* acts by inducing mesodermal expression of *Lmx-1.*[97,98] Like *apterous, Lmx-1* is a related member of the *LIM*-homeodomain family of transcription factors. As with *apterous (Drosophila), Lmx-1* (vertebrates) expression defines a dorsal compartment, being expressed early throughout the dorsal half of the limb bud, and it is sufficient to convey dorsal cell fate.[97,98]

The simplest phylogenetic implication to draw from these comparisons is that individual genes that are expressed in the three orthogonal axes are more ancient than either insect or vertebrate limbs (Fig. 6). Indeed, several of the regulatory systems seen in arthropod and vertebrate limbs are also involved in the development of other organs in a variety of taxa. The phylogenetic distribution of regulatory circuits and morphological structures presents two major interpretations: either similar genetic circuits were convergently recruited to make the limbs of different taxa or

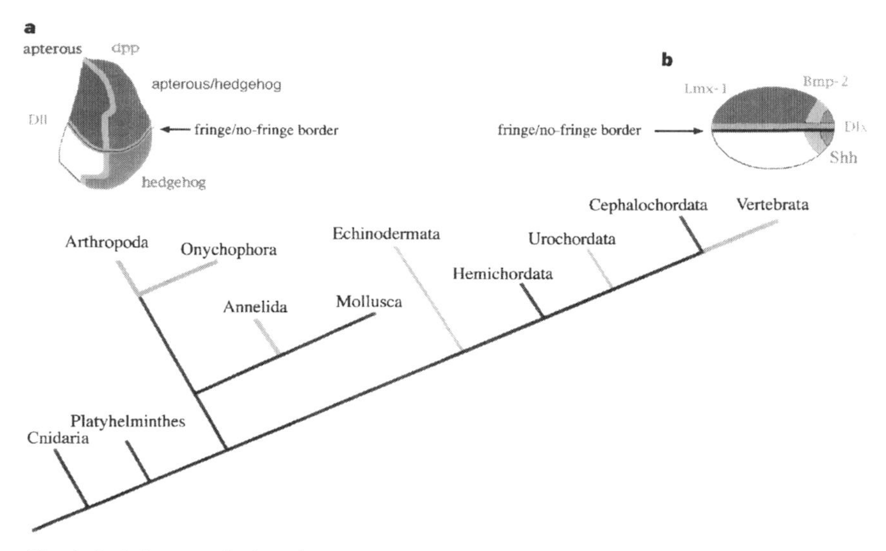

Fig. 6 A cladogram of selected metazoans shows the distribution of major genes involved with appendage development. Homologous signals are deployed in similar locations in the limb primordia of arthropods *(a, Drosophila)* and vertebrates (*b,* chick). Equivalent orientations (dorsal up, anterior left) of a chick left wing bud and a *Drosophila* left wing imaginal disk are shown. *Sonic hedgehog* in the chick, and its homologue *hedgehog* in the fly are produced in a posterior domain. These factors induce the expression of secondary patterning signals in the appendages: overlapping expression of *Bmp-2* in the chick and adjacent expression of *dpp,* the *Bmp-2* homologue, in the fly. Dorsal cell fates are specified in both systems by LIM homeodomain transcription factors expressed throughout the dorsal half of the appendage primordia: *Lmx-1* in the chick, and *apterous* in the fly. The outgrowth of both appendages is driven by a specialized group of cells (the AER in the chick and the wing margin in the fly) running along the anteroposterior axis at the junction of the dorsal and ventral compartments. These key groups of cells are specified in both the chick and the fly by the border between dorsal cells expressing the gene *fringe* and ventral cells not expressing *fringe.* For simplicity in viewing, conserved genes in signal transduction (such as the *hedgehog* receptor, *patched*) and other parallels between the two systems (such as the expression of *Wnt* genes) are not shown. *Distal-less* orthologues (*Dll* in *Drosophila* and *Dlx* in the chick) are expressed in a wide variety of animal appendages, including the lobopods of onychophorans, the tube feet of echinoderms, and the wings of birds and flies *(light shading).* The limbs of these taxa are not homologous as appendages because phylogenetically intermediate groups do not possess comparable structures. This suggests at least two phylogenetic possibilities: either similar genetic circuits were convergently recruited to make the limbs of different taxa, or these signalling and regulatory systems are ancient and patterned a different structure (presumably another type of outgrowth) in the common ancestor of protostomes and deuterostomes.[99] (A color version of this figure is available on-line at <http://www.press.uchicago.edu/books/gee>.)

a set of these signalling and regulatory systems are ancient and patterned a structure in the common ancestor of protostomes and deuterostomes.[99]

The first model holds that genes and/or genetic circuits were convergently recruited for limb development during the evolution of vertebrates and arthropods. These genes would not be involved in appendage development in the common ancestor of vertebrates and arthropods; each gene or circuit was involved in other developmental events. This notion would require the parallel co-option of members of similar gene families, acting along different developmental axes to pattern an outgrowth of the body wall in at least two taxa. The evolution of limbs in each group would, then, have involved the convergent recruitment of numerous genes to define similar developmental axes. If confirmed, this hypothesis would provide a stunning case of convergent evolution.

The second model is that some of these genes or circuits were components of an ancestral genetic regulatory system that was used to pattern a structure in the common ancestor of vertebrates and arthropods. This ancestral structure need not have been homologous to arthropod or vertebrate limbs; the regulatory system could have originally patterned any one of a number of outgrowths of the body wall in a primitive bilaterian for example. The genes themselves were initially involved in other developmental events; the key step in animal limb evolution was the establishment of an integrated genetic system to promote and pattern the development of certain outgrowths. Once established, this system provided the genetic and developmental foundation for the evolution of structures as diverse as wings, fins, antennae and lobopodia.

The evaluation of these two alternative models requires consideration of several factors that could affect the identity and deployment of appendate-patterning genes. First, given the independent histories spanning more than 500 Myr of the lineages being compared, one should expect regulatory differences among patterning systems. Even the insect wing and leg, which are both probably derived from an ancestral polyramous appendage, have acquired different patterning mechanisms. Because individual regulatory components appear to have been gained, lost and modified during insect wing and leg evolution, there is no *a priori* reason to expect that either structure would be more genetically similar to vertebrate limbs. Second, one member of a gene family may substitute for another during normal development. Indeed, some of the genes that are deployed similarly in arthropods and vertebrates are not strictly orthologous. Such substitution could be the product of either convergent evolution or descent with modification (by substitution) among redundant genes. Third, it could be argued that the presumed inversion of the DV body axis during deuterostome evolution would imply a corresponding inversion of appendage DV axis patterning signals, an expectation that is not met by the observed expression of *apterous*- and *fringe*-related genes in vertebrate limb buds. A parallel in reversal of the DV axes of the body and the limbs would be expected if vertebrate and insect limbs were structurally homologous to a common ancestral appendage which predated the reversal. However, we know that the modern vertebrate limb evolved after the DV inversion of the body axis. Thus a DV inversion in the patterning of the limbs would not necessarily be predicted, whether the axis patterning genes were independently co-opted for appendage formation according to the first model, or whether they were co-opted as a unit, but regulated independently of the body axis, according to the second model.

One can argue many ways from the comparison of only two taxa: the alternative phylogenetic hyopotheses need to be tested by additional comparative data. Evidence in support of an ancient common mechanism for the formation of outgrowths of the body wall comes from phylogenetic comparison of the expression of the transcription factor *Distal-less (Dll)*[100] (Fig. 6). *Dll* is expressed at the distal end of growing insect limbs,[55,60,61] and is essential for appendage outgrowth.[59] *Dll* orthologues are expressed in the distal portion (AER) of the embryonic limb buds of vertebrates, the ampullae and siphons of tunicates, the tube feet of echinoderms, the parapodia of annelids, as well as onychophoran lobopodia.[100] The expression of *Dll*-related genes could represent convergent utilization of the gene. However, the fact that out of the hundreds of transcription factors that potentially could have been used, *Dll* is expressed in the distal portions of appendages in six coelomate phyla makes it more likely that *Dll* was already involved in regulating body-wall outgrowth in a common ancestor of these taxa (Fig. 6). The additional parallels between vertebrate and arthropod limbs suggest that this ancestral outgrowth may have also been patterned along the three orthogonal axes.

If a conserved outgrowth patterning system was available for co-option in the evolution of vertebrate limbs, then it must have been used in patterning non-limb outgrowths in basal taxa. Genetic studies provide an example of at least one secondary outgrowth patterned along these axes that predated the evolution of vertebrate limbs: the branchial arches. As the branchial arches grow out from the cranial region of the chick embryo, they express important components of the limb patterning system in similar developmental regions.[101,102] Like arthropod and vertebrate limbs, the branchial arches contain localized, posterior expression of a *hedgehog* gene, in this case *Shh.*[101,102] Furthermore, *Shh* is coexpressed with *Bmp* posteriorly.[101] Yet more similarities lie in the DV and AP axes: *fringe*-expressing cells are initially confined to the dorsal ectoderm and later are restricted to distal regions of the outgrowth where *Distal-less* orthologues are also expressed.[93] The ectopic deployment and modification of an existing patterning program, such as that of the branchial arches, may have given rise to the predecessors of vertebrate appendages.

Determination of whether two structures are homologous depends on the hierarchical level at which they are compared.[103–105] For example, bird wings and bat wings are analogous as wings, having evolved independently for flight in each lineage. However, at a deeper hierarchical level that includes all tetrapods, they are homologous as forelimbs, being derived from a corresponding appendage of a common ancestor. Similarly, we suggest that whereas vertebrate and insect wings are analogous as appendages, the genetic mechanisms that pattern them may be homologous at a level including most protostomes and deuterostomes. Furthermore, we propose that the regulatory systems that pattern extant arthropod and vertebrate appendages patterned an ancestral outgrowth and that these circuits were later modified during the evolution of different types of animal appendages. Animal limbs would be, in a sense, developmental "paralogues" of one another; modification and redeployment of this ancient genetic system in different contexts produced the variety of appendages seen in Recent and fossil animals.

NEIL SHUBIN is in the Department of Biology, University of Pennsylvania, Philadelphia, Pennsylvania 19104, USA; Cliff Tabin is in the Department of Genetics, Harvard Medical School, 200 Longwood Avenue, Boston, Massachusetts 02115, USA; Sean Carroll is at the HHMI and Laboratory of Molecular Biology, University of Wisconsin, 1525 Linden Drive, Madison, Wisconsin 53706, USA.

References

1. Müller, G. B. & Wagner, G. P. Novelty in evolution: Restructuring the concept. *Ann. Rev. Ecol. Syst.* **22,** 229–256 (1991).
2. Coates, M. I. The origin of vertebrate limbs. *Development* (suppl.) 169–180 (1994).
3. Coates, M. I. Fish fins or tetrapod limbs: A simple twist of fate? *Curr. Biol.* **5,** 844–848 (1995).
4. Shubin, N. The evolution of paired fins and the origin of tetrapod limbs. *Evol. Biol.* **28,** 39–85 (1995).
5. Coates, M. I. The Devonian tetrapod *Acanthostega gunnari* Jarvik: Postcranial anatomy, basal tetrapod interrelationships and patterns of skeletal evolution. *Trans. R. Soc. Edinb.* **87,** 363–421 (1996).
6. Johnson, R. & Tabin, C. The long and short of *hedgehog* signaling. *Cell* **81,** 313–316 (1995).
7. Nelson, C. E. *et al.* Analysis of *Hox* gene expression in the chick limb bud. *Development* **122,** 1449–1466 (1996).
8. Tabin, C. J. & Laufer, E. *Hox* genes and serial homology. *Nature* **361,** 692–693 (1993).
9. Sordino, P., van der Hoeven, F. & Duboule, D. *Hox* gene expression in teleost fins and the origin of vertebrate digits. *Nature* **375,** 678–681 (1995).

10. Kessel, M. & Gruss, P. Murine developmental control genes. *Science* **249,** 374–379 (1990).
11. Mackem, S., Ranson, M. & Mahon, K. Limb-type differences in expression domains of certain chick *Hox-4* genes and relationship to pattern modification for flight. *Prog. Clin. Biol. Res.* **383** A, 21–30 (1993).
12. Peterson, R. J., Papenbrock, T., Davada, M. M. & Awgulewitsch, A. The murine *Hoxc* cluster contains five neighboring *abdB*-related *Hox* genes that show unique spatially coordinated expression in posterior embryonic subregions. *Mech. Dev.* **47,** 253–260 (1994).
13. Gibson-Brown, J. J. *et al.* Evidence of a role for *T-box* genes in the evolution of limb morphogenesis and the specification of forelimb/hindlimb identity. *Mech. Dev.* **56,** 93–101 (1996).
14. Davis, A. P., Witte, D. P., Hsieh-Li, H. M., Potter, S. S. & Capecchi, M. R. Absence of radius and ulna in mice lacking *hoxa-11* and *hoxd-11. Nature* **375,** 791–795 (1995).
15. Vorobyeva, E. & Hinchliffe, J. R. From fins to limbs. *Evol. Biol.* **29,** 263–311 (1996).
16. Hinchliffe, J. R. & Johnson, D. R. *The Development of the Vertebrate Limb* (Clarendon, Oxford, 1980).
17. Holmgren, N. On the origin of the tetrapod limb. *Acta Zoologica* **14,** 185–295 (1933).
18. Holmgren, N. Contribution on the question of the origin of the tetrapod limb. *Acta Zoologica* **20,** 89–124 (1939).
19. Watson, D. M. S. On the primitive tetrapod limb. *Anat. Anzeiger* **44,** 24–27 (1913).
20. Gregory, W. K. & Raven, H. C. Studies on the origin and early evolution of paired fins and limbs. *Ann. N.Y. Acad. Sci.* **42,** 273–360 (1941).
21. Sordino, P. & Duboule, D. A molecular approach to the evolution of vertebrate paired appendages. *Trends Ecol. Evol.* **11,** 114–119 (1996).
22. Ahlberg, P. E. & Milner, A. R. The origin and early diversification of tetrapods. *Nature* **368,** 507–512 (1994).
23. Yokouchi, Y. *et al.* Homeobox gene expression correlated with the bifurcation process of limb cartilage development. *Nature* **353,** 443–445 (1991).
24. Gerard, M., Duboule, D. & Zakany, J. C. Cooperation of regulatory elements involved in the activation of the *Hoxd-11* gene. *Compt. R. Acad. Sci.* III **316,** 985–994 (1993).
25. Beckers, J., Gerard, M. & Duboule, D. Transgenic analysis of a potential *Hoxd-11* limb regulatory element present in tetrapods and fish. *Dev. Biol.* **180,** 543–553 (1996).
26. van der Hoeven, F., Zakany, J. & Duboule, D. Gene transpositions in the *HoxD* complex reveal a hierarchy of regulatory controls. *Cell* **85,** 1025–1035 (1996).
27. Shubin, N. & Alberch, P. A morphogenetic approach to the origin and basic organization of the tetrapod limb. *Evol. Biol.* **20,** 318–390 (1986).
28. Tabin, C. J. Why we have (only) five fingers per hand: *hox* genes and the evolution of paired limbs. *Development* **116,** 289–296 (1992).
29. Holder, N. Developmental constraints and the evolution of vertebrate digit patterns. *J. theor. Biol.* **104,** 451–471 (1983).
30. Morse, E. On the tarsus and carpus of birds. *Ann. Lyc. Nat. Hist.* **10,** 141–158 (1872).

31. Shubin, N., Crawford, A. & Wake, D. Morphological variation in the limbs of *Taricha granulosa* (Caudata: Salamandridae): Evolutionary and phylogenetic implications. *Evolution* **49,** 874–884 (1995).
32. Greer, A. Limb reduction in the Scincid lizard genus *Lerista.* 2. Variation in the bone complements of the front and rear limbs and the number of postsacral vertebrae. *J. Herpetol.* **24,** 142–150 (1980).
33. Lande, R. Evolutionary mechanisms of limb loss in tetrapods. *Evolution* **32,** 73–92 (1978).
34. Gauthier, J. Saurischian monophyly and the origin of birds. *Mem. Calif. Acad. Sci.* **8,** 1–55 (1986).
35. MacFadden, B. J. *Fossil Horses* (Cambridge Univ. Press, Cambridge, 1992).
36. Davis, A. P. & Capecchi, M. R. Axial homeosis and appendicular skeleton defects in mice with a targeted disruption of *hoxd-11. Development* **120,** 2187–2198 (1994).
37. Davis, A. P. & Capecchi, M. R. A mutational analysis of the 5′ *HoxD* genes: Dissection of genetic interactions during limb development in the mouse. *Development* **122,** 1175–1185 (1996).
38. Favier, B. *et al.* Functional cooperation between the non-paralogous genes *Hoxa-10* and *Hoxd0-11* in the developing forelimb and axial skeleton. *Development* **122,** 449–460 (1996).
39. Dollé, P. *et al.* Disruption of the *Hoxd-13* gene induces localized heterochrony leading to mice with neotenic limbs. *Cell* **75,** 431–441 (1993).
40. Favier, B., LeMeur, M., Chambon, P. & Dollé, P. Axial skeleton homeosis and forelimb malformations in *Hoxd-11* mutant mice. *Proc. Natl. Acad. Sci. USA* **92,** 310–314 (1995).
41. Capecchi, M. R. Function of homeobox genes in skeletal development. *Ann. N.Y. Acad. Sci.* **97,** 34–37 (1996).
42. Wigglesworth, V. B. Evolution of insect wings and flight. *Nature* **246,** 127–203 (1973).
43. Budd, G. The morphology of *Opabinia regalis* and the reconstruction of the arthropod stem-group. *Lethaia* **29,** 1–14 (1996).
44. Hou, X. G. & Bergström, J. Cambrian lobopodians—ancestors of extant onychophorans? *Zool. J. Linn. Soc. Lond.* **114,** 3–19 (1995).
45. Simonetta, A. M. & Delle Cave, L. in *The Early Evolution of Metazoa and the Significance of Problematic Taxa* (eds. Simonetta, A. M. & Conway Morris, S.) 189–244 (Cambridge Univ. Press, Cambridge, 1991).
46. Budd, G. A Cambrian gilled lobopod from Greenland. *Nature* **364,** 709–711 (1993).
47. Chen, J. Y., Ramsköld, L. & Zhou, G. Q. Evidence for monophyly and arthropod affinity of Cambrian giant predators. *Science* **264,** 1304–1308 (1994).
48. Carroll, S. B. Homeotic genes and the evolution of arthropods and chordates. *Nature* **376,** 479–485 (1995).
49. Struhl, G. Genes controlling segmental specification in the *Drosophila* thorax. *Proc. Natl. Acad. Sci. USA* **79,** 7380–7384 (1982).
50. Struhl, G. A homeotic mutation transforming leg to antenna in *Drosophila. Nature* **292,** 635–638 (1981).
51. Gibson, G. & Gehring, W. J. Head and thoracic transformations caused by ectopic expression of *Antennapedia* during *Drosophila* development. *Development* **102,** 657–675 (1988).

52. Stuart, J., Brown, S., Beeman, R. & Denell, R. A deficiency of the homeotic complex of the beetle *Tribolium. Nature* **350,** 72–47 (1991).
53. Averof, M. & Akam, M. *Hox* genes and the diversification of insect-crustacean body plans. *Nature* **376,** 420–423 (1995).
54. Vachon, G. *et al.* Homeotic genes of the Bithorax complex repress limb development in the abdomen of the *Drosophila* embryo through the target gene. *Cell* **71,** 437–450 (1992).
55. Panganiban, G. *et al.* The development of crustacean limbs and the evolution of arthropods. *Science* **270,** 1363–1366 (1995).
56. Manton, S. M. *Mandibular Mechanisms and the Evolution of Arthropods* Vol. 247 (British Museum and Queen Mary College, London, 1964).
57. Wheeler, W. C., Cartwright, P. & Hayashi, C. Y. Arthropod phylogeny: A combined approach. *Cladistics* **9,** 1–39 (1993).
58. Boore, J. L., Collins, T. M., Stanton, D., Daehler, L. L. & Brown, W. M. Deducing the pattern of arthropod phylogeny from mitochondrial DNA rearrangements. *Nature* **376,** 163–165 (1995).
59. Cohen, S. M. & Jürgens, G. Proximal-distal pattern formation in *Drosophila:* Cell autonomous requirement for *Distal-less* gene activity in limb development. *EMBO J.* **8,** 2045–2055 (1989).
60. Cohen, S. *et al. Distal-less* encodes a homeodomain protein required for limb development in *Drosophila. Nature* **338,** 432–434 (1989).
61. Panganiban, G., Nagy, L. & Carroll, S. B. The development and evolution of insect limb types. *Curr. Biol.* **4,** 671–675 (1994).
62. Popadic, A., Rusch, D., Peterson, M., Rogers, B. T. & Kaufman, T. C. Origin of the arthropod mandible. *Nature* **380,** 395 (1996).
63. Jeram, A. J., Selden, P. A. & Edwards, D. Land animals in the Silurian: Arachnids and myriapods from Shropshire, England. *Science* **250,** 658–661 (1990).
64. Kukalová-Peck, J. *The Insects of Australia* 2nd edn. (Cornell Univ. Press, Ithaca, NY, 1991).
65. Snodgrass, R. *Principles of Insect Morphology* (McGraw-Hill, New York, 1935).
66. Kukalová-Peck, J. Origin and evolution of insect wings and their relation to metamorphosis, as documented from the fossil record. *J. Morphol.* **156,** 53–126 (1978).
67. Marden, J. H. & Kramer, M. G. Surface-skimming stoneflies: A possible intermediate stage in insect flight evolution. *Science* **266,** 427–430 (1994).
68. Cohen, B. *et al.* Allocation of the thoracic imaginal primordia in the *Drosophila* embryo. *Development* **117,** 597–608 (1993).
69. Diaz-Benjumea, F. & Cohen, S. M. Interaction between dorsal and ventral cells in the imaginal disc directs wing development in *Drosophila. Cell* **75,** 741–752 (1993).
70. Carroll, S. B. *et al.* Pattern formation and eyespot determination in butterfly wings. *Science* **265,** 109–114 (1994).
71. Averof, M. & Cohen, S. M. Evolutionary origin of insect wings from ancestral gills. *Nature* **385,** 627–630 (1997).
72. Lee, J. J. *et al.* Secretion and localized transcription suggest a role in positional signaling for products of the segmentation gene *hedgehog. Cell* **71,** 33–50 (1992).
73. Tabata, T. *et al.* The *Drosophila hedgehog* gene is expressed specifically in poste-

rior compartment cells and is a target of *engrailed* regulation. *Genes Dev.* **6,** 2635–2645 (1992).

74. Basler, D. & Struhl, G. Compartment boundaries and the control of *Drosophila* limb pattern by *hedgehog* protein. *Nature* **368,** 208–214 (1994).
75. Posakony, L., Raftery, L. & Gelbart, W. Wing formation in *Drosophila melanogaster* requires *decapentaplegic* gene function along the anterior-posterior compartment boundary. *Mech. Dev.* **33,** 69–82 (1991).
76. Capdevila, J. & Guerrero, I. The *Drosophila* segment polarity gene *patched* interacts with *decapentaplegic* in wing development. *EMBO J.* **6,** 715–729 (1994).
77. Sanicola, M., Sekelsky, J., Elson, S. & Gelbart, W. M. Drawing a stripe in *Drosophila* imaginal discs: Negative regulation of *decapentaplegic* and *patched* expression. *Genetics* **139,** 745–756 (1995).
78. Nellen, D., Burke, R., Struhl, G. & Basler, K. Direct and long-range actions of a *Dpp* morphogen gradient. *Cell* **85,** 357–368 (1996).
79. Lecuit, T. *et al.* Two distinct mechanisms for long-range patterning by *Decapentaplegic* in the *Drosophila* wing. *Nature* **381,** 387–393 (1996).
80. Echelard, Y. *et al. Sonic hedgehog,* a member of a family of putative signaling molecules, is implicated in the regulation of CNS polarity. *Cell* **75,** 1417–1430 (1993).
81. Krauss, S., Concordet, J. P. & Ingham, P. W. A functionally conserved homolog of the *Drosophila* segment polarity gene *hh* is expressed in tissues with polarizing activity in zebrafish embryos. *Cell* **75,** 1431–1444 (1993).
82. Chang, D. T. *et al.* Products, genetic linkage and limb patterning activity of a murine *hedgehog* gene. *Development* **120,** 3339–3353 (1994).
83. Riddle, R. D. *et al. Sonic hedgehog* mediates the polarizing activity of the ZPA. *Cell* **75,** 1401–1416 (1995).
84. Tickle, C. Genetics and limb development. *Dev. Genet.* **19,** 1–8 (1996).
85. Irvine, K. & Weischaus, E. *fringe,* a boundary-specific signaling molecule, mediates interactions between dorsal and ventral cells during *Drosophila* wing development. *Cell* **79,** 595–606 (1994).
86. Spreicher, S., Thomas, U., Hinz, U. & Knust, E. The *Serrate* locus of *Drosophila* and its role in morphogenesis of imaginal discs: Control of cell proliferation. *Development* **120,** 535–544 (1994).
87. Kim, J., Irvine, K. & Carroll, S. Cell recognition, signal induction, and symmetrical gene activation at the dorsal-ventral boundary of the developing *Drosophila* wing. *Cell* **82,** 795–802 (1995).
88. Couso, J. P., Knust, E. & Martinez Arias, A. *Serrate* and *wingless* cooperate to induce vestigial gene expression and wing formation in *Drosophila. Curr. Biol.* **5,** 1437–1448 (1995).
89. Diaz-Benjumea, F. J. & Cohen, S. *Serrate* signals through *Notch* to establish a *Wingless*-dependent organizer at the dorsal/ventral compartment boundary of the *Drosophila* wing. *Development* **121,** 4215–4225 (1995).
90. Kim, J. *et al.* Integration of positional signals and regulation of wing formation and identity by *Drosophila* vestigial gene. *Nature* **382,** 133–138 (1996).
91. Todt, W. L. & Fallon, J. F. Development of the apical ectodermal ridge in the chick wing bud. *J. Embryol. Exp. Morphol.* **80,** 21–41 (1984).
92. Rodriguez-Estaban, C. *et al. Radical fringe* positions the apical ectodermal ridge at the dorsoventral boundary of the vertebrate limb. *Nature* **386,** 360–361 (1997).

93. Laufer, E. *et al.* Expression of *Radical fringe* in limb-bud ectoderm regulates apical ectodermal ridge formation. *Nature* **386,** 366–373 (1997).
94. Williams, J. A., Paddock, S. W. & Carroll, S. B. Pattern formation in a secondary field: A hierarchy of regulatory genes subdivides the developing *Drosophila* wing disc into discrete sub-regions. *Development* **117,** 571–584 (1993).
95. Couso, J. P., Bate, M. & Martinez-Arias, A. A *wingless*-dependent polar coordinate system in *Drosophila* imaginal discs. *Science* **259,** 484–489 (1993).
96. Parr, B. A. & McMahon, A. P. Dorsalizing signal *Wnt-7a* required for normal polarity of D-V and A-P axes of mouse limb. *Nature* **374,** 350–353 (1995).
97. Riddle, R. D. *et al.* Induction of the LIM homeobox gene *Lmx-1* by *Wnt-7a* establishes dorsoventral pattern in the vertebrate limb. *Cell* **83,** 631–640 (1995).
98. Vogel, A. *et al.* Dorsal cell fate specified by chick *Lmx1* during vertebrate limb development. *Nature* **378,** 716–720 (1995).
99. Raff, R. *The Shape of Life* (Univ. Chicago Press, Chicago, 1996).
100. Panganiban, G. *et al.* The origin and evolution of animal appendages. *Proc. Natl. Acad. Sci. USA* **94,** 5162–5166 (1997).
101. Wall, N. A. & Hogan, B. L. M. Expression of *bone morphogenetic protein-4 (BMP-4), bone morphogenetic protein-7 (BMP-7), fibroblast growth factor-8 (FGF-8)* and *Sonic hedgehog (SHH)* during branchial arch development in the chick. *Mech. Dev.* **53,** 383–392 (1995).
102. Marigo, V., Scott, M. P., Johnson, R. L., Goodrich, L. V. & Tabin, C. J. Conservation in *hedgehog* signaling: Induction of a chicken *patched* homolog by *Sonic hedgehog* in the developing limb. *Development* **122,** 1225–1233 (1996).
103. Roth, V. L. Homology and hierarchies: Problems solved and unresolved. *J. Evol. Biol.* **4,** 167–194 (1991).
104. Wagner, G. P. The origin of morphological characters and the biological basis of homology. *Evolution* **43,** 1157–1171 (1989).
105. Bolker, J. A. & Raff, R. A. Developmental genetics and traditional homology. *BioEssays* **18,** 489–494 (1996).
106. Carroll, R. L. *Vertebrate Paleontology* (Freeman, San Francisco, 1988).
107. Jarvik, E. *The Structure and Evolution of the Vertebrates* Vol. 1 (Academic, New York, 1980).
108. Jarvik, E. The Devonian tetrapod *Ichthyostega. Fossils and Strata* **40,** 1–213 (1996).
109. Fromental-Ramain, C. *et al.* Specific and redundant functions of the paralogous *Hoxa-9* and *Hoxd-9* genes in forelimb and axial skeleton patterning. *Development* **122,** 461–472 (1996).
110. Mortlock, D. P., Post, L. C. & Innis, J. W. The molecular basis of hypodactyly (Hd): A deletion in *Hoxa13* leads to arrest of digital arch formation. *Nature Genet.* **13,** 284–289 (1996).
111. Mortlock, D. P. & Innis, J. W. Mutation of *HOXA13* in hand-foot-genital syndrome. *Nature Genet.* **15,** 179–181 (1997).
112. Saunders, J. The proximo-distal sequence of origin of the parts of the chick wing and the role of the ectoderm. *J. Exp. Zool.* **108,** 363–403 (1948).
113. Summerbell, D., Lewis, J. H. & Wolpert, L. Positional information in chick limb morphogenesis. *Nature* **244,** 492–496 (1973).
114. Lebedev, O. A. & Coates, M. I. The postcranial skeleton of the Devonian tetrapod *Tulerpeton curtum* Lebedev. *Zool. J. Linn. Soc.* **113,** 302–348 (1995).

115. Hou, X. G., Bergström, J. & Ahlberg, P. *Anomalocaris* and other large animals in the Lower Cambrian Chengjiang fauna of southwest China. *Geol Forening. Forhandling.* **117,** 163–183 (1995).

Acknowledgments

We thank P. Ahlberg, G. Budd, A. C. Burke, M. Coates, A. Meyer, G. Panganiban, P. Sniegowski, D. Wake, R. S. Winters, L. Wolpert and members of our laboratories for their critiques of drafts of this manuscript. S.B.C. is an investigator of the HHMI. C.T. is supported by grants from the NIH and the American Cancer Society. N.S. is supported by grants from the NSF, from the National Geographic Society and from the Research Foundation of the University of Pennsylvania.

Correspondence and requests for materials should be addressed to N.S. (e-mail: nshubin@sas.upenn.edu).

S. Conway Morris

THE FOSSIL RECORD AND THE EARLY EVOLUTION OF THE METAZOA

The appearance of the multicellular animals, or Metazoa, in the fossil record about 600 million years ago marks a revolution in the history of life. Molecular biology is continuing to increase our understanding of metazoan evolution, yet information from fossils is still an important component in deciphering metazoan phylogeny, and data on rapidly radiating animal groups place early metazoan evolution in a new perspective.

Would you like a game of "Metazoan Phylogeny"? The prospect is daunting. The board encompasses the entire earth: the only game ever attempted has already taken almost a billion years and shows no signs of a conclusion. The number of players varies (around 35 at the last count), and the strategic implications of such rules as may exist have not been fully worked out. Ideas about metazoan phylogeny are legion and often contradictory.[1–4] A few principles are widely, but not universally, accepted, but no coherent phylogeny for the roughly 35 metazoan phyla exists. Speculation has been based mostly on anatomy, both adult and larval; meanwhile ultrastructural studies, immunology and biochemistry have added to the debate, and sometimes the confusion. Now, the picture has changed for ever. The true outlines of metazoan phylogeny seem to be emerging. This is due to advances in molecular biology, most notably data from ribosomal RNA.[5–8]

If the broad outline of metazoan phylogeny is now becoming clear, then surely all we need to do is commiserate with the authors of innumerable failed schemes constructed over the past century and promptly move on to more interesting problems. This would be a mistake, for three reasons.

First, the power of molecular phylogeny is not unlimited. If diversification is rapid, then the precise order of branching is very difficult to resolve.[8,9] At present, unresolved branchings (polychotomies) persist, especially in triploblastic protostomes.[6]

Second, the new schemes of metazoan phylogeny tell us nothing about the actual anatomical and functional transitions between related phyla. For instance, molecular evidence[10] suggests a close alliance between molluscs and annelids, but it is hard to imagine how an animal like the most

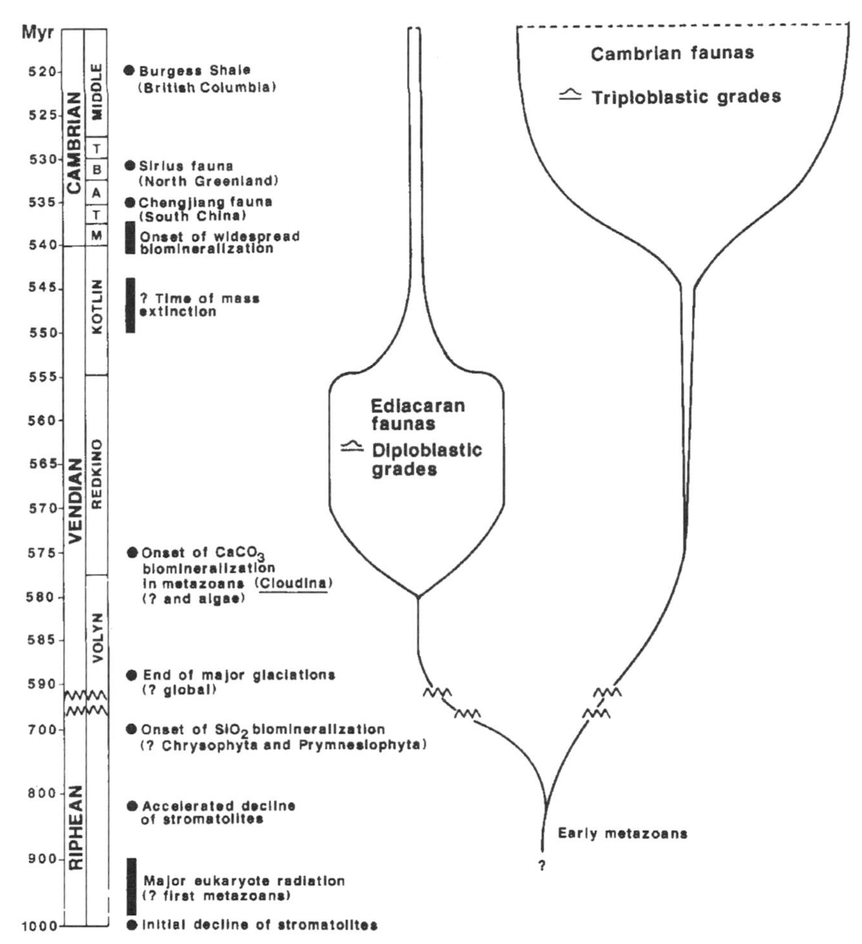

Fig. 1 The geological framework of early metazoan evolution and related biological events. The geochronological scale takes the Precambrian-Cambrian boundary at ~540 Myr ago[112] and the Ediacaran faunas as ~580–560 Myr ago. The stratigraphic divisions of the Vendian and Lower Cambrian (*M,* Manykay; *T,* Tommotian; *A,* Atdabanian; *B,* Botomian; *T,* Toyonian) follow the Russian and Siberian standards. Metazoans are hypothesized to have arisen as part of a major eukaryotic radiation, 800–1,000 Myr ago.[13,23] Tangible fossil evidence includes the Ediacaran radiations of mostly cnidarian grade, but also stem protostomes and possible deuterostomes. The "Cambrian explosion" is largely a triploblastic radiation, but diploblasts and triploblasts may have diverged substantially before Ediacaran times.

primitive known annelid (an archiannelid?) was transformed into a primitive mollusc (a chiton?). The morphological gaps that, by definition, separate phyla, remain inviolate. We remain uninformed both about the now-extinct intermediates and the evolutionary processes that would have been responsible for the diversification of early multicellular animals into what we now perceive as distinct phyla, each with its own body plan.

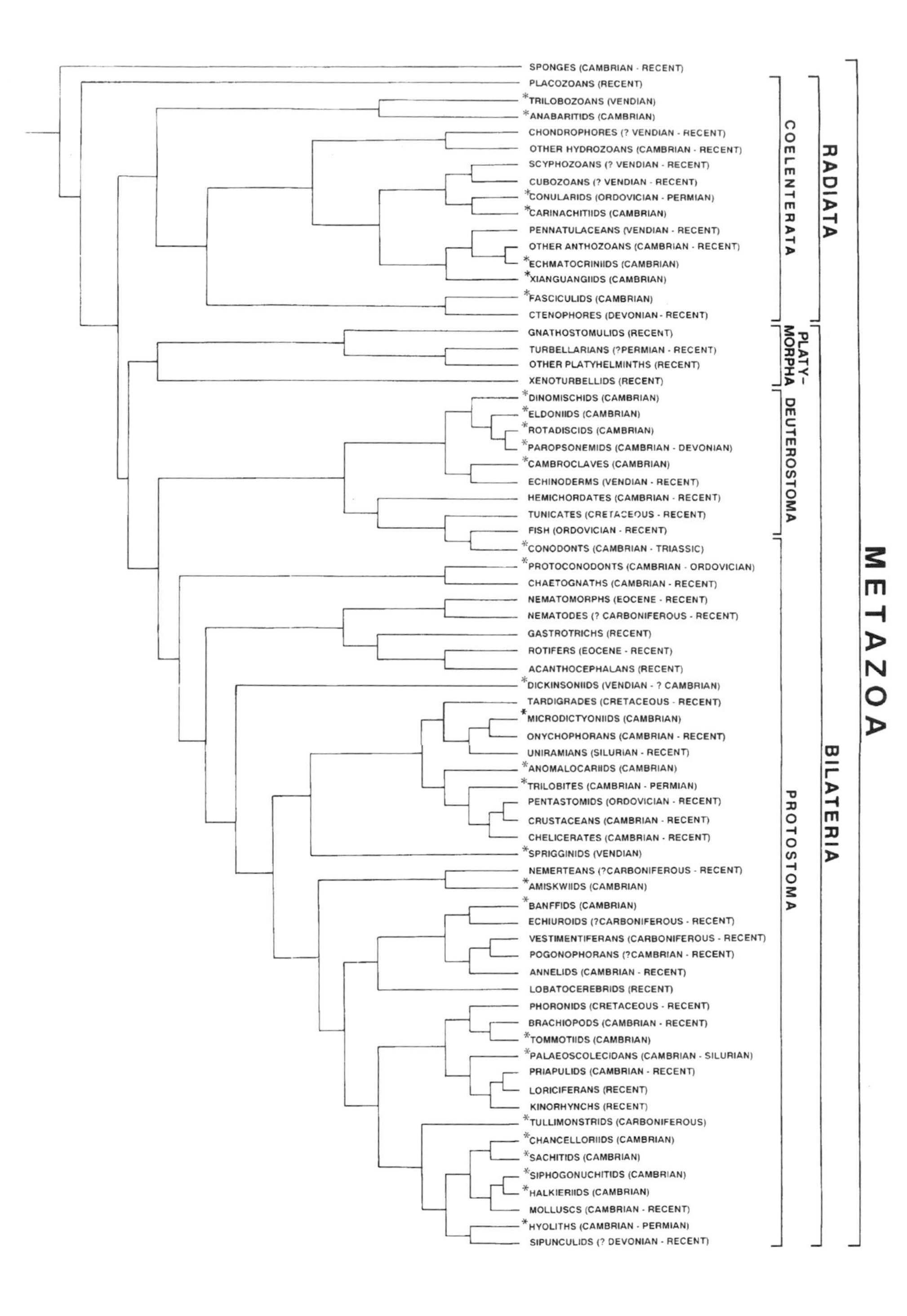
SPONGES (CAMBRIAN - RECENT)
PLACOZOANS (RECENT)
*TRILOBOZOANS (VENDIAN)
*ANABARITIDS (CAMBRIAN)
CHONDROPHORES (? VENDIAN - RECENT)
OTHER HYDROZOANS (CAMBRIAN - RECENT)
SCYPHOZOANS (? VENDIAN - RECENT)
CUBOZOANS (? VENDIAN - RECENT)
*CONULARIDS (ORDOVICIAN - PERMIAN)
*CARINACHITIIDS (CAMBRIAN)
PENNATULACEANS (VENDIAN - RECENT)
OTHER ANTHOZOANS (CAMBRIAN - RECENT)
*ECHMATOCRINIIDS (CAMBRIAN)
*XIANGUANGIIDS (CAMBRIAN)
*FASCICULIDS (CAMBRIAN)
CTENOPHORES (DEVONIAN - RECENT)
GNATHOSTOMULIDS (RECENT)
TURBELLARIANS (?PERMIAN - RECENT)
OTHER PLATYHELMINTHS (RECENT)
XENOTURBELLIDS (RECENT)
*DINOMISCHIDS (CAMBRIAN)
*ELDONIIDS (CAMBRIAN)
*ROTADISCIDS (CAMBRIAN)
*PAROPSONEMIDS (CAMBRIAN - DEVONIAN)
*CAMBROCLAVES (CAMBRIAN)
ECHINODERMS (VENDIAN - RECENT)
HEMICHORDATES (CAMBRIAN - RECENT)
TUNICATES (CRETACEOUS - RECENT)
FISH (ORDOVICIAN - RECENT)
*CONODONTS (CAMBRIAN - TRIASSIC)
*PROTOCONODONTS (CAMBRIAN - ORDOVICIAN)
CHAETOGNATHS (CAMBRIAN - RECENT)
NEMATOMORPHS (EOCENE - RECENT)
NEMATODES (? CARBONIFEROUS - RECENT)
GASTROTRICHS (RECENT)
ROTIFERS (EOCENE - RECENT)
ACANTHOCEPHALANS (RECENT)
*DICKINSONIIDS (VENDIAN - ? CAMBRIAN)
TARDIGRADES (CRETACEOUS - RECENT)
*MICRODICTYONIIDS (CAMBRIAN)
ONYCHOPHORANS (CAMBRIAN - RECENT)
UNIRAMIANS (SILURIAN - RECENT)
*ANOMALOCARIIDS (CAMBRIAN)
*TRILOBITES (CAMBRIAN - PERMIAN)
PENTASTOMIDS (ORDOVICIAN - RECENT)
CRUSTACEANS (CAMBRIAN - RECENT)
CHELICERATES (CAMBRIAN - RECENT)
*SPRIGGINIDS (VENDIAN)
NEMERTEANS (?CARBONIFEROUS - RECENT)
*AMISKWIIDS (CAMBRIAN)
*BANFFIDS (CAMBRIAN)
ECHIUROIDS (?CARBONIFEROUS - RECENT)
VESTIMENTIFERANS (CARBONIFEROUS - RECENT)
POGONOPHORANS (?CAMBRIAN - RECENT)
ANNELIDS (CAMBRIAN - RECENT)
LOBATOCEREBRIDS (RECENT)
PHORONIDS (CRETACEOUS - RECENT)
BRACHIOPODS (CAMBRIAN - RECENT)
*TOMMOTIIDS (CAMBRIAN)
*PALAEOSCOLECIDANS (CAMBRIAN - SILURIAN)
PRIAPULIDS (CAMBRIAN - RECENT)
LORICIFERANS (RECENT)
KINORHYNCHS (RECENT)
*TULLIMONSTRIDS (CARBONIFEROUS)
*CHANCELLORIIDS (CAMBRIAN)
*SACHITIDS (CAMBRIAN)
*SIPHOGONUCHITIDS (CAMBRIAN)
*HALKIERIIDS (CAMBRIAN)
MOLLUSCS (CAMBRIAN - RECENT)
*HYOLITHS (CAMBRIAN - PERMIAN)
SIPUNCULIDS (? DEVONIAN - RECENT)
COELENTERATA
PLATY- MORPHA
DEUTEROSTOMA
PROTOSTOMA
RADIATA
BILATERIA
METAZOA

Third, the new view of metazoan phylogeny requires a radical review of existing data, especially about ultrastructure, functional anatomy and biomineralization. Convergent evolution is likely to have been rampant.

If these problems can be solved, we would have an enhanced understanding of the constraints that guide evolutionary processes, the role of intrinsic and extrinsic factors in diversification, and a new appreciation of metazoan diversity: in short, the rules of the game.

My purpose here is to review how the fossil record of early metazoans might be integrated with the new phylogeny. Three types of fossil deposit yield relevant information: Ediacaran-type assemblages of soft-bodied animals from the latest Precambrian (Vendian), typified by the Ediacara fauna of South Australia; the earliest assemblages of animals with hard parts, from the Lower Cambrian; and the distinctive faunas from the Lower and Middle Cambrian, exemplified by the Burgess Shale of western Canada (Fig. 1). All these deposits have attracted wide attention because of the apparent abundance of novel body plans. These so-called "bizarre" fossils may help to unravel the basic structure of metazoan evolution (Fig. 2).

Pre-Ediacaran Metazoans?

When did metazoans appear? The objective fossil record starts with the first Ediacaran assemblages (~560–600 million years [Myr] ago), but some evidence suggests that metazoans were already in existence as early as ~800–1000 Myr ago. Several lines of evidence indicate that metazoans originated at around that time, such as a decline in the diversity of stromatolites, possibly indicative of grazing and burrowing of the microbial-mat communities by metazoans.[11] Other explanations for stromatolite decline,

Fig. 2 A provisional phylogeny of metazoan relationships. The backbone of relationships between extant phyla is based on data from molecular biology, especially the phylogeny of Lake (ref. 7). This phylogeny is chosen because some evidence from the fossil record may be in agreement. This includes (1) the early appearance of cnidarians, including many Ediacaran taxa, (2) the dickinsoniids as a stem-protostome group, and (3) sprigginids (? and *Redkinia*) as stem arthropods. Alternative cladograms based on various analyses of ribosomal RNA are available,[5,8,97] and while having features in common, differ in other respects, especially with respect to protostome relationships. The provisional phylogeny used here seeks to do two things. First, it emphasizes the importance of extinct groups (n = 29, marked with an asterisk), many of which are generally regarded as major body plans, but may have morphological characters that could reveal more clearly relationships between disparate extant phyla. Second, this cladogram is an exercise in inference because many of the extinct groups remain poorly documented; the text offers justification for many of the placements. Note that the levels of branching are arbitrary and no precise metric is applied to distance between the nodes. A number of clades are still too poorly known to be included. These include mobergellids, rhombocorniculids, nectocariids, cribricyathids, radiocyathids, escumasiids, myoscolexids, coleoliids, agmatans, typhloesids and ainiktozoonids.

however, are also plausible. These include changes in ocean chemistry[12] or the microbial mats adapting to the evolution and arrival of a wide range of new types of protistan.[13] Second, trace fossils, including what may be metazoan fecal pellets,[14] are found in rocks of this age. But protistans can produce structures similar to metazoan fecal pellets,[15] and no systematic survey for pre-Ediacaran traces, including bioturbation fabrics[16,17] and geochemical anomalies around possible burrows,[18,19] has been undertaken. Third, evidence from molecular biology[20,21] indicates that the major lineages of Metazoa were distinct at least 700 Myr ago (suggestive of even earlier origins), but this too is disputed.[22]

Molecular evidence indicates a major diversification of eukaryotes[8] perhaps ~1,000 Myr ago.[13,23] Were metazoans part of that evolutionary event? Millimeters in size, perhaps interstitial in sediments, their inability to fossilize or leave obvious traces could explain their absence from Proterozoic rocks. The reconciliation of this idea with what is known of subsequent evolutionary events remains problematic (Fig. 1). If triploblasts (metazoans with three germ layers) occur in Ediacaran assemblages, then perhaps they too had an extended history in parallel with early diploblasts (two-layered animals such as modern coelenterates). Why the first, albeit hypothetical, metazoans never grew beyond a few millimeters in size is not understood. A persistently low concentration of atmospheric oxygen is one possible explanation, but alternatives include the late acquisition of collagen (not unconnected with the oxygen-concentration idea[24]).

Ediacaran Metazoans

The soft-bodied Ediacaran animals occupied a wide range of Vendian (latest Proterozoic) marine environments and are now known from all over the world. Many of the fossils are generally interpreted as coelenterates, and some segmented forms as arthropods, annelids or other triploblasts. Nevertheless, Ediacaran fossils may not be representative of all Vendian metazoans. A newly described assemblage[25] from black shales in the Doushantuo Formation (Yangtze Gorges) in China includes tubular fossils that might have been occupied by metazoans. The presence of fossils of seaweeds from the same deposits suggests that the remains of soft-bodied animals, not just their traces, will be found there in the future.

Molecular evidence is also consistent with the early appearance of coelenterates (cnidarians and ctenophores),[6] but probably not independently of other metazoans.[26] Only sponges[8] and the microscopic placozoans may have branched off earlier. The chance of placozoans fossilizing is minuscule, but the apparent absence of Precambrian sponges is puzzling. Siliceous spicules of demosponges occur in strata in Iran[27] that may be of Ediacaran age, and claims for Precambrian sponge spicules[28] need reexamination, especially as protistan biomineralization,[29] possibly of silica, is known from the preceding Riphean period (Fig. 1).[30]

The primitive position of coelenterates[4–8] accords well with the domination of the Vendian radiation by diploblastic animals. This supposition[31] has been overshadowed by the "Vendobionta" hypothesis,[32,33] in which Ediacaran animals are thought to constitute an extinct clade of multicellular eukaryotes, neither diploblast nor triploblast, with a unique construction of tough cuticle and mattress-like body, and possibly the absence of alimentary tract, musculature and nervous tissue. This concept is open to debate, in that at least some Ediacaran fossils can be compared with known metazoans.[31,34] *Inaria,* for example, may be an actinian coral,[35] and the fivefold arrangement of putative feeding grooves in the minute epibenthic *Arkurua* suggests an affinity with echinoderms.[36] Other organisms show evidence for muscular activity[34] and so presumably a nervous system, as well as the inferred presence of a circulatory system. An analogue of Ediacaran organization may be found in the modern deep-water scleractinian coral *Leptoseris fragilis:* instead of tentacles, it uses its ciliated surface as a food-gathering organ, and has an internal canal-like gastrovascular system that communicates with the exterior through pores.[37] So is the Vendobionta concept completely redundant? Not necessarily, because the Proterozoic may have witnessed the achievement of the multicellular grade of organization several times, and in quite unrelated lineages.

As for coelenterates, though, several arguments suggest that all four extant cnidarian classes can be recognized in early metazoan assemblages. Anthozoans are represented by pennatulacean-like fossils[38] such as the Ediacaran *Charniodiscus.* Relatives in the Burgess Shale show what may be zooids.[39] Actinians may have occupied bowl-like fossils such as the Vendian *Beltanelliformis*[40,41] and the Cambrian *Berguaria.*[42] Biomineralization of cnidarians produced a variety of primitive coral-like fossils.[43] *Kimberella* may be a cubozoan,[38] represented today by the venomous box-jellies, whereas forms such as *Ovatoscutum*[38] may represent the floats of hydrozoans, as in the present-day Portuguese man-of-war. Jellyfish-like Ediacaran forms have been compared with "true" jellyfish, or scyphozoans, but there the evidence is more tenuous.

Probably related to the scyphozoans are the enigmatic conulariids, with ribbed phosphatic tests and a characteristic tetraradial symmetry. The Lower Cambrian carinachitiids (Fig. 3f) and hexagulaconulariids[44] are probably conulariids, and the Ediacaran *Conomedusites*[45] may also belong in this group. The xianguangiids,[46] from the Burgess Shale–type fauna at

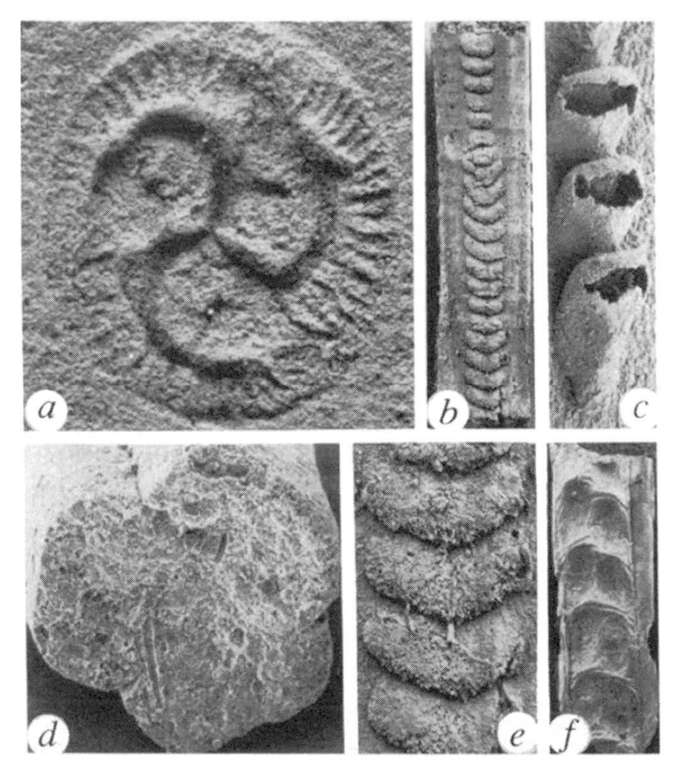

Fig. 3 Vendian and Cambrian metazoans. *a, Tribrachidium heraldicum,* a discoidal Ediacaran fossil with three "arms" from the Pound Subgroup, South Australia; ×2.9. *Tribrachidium* has been attributed to the trilobozoans (a possible group of early cnidarians) or alternatively a proto-echinoderm with the triradial symmetry preceding the penta-radial arrangement of most later echinoderms. *b, e, Anabarites compositus* from the Lower Cambrian of northern Siberia showing steinkern of tube (*b;* ×50) and detail of porous blade-like extensions (*e;* ×200) that arise in three rows along the interior wall of the tube. (Specimen kindly made available by V. V. Missarzhevksy.) *c, Deltaclavus graneus,* a series of articulated sclerites from the Lower Cambrian of Hubei, China, ×90. More complex arrays of sclerites have an "arm-like" structure that could indicate a relationship to echinoderms (see Fig. 2). *d, Anabarites trymatus* from the Lower Cambrian of South Australia, steinkern showing triradial symmetry of tube; ×150. *f, Carinachites spinatus,* an early ?conulariid (Cnidaria) from the Lower Cambrian of Shaanxi, China; ×36.

Chengjiang in China (Fig. 1), are interpreted as anthozoan-like cnidarians on account of a basal disk, polyp-like body with possible septal impressions, and a distal crown of tentacles. The tentacles, however, are unique in bearing closely spaced pinnules, a feature overlooked in the original description.[46] One group of Ediacaran medusiforms is characterized by triradial symmetry. Cambrian descendants of these so-called trilobozoans (such as *Albumares* and *Tribrachidium;* Fig. 3a) may be represented by the anabaritids;[47,48] their calcareous tubes are triradial (Fig. 3d), as defined by internal keels (Fig. 3b, e). Rare branching in anabaritid tubes[49] is consistent with a cnidarian grade. Cnidarians may also be represented by the tube-like byroniids,[48] the possibly tentacle-bearing *Cambrorhytium,*[50] the paiutiids with septate tubes,[51] and conceivably the tubular cloudinids. The last constitute an important Vendian group,[52] and are among the first metazoans to have acquired hard parts. Another candidate is the Burgess Shale *Echmatocrinus,* currently interpreted not as a cnidarian at all, but as an echinoderm, the earliest crinoid:[53] the supposed tube feet are very large and could instead be anthozoan-like tentacles. In contrast, *Echmatocrinus* had plated structures on its polyp and tentacles but, unlike the co-occurring eocrinoids,[53] the plates appear not to show a stereom structure.

The sister group of the Cnidaria is the Ctenophora, the comb-jellies. A pelagic habit and a delicate, gelatinous body account for their poor fossil record. *Fasciculus,*[54] from the Burgess Shales and adjacent localities, is an early ctenophore, but differs from extant and Devonian ctenophores[55] in the number of comb-rows.

Triploblasts Inherit the Earth

If Ediacaran assemblages are dominated by coelenterates, the ensuing Cambrian radiations are largely the irruption of triploblastic phyla. There are essentially two kinds of coelomate triploblastic metazoans: the protostomes and the deuterostomes. Protostomes, such as annelids and arthropods, are usually regarded as having spiral, determinate embryonic cleavage, and the embryonic blastopore is situated at the rear, at or near the site of the future mouth. Deuterostomes, in contrast (principally echinoderms and chordates) generally have radial, indeterminate cleavage, and the blastopore is associated with the position of the adult anus.[4]

Fossil evidence for the "Cambrian explosion"[56–59] is evident from the widespread rise of animal skeletons,[48,49] the major diversification of trace fossils[60] and the Burgess Shale–type faunas.[61] But it would be unwise to draw too strong a line between Ediacaran and Cambrian faunas. The Ediacaran *Dickinsonia* may be a stem-group protostome: it is segmented, bilaterally symmetrical and has a clear anteroposterior axis.[31,34] Other Ediacaran taxa could also be protostomes. *Spriggina* (Fig. 4e) could be a stem arthropod, and *Redkinia*[59] could represent mandible-like jaws: and if *Arkurua*[36] is interpreted as an echinoderm, the deuterostome lineage is also present in Ediacaran times.

Triploblastic diversification was rapid on a geological timescale,[6,8] and resolution of the branching order is consequently difficult. Refined stratigraphy, including recent advances in chemo- and magnetostratigraphy,[62]

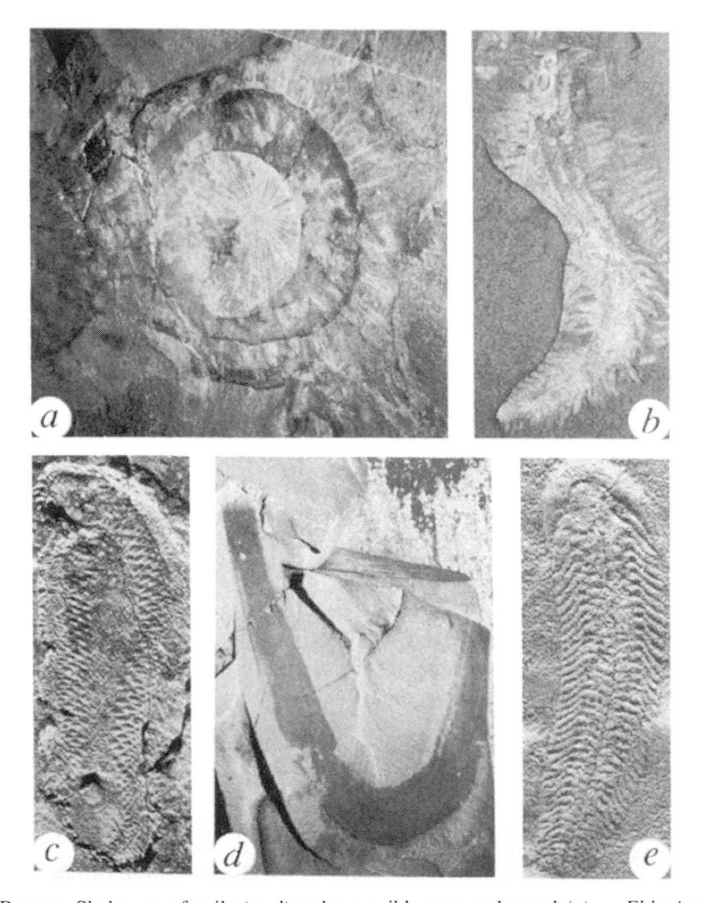

Fig. 4 Burgess Shale–type fossils *(a–d)* and a possible stem arthropod *(e)*. *a, Eldonia ludwigi,* an early deuterostome that may show features of pre-echinoderms; ×1.1. Specimen from Middle Cambrian (Burgess Shale), British Columbia. *b,* Undescribed worm from Middle Cambrian (Burgess Shale), British Columbia, possibly representing an early annelid; ×3.2. *c,* Articulated halkieriid from the Lower Cambrian (Buen Formation) of the Sirius Passet fauna, north Greenland; ×1.5. Halkieriids appear to represent a very early stage of mollusc evolution. *d,* The priapulid worm *Louisella pedunculata* from the Middle Cambrian (Burgess Shale), British Columbia; ×0.85. The trunk bears rows of sclerites, closely similar in arrangement to those of palaeoscolecidans. *e, Spriggina floundersi,* with segmented body and cephalic shield, a possible member of the arthopod stem group from the Vendian Pound Subgroup, South Australia; ×2.3.

may resolve some orders of appearance on a global scale. As far as fossils go, though, local taphonomic bias in favor of the preservation of either articulated scale-like armor,[63] phosphatized skeletal fragments[48,49] or soft parts[61] leads to a sporadic and incomplete record.

The "Cambrian explosion" is a real evolutionary event, but its origins are obscure. At least 20 hypotheses have been proposed, and although arguments linking diversification to oxygen levels, predation, faunal provinciality and ocean chemistry all attract support, it is the case that "the emergence of Metazoa remains the salient mystery in the history of life" (ref. 58, p. 17).

Deuterostome Radiations

Molecular data[6–8] suggest a profound evolutionary gulf between deuterostomes and protostomes. Among the Cambrian deuterostomes, the echinoderms have a good fossil record, and the interrelationships between many of the supposedly disparate classes are becoming clearer.[64] If the characteristically echinoderm skeleton of calcareous "stereom" was acquired before the equally distinctive water-vascular system, then the aberrant cornutes (which have the first, but not the second) are a very early branch of echinoderms. The chordate-like features of cornutes, which have been proposed to indicate a close relationship with chordates,[65] arose by convergence. The most primitive echinoderm with a water-vascular system appears to have been *Helicoplacus,* notable in having three rather than five ambulacra.[64] These observations need to be reconciled with the pentaradial arrangement in the Ediacaran *Arkurua,*[36] and also suggest an alternative assignment for trilobozoans (Fig. 3a). The cambroclaves may be allied to echinoderms, albeit more tentatively, because of their arm-like arrays[66,67] of articulated sclerites (Fig. 3c). These structures seem to suggest a sessile, echinoderm-like form rather than the creeping, slug-like habit previously suggested for these animals.[67]

Chordates appear with *Pikaia* from the Burgess Shale.[61] This creature superficially resembled the modern amphioxus and had a bilobed head and pair of tentacles, as well as myotomes and a notochord which, unlike the amphioxus, appears not to have extended to the anterior. Gill slits may have been present, but are hard to identify with certainty in the compressed material available. Apart from conodonts,[68] unequivocal evidence for fish does not occur until the Ordovician.[69] In particular, the fossil *Anatolepis,* which first occurs in the Upper Cambrian and has been interpreted as the dermal scales of a heterostracan fish,[70] is more likely to have belonged to an arthropod.[71]

The record of hemichordates is moderately good;[72,73] the organic tubes of rhabdopleurid pterobranchs have a fair preservation potential, and the Burgess Shale *"Ottoia" tenuis* may be an enteropneust similar to the extant acorn-worm *Balanoglossus* (unpublished observations by S.C.M.).

The nature of the deuterostome ancestor is speculative. The eldoniids,[74] a group of medusiform animals each with a prominent coiled gut (Fig. 4a), may repay consideration in this regard. They could represent a group of pre-echinoderm deuterostomes, as neither stereom nor a true water-vascular system appears to be present. Groups that may be related to the eldoniids include rotadiscids,[75] paropsonemids[76] and *Velumbrella.*[77]

Protostome Radiations

Platyhelminthes appear to be the most primitive triploblasts but, apart from possible trace fossils,[78] their fossil record is almost non-existent. Again, fossils have little to say about the aschelminthes (nematodes, rotifers and others) which may, in any case, be a polyphyletic grade. The nematodes and the related nematomorphs may be relatively primitive, but the present proximity of groups such as the rotifers and the related acanthocephalans[79] could require revision.

The branching pattern of the remaining protostomes is just as contentious. Lake[7] upset established orthodoxy by arguing that the arthropods are paraphyletic, arising before annelids and molluscs. In contrast, R. A. Raff (personal communication) regards arthropods as the sister group of molluscs–annelids–brachiopods–pogonophorans, the latter assemblage arising as an unresolved polychotomy.

Arthropods and Pre-Arthropods

The fossil record of Cambrian arthropods is moderately good, and even the riot of supposedly disparate forms from the Burgess Shales seem to fall into a phylogenetically coherent scheme. The available cladograms[80] are as yet tentative, but as information about arthropods from the Burgess Shale–like Chengjiang[81] and the Sirius Passet fauna from northern Greenland becomes available, cladistic analyses should improve.

More important is the nature of the pre-arthropod stocks. Apart from *Spriggina* (and possibly *Redkinia*), the anomalocarids[82] currently attract much interest. They are very diverse, but they all have prominent lobate appendages, a well-developed head with eyes, and sometimes a tail fan. Undescribed material from localities adjacent to the Burgess Shale (D. Collins, personal communication) and the Sirius Passet fauna will help to make sense of this intriguing group. Anomalocarids may achieve importance, not as a supposedly unique clade of extinct animals, but as a group close to the ancestry of modern arthropods.

The onychophorans, represented today by *Peripatus* of southern tropical forests and its relatives, is now known to have enjoyed considerable success in the Cambrian. These early, marine forms included the hitherto enigmatic *Hallucigenia,*[83] as well as *Microdictyon,*[84] which before its discovery in the Chengjiang fauna (Fig. 1) was known only from dispersed phosphatic sclerites. The Chengjiang fauna has yielded several other onychophores,[85,86] and other examples include *Xenusion* from Scandinavia.[87] *Facivermis*[88] is peculiar because the anterior region bears five pairs of lobopod-like appendages, whereas the rest of the body is smooth. Sclerotization of such an animal, with jointed limbs replacing lobopods, could lead to an animal similar to phosphatized Ordovician fossils from Öland,[89] regarded as primitive pentastomids (still extant as parasitic animals whose postulated position in the crustaceans[90] has been recently confirmed by molecular evidence[91]). There is insufficient information to resolve the phylogenetic relationships between anomalocarids, onychophores and arthropods exactly, but the first group may be a glimpse of what early arthropod-like protostomes were really like. Recent molecular analysis of extant onychophorans supports their place in the arthropods, but questions their primitive status.[92]

Annelids, Molluscs and Relatives

Further protostome diversification led to a plexus of annelids, molluscs and near relatives. If the scheme of Lake[7] is followed, then the annelids arose next. In the Burgess Shale there is an undescribed worm (Fig. 4b) with prominent lateral extensions that may be close to the basal annelid stock. The first definitive record of the annelids is as polychaetes,[93] which are diverse in the Burgess Shale: their absence from the Chengjiang fauna may not be significant because they are present in the slightly younger Sirius Passet fauna. There is general agreement that pogonophorans are related to annelids.[94] The elongate, organic-walled tubes of sabelliditids, abundant in the Vendian and lower Cambrian, may have housed pogonophorans. There are, however, differences in wall ultrastructure between these and modern pogonophoran tubes.[95,96]

Molecular data suggest that the nemerteans, traditionally placed close to the platyhelminths, are coelomate protostomes, perhaps related to the annelids,[97] but with a distinctive coelom (rhynchocoel surrounding the proboscis). The nemertean fossil record is very meager, but amiskwiids[98] (represented by *Amiskwia* from the Burgess Shale) may represent an early stage of their divergence. Molecular evidence likewise places the brachiopods close to the annelids, and the chaetae in both groups are similar.[99] Understanding of the initial radiation of brachiopods is improving, especially with the description of supposedly enigmatic groups.[48,58] But pitfalls remain; for example, a putative brachiopod-like shell from the Lower Cambrian of Irkutsk[100] could be a percussion fracture formed when the drill core was broken open, a telling illustration of the difficulties sometimes incurred in interpreting problematic fossils. Tommotiids, only known from dispersed phosphatic sclerites, have been reconstructed as slug-like animals.[101] Some sclerites, however, recall brachiopods, and comparison of shell ultrastructure[102] supports the possibility that tommotiids and brachiopods are related.[56]

A relationship between annelids and brachiopods may be difficult to reconcile with data from hemerythrin sequences that indicate a close relationship between priapulids and the inarticulate brachiopod *Lingula.*[103] Nevertheless, current evidence suggests that priapulids (and the closely related kinorhynchs and loriciferans, both without a fossil record), are part of the coelomate protostome radiation. Whether Cambrian priapulids[50,104] will throw light on early protostome relationships is difficult to judge. Fossils called palaeoscolecidans, known to bear phosphatic sclerites,[105] may be of interest here: specimens from the Sirius Passet fauna display what look like spiny snouts similar to those that distinguish priapulids, suggestive of a close relationship. Conversely, rows of phosphatic sclerites similar to those of palaeoscolecidans have been found in a Burgess Shale priapulid[104] (Fig. 4d). Of potential significance also is the Chengjiang priapulid *Cricocosmia,*[106] which has two rows of convex shell-like sclerites on its trunk.

Molluscs may have evolved from a flatworm-like ancestor independently of the annelids,[4] but molecular evidence[10] supports the possibility of an annelid link. The fossil record is silent on this transition, but the earlier stages of mollusc evolution are becoming clear. Cardinal evidence comes from the halkieriids[63] (Fig. 4c), which had an articulated armor of sclerites together with prominent anterior and posterior shells. A number of mollusc-like shells are known from the Cambrian and it is likely that some derive from halkieriid or related scleritomes. *Triplicatella,*[48] from South Australia and possibly southern Siberia,[107] very probably comes from the scleritome of the halkieriid *Thambetolepis.* Prominent folds on the shell margin of *Triplicatella* may mark sites of exhalent and inhalent water currents,[48] perhaps connected with gills similar to molluscan ctenidia.

In turn, similarity of sclerite structure suggests a link between halkieriids and the Coeloscleritophora,[48] a diverse Cambrian group that includes siphogonuchitids, sachitids and chancelloriids. The first of these probably had a scleritome similar to halkieriids, with at least one shell and two types of sclerite.[108] Chancelloriid sclerite structure does not support their earlier assignment to sponges, but overall body shape is consistent with a sessile habit. Chancelloriid sclerites were evidently embedded in a resistant cuticle (Hou Xianguang, personal communication). Sachitids may be intermediate in form between chancelloriids and halkieriids.[48]

Trace Fossils

Even if no soft-bodied fossils had been preserved in the Cambrian, their adaptive radiations would be evident from the record of trace fossils.[58,60] Although some traces, such as scratch marks, can be attributed to arthropods (or perhaps the related anomalocarids), in general tying traces to their makers is difficult. Trace fossils have consequently played little part in discussions of metazoan phylogeny, but they may nevertheless contribute useful information. Although many kinds of trace fossil are known from a long geological time span, some are confined to the Cambrian and could record extinct body plans, as well as details of early metazoan activities such as locomotion and feeding. For example, the giant trace *Climactichnites* from the Upper Cambrian of North America is believed to have been constructed by an otherwise unknown metazoan of novel appearance.[109]

Phyla and the Role of Problematic Taxa

Schemes of metazoan phylogeny based on molecular evidence but which have been broadened to include information from fossils have been presented before.[110,111] I have taken it a stage further with the inclusion of many groups usually seen as problematic, linking them with more familiar body plans. But this is not meant to belittle the magnitude of the adaptive radiations that took place in the Vendian and Cambrian periods. Within a period of about 20 Myr (taking the Ediacaran faunas as ~560 Myr old and the base of the Cambrian as ~540 Myr old[112]), the oceans changed from habitats housing a rich but effectively microscopic biota,[13] to one teeming with macroscopic animals engaged in a wide range of ecologies and presumably showing a degree of behavioral sophistication. Is it realistic to talk of a multiplicity of body plans in the Cambrian, far exceeding that of the present day?[113] At the other extreme, the argument that extant phyla maintain their identity back to the Cambrian[114] offers no more than a tautology. Such phyla so persist because the only way they can be recognized is by reference to themselves. Consider the potential disagreements that surround the interpretation of various Cambrian fossils: are the pseudobrachiopods[48] "truly" brachiopods? Are the anomalocarids[82] "really" arthropods? Are hyoliths or stenothecoids "actually" molluscs? Definitions of major groups are less secure than sometimes imagined.

I argue that the supposed problematic taxa, now in imminent danger of elevation to a classic status as evolutionary enigmas,[113] hold the key to understanding many aspects of early metazoan evolution. This view contrasts with that of a "pool" of generalized coelomate ancestors from which at least the main phyla of protostomes were derived.[114,115] In such models the problematic taxa are further and separate derivatives, useful for documenting the range of metazoan morphospace but irrelevant for establishing relationships at the base of metazoan evolution.

Conclusion

Future advances will depend largely on molecular biologists and paleontologists, although no data source should be neglected.[116] The former need to tackle new gene sequences and extend the database. It is necessary also to enlarge our understanding of developmental mechanisms, and there is growing interest in the widespread occurrence of at least some regulatory genes in disparate phyla.[117,118] What is equally important to know, however, is how these genes have either been co-opted or have changed their function, not only among metazoan phyla but in their protistan ancestors.

What can paleontologists contribute? The discovery of Cambrian faunas similar to that of the Burgess Shale is likely.[61] Further preparation of collections of calcareous and phosphatic shelly Cambrian fossils may well produce dividends, and such disarticulated remains that are etched out of the rock should be held up against articulated scleritomes, either actual or hypothesized. The diversification of trace fossils still lacks a biological context, and there is an urgent need to test whether at least some Ediacaran fossils could be assigned to the Vendobionta, rather than to Metazoa.[39]

And yet, at the heart of the matter, a coherent explanation for the origin and scope of the early metazoan radiations is still missing.[58] Family trees based on molecular data can be festooned with fossils but this is perhaps rather premature. Instead, we should be asking how the diversity of body plans so evident in the Vendian and Cambrian affected the shape of things to come. Evolutionary innovations such as nerve tracts, seriation, mesoderm, spacious body cavities and circulatory systems, all implied by the fossils already available, must have had profound phylogenetic consequences. Again, the causes of this diversification still remain a mystery, although changes in the concentration of atmospheric oxygen,[119] trophic resources, and ecological response, especially to predation, may have all played a part. Certainly the acquisition of hard parts as a deterrent to predators is a compelling hypothesis.[120] Increasing evidence for major changes in the environment during the Vendian and Cambrian[121] also require further assessment. The once stagnant field of metazoan phylogeny is being rejuvenated by new discoveries that promise to call on a wide range of scientific disciplines. "Metazoan Phylogeny" is exciting to play, and we may, at last, be on the verge of working out some of the rules of the game.

SIMON CONWAY MORRIS is in the Department of Earth Sciences, University of Cambridge, Cambridge CB2 3EQ, UK.

References

1. Salvini-Plawen, L. *Z. zool. System. Evolut-forsch.* **16,** 40–88 (1978).
2. Schram, F. R. in *The Early Evolution of Metazoa and the Significance of Problematic Taxa* (eds. Simonetta, A. M. & Conway Morris, S.) 35–46 (Cambridge Univ. Press, Cambridge, 1991).
3. Ax, P. in *The Hierarchy of Life, Molecules and Morphology in Phylogenetic Analysis* (eds. Fernholm, B., Bremer, K. & Jörnval, H.) 229–245 (Excerpta Medica, Amsterdam, 1989).
4. Willmer, P. *Invertebrate Relationships: Patterns in Animal Evolution,* xiii, 1–400 (Cambridge Univ. Press, Cambridge, 1990).
5. Patterson, C. in *The Hierarchy of Life, Molecules and Morphology in Phylogenetic Analysis* (eds. Fernholm, B., Bremer, K. & Jörnval, H.) 471–488 (Excerpta Medica, Amsterdam, 1989).
6. Field, K. G. *et al. Science* **239,** 748–753 (1988).
7. Lake, J. A. *Proc. natn. Acad. Sci. USA* **87,** 763–766 (1990).
8. Christen, R. *et al. EMBO J.* **10,** 499–503 (1991).
9. Erwin, D. H. *Trends Ecol. Evol.* **6,** 131–134 (1991).
10. Ghiselin, M. T. *Oxford Surv. Evol. Biol.* **5,** 66–95 (1988).
11. Walter, M. R. & Heys, G. R. *Precambrian Res.* **29,** 149–174 (1985).
12. Grotzinger, J. P. *Am. J. Sci.* **290** A, 80–103 (1990).
13. Knoll, A. H. *Science* **256,** 622–627 (1992).
14. Robbins, E. I., Porter, K. G. & Haberyan, K. A. *Proc. natn. Acad. Sci. USA* **82,** 5809–5813 (1985).
15. Nöthig, E.-M. & Bedungen, B. V. *Mar. Ecol. Prog. Ser.* **56,** 281–289 (1989).
16. O'Brien, N. R. *J. sedim. Petrol.* **57,** 449–455 (1987).
17. Reichelt, A. C. *Symp. zool. Soc. Lond.* **63,** 33–52 (1991).
18. Harding, S. C. & Risk, M. J. *J. sedim. Petrol.* **56,** 684–696 (1986).
19. Aller, R. C. & Yingst, J. Y. *J. mar. Res.* **36,** 201–254 (1978).
20. Runnegar, B. *Lethaia* **15,** 199–205 (1982).
21. Runnegar, B. *J. molec. Evol.* **22,** 141–149 (1985).
22. Erwin, D. H. *Lethaia* **22,** 251–257.
23. Sogin, M. L. *Curr. Opin. genet. Dev.* **1,** 457–463 (1991).
24. Towe, K. M. *Proc. natn. Acad. Sci. USA* **65,** 781–788 (1970).
25. Chen M. & Xiao, Z. *Sci. Geol. Sinica* **4,** 317–324 (1991).
26. Field, K. G. *et al. Science* **243,** 550–551 (1989).
27. Brasier, M. D. *J. geol. Soc. Lond.* **149,** 621–629 (1992).

28. Dunn, P. R. *J. geol. Soc. Aust.* **11,** 195–197 (1964).
29. Allison, C. W. & Hilgert, J. W. *J. Paleont.* **60,** 973–1015 (1986).
30. Kaufman, A. J., Knoll, A. H. & Awramik, S. M. *Geology* **20,** 181–185 (1992).
31. Glaessner, M. F. *The Dawn of Animal Life: A Biohistorical Study* Vol. 11, 1–244 (Cambridge Univ. Press, Cambridge, 1984).
32. Seilacher, A. *Lethaia* **22,** 229–239 (1989).
33. Seilacher, A. *J. geol. Soc. Lond.* **149,** 607–613 (1992).
34. Gehling, J. G. *Mem. geol. Soc. India* **20,** 181–224 (1991).
35. Gehling, J. G. *Alcheringa* **12,** 299–314 (1988).
36. Gehling, J. G. *Alcheringa* **11,** 337–345 (1987).
37. Schlichter, D. *Helgolander wiss Meeresunters* **45,** 423–443 (1991).
38. Jenkins, R. J. F. *Mem. Ass. Australas. Palaeontol.* **8,** 307–317 (1989).
39. Conway Morris, S. *Palaeontology* **36,** 593–635 (1993).
40. Fedonkin, M. A. in *The Vendian System: I. Paleontology* (eds. Sokolov, B. S. & Iwanowski, A. B.) 71–120 (Springer, Berlin, 1990).
41. Narbonne, G. M. & Hofmann, H. J. *Palaeontology* **30,** 647–676 (1987).
42. Alpert, S. P. *J. Paleont.* **47,** 919–924 (1973).
43. Lafuste, J., Debrenne, F., Gandin, A. & Gravestock, D. *Geobios* **24,** 697–718 (1991).
44. Conway Morris, S. & Chen, M. *J. Paleont.* **66,** 384–406 (1992).
45. Glaessner, M. F. *Paläont. Z.* **45,** 7–17 (1971).
46. Chen Junyuan & Erdtmann, B.-D. in *The Early Evolution of Metazoa and the Significance of Problematic Taxa* (eds. Simonetta, A. M. & Conway Morris, S.) 57–76 (Cambridge Univ. Press, Cambridge, 1991).
47. Conway Morris, S. & Chen, M. *Geol. Mag.* **126,** 615–632 (1989).
48. Bengtson, S., Conway Morris, S., Cooper, B. J., Jell, P. A. & Runnegar, B. N. *Mem. Ass. Australas. Palaeontol.* **9,** 1–364 (1990).
49. Missarzhevsky, V. V. *Oldest Skeletal Fossils and Stratigraphy of Precambrian and Cambrian Boundary Beds* (Nauka, Moscow, 1989).
50. Conway Morris, S. & Robison, R. A. *Univ. Kansas Paleont. Contrib. Pap.* **122,** 1–48 (1988).
51. Tynan, M. C. *J. Paleont.* **57,** 1188–1211 (1983).
52. Grant, S. W. F. *Am. J. Sci.* **290** A, 261–294 (1990).
53. Sprinkle, J. *Spec. Publ. Mus. comp. Zool. Harvard* 1–283 (1973).
54. Simonetta, A. & Cave, L. D. *Atti Soc. tosc. Sci. nat. Mem.* **85** A, 45–49 (1978).
55. Stanley, G. D. & Stürmer, W. *Nature* **327,** 61–63 (1987).
56. Brasier, M. D. in *The Origin of Major Invertebrate Groups* (ed. House, M. R.) 103–159 (Academic, London, 1979).
57. Valentine, J. W., Awramik, S. M., Signor, P. W. & Sadler, P. M. *Evol. Biol.* **25,** 279–356 (1991).
58. Lipps, J. H. & Signor, P. W. (eds.) *Origin and Early Evolution of Metazoa* (Plenum, New York, 1992).
59. Schopf, J. W. & Klein, C. (eds.) *The Proterozoic Biosphere: A Multidisciplinary Study* (Cambridge Univ. Press, Cambridge, 1992).
60. Crimes, T. P. in *The Precambrian-Cambrian Boundary* (eds. Cowie, J. W. & Brasier, M. D.) 166–185 (Clarendon, Oxford, 1989).
61. Conway Morris, S. *Science* **246,** 339–346 (1989).

62. Kirschvink, J. L., Magaritz, M., Ripperdan, R. L., Zhuravlev, A. Yu. & Rozanov, A. Yu. *Geol. Soc. Am. Today* **1,** 69–71, 87, 91 (1991).
63. Conway Morris, S. & Peel, J. S. *Nature* **345,** 802–805 (1990).
64. Smith, A. B. in *Major Evolutionary Radiations* (eds. Taylor, P. D. & Larwood, G. P.) 265–286, *Syst. Ass. Spec.* Vol. 42 (Clarendon, Oxford, 1990).
65. Jefferies, R. P. S. *The Ancestry of the Vertebrates* (British Museum [Natural History] and Cambridge Univ. Press, Cambridge, 1986).
66. Yue, Z. *Kexue Tongbao* 47–50 (1991).
67. Conway Morris, S. and Chen, M. *Palaeontology* **34,** 357–397 (1992).
68. Sansom, I. J., Smith, M. P., Armstrong, H. A. & Smith, M. A. *Science* **256,** 1308–1311 (1992).
69. Blieck, A. *Geobios* **25,** 101–113 (1992).
70. Repetski, J. E. *Science* **200,** 529–531 (1978).
71. Peel, J. S. *Rapp. Grønlands geol. Unders.* **91,** 111–115 (1979).
72. Bengtson, S. & Urbanek, A. *Lethaia* **19,** 293–308 (1985).
73. Urbanek, A., Mierzejewski, P. & Bengtson, S. *Lethaia* **25,** 349–350 (1992).
74. Durham, J. W. *J. Paleont.* **48,** 750–755 (1974).
75. Sun, W. & Hou, X. *Acta Palaeont. Sin.* **26,** 257–271 (1987).
76. Clarke, J. M. *Bull. N.Y. State Mus.* **39,** 172–178 (1900).
77. Dzik, J. in *The Early Evolution of Metazoa and the Significance of Problematic Taxa* (eds. Simonetta, A. M. & Conway Morris, S.) 47–56 (Cambridge Univ. Press, Cambridge, 1991).
78. Alessandrello, A., Pinna, G. & Teruzzi, G. *Atti Soc. ital. Sci. nat. Museo civil. storia natur.* **129,** 139–145 (1988).
79. Lorenzen, S. in *The Origins and Relationships of Lower Invertebrates* (eds. Conway Morris, S., George, J. D., Gibson, R. & Platt, H. M.) 210–223 (Clarendon, Oxford, 1985).
80. Briggs, D. E. G., Fortey, R. A. & Wills, M. A. *Science* **256,** 1670–1673 (1992).
81. Hou, X. & Bergström, J. in *The Early Evolution of Metazoa and the Significance of Problematic Taxa* (eds. Simonetta, A. M. & Conway Morris, S.) 179–187 (Cambridge Univ. Press, Cambridge, 1991).
82. Whittington, H. B. & Briggs, D. E. G. *Phil. Trans. R. Soc.* B **309,** 569–609 (1985).
83. Ramsköld, L. & Hou, X. *Nature* **351,** 225–228 (1991).
84. Chen, J., Hou, X. & Lu, H. *Acta Palaeont. Sin.* **28,** 1–16 (1989).
85. Hou, X., Ramsköld, L. & Bergström, J. *Zool. Scripta* **20,** 395–411 (1991).
86. Ramsköld, L. *Lethaia* **25,** 443–460 (1992).
87. Dzik, J. & Krumbiegel, G. *Lethaia* **22,** 169–182 (1989).
88. Hou, X. & Chen, J. *Acta Palaeont. Sin.* **28,** 32–41 (1989).
89. Andres, D. *Berl. geowiss. Abh.* A **106,** 9–19 (1989).
90. Riley, J., Banaja, A. A. & James, J. L. *Int. J. Parasit.* **8,** 245–254 (1978).
91. Abele, L. G., Kim, W. & Felgenhauer, B. E. *Molec. Biol. Evol.* **6,** 685–691 (1989).
92. Ballard, J. W. O. *et al. Science* **258,** 1345–1348 (1992).
93. Conway Morris, S. *Phil. Trans. R. Soc.* B **285,** 227–274 (1979).
94. Southward, E. C. *Symp. zool. Soc. Lond.* **36,** 235–251 (1975).
95. Urbanek, A. & Mierzejewska, G. in *Upper Precambrian and Cambrian Palaeontology of the East-European Platform* (eds. Urbanek, A. & Rozanov, A. Yu.) 100–111 (Wydawnictwa Geologicze, Warsaw, 1983).
96. Ivantsov, A. Yu. *Paleont. Zh.* 125–128 (1990).

97. Turbeville, J. M., Field, K. G. & Raff, R. A. *Molec. Biol. Evol.* **9,** 235–249 (1992).
98. Conway Morris, S. *Paläont. Z.* **51,** 271–287 (1977).
99. Gustus, R. M. & Cloney, R. A. *Acta Zool. Stockh.* **53,** 229–233 (1972).
100. Galimova, V. S. in *Cambrian of the Siberia and the Middle Asia* (eds. Zhuravleva, I. T. & Repina, L. N.) 185–190 (Nauka, Moscow, 1988).
101. Evans, K. R. & Rowell, A. J. *J. Paleont.* **64,** 692–700 (1990).
102. Conway Morris, S. & Chen, M. *J. Paleont.* **64,** 169–184 (1990).
103. Runnegar, B. & Curry, G. B. *Abstr. Int. Geol. Congress, Kyoto* (1992).
104. Conway Morris, S. *Spec. Pap. Palaeont.* **20,** 1–95 (1977).
105. Hinz, I., Kraft, P., Mergl, M. & Müller, K. J. *Lethaia* **23,** 217–221 (1990).
106. Hou, X. & Sun, W. *Acta Palaeont. Sin.* **27,** 1–12 (1988).
107. Khomentovskii, V. V. & Karlova, G. A. in *Late Precambrian and Early Palaeozoic of Siberia* (ed. Khomentovskii, V. V.) 3–40 (Novosibirsk, 1991).
108. Bengtson, S. *Lethaia* **25,** 401–420 (1992).
109. Yochelson, E. L. & Fedonkin, M. A. *Spec. Publ. Paleont. Soc.* **6,** 321 (1992).
110. Valentine, J. W. in *Constructional Morphology and Evolution* (eds. Schmidt-Kittler, N. & Vogel, K.) 389–397 (Springer, Berlin 1991).
111. Runnegar, B. in *Major Events in the History of Life* (ed. Schopf, J. W.) 65–93 (Jones & Bartlett, Boston, 1992).
112. Compston, W., Williams, I. S., Kirschvink, J. L., Zhang, Z. & Ma, G. *J. geol. Soc. Lond.* **149,** 171–184 (1992).
113. Gould, S. J. *Wonderful Life: The Burgess Shale and the Nature of History* (Norton, New York, 1989).
114. Bergström, J. *Lethaia* **22,** 259–269 (1989).
115. Valentine, J. W. *Syst. Zool.* **22,** 97–102 (1973).
116. Willmer, P. G. & Holland, P. W. H. *J. zool. Lond.* **224,** 689–694 (1991).
117. Akam, M. *Cell* **57,** 347–349 (1989).
118. Schierwater, B., Murtha, M., Dick, M., Ruddle, F. H. & Buss, L. W. *J. exp. Zool.* **260,** 415–416 (1991).
119. Derry, L. A., Kaufman, A. J. & Jacobsen, S. B. *Geochim. cosmochim. Acta* **56,** 1317–1329 (1992).
120. Vermeij, G. J. *Palaios* **4,** 585–589 (1990).
121. Knoll, A. H. & Walter, M. R. *Nature* **356,** 673–678 (1992).

Acknowledgments

I thank R. A. Wood, D. Erwin and N. J. Butterfield, S. J. Last for typing, K. Harvey for photography and H. Alberti for drafting. Research for this work has been supported by the NERC, the Carlsberg Foundation, the Royal Society and St. John's College, Cambridge.

PART 3

SEEING THE TREE FOR THE WOODS: THE GLOBAL CONTEXT

How did the tree of life get its shape? Did it run wild, or was it pruned from time to time? If so, was the pruning done with a topiarist's skill, or with a rampaging chainsaw? Understanding the context in which the tree of life grew is clearly vital to its greater appreciation. This is the theme that unites the otherwise very different contributions in this section.

The first, from Andrew Knoll and Malcolm Walter, looks at the geochronological context of the increasingly well known Proterozoic eon, whose end saw the fantastic flowering of multicellular animal life. What were the conditions that set the stage? When did each occur, and in what order? This interval was, as Knoll and Walter say, "a time of pronounced biological, biogeochemical, climatic and tectonic change." Ice ages came and went; groups of algae underwent seemingly sudden extinctions; partial pressures of oxygen rose, the chemistry of the sea underwent profound changes and a variety of biogeochemical cycles switched gear.

Some or all of these factors contributed to the rise of multicellular life, but it is possible that life itself helped promote and reinforce these changes. Knoll and Walter record that the earliest fossils are more than 3.5 billion years old.[1] Isotopic evidence for life—although no actual fossils—comes from rocks 3.8 billion years old,[2,3] almost back to the time when the Earth was regularly bombarded with planet-sized impactors that made the end-Cretaceous bolide, 65 million years ago, seem like a mote of fluff. Life, once it appeared, seems to have been remarkably tenacious.

For most of the Precambrian, the period between Earth's formation and 543 million years ago—the beginning of the Cambrian period—life consisted of various kinds of prokaryote.[4] Molecular and some fossil evidence suggests that the origin of eukaryotes—cells with membrane-shrouded nuclei—happened around 2 billion years ago, if not earlier.[5,6] Nevertheless, some molecular work has suggested, controversially, a divergence between prokaryotes and eukaryotes of around 2 billion years ago.[7,8]

In this context, the interval between about a billion and 543 million years ago[9,10] was crucial in evolution. At the start the world was clothed in bacteria and algae: at the end, the Metazoa were well on the way to becoming established. The oxygen that might have allowed animals to assume large sizes was assuredly a biogenic product. Once evolved, marine animals large enough to drop fecal pellets to the bottom of the sea, rather than distribute suspended excreta, may have cleared the oceans of an anoxic zone that spread almost to the surface, thereby opening up the plankton and nekton for colonization.[11,12] A wealth of literature exists on the end-Proterozoic transition (see, for example, refs. 11–17), but as Knoll and Walter stress, we can really only unravel the complex events that ushered in multicellular life once a reliable geochronological framework is in place.

Once metazoan life had appeared in the sea, what then? The land still remained ripe for colonization. But as William Shear shows in the second item in this section, the development of the first terrestrial ecosystems was not the simple tale of steady advance that one might imagine, by looking at the modern ecosystem and working backward from there. Instead, argues Shear, paleontologists are learning that it is dangerous to draw too close an analogy between the ecology of Paleozoic organisms and their modern counterparts, as the earliest terrestrial ecosystems could have been very different. A marked difference, for example, was the almost total absence of herbivory for many millions of years after the land was first invaded. Today, it is hard to imagine a terrestrial ecosystem in which the consumption of living plant matter is not an integral part of the food web.

Shear's review is the oldest in this book, published in 1991. I include it here as it is still unmatched as a brief, general overview. See refs. 17–27 for a range of published perspectives on various aspects of parts of the story.

After all this burgeoning life it is no surprise to learn that the Reaper is not far behind. Extinction is a normal part of life, yet on at least five occasions in the Phanerozoic eon (the name for the past 543 million years, since the beginning of the Cambrian period), the rate of extinction has proceeded at a rate far enough above the "background rate" that researchers recognize a so-called "mass-extinction" event.

In the third contribution to this section, Douglas Erwin looks at our increasing understanding of the largest of these events, which occurred at the end of the Permian period, 251 million years ago. Erwin has also delved into the issue at book length.[28] "Killing over 90% of the species in the oceans and about 70% of vertebrate families on land is remarkably difficult," says Erwin, so how did it happen? Our own difficulty, trying to understand such events, stems from trying to put the blame on any one cause, when we should rather be looking for a "tangled web" of interlinked factors.

The past few years has seen greater refinement in our understanding of the various factors at work whose interplay could have triggered the extinction. Absolute chronology shows that whatever happened, and how, it was remarkably quick, occurring in 500,000 years or less,[29] and might have been related to massive volcanic activity in what is now Siberia.[30] Even though mortality was high, one can see that extinction struck some forms more than others, a selectivity that could be consistent with the release of carbon dioxide into the atmosphere, triggered by the rapid overturn of unusually anoxic (that is, oxygen-starved) seawater.[31] Although there is evidence for extensive deep-sea anoxia at the time,[32] the mechanism behind any oceanic overturn is not clear. In any event, a sudden "spike" of fungi, or possibly mosses, at the boundary, indicates widespread terrestrial distress.[33] Clearly—something happened, and a small industry is at work untangling Erwin's "tangled web."

Although each mass-extinction event is different, one could apply this lesson of caution to the end-Cretaceous extinction, 65 million years ago. This event has grabbed the popular imagination thanks to compelling evidence for the impact of a large bolide, and the demise of the dinosaurs at around the same time. The title of Walter Alvarez's page-turning account[34] of research into the end-Cretaceous impact—*T. rex and the Crater of Doom*—may appeal to popular taste, but does little to promote the idea of extinctions as the results of many causes, rather than just one.

Recent research suggests that the biotic response to the end-Cretaceous crisis varied according to the group under study: dinosaurs aside, the fates of the organisms involved was less dramatic overall, and dependent on the complexities of their own ecologies.[35,36]

Sadly, news that downplays drama and increases complexity is unlikely to make headlines anywhere near as huge as one promising a story that *T. rex* died from an invader from outer space. Yet very few dinosaurs were actually hit on the head by the impactor—most would have died from the secondary effects of the impact, which could have been many and varied, ranging from acid rain to "nuclear winter." As an aside, Mike Benton has compiled a bibliography of ideas, ranging from serious to silly, purporting to "explain" the extinction of the dinosaurs.[37] Three symposium volumes[38–40] provide valuable compendia of thought on the topic of mass extinctions generally.

Perhaps unexpectedly, cladistics has a key role to play in the study of mass extinctions. Patterson and Smith[41] have argued that no work on, say, the biotic selectivity of extinction events can be meaningful unless and until the phylogenies of the organisms concerned are properly understood.

References

1. J. W. Schopf. Microfossils of the early Archean Apex chert: New evidence of the antiquity of life. *Science* 260 (1993): 640–646.
2. S. J. Mojzsis, G. Arrhenius, K. D. McKeegan, T. M. Harrison, A. P. Nutman and C. R. L. Friend. Evidence for life on Earth before 3,800 million years ago. *Nature* 384 (1996): 55–59.
3. J. M. Hayes. The earliest memories of life on Earth. *Nature* 384 (1996): 21–22.
4. J. W. Schopf, ed. *Earth's earliest biosphere: Its origin and evolution.* Princeton: Princeton University Press, 1983.
5. A. H. Knoll. The early evolution of eukaryotes: A geological perspective. *Science* 256 (1992): 622–627.
6. T. M. Han and B. Runnegar. Megascopic eukaryotic algae from the 2.1-billion-year-old Negaunee Iron Formation, Michigan. *Science* 257 (1992): 232–235.
7. R. F. Doolittle, D.-F. Sheng, S. Tsang, G. Cho and E. Little. Determining divergence times of the major kingdoms of living organisms with a protein clock. *Science* 271 (1996): 470–477.
8. A. Ø. Mooers and R. J. Redfield. Digging up the roots of life. *Nature* 379 (1996): 587–588.
9. S. A. Bowring, J. P. Grotzinger, C. E. Isachsen, A. H. Knoll, S. M. Pelechaty and P. Kolosov. Calibrating rates of early Cambrian evolution. *Science* 261 (1993): 1293–1298.
10. J. P. Grotzinger, S. A. Bowring, B. Z. Saylor and A. J. Kaufman. Biostratigraphic and geochronologic constraints on early animal evolution. *Science* 270 (1995): 598–604.
11. G. A. Logan, J. M. Hayes, G. B. Hieshma and R. E. Summons. Terminal Proterozoic reorganization of biogeochemical cycles. *Nature* 376 (1995): 53–56.
12. M. Walter. Faecal pellets in world events. *Nature* 376 (1995): 16–17.
13. A. H. Knoll and M. R. Walter, eds. *Neoproterozoic stratigraphy and earth history.* Special issue of *Precambrian Research* 73 (1995): 1–298.

14. D. R. Canfield and A. Teske. Late Proterozoic rise in atmospheric oxygen concentration inferred from phylogenetic and sulphur isotope studies. *Nature* 382 (1996): 127–132.
15. A. J. Kaufman *et al.* Isotopes, ice ages, and terminal Proterozoic Earth history. *Proceedings of the National Academy of Sciences, USA* 94 (1997): 6600–6605.
16. J. H. Lipps and P. W. Signor, eds. *Origin and early evolution of the Metazoa.* New York: Plenum, 1992.
17. A. K. Behrensmeyer et al., eds. *Terrestrial ecosystems through time.* Chicago: University of Chicago Press, 1992.
18. W. A. Shear and J. Kukalová-Peck. The ecology of Paleozoic terrestrial arthropods: The fossil evidence. *Canadian Journal of Zoology* 68 (1990): 1807–1834.
19. W. A. Shear. Les premiers ecosystems terrestres. *La Recherche* 23 (1992): 1258–1267.
20. J. Gray and W. A. Shear. Early life on land. *American Scientist* 80 (1992): 444–456.
21. D. Edwards and P. A. Selden. The development of early terrestrial ecosystems. *Botanical Journal of Scotland* 46 (1993): 337–366.
22. A. C. Scott *et al.* Interaction and coevolution of plants and arthropods during the Palaeozoic and Mesozoic. *Philosophical Transactions of the Royal Society* B 335 (1992): 129–165.
23. K. McNamara and P. A. Selden. Strangers on the shore. *New Scientist* 139 (1993): 23–27.
24. D. Edwards *et al.* Coprolites as evidence of plant-animal interaction in Siluro-Devonian terrestrial ecosystems. *Nature* 377 (1995): 329–331.
25. A. C. Scott. Evidence for plant-arthropod interactions in the fossil record. *Geology Today* (March–April 1991): 58–61.
26. J. Gray and A. J. Boucot. Early Silurian nonmarine animal remains and the nature of the early continental ecosystem. *Acta Palaeontologica Polonica* 38 (1994): 303–328.
27. E. N. K. Clarkson, A. L. Panchen and W. D. I. Rolfe, eds. Volcanism and early terrestrial biotas. *Transactions of the Royal Society of Edinburgh Earth Science* 84 (1994): 175–464.
28. D. H. Erwin. *The great Paleozoic crisis: Life and death in the Permian.* New York: Columbia University Press, 1993.
29. S. A. Bowring, D. H. Erwin, Y. G. Jin, M. W. Martin, K. Davidek and W. Wang. U/Pb geochronology and tempo of the end-Permian mass extinction. *Science* 280 (1998): 1039–1045.
30. P. Renne *et al.* Synchrony and casual relations between Permian-Triassic boundary crises and Siberian Flood volcanism. *Science* 269 (1995): 1413–1416.
31. A. H. Knoll, R. K. Bambach, D. E. Canfield and J. P. Grotzinger. Comparative Earth history and late Permian mass extinction. *Science* 273 (1996): 453–457.
32. Y. Isozaki. Permo-Triassic boundary superanoxia and stratified superocean: Records from the lost deep sea. *Science* 276 (1997): 235–238.
33. H. Visscher *et al.* Terminal Paleozoic fungal event: Evidence of terrestrial ecosystem destabilization and collapse. *Proceedings of the National Academy of Sciences USA* 93 (1996): 2155–2158.
34. W. Alvarez. *T. rex and the crater of doom.* Princeton: Princeton University Press, 1997.

35. N. Macleod *et al.* The Cretaceous-Tertiary biotic transition. *Journal of the Geological Society, London* 154 (1997): 265–292.
36. A. B. Smith and C. H. Jeffrey. Selectivity of extinction among sea urchins at the end of the Cretaceous Period. *Nature* 392 (1998): 69–71.
37. M. J. Benton. Scientific methodologies in collision: The history of the study of the extinction of the dinosaurs. *Evolutionary Biology* 24 (1990): 371–400.
38. W. G. Chaloner and A. Hallam, eds. *Evolution and extinction.* London: Royal Society, 1989.
39. S. K. Donovan, ed. *Mass extinctions: Processes and evidence.* New York: Columbia University Press, 1989.
40. G. P. Larwood, ed. *Extinction and survival in the fossil record.* Systematics Association Special Volume 24. Oxford: Clarendon Press, 1988.
41. C. Patterson and A. B. Smith. Is the periodicity of mass-extinctions a taxonomic artefact? *Nature* 330 (1987): 248–252.

Andrew H. Knoll and Malcolm R. Walter

LATEST PROTEROZOIC STRATIGRAPHY AND EARTH HISTORY

The end of the Proterozoic eon was a time of pronounced biological, biogeochemical, climatic and tectonic change. New bio- and chemostratigraphic data provide an improved framework for stratigraphic correlation, making possible a deeper understanding of latest Proterozoic Earth history and providing tools for a chronostratigraphic division of late Proterozoic time.

Like the Bible, Earth history is conventionally divided into two segments of markedly different length and character. Earth's "Old Testament" comprises the Archean and Proterozoic eons: the immense span of Precambrian time that extends from the earliest known rocks to about 540 million years (Myr) ago. It is separated from geology's "New Testament" (the Phanerozoic eon) by a discrete event: a unique radiation of eucoelomate animals that permanently changed the nature of ecological and sedimentary systems.

The diverse skeletons that mark the base of the Cambrian System have long presented paleontologists with an evolutionary problem of the first order. Darwin was keenly aware of it and urged that the sudden appearance of animals in the fossil record was a valid argument against his view of evolution by natural selection.[1] He speculated that the Precambrian earth "swarmed with living creatures," having in mind principally the animal ancestors of trilobites and other early Cambrian fauna. Darwin would no doubt be pleased by evidence that microbial life goes back 3,500 Myr or more.[2] But he would be unsettled by the lack of a long Proterozoic record of animal evolution; unequivocal metazoans occur only in the final 40 Myr of the eon, and most of the eucoelomate phyla that populate Phanerozoic seas have no Proterozoic record. The end of the Proterozoic eon was truly a period of remarkable evolutionary innovation. Less widely appreciated, these biological events are rooted in a broader framework of tectonic, biogeochemical and climatic change that altered the earth's surface and, in doing so, ushered in the Phanerozoic world.[3]

The idea that evolution and environmental change are intertwined is not new, but empirical support for Proterozoic interactions has been difficult to muster because it requires the reliable correlation of physical and biological events documented in the rock record. Students of Phanerozoic history can debate whether extinctions at the Cretaceous-Tertiary boundary took place in a geological instant, precipitated by the impact of a giant bolide. They can link Tertiary vegetational change in North America to climatic deterioration induced by the opening of the Southern Ocean, or ask about the biological consequences of oxygen depletion in late Devonian ocean basins. By contrast, it is not always easy to establish the synchroneity of Archean or Proterozoic events to within 100 Myr.

The far better resolution of Phanerozoic history does not stem from a denser framework of radiometric dates. As in the Proterozoic, geochronometry provides a coarse means of correlating Phanerozoic rocks.[4] Fossils and other chronostratigraphic indicators make the difference (see Box 1). The critical importance of chronostratigraphy is most clear in attempts to understand the great events of the late Proterozoic eon. Important advances have been made in unravelling regional geological histories using geochronometric data,[5] but many sedimentary successions are not easily dated, and even well-dated sections cannot be correlated with anything like biostratigraphic precision. In the absence of reliable tools for interbasinal correlation, only a broad understanding of late Proterozoic history has been attainable. Fortunately, the paleontological and geochemical records that document this history provide the raw materials for improved

BOX 1 Of time and stratigraphy

Geochronology is the discipline in which the ages of rocks are estimated. Estimates can be made in two different ways, and time divisions are, correspondingly, of two sorts. In the chronometric scale, the basic unit is the year and radiometric dating provides the means by which rocks are related to the scale. Chronostratigraphic scales are based on a series of reference points in rock successions, and stratigraphy permits the estimation of age. In the global chronostratigraphic scale, the beginning of each division is defined by a global stratotype section and point (a point in a stratigraphic section chosen by convention to define the boundary between the division in question and the preceding interval [popularly known as the "golden spike"]). Chronostratigraphic age can be known with certainty only in the stratotype section; the ages of all other strata are interpretations based on correlation, the use of the physical, chemical and/or biological features of rocks to establish their age relationships. Correlation may be based on overall characteristics (lithostratigraphy), the specific geometric relationships of sedimentary rock packages (sequence stratigraphy), fossils (biostratigraphy), chemical features (chemostratigraphy) and/or magnetic features (magnetostratigraphy).[65] Ideally, correlation relies on the recognition in different successions of geological signatures judged to be synchronous (of the same age wherever detected). In fact, many signatures, including most biostratigraphic and climatological events, are at least slightly diachronous; that is, they differ in age from place to place. The degree to which diachroneity compromises correlation depends on the stratigraphic resolution required. Like phylogenetic hypotheses in biology, hypotheses of correlation may be well or poorly corroborated. The well-known geological timescale for the Phanerozoic eon is a chronostratigraphic scale that has been integrated with the chronometric scale by means of geochronological calibration.

stratigraphic correlation. We argue here that, in conjunction with improved geochronometric data, protistan microfossils and isotope geochemistry are likely to provide the means by which the great physical and biological events of the latest Proterozoic eon can be correlated and, eventually, integrated.

Tradition in Neoproterozoic Stratigraphy

Thick sedimentary successions occur immediately beneath Lower Cambrian rocks in many places. Originally referred to as "Sinian" after exposures along the Yangtze River,[6] these rocks have conventionally been correlated on the basis of two features: glaciogenic sediments and Ediacaran animal remains.[4,7] Both define broad stratigraphic bands, but each is attended by significant uncertainties.

Paleozoology

Ediacaran animals comprise a morphologically distinctive fauna of architecturally simple, unskeletalized invertebrates known from over 24 localities on six continents[8] (Fig. 1b). There is general agreement that Ediacaran assemblages occur only within a discrete interval of latest Proterozoic time, but it is unclear whether all assemblages are strictly coeval, whether all or some taxa extend up to the Proterozoic-Cambrian boundary, or whether distinct biozones can be recognized within the interval. It is likely that assemblage zones will eventually be established, but for the present, such questions can be addressed most effectively using stratigraphic tools other than the fossils themselves. Taphonomic, paleoenvironmental and, possibly, biogeographical variations further complicate the picture in ways that remain poorly understood.[8] Perhaps most limiting, Ediacaran body fossils are relatively rare.

Trace fossils provide additional and largely independent evidence of Proterozoic animal evolution (Fig. 1c). The assemblages of simple tracks, traces and burrows found in Proterozoic rocks are distinct from those in basal Cambrian and younger deposits, and they seem to be recognizable globally in siliciclastic sediments.[9] Thus, trace fossils provide useful markers of the Proterozoic-Cambrian boundary and may allow as many as three uppermost Proterozoic biozones to be recognized[10–12] (Fig. 2).

Conventionally, the evolution of skeletons is seen as the defining event of the Proterozoic-Cambrian boundary, but in Namibia, carbonates interbedded with sandstones bearing Ediacaran fossils contain complex tubular skeletons made of calcite[13,14] (Fig. 1a). Indeed, architecturally similar skeletons occur throughout the world in carbonates that are definitely or probably of latest Proterozoic age.[14,15] The degree of stratigraphic overlap

Fig. 1 Latest Proterozoic metazoans. *a, Cloudina hartmanni,* a skeletalized animal from the Nama Group, Namibia, in longitudinal and transverse sections; ×5 (photograph by Stephen Grant). *b, Dickinsonia costata,* a nonskeletal, Ediacara-grade metazoan from the Pound Subgroup, Adelaide Geosyncline, South Australia; ×1.3 (photograph by R. J. F. Jenkins). *c, Planolites ballandus,* a trace fossil from the Elkera Formation, Georgina Basin, Australia; ×1.

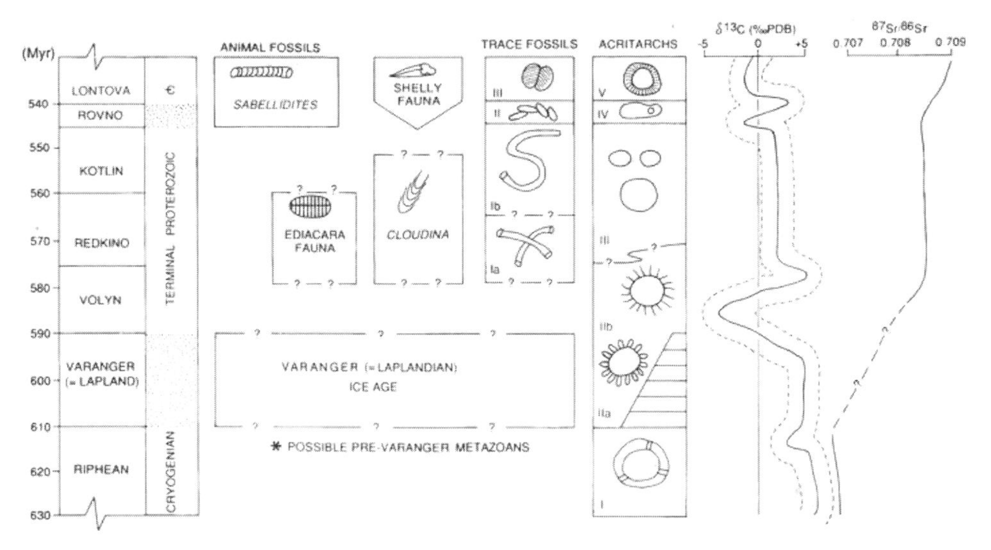

Fig. 2 Summary of terminal Proterozoic biostratigraphy and chemostratigraphy, showing the position of the Varanger ice age. Stratigraphic names are modified from Sokolov;[62] the shaded areas between periods indicate that global stratotype sections and boundary points have yet to be ratified by the International Union of Geological Sciences. Geochronometric estimates are best guesses, and are subject to change as data improve. The numbered vertical sequences of trace fossils and acritarchs indicate biostratigraphic zones that can be used for the intercontinental correlation of sedimentary successions. C and Sr isotopic curves are based on the analysis of carbonate rocks. See text for discussion and data sources.

between Ediacaran fossils and these early skeletons is uncertain. Chitinous sabelliditid worms and the first biomineralized shells of Cambrian aspect appear at or just below the Proterozoic-Cambrian boundary.[16]

Invertebrate animals provide a reliable guide to the broad recognition of uppermost Proterozoic strata, but because we can at present correlate using only a grade of evolution rather than well-defined species ranges, invertebrate biostratigraphy remains a fairly coarse means of correlating Neoproterozoic strata. (A recent report by Hofmann *et al.*[17] of possible metazoan impressions in older rocks opens up the possibility that an earlier stratigraphic interval may be defined by simple animals, but it does not alter the general conclusion.)

Neoproterozoic Ice Ages

Tillites and associated glaciogenic strata occur throughout the world in Upper Proterozoic successions.[18] In 1964, Harland[19] hypothesized that these record a great late Proterozoic ice age of global extent. It has now become clear that the tillites vary in age between about 850 and 590 Myr.[18]

Opinion has varied as to the discreteness of Neoproterozoic ice ages; at present, many would agree that there were at least two major glacial periods and probably more. Within each ice age, deposits are broadly but not necessarily strictly coeval. Of particular interest has been the Varanger (= Laplandian; ~610–590 Myr) ice age, probably the most widespread of the Neoproterozoic glaciations. Varanger tillites occur stratigraphically below all known diverse Ediacaran faunal assemblages. As with metazoan fossils, the extent of diachroneity among Neoproterozoic glaciogenic rocks must be evaluated using independent methods of correlation.

Stromatolites

Like invertebrates, microbial communities have left a record of trace fossils, in this case the distinctively laminated sedimentary structures known as stromatolites. Coarse patterns of temporal change have been recognized in the Proterozoic stromatolite record,[20] and recent observations continue to suggest that these structures can have a useful auxiliary role in stratigraphic correlation.[16,21–23] But there is no evidence that stromatolites can resolve the stratigraphic uncertainties noted above or match the resolution discussed below.[24]

Emerging Methodologies

Fossil Protists and Prokaryotes

Photosynthetic organisms are abundantly represented in uppermost Proterozoic rocks, but their stratigraphic potentials vary widely. Seaweeds can be morphologically complex, but their low probability of preservation results in a patchy record of uncertain stratigraphic usefulness. One possible exception may be the problematic organic ribbons known as vendotaenids, which regularly occur in uppermost Proterozoic sediments.[25] But even these define a broad stratigraphic band whose boundaries remain uncertain. Cyanobacteria and cyanobacteria-like microfossils are widely distributed in uppermost Proterozoic rocks, but virtually all have close similarities both to older fossil and to living taxa. Most distinctive is the helical *Obruchevella,* which has a distinct (but stratigraphically broad) acme in uppermost Proterozoic and Lower Cambrian rocks.

The most promising fossils for Neoproterozoic biostratigraphy are the acritarchs, organic-walled microfossils produced for the most part by phytoplanktonic protists. Acanthomorphic and other morphologically complex acritarchs have been useful in correlating Phanerozoic successions but, until recently, most paleontologists considered such fossils to be rare or absent from Proterozoic successions. Now, over 24 assemblages of complex acritarchs have been described from Neoproterozoic rocks, many of these fossils being extremely large relative to younger examples (Fig. 3). Although the total number of assemblages remains small, it is becoming clear that several biostratigraphic zones can be recognized (Fig.

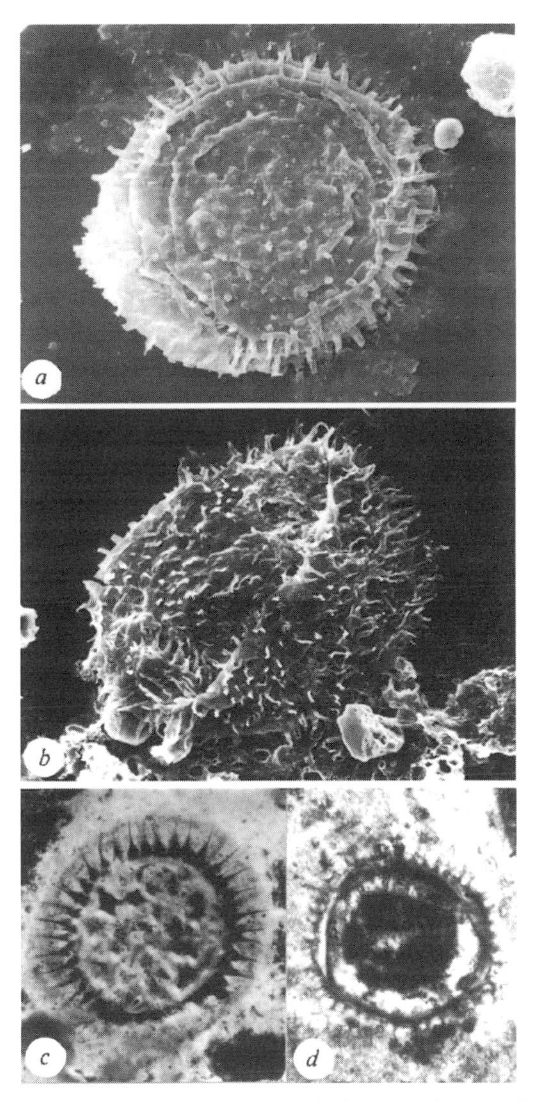

Fig. 3 Morphologically complex acritarchs from terminal Proterozoic successions. *a, b,* From the Pertatataka Formation, Amadeus Basin, Australia; vesicle diameters 155 and 200 µm, respectively (photographs by Zang Wenlong). *c,* From the Doushantuo Formation, China; vesicle diameter 200 µm (material from Zhang Yun). *d,* From greenschist-facies metasediments of the Scotia Formation, Prins Karls Froland, Svalbard; vesicle 155 µm.

2). Varanger and immediately post-Varanger acritarchs can be distinguished from earlier assemblages,[26–30] and further subdivision may be possible. Most species of large, morphologically complex acritarchs disappeared at about the time of the Ediacaran faunal radiation, leaving very latest Proterozoic assemblages characteristically simple and low in diversity.[16] Several distinctive taxa appeared near the very end of the Proterozoic, and a major radiation of acritarchs marks the beginning of the Cambrian.[16,31] Thus, microfossil assemblages provide an independent framework for evaluating the stratigraphic ranges of early animals and for correlating the many Neoproterozoic successions in which no metazoan fossils are known.

Isotopic Chemostratigraphy

Secular variations in the isotopic composition of seawater are useful in the correlation of Phanerozoic successions. For example, oxygen isotopic variation linked to the growth and decay of ice sheets has been critically important in unravelling Pleistocene history. A pronounced and essentially monotonic increase in $^{87}Sr/^{86}Sr$ over the past 100 Myr provides stratigraphic resolution that rivals biostratigraphy. More broadly, the C and Sr isotopic compositions of carbonates and the S isotopic composition of sulfates exhibit stratigraphically useful patterns of secular variation throughout the Phanerozoic.[32]

A common prejudice is that, although Phanerozoic carbonates may be little disturbed, their Proterozoic counterparts are almost invariably altered. There is little empirical evidence to support this supposition. On the contrary, a number of studies have demonstrated that depositional C and Sr isotopic signatures can be retained in micritic carbonates and oolites bound by fine-grained synsedimentary cements.[33–41] A principal finding is that these signatures varied markedly during Neoproterozoic time, providing a secular pattern of strong stratigraphic potential (Fig. 2). Detailed isotopic curves for the Neoproterozoic eon are being constructed from analyses of numerous rock samples collected at close intervals in individual measured sections, with sedimentological and petrographic control.

The carbon isotopic composition of carbonates and organic matter formed in seawater reflects both the degree of isotopic fractionation by photoautotrophs and carbon fluxes into and out of the ocean. Rates of organic carbon burial exert a primary influence on these ratios, with higher $\delta^{13}C$ indicating greater rates of burial; ocean mixing can also be significant. At the close of the Varanger ice age, both carbonate and organic carbon isotopic ratios plunged to distinctly light values ($\delta^{13}C_{CARB} = -2$ to -4% relative to the PDB Standard).[33–35] A subsequent positive excursion resulted in a short-lived peak of about +4 to +5%, after which $\delta^{13}C_{CARB}$ remained at about +2% in latest Proterozoic time.[35,36] Profiles from Siberia and elsewhere indicate another rapid shift to negative and then strongly positive values just below the Proterozoic-Cambrian boundary, with basal Cambrian $\delta^{13}C_{CARB}$ running to about −1% (ref. 42).

Fewer pre-Varanger profiles have been examined, but successions in Svalbard, East Greenland, northwestern Canada and Namibia show similarly distinctive patterns of secular variation.[33,35,40] Many 850–610 Myr carbonates contain carbon that is isotopically heavy ($\delta^{13}C_{CARB} \geq 5\%$) and matched by correspondingly heavy organic carbon; this signal is punctuated by several shorter intervals of isotopically light carbonate (about 0 to −2%). It might be argued that these negative excursions are of diagenetic origin; for example, carbonates exposed subaerially during low sea-level stands commonly become altered by isotopically light groundwaters.[43] But known isotopic excursions are associated with transgression rather than regression, and organic carbon signatures co-vary smoothly with the carbonates, an unlikely circumstance if the signals were produced diagenetically. Several and perhaps all of the negative excursions fall at the top of glaciogenic horizons.[35] It seems, therefore, that the marked isotopic variations of the interval 850–610 Myr reflect changes in the world ocean and as such provide a useful stratigraphic signal.

The Neoproterozoic carbon isotopic record thus forms an oscillating pattern that, like the Pleistocene oxygen isotope curve, must be paired with other stratigraphic tools to achieve its full potential. Accumulating evidence also makes it clear that despite the common preservation of near primary carbon isotopic ratios in Neoproterozoic carbonates and organic matter, every sample must be screened for possible diagenetic alteration.[37]

A second isotopic system of stratigraphic value is that of strontium. The Sr isotopic composition of seawater (and of carbonates precipitated from seawater) reflects the relative contributions of continental erosion (generally high $^{87}Sr/^{86}Sr$, but varying as a function of source rock) and hydrothermal input (typically low $^{87}Sr/^{86}Sr$). Like C isotopes, the Sr isotopic compositions of Neoproterozoic carbonates define a curve of great geochemical interest and stratigraphic usefulness (Fig. 2). As first argued by Veizer *et al.*,[44] $^{87}Sr/^{86}Sr$ is remarkably low in ~850–800 Myr carbonates; values as low as 0.7055 reflect a relative hydrothermal contribution greater than anything during the succeeding 800 Myr.[40] Ratios then increased to ~0.7070, where they remained until the Varanger glaciation.[39] By the end of the Varanger ice age, $^{87}Sr/^{86}Sr$ had increased to 0.7080, and by the time that Ediacaran metazoans diversified, the ratio stood at 0.7085 (refs. 36, 44). Strontium isotopic ratios remained at about that level until near the Proterozoic-Cambrian boundary, when another increase lifted $^{87}Sr/^{86}Sr$ to values approached only by the present.[36,45,46] Thus, secular change in the Sr isotopic composition of seawater provides another chemostratigraphic marker, and one whose stratigraphic sensitivity is potentially fine for certain periods, particularly the intervals near the Varanger glaciation and the Proterozoic-Cambrian boundary.

The isotopic composition of sulfur in Neoproterozoic sulfates is also of potential stratigraphic interest insofar as a major shift in $\delta^{34}S$ occurred near the end of the Proterozoic.[47] Unfortunately, the stratigraphic sampling of Neoproterozoic sulfates is patchy so that this pattern remains rather broadly defined. Sulfide S is ubiquitous in Neoproterozoic sedimentary rocks, but its isotopic composition reflects local sulfate availability and bears no simple relationship to seawater composition. This problem may potentially be overcome by limiting stratigraphic interpretation to those sulfides that reflect maximum isotopic fractionation (see, for example, ref. 48), but few systematic data are available.

Sequence Stratigraphy

Stratigraphers have long used transgressive and regressive facies in lithostratigraphic correlation. In recent years, a more rigorously articulated analysis of sedimentary sequences has provided potentially fine chronostratigraphic information for the correlation of Phanerozoic successions; sequence boundaries seem to be synchronous on a regional scale, at the resolution of other stratigraphic techniques.[49] Although it has been proposed that eustatic change of sea level exerts a primary influence on sequence development, the controls are more complex and include local tectonism.

Packages of genetically related sedimentary facies bounded by unconformities or their correlative conformable surfaces can be recognized, as well, in Proterozoic successions.[50–55] Because sequence boundaries share sets of common characteristics that recur through time, independent criteria must be used to enable (or at least support) the interregional correlation of sequences. There has been far too little research to allow the confident recognition of Proterozoic sequences likely to reflect eustatic sea-level change; however, there are intriguing hints. Glaciations inevitably lead to changes of sea level, and ice-age-related sequences that may well be of global significance have been recognized. For example, the sequence overlying the upper (Marinoan, probably equivalent to Varanger) glacial beds in the Adelaide Geosyncline, Australia, is lithologically distinctive[52] and has long been correlated across the continent (for example ref. 56) on the basis of stratigraphic position and a distinctive marker bed (the "upper marker cap dolomite"). Marker beds that are the product of unusual biological and/or environmental conditions may enable otherwise similar sequences to be distinguished.

A special complication arises in Proterozoic analyses because many historically important successions formed in intracratonic settings. In such settings, depositional space is available only intermittently, resulting in a very incomplete sedimentary record that may be difficult to correlate between regions.

Although it may be premature to correlate sequence boundaries among continents, sequence analysis within basins is necessary for the evaluation of bio- and chemostratigraphic data, because stratigraphic patterns in basins will be influenced by regional patterns of non-deposition or erosion, as well as by global secular changes.

Geochronometry

The International Commission on Stratigraphy recommends that newly selected stratotypes for system boundaries be amenable to radiometric dating. This is clearly desirable, as is the dating of significant events within systems, but it is rarely easy to achieve. The history of attempts to date the initial Cambrian boundary illustrates this. For the past 20 years, age estimates have ranged from 530 to 600 Myr, an uncertainty equal to the full duration of the Ordovician period.[57] The older estimates come from various techniques for the direct dating of sedimentary rocks, including K/Ar on glauconite and illite and Rb/Sr on shale, and to a lesser extent from K/Ar and Rb/Sr determinations on igneous rocks. Dating igneous rocks can provide precise and accurate results, but it can be difficult to establish the relationship of the dated event to sedimentary rocks of interest. The best lithologies for the calibration of Neoproterozoic time are volcanic ash beds, rocks that occur in a clear stratigraphic context and contain minerals such as zircon that can be dated precisely by U/Pb techniques.

Even zircon dating requires care, because zircon is a recalcitrant mineral that can be weathered from its host rock, redeposited and later metamorphosed, all without disturbing its isotopic composition and thus the age it will yield. Both sedimentary and igneous rocks commonly contain several generations of zircon grains, and these must be dated separately if a meaningful age is to be obtained. A powerful new approach has been made possible by the development of ion microprobes that can analyze small parts of single zircon grains.[58] Individual zircon populations can be identified, permitting meaningful and precise age determinations. Even so, if only a single zircon population is present in a sample, it can be difficult or impossible to determine whether it should be considered pristine or inherited, and thus what any resulting date might signify. Radiometric dates on igneous rocks may be supplemented usefully by U/Pb and/or Pb/Pb dates on early diagenetic phosphorites and carbonates.

At present, evidence favors a Proterozoic-Cambrian boundary at ~540 Myr[58–60] and a range for the Varanger ice age of about 610–590 Myr.[4,60]

A Timescale for the Neoproterozoic Era

The Phanerozoic geological timescale both reflects and is characterized by historical events, chiefly biological. By contrast, a recently ratified convention imposes a strictly chronometric basis for subdividing the Archean and Proterozoic eons.[61] This refinement of the timescale provides coarse stratigraphic tools, but the study of Earth history demands a well-honed scalpel. The significant biological, climatic and chemostratigraphic events of the Neoproterozoic era suggest that the chronostratigraphic basis of Phanerozoic correlation can be extended downward to cover at least the last 300 Myr of the Proterozoic eon.

In chronostratigraphic timescales, a division is defined by a point in a global stratotype (see Box 1). Boundary points are chosen by convention, but their placement depends on conceptions of what they are meant to bound and the prospects for correlation from the stratotype section to other parts of the globe. Several chronostratic divisions have been proposed for latest Proterozoic time. Sokolov[62] proposed a Vendian period, with an initial boundary originally based on a widespread transgressive sequence in Ukraine that bears an Ediacara fauna, but later broadened to include Varanger tillites. Jenkins[63] and Cloud and Glaessner[7] independently proposed an Ediacaran (or Ediacarian) period, with initial boundary stratotypes in South Australia placed to reflect the first appearance of Ediacaran metazoans and the postglacial (Marinoan, probably equivalent to Varanger) transgression, respectively. Harland and Herod[64] advocated a Vendian period containing an early Varanger and later Ediacara epoch (terminology as in ref. 4).

We agree that latest Proterozoic Earth history may best be characterized in terms of ice ages and animal fossils, but argue that the stratigraphic tools that will unlock this history are microfossils and stable isotopes. Used together, acritarchs, C isotopes and Sr isotopes permit correlation between Neoproterozoic successions in Svalbard and western Canada at the formation level.[40] Similarly fine correlations are proving possible among uppermost Proterozoic basins as widely scattered as Australia, Siberia, southern China, Africa and the lesser Himalaya. As paleontological and isotopic data are collected simultaneously and evaluated in light of the sequence stratigraphy of basins under consideration, Neoproterozoic correlation will come of age.

The potential for improved correlation has implications for the choice of stratotype section and boundary point for a terminal Proterozoic division. As noted above, several proposals place the beginning of this unit at the onset of Varanger tillite deposition. But the beginning of this ice age is difficult to identify on grounds other than the appearance of glaciogenic sediments. Given the uncertainties regarding tillite synchroneity, global correlation from a stratotype section so chosen is likely to prove difficult. By contrast, distinctive microfossil assemblages, a pronounced transgression, a major decrease in $\delta^{13}C$, and the rise of $^{87}Sr/^{86}Sr$ to 0.7080 all fall at or near to the end of the Varanger ice age. We therefore suggest that the initial boundary of a terminal Proterozoic division be defined by a boundary stratotype that reflects these attributes. This has the advantage of a chronostratic definition based on events that can be identified globally in a wide range of lithologies. The best stratotype section is yet to be determined, but candidates include the Adelaide Geosyncline, Australia; the Windermere Supergroup, northwestern Canada; the Doushantuo/Denying succession, China; the Krol succession in the lesser Himalaya; the Witlvei and Nama groups, Namibia; and several successions in Siberia. By definition, the end of this interval coincides with the beginning of the Cambrian period.

A second advantage of such a definition is that it would make the Varanger glaciation the final event of a previous division. Ice ages, iron formation, distinctive protistan fossils and stromatolites, unusual C isotope profiles, and unique Sr isotopic ratios collectively mark the time interval from ~850 to 590 Myr as one of the most distinctive in the past 2,000 Myr. The International Commission on Stratigraphy[60] has recognized a Cryogenian period defined chronometrically as beginning 850 Myr ago. This definition needs to be complemented by a chronostratic scale based on microfossils and isotopic geochemistry, with a major effort made to tie such data (as well as those of the terminal Proterozoic interval) into the broader chronometric framework of the Proterozoic timescale.

Conclusion

We stand at the edge of a new understanding of late Proterozoic Earth history. Previous research has demonstrated that this was a remarkable interval characterized by profound biological, tectonic and environmental events. The challenge now is to use new methods of stratigraphic correlation to integrate these disparate records so that we may learn how the foundations for Phanerozoic physical and biological evolution were established. The emerging stratigraphic framework promises a new exegesis of the eventful final chapters in Earth's Old Testament.

ANDREW H. KNOLL is at the Botanical Museum, Harvard University, Cambridge, Massachusetts 02138, USA; Malcolm R. Walter is at the School of Earth Sciences, Macquerie University, New South Wales 2109, Australia.

References

1. Darwin, C. R. *The Origin of Species* (Murray, London, 1859).
2. Schopf, J. W. (ed.) *Earth's Earliest Biosphere: Its Origin and Evolution* (Princeton Univ. Press, Princeton, NJ, 1983).
3. Knoll, A. H. *Scient. Am.* **265,** 65–73 (1991).
4. Harland, W. B. *et al. A Geological Time Scale 1989* (Cambridge Univ. Press, Cambridge, 1990).
5. Krogh, T. E., Strong. D. F., O'Brien, S. J. & Papezik, V. S. *Can. J. Earth Sci.* **25,** 442–453 (1988).
6. Wills B., Blackwelder, E. & Sargent, R. H. *Research in China* (Carnegie Institute of Washington, 1907).
7. Cloud, P. & Glaessner, M. F. *Science* **217,** 783–792 (1982).
8. Glaessner, M. F. *The Dawn of Animal Life* (Cambridge Univ. Press, Cambridge, 1984).
9. Crimes, T. P. *Geol. Mag.* **124,** 97–119 (1987).
10. Fedonkin, M. A. in *The Vendian System:* Vol. 1 *Paleontology* English edn. (eds. Sokolov, B. S. & Iwanowski, A. B.) 112–117 (Springer, Berlin, 1990).
11. Narbonne, G. M. & Myrow, P. *N.Y. St. Mus. Bull.* **463,** 72–76 (1988).
12. Walter, M. R., Elphinstone, R. & Heys, G. R. *Alcheringa* **13,** 209–256 (1989).
13. Germs, G. J. B. *Am. J. Sci.* **272,** 752–761 (1972).
14. Grant, S. *Am. J. Sci.* **290,** 261–294 (1990).
15. Conway Morris, S., Mattes, B. W. & Chen, M. *Am. J. Sci.* A **290,** 245–260 (1990).
16. Sokolov, B. S. & Iwanowski, A. B. (eds.) *The Vendian System:* Vol. 1 *Paleontology* English edn. (Springer, Berlin, 1990).
17. Hofmann, H., Narbonne, G. M. & Aitken, J. D. *Geology* **18,** 1199–1202 (1990).
18. Hambrey, M. B. & Harland. W. B. (eds.) *Earth's Pre-Pleistocene Glacial Record* (Cambridge Univ. Press, Cambridge, 1981).
19. Harland, W. B. *Geol. Rundschau* **54,** 45–61 (1964).
20. Walter, M. R. (ed.) *Stromatolites* (Elsevier, Amsterdam, 1976).
21. Bertrand-Sarfati, J. & Walter, M. R. *Precambrian Res.* **15,** 353–371 (1981).
22. Walter, M. R. & Heys, G. R. *Precambrian Res.* **29,** 149–175 (1985).
23. Hofmann, H. J. *Geosci, Canada* **14,** 135–154 (1987).

24. Khomentovsky, A. A. in *The Vendian System:* Vol. 2 *Regional Geology* English edn. (eds. Sokolov, B. S. & Fedonkin, M. A.) 102–183 (Springer, Berlin, 1990).
25. Gnilovskaya, M. B. (ed.) *Vendotaenids of the East European Platform* (Nauka, Leningrad, 1988).
26. Yin, L. *Palaeontologia Cathayana* **2,** 229–249 (1985).
27. Zang, W. & Walter, M. R. *Nature* **337,** 632–645 (1989).
28. Knoll, A. H. & Butterfield, N. J. *Nature* **337,** 602–603 (1989).
29. Vidal, G. *Palaeontology* **33,** 287–298 (1990).
30. Knoll, A. H. *Palaeontology* (in the press).
31. Moczydlowska, M. *Fossils Strata* **29,** 1–127 (1991).
32. Holser, W. T. in *Patterns of Change in Earth Evolution* (eds. Holland, H. D. & Trendall, A. F.) 123–143 (Springer, Berlin, 1984).
33. Knoll, A. H., Hayes, J. M., Kaufman, A. J., Swett, K. & Lambert, I. B. *Nature* **321,** 832–838 (1986).
34. Fairchild, I. R. & Spiro, B. *Sedimentology* **34,** 973–988 (1987).
35. Kaufman, A. J., Hayes, J. M., Knoll, A. H. & Germs, G. J. B. *Precambrian Res.* **49,** 301–327 (1991).
36. Lambert, I. B., Walter, M. R., Zang, W., Lu. S. & Ma, G. *Nature* **325,** 140–142 (1987).
37. Fairchild, I. R., Marhall, J. D. & Bertrand-Sarfati, J. *Am. J. Sci.* A **290,** 46–79 (1990).
38. Aharon, P., Schidlowski, M. & Singh, I. B. *Nature* **327,** 699–702 (1987).
39. Derry, L. A., Keto, L. S., Jacobsen, S. B., Knoll, A. H. & Swett, K. *Geochim. cosmochim. Acta* **53,** 2331–2339 (1989).
40. Asmerom, Y., Jacobsen, S. B., Knoll, A. H., Butterfield, N. J. & Swett, K. *Geochim. cosmochim. Acta* **55,** 2883–2894 (1991).
41. Derry, L. A., Kaufman, A. J. & Jacobsen, S. B. *Geochim. cosmochim. Acta* (in the press).
42. Kirschvink, J. L., Magaritz, M., Ripperdam, R. L., Zhuravlev, A. Y. & Rozanov, A. Y. *GSA Today* **1,** 69–91 (1991).
43. Beanus, M. A. & Knauth, L. P. *Geol. Soc. Am. Bull.* **96,** 737–745 (1985).
44. Veizer, J., Compston, W., Clauer, N. & Schidlowski, M. *Geochim. cosmochim. Acta* **47,** 295–302 (1983).
45. Donnelly, T. H., Shergold, J. H., Southgate, P. N. & Barnes, C. N. in *Phosphorite Research and Development* (ed. Notholt, A. J. G. & Jarvis, I.) *Geol. Soc. Spec. Publ.* **52,** 273–287 (1990).
46. Kaufman, A. J., Jacobsen, S. B. & Knoll, A. H. *Earth planet. Sci. Lett.* (in the press).
47. Claypool. G. E., Holser, W. T., Kaplan, I. R., Sakai, H. & Zak, I. *Chem. Geol.* **28,** 199–260 (1980).
48. Ross, G. M., Bloch, J. D. & Krouse, H. R. *Abstr. Geol. Assoc. Canada/Miner. Assoc. Canada Ann. Mtg.* **16,** 108 (1991).
49. Christie-Bick, N. *Mar. Geol.* **97,** 35–56 (1991).
50. Lindsay, J. F. *Am. Assoc. Petrol. Geol. Bull.* **71,** 1387–1403 (1987).
51. Lindsay, J. F. & Korsch, R. J. *Basin Res.* **2,** 3–25 (1989).
52. von der Borch, C. C., Christie-Blick, N. & Grady, A. E. *Austral. J. Earth Sci.* **35,** 59–72 (1988).
53. Christie-Blick, N., Grotzinger, J. P. & von der Borch, C. C. *Geology* **16,** 100–104 (1988).

54. Christie-Blick. N., von der Borch. C. C. & DiBona, P. A. *Am. J. Sci.* A **290,** 295–332 (1990).
55. Ross, G. M. & Murphy, D. C. *Geology* **16,** 139–143 (1988).
56. Preiss, W. V., Walter, M. R., Coats, R. P. & Wells, A. T. *BMR J. Aust. Geol. Geophys.* **3,** 43–53 (1978).
57. Cowie, J. W. & Harland, W. B. in *The Precambrian-Cambrian Boundary* (eds. Cowie, J. R. & Brasier, M. D.) 186–204 (Clarendon, Oxford, 1989).
58. Compston, W. *et al. Res. Sch. Earth Sci. Aust. Nat. Univ. A. Rep.* **1990,** 26–27 (1990).
59. Jenkins, R. J. F. *Geol. Mag.* **121,** 635–643 (1984).
60. Conway Morris, S. in *The Precambrian-Cambrian Boundary* (eds. Cowie, J. R. & Brasier, M. D.) 7–39 (Clarendon, Oxford, 1989).
61. Plumb, K. *Episodes* **14,** 139–140 (1991).
62. Sokolov, B. S. *Izv. Akad. nauk SSSR ser. geol.* **5,** 21–31 (1952).
63. Jenkins, R. J. F. *Trans. R. Soc. S. Aust.* **105,** 179–194 (1984).
64. Harland, W. B. & Herod, K. M. in *Ice Ages: Ancient and Modern* (eds. Wright, A. E. & Moseley, F.) 189–216 (Steel Horse, Liverpool, 1975).
65. Whittaker, A. *et al. J. geol. Soc. Lond.* **148,** 813–824 (1991).

Acknowledgments

We thank L. Smith, S. Grant, S. Conway Morris, P. Link, W. B. Harland, J. Veevers and R. Jenkins for comments. This work was supported by NASA, NSF and a Visiting Fellowship from Gonville and Caius College, Cambridge (A.H.K.); and the Australian Research Council (M.R.W.).

William A. Shear

THE EARLY DEVELOPMENT OF TERRESTRIAL ECOSYSTEMS

New work on the fossil record of early terrestrial ecosystems is challenging the validity of the accepted picture of their development. In particular, paleoecologists are learning of the dangers of drawing analogies between Paleozoic organisms and their modern counterparts, and are beginning to rely more on direct inferences from the fossils themselves.

The first terrestrial ecosystems are now thought to have been based on plants of a moss-like grade of organization, mats of algae, crusts of lichens and cyanobacteria, or even on plants unlike any alive today. These mats, crusts and clumps were probably colonized by small, air-breathing arthropods long before the arrival of vascular plants. In other words, some of the animal components of terrestrial ecosystems may have been on land waiting for the major primary producers of today to arrive. By the time vascular plants appeared, their productivity was available to animals only after the plants (or parts of them) had died. Carnivores and detritivores, not herbivores, dominated terrestrial animal communities for tens of millions of years before animals that could eat the living tissues of higher plants appeared. During the late Paleozoic, dominance at the level of primary production in terrestrial communities was evidently shared by a wider range of higher taxa than is now the case; the tree form was assumed by members of every major vascular plant clade.

The Origins of Terrestrial Life

Examining the colonization of the land by single species[1,2] illuminates the individual physiological adaptations developed in the process, but the pioneering of new habitats is always a function of a community of species. For this reason, the community will be the focus of this review.

Beerbower[3] has discussed in detail the conditions that may have been present on the abiotic continents. He emphasized the difference in the rate of chemical weathering and the release of mineral nutrients in the absence of macrophytic plants. A practical absence of organic detritus would have lowered the partial pressure of CO_2 in groundwater, reducing the production of carbonic acids. Fungi would have largely lacked a substrate, so even if present, they would have contributed to leaching only in a surface film.[4] Without leaf cover or root mats to slow the movement of water through the sediments, hydrological regimes would have been "flashy," with heavy flooding during storms, and rapid drying afterward.[5] The water in streams and lakes would have been low in mineral nutrients because of reduced chemical weathering, and would probably have carried a heavier load of suspended sediments. Lack of the binding effects of macrophyte cover would also have resulted in rapid physical erosion and instability of the surface. Temperatures at the surface and in the soil would have fluctuated violently with changes in insolation.

On a worldwide basis, the absence of evapotranspiration and a more uniform planetary albedo would have inhibited the transfer of atmospheric moisture, so that the interiors of continents would probably have had unstable rainfall regimes. But it seems that short-wave ultraviolet radiation, often cited as an environmental problem for early land life, would have been strongly reduced if the partial pressure of atmospheric oxygen were as little as 1% of the present value.[6] Long-wave ultraviolet, far less damaging to life, may still have presented problems in the early Paleozoic.

The most favorable environments for the invasion of land would have been coastal lowlands,[3] not only because of their proximity to the sources of colonists, but because most of the factors mentioned above would be ameliorated by low relief and proximity to large bodies of water. Such habitats were common from the Cambrian onward.

The earliest terrestrial communities were probably microbial crusts and mats,[7] which may have appeared on damp substrates even in the Precambrian, leaving as traces paleosols enriched in organic matter.[8] Intertidal cyanobacteria would have been preadapted for the formation of these crusts, as they would already have been able to resist alternate wetting and drying and fluctuating salinities. One well-studied extant microbial crust consists mainly of two species of cyanobacteria, one normally freshwater and the other marine; this example occurs in a desert environment, a strong indication of the potential adaptability of such communities.[8] Only regions unsuited for colonization by vascular plants are available for microbial crusts today, but such communities may have been varied and widespread from Precambrian times[9] until the appearance of macrophytes in the Ordovician or Silurian (see Box 1). Few well-studied fossil examples are available. Kobluk *et al.*[10] studied a karstic surface in the earliest Gedinnian (Devonian) in Ontario and suggested that algal, lichen or moss cover was present. The oldest known ascomycte fungi were found in Lud-

BOX 1 Divisions of geological time

Guide to some of the divisions of geological time mentioned in the text. The subdivisions of the Ordovician and Silurian periods are considered epochs, while those of the Devonian are ages. The epochs of the Devonian are prosaically named early (Gedinnian, Siegenian, and Emsian), middle (Eifelian and Givetian) and late (Frasnian and Famennian). The Namurian, Westphalian and Stephanian in the Carboniferous are each divided into A, B and C, going from oldest to youngest. The subperiods Mississippean and Pennsylvanian are used primarily in North America. The dates are subject to periodic revision and are in millions of years ago. The Brigantian (also mentioned in the text) is at the top of the Visean (latest Visean).

Palaeozoic	Permian		Late	248 Myr
			Early	258
	Carboniferous	Pennsylvanian	Stephanian	286
			Westphalian	296
			Namurian	316
		Mississippian	Viséan	333
			Tournaisian	352
	Devonian		Famennian	360
			Frasnian	367
			Givetian	374
			Eifelian	380
			Emsian	387
			Siegenian	394
			Gedinnian	401
	Silurian		Pridoli	408
			Ludlow	414
			Wenlock	421
			Llandovery	428
	Ordovician		Ashgill	438
			Caradoc	448
			Llandeilo	458
			Llanvirn	468
			Arenig	478
			Tremadoc	488
				505

lovian (Upper Silurian) rocks in Sweden.[11] Ascomycetes are components of all but a small percentage of lichen symbioses; lichens are important pioneers of abiotic environments. The Swedish specimens also included what were interpreted as microarthropod fecal pellets, and "animal-type" cuticles, indicating that early microbial/lichen communities, like their extant analogues, probably harbored populations of tiny herbivores, detritivores and their predators.[3]

Development of an organic soil (humus) under such mats would have been extremely slow, especially if large burrowing animals, which mix surface organic material and underlying mineral particles, were absent.[7] Burrows have been described in an Ordovician paleosol[12] which may be attributable to large annelids or arthropods; their activities would have aided in mixing and their feces would have enriched the soil in available nitrates and phosphates.[13] In addition to their support of animal communities, early mats of microbes and/or lichens would also have accelerated chemical weathering of mineral substrates. Oxalic and other acids produced by lichens and fungi are of great importance in dissolving minerals from rock.[14] The activity of these communities occurring in already favorable habitats would have—in the absence of disturbance—set up a closed positive feedback loop in which increased biological activity would have made the site even more suitable for life. Eventually, the accumulation of organic soil components and a greater range of animal life may have created conditions favorable for invasion by, or *in situ* development of, more complex photosynthetic forms. These forms may have been early eukaryotic green algae, charophytes, or even their close relatives, the embryophytic plants.

Embryophytes (liverworts, hornworts, mosses and vascular plants) may have appeared as early as the middle Ordovician.[15–18] The recent debate on the timing of the advent of such plants on land has been instructive. The old paradigm suggested that plants developed most of their terrestrial adaptations (conducting strands, cuticle, stomates, resistant spores, gametophyte and sporophyte life-cycle phases) synchronously and probably in an aquatic phase. The sudden appearance in the fossil record of vascular plants also engendered an emphasis on macrofossil evidence and a persistent confusion between the category "land plants," an ecological grouping, and vascular plants, a morphologically defined taxon.[19,20] The failure to find a conducting strand in early specimens of *Cooksonia,*[21] long regarded as the earliest vascular plant, opened the door to the more realistic model in which the main adaptations of land plants arose in at least two phases, if not in a mosaic fashion.[22,23]

Gray[16] has maintained that the initial preadaptation leading to the colonization of land was the development of the resistant spore, probably in a *Coleochaete*-like green alga. It now seems irrefutable that embryophytes sprang from such an ancestor, and that the vascular plants are paraphyletically nested within the group conventionally called the bryophytes, having the mosses as their closest relatives.[24]

The earliest meiospores attributed by Gray to embryophytes or embryophyte progenitors are found in Llanvirn-Llandeilo (middle Ordovician) sediments in Arabia.[18] They consist of obligate tetrahedral tetrads of

Fig. 1 Tetrad of meiospores from early Silurian rocks in Arabia. Scanning electron micrograph courtesy of Jane Gray.

smooth-walled spores, often enclosed in a common outer wall probably contributed by the mother cell (Fig. 1). Such spores are very close in form to those produced by living primitive liverworts.[25] The earliest macrofossils of liverworts are lowermost middle Devonian or late Silurian.[26] The ecology of certain extant bryophytes suggests that Ordovician or early Silurian hepaticoid plants may have been able to withstand severe environmental stress by means of an ephemeral lifestyle—colonizing a temporarily favorable habitat and maturing and producing propagules (resistant spores) before conditions deteriorated. Yet other bryophytes are extremely tolerant of desiccation, surviving for long periods of time in a dry state, to "come back to life" when moisture is once again available.[27] The production of tetrads of spores is an advantageous strategy for plants with unisexual gametophytes (which occur in living hepatics with obligate dispersed tetrads) and in which the sex of the gametophyte is controlled by a chromosomal mechanism. Each tetrad has the potential to produce two male and two female gametophytes, increasing chances for fertilization and providing some limit, however small, to inbreeding.[28]

In the later early Silurian, single dispersed spores (Fig. 2) appear for the first time.[17,18] These spores bear the characteristic triradiate mark found only in meiospores with a resistant wall, which are today known from less than 3% of living moss genera,[29] and from all vascular plants except those that flower under water and have secondarily lost the resistant wall.[17] Extant homosporous vascular plants (that is, ferns) have gametophytes which are at least potentially hermaphroditic and self-fertilizing. The sporophytes of such plants achieve rapid colonization and domination of a

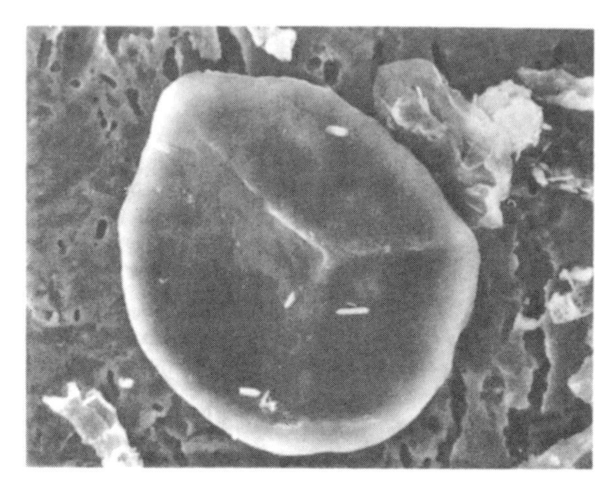

Fig. 2 Single dispersed spore with triradiate mark, isolated from Llandoverian sediments in Pennsylvania, USA; about 430 Myr old. Scanning electron micrograph courtesy of Jane Gray.

pristine habitat by vegetative reproduction; inbreeding to preserve adaptive complexes allowing survival in restrictive habitats would be advantageous. But as density increases, the environment may become more stabilized. As competition between individuals increases, so might the advantages of outbreeding, which would occur as a natural correlate of a higher density of gametophytes. The combination of rapid vegetative reproduction and the potential for greater heterozygosity would have allowed early vascular plants to dominate most environments rapidly, perhaps even by the early Silurian, driving the hepatics and other bryophyte-grade plants into their present limited ecological niches.[15–17] There was an increase in spore size during this interval, so that some of the single trilete spores found in the early Silurian are larger than earlier entire tetrads.[15] Small differences in spore size can make a great difference in dispersal potential, with large spores representing an adaptation to limited dispersal,[30] permitting a build-up of local populations in favorable habitats. With the origin of vascular tissue, organ differentiation in sporophytes was accelerated, particularly in the early and middle Devonian.[22] Around this time, vascular plants might also have formed mycorrhizal relationships that would have greatly increased their water and mineral absorbing capabilities.[31] Sporophyte-dominant lines took over the richest habitats; gametophyte-dominant plants remained in low-resource, stressful habitats.

Isolated scraps of peculiar, characteristic cuticle and both smooth and banded tubes that closely resemble vascular plant tracheids coincide at some sites with the spores Gray describes.[15,32,33] The tracheid is the most characteristic feature of a vascular plant, being a tubular cell, non-living at functional maturity, whose walls are strengthened by thick spiral or circular bands containing lignin. The cuticles, which first appear in middle Ordovician spore macerations,[15] continue with little variation at least into the later middle Devonian. Several distinct types of cuticle have been recognized,[33] and there is strong evidence[34] for association with enigmatic extinct plants known as nematophytes, whose bodies consisted of clusters of banded and smooth tubes, often of at least two distinct sizes. These plants seem to have been quite diverse, and included (in *Prototaxites*) the largest land organisms of their time, trunklike structures up to a meter in diameter and several meters long.[35,36] Quite unlike any extant plants but almost certainly terrestrial and photosynthetic, nematophytes were evidently important components of early land communities. Lignin or lignin-like compounds were isolated[37] from one of the earliest occurrences of nematophytes in the Llandovery (Silurian) of Virginia,[38] where they are found with land-plant spores. Such compounds have also been found in mosses and in the charophyte alga *Coleochaete,* thought to be the closest relative of embryophytes.[39] No *in situ* spores have as yet been associated with nematophytes.

Do at least some of Gray's spore tetrads, cuticles and banded tubes derive from nematophytes, or are they evidence of early vascular plants as yet unknown from macrofossils? The earliest land-plant macrofossil remains are the tiny (2–4 cm tall) fertile axes of *Cooksonia,* found in the Welsh Upper Wenlock (early in the late Silurian).[40] No tracheids have ever been demonstrated in these early finds, though short sterile axes with tracheids co-occur with *Cooksonia* in later, Ludlow, sediments.[41] By late Silurian time, diverse assemblages of land plants, including probable vascular plants, bryophytes, hepatic-like thalli and nematophytes are found.[26] This evident diversity is reflected in the dispersed spore record, with many new spore morphologies, including elaborations of the spore coat.[15,16]

The adaptations of vascular plants probably arose only after their ancestors were subjected to the strong selection pressures of the terrestrial environment.[15,16,22] The earliest of these pressures were nonbiological; only later did competition between plants become important, leading primarily to a "stature race," the taller winners gaining more access to light. Probably many of the plant species present in the late Silurian were ecologically interchangeable, with little distinction between species in habitat preferences.[42] Diversity in local areas was low, despite increasing taxonomic diversity; vegetated areas consisted of dense strands of one or a few species of plants.[43]

There is little evidence of animal life on land during this interval. Burrows from the Ordovician of Pennsylvania[12] remain a subject of debate. Could microbial mats or clumps of hepatics have provided a base for relatively large herbivores resident in burrows? But even the earliest microbial mats could have supported populations of animals occupying at least two higher trophic levels.[3] The coprolite evidence[11] from the late Silurian of Sweden is complemented by the scattered, rather poorly preserved fossils of millipedes and millipede-like animals from the Old Red Sandstone of Britain.[44] Recently paleozoologists have adopted some of the paleobotanical techniques used to isolate microfossils from shales—by dissolving the rock itself in hydrofluoric acid. Bits of plants and animals, consisting largely of highly reduced and polymerized organic matter, resist the acid and may be picked from the sample.[44] A newly reported fauna, as yet not studied in detail, has been recovered in this way from Ludlow rocks (early Upper Silurian; about 429 million years [Myr] old) in Wales.[45] This discovery is the oldest known for body fossils of terrestrial animals. Abundant animal remains had already been obtained by this means from a few Devonian localities (see below). Remarkably, the Ludlow fauna seems to share many evolutionarily advanced elements (trigonotarbids, centipedes) with the middle Devonian Gilboa fauna, 40–45 Myr younger.

Modernization in the Devonian

By the end of the Devonian, there existed terrestrial ecosystems that were at least architecturally modern.[42] Nearly all the main adaptations of vascular plants except flowers and fruit were complete, and all the main clades (lycopods, sphenopsids, ferns, seed plants) has emerged.[22,26,42,47] As abiotic selective factors continued to decline in importance in colonized habitats, competition between major clades of plants produced rapid architectural (and probably physiological) diversification. By the end of the Devonian, recognizable forests dominated by progymnosperm, lycopsid and sphenopsid trees, with a probable shrubby understory and ground cover of ferns, herbaceous lycopods and small seed plants, had appeared.[42] Competition would also lead to more sharply defined habitat preferences. Particular strategies seem, however, to have been characteristic of only one or a few species in each case, probably making late Devonian plant communities less stable than modern ones with greater "guild depth."[42] In addition, there is an emerging consensus regarding considerable biogeographic diversity in floras.[47]

Body fossil evidence of animals in a community setting comes from early and middle Devonian localities in North America and Europe. The oldest of these—the Rhynie Chert of Scotland—is Pragian, 400–410 Myr old.[48] As with all early to middle Devonian fossil faunas, it consists entirely of small arthropods. Much better understood and evidently far more diverse is the Givetian Gilboa fauna of New York, probably 30 Myr younger.[45] The faunal assemblage at Gilboa is strongly predator-dominated, both in terms of individuals and species (Fig. 3). The few non-predator species present were probably detritivores, and there is no evidence for direct consumption of live plant material. The predator populations must have supported themselves on abundant detritivore species, as in today's litter faunas, but there is no obvious taphonomic reason why the remains of the detritivores would not have been preserved. It may be that the hydrofluoric acid extraction selectively destroys some kinds of fossils, but this would be very hard to check. Although many long-extinct groups are represented, in functional terms the individual species are fully modern, and in a few cases it would be no surprise to find them among

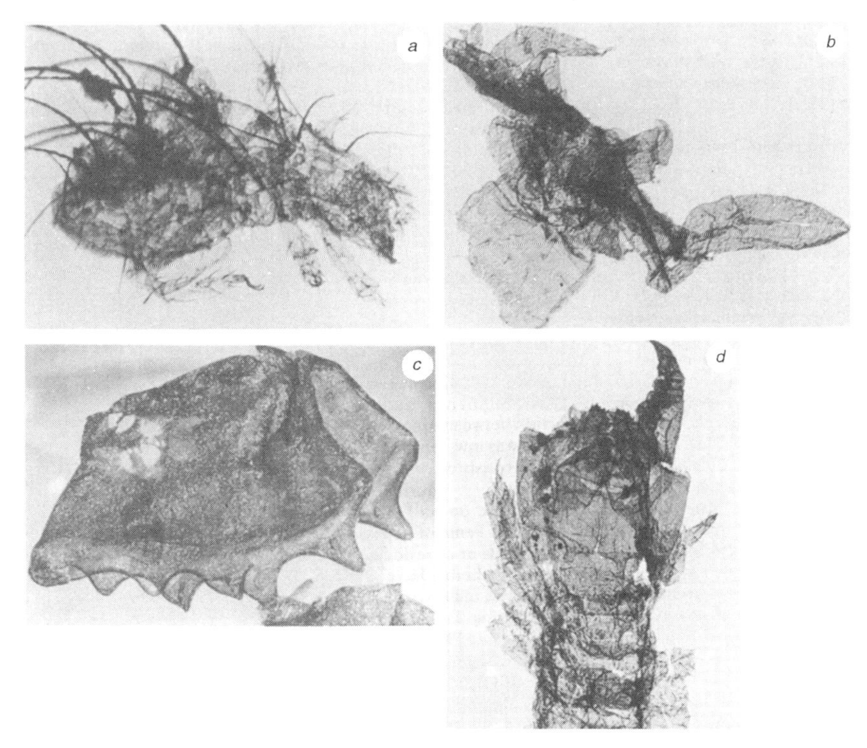

Fig. 3 Identifiable arthropod cuticles isolated from Middle Devonian rocks near Gilboa, New York. *a, Devonacarus sellnicki,* a primitive oribatid mite. *b, Dracochela deprehendor,* the earliest known false scorpion. *c,* Carapace of *Gelasinotarbus reticulatus,* a trigonotarbid arachnid, illuminated partially with incident light to show the slightly metallic luster of the Gilboa fossils. *d, Devonobius delta,* the earliest known centipede and the type of a new order of centipedes.

the extant litter fauna.[49–54] Given classic assumptions about rates of evolution, the modern aspect of these animals and the occurrence of similar forms in the late Silurian Ludlow fauna[46] suggests that their ancestors had already spent a long period adapting to the land habitat, and leads one to predict that steadily older, similar faunas should be found, extending back at least into the early Silurian. If, however, adaptation and diversification were initially rapid, followed by essential stasis and selective winnowing, the earliest invasions may have taken place as late as the middle Silurian.

It seems likely that transfer of primary productivity to higher trophic levels in early ecosystems based on algae, lichens and mosses or moss-like plants took place through microherbivores as well as detritivores. But with the dominance of vascular plants, the direct link between primary producers and consumers is very much weakened, if it does not disappear altogether.[42] A possible explanation for this decoupling of plant and animal worlds, which lasted even through the early Carboniferous, may lie in the structural molecule lignin, abundant only in vascular plants. The synthesis of this compound from phenolic precursors results in a number of toxic by-products, which, as plants are unable to excrete, must be stored in cell walls and dead tissues.[55,56] These poisonous secondary compounds, coupled with the low nutritive value of fresh vegetative parts of plants, may have deterred animal herbivory for a long time. Although initially fortuitous, the production of these toxins by plants would have responded to natural selection through refinement of their toxic effects, and the development of new biosynthetic pathways producing unrelated toxins. Only after plant litter had been partially autolyzed, and decomposed by bacteria and fungi, would it have been attacked by animals. The vegetative parts of plants are not nutrient-rich, being especially low in sodium and aromatic amino acids, and the complex carbohydrate polymers making up cell walls resist digestion by animals.[55] Hence the early emphasis on detritivory—feeding on plant matter that had been partially autolyzed and further broken down by microbes and fungi. The latter improve the nutritional quality of litter, and may have been the main foodstuff of many of the small arthropods.[11]

Thus, the modernization of the architecture of plant communities was not matched by the development of plant-animal interactions, largely absent through much of the middle to early late Paleozoic, nor by the evolution of animal food webs based on herbivory.[42] Among arthropods and vertebrates, two main strategies have made herbivory possible. The exploitation of microbial, protoctistal or fungal mutualists, both internal and external, may have provided some detoxification and greatly enhanced the available calorific and nutritional value of plant material.[56] Likewise, the consumption of more ephemeral plant parts such as spores, ovules and seeds, or young leaves, before they accumulated significant concentrations of toxins, was adopted early on by insects; the stored foods in reproductive structures make them (at least in calorific terms) a much richer food source than vegetative parts.[57] In addition, insects and some vertebrate herbivores possess enzyme complexes that act as general detoxification agents, and some species sequester plant toxins, even using them in their own defense.[55,56]

Insects and Tetrapods Rise to Dominance

Insects represent an extraordinarily successful line within the myriapod-hexapod assemblage. There is still no fossil evidence bearing on their origin. The sister group of the Insecta, Parainsecta (Collembola), first appears at Rhynie with two or more species so modern in aspect that they may be assignable to extant families;[58] obviously the origin of this group and its dichotomy with Insecta lies in the more distant past. Parainsectans are adapted for life in confined spaces; their entognathous jaws are enclosed between the labrum (upper lip) and side lobes of the head. Parainsects initially used hiding as their main defensive adaptation, whereas true insects relied upon jumping and running to escape predators.[57] Later, Insecta would split again along these lines and the entognath condition would arise once more independently. The first true insect, *Gaspea palaeoentognatha,* is known from a single specimen unaccompanied by any other terrestrial animal fossils. It occurred in paleobotanical macerates of Lower Devonian rocks from Quebec.[59] The lack of associated animal fossil, diagenetic changes and ambiguities in the anatomical interpretation of *Gaspea* have led to doubts about its being a fossil.[46] Remains of related animals, wingless archaeognathan machilids, occur at Gilboa.[45] Following these occurrences, there is a long gap covering most of the Devonian and early Carboniferous, until in the early Upper Carboniferous, winged insects seem suddenly to appear at a moderate, but rapidly increasing, level of diversity.[60] But few localities are known; the Namurian B deposit at Hagen-Vorhalle, Germany, now shows all palaeopterous orders as well as diverse hemipteroids, suggesting that the "explosion" of insect diversity in the late Carboniferous may only be apparent.[57] Any increase in insect diversity in the late Carboniferous and early Permian must have been linked to the increasing diversity of vascular plants during the same period, though whether the numbers of available niches expanded gradually or remained essentially stable after a rapid, early diversification remains to be determined for insects and Paleozoic plant life.[61,62] Shear and Kukalová-

Peck[57] favor the latter alternative; the palaeodictyopteroid orders remained stable through the Carboniferous and disappeared abruptly with the old Carboniferous flora of seed ferns, lycopod trees and cordaites.

Fossil evidence for Carboniferous insects feeding by chewing on living vegetative parts of plants is weak. Scott and Taylor[63] looked for reports of "bite marks" on *Neuropteris* leaves, and examined collections in the Field Museum (Chicago). They concluded that such marks were "quite common," but when they examined a collection of one hundred randomly chosen leaves from Pit 11 at Mazon Creek (late Carboniferous of Illinois), only four could be identified as bitten or chewed. The problem of demonstrating that the leaf, stem or wood was alive or part of a living plant when attacked further complicates such interpretations. How does one differentiate a bitten dead leaf from a bitten living leaf on the basis of fossil evidence? In analyzing stems, signs of healing or of reaction tissue are of aid. The same arguments apply to wood-boring (reported from the Carboniferous[64]) though, as a dead tissue, wood cannot be expected to heal. Leaf-mining, know only from a few fossil examples, may be more reliable.[62,65]

Among the earliest winged insects, members of the palaeodictyopteroid orders had specialized mouthparts used for tearing apart loosely organized, primitive cones (Fig. 4), or piercing ovules and sucking up the contents; their guts are sometimes crammed with spores[66] (Fig. 5). These short-lived plant parts would have been nutritious and low in toxins. A more complete analysis of insect mouthpart evolution, recently completed by Labandeira,[62] may shed more light on the early development of insect herbivory, but for the time being it seems unlikely that feeding on the leafy part of plants, at last by early Carboniferous insects, was widespread. But there is convincing indirect evidence that many Paleozoic hemipteroids fed by sucking: triangular stylet-like or bristle-like mouthparts and a pronounced, swollen pumping apparatus on the front of the head.[57,62] Perhaps the original foods of these groups included plant cell contents and semi-liquid decaying matter. If phloem was a target, they would have required a mechanism to eliminate the excess water.

Pollination, another insect-plant interaction, may also have originated in the Carboniferous,[57,62,67] though the evidence is circumstantial; medullosan seed fern pollen, for example, is often so large that wind pollination is not likely.[63] Megasecopteran and diaphanopteran insects also bore structures which, among other functions, may have served to collect and dispense pollen.[57]

Predatory insects of the Carboniferous were typified by the often gigantic Protodonata, relatives of today's dragonflies. They were probably spe-

Fig. 4 Reconstruction of *Homaloneura lehmani* feeding on a *Cordaites* cone. The wing-color pattern, frequently preserved in Paleozoic insect fossils, may have been disruptive or served for intraspecies signalling. Drawing courtesy of Jarmila Kukalová-Peck.

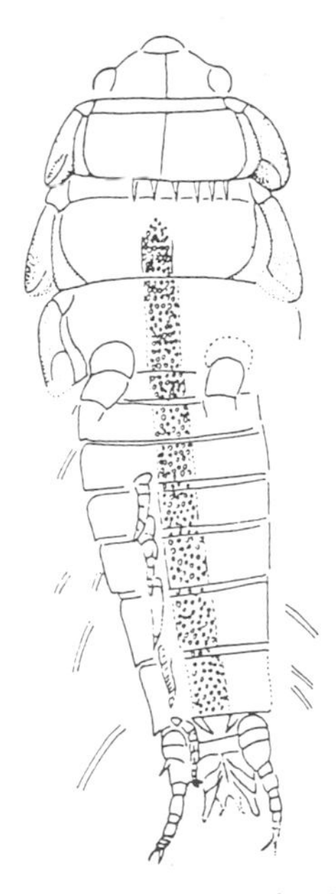

Fig. 5 Many Upper Carboniferous insects fed from plant fructifications or other ephemeral tissues, perhaps to avoid toxins, and because spores are relatively high in nutrients. Nymph of a fossil diaphanopteroid insect from the Upper Carboniferous of Illinois; the gut is packed with spores. Drawing courtesy of Jarmila Kukalová-Peck.

cialists in plucking other large insects from the trunks of lycopod trees in the coal swamps. Unable to fold their wings, many groups of Carboniferous insects would not have been able to walk through tangled vegetation and indeed would have been restricted to the open structure of the coal swamp "pole-forest." These predators and their prey were probably participants in a size-based arms race, as prey evolved to avoid predation by becoming larger, matched by increases in size on the part of the predator. By the mid-late Carboniferous, the largest insects that ever lived exhibited wingspans of more than 60 cm.[57]

The early evolution of communities of tetrapods is analogous to that of arthropods. The earliest evidence from body fossils appears in the late Devonian of Greenland, Australia, and the western Soviet Union.[68] The skeletons of *Ichthyostega* and *Acanthostega* from Greenland are already quite modern,[69] but show some characters that might exclude them from the ancestry of extant tetrapods.[70,71] *Acanthostega* in particular appears to have been entirely aquatic.[72] Ecologically, many of the diverse lines of amphibians that had appeared by the early Carboniferous, some of them very large animals, bridged aquatic and terrestrial environments. More terrestrially adapted forms (anthracosaurs and temnospondyls) also appeared very early. All the late Devonian and early Carboniferous tetrapods were predatory, feeding on fish, one another, and, most important for terrestrial ecosystems, insects. No potentially herbivorous tetrapods appear in the fossil records until the Stephanian (latest late Carboniferous), when functional morphology of the jaws of edaphosaurs (pelycosaurid reptiles) and diadectid amphibians suggests possible adaptations for feeding on vegetation, though large, hard seeds or shelled invertebrates would have been other possibilities. By the early Permian, the caseid pelycosaurs had developed an obvious suite of dental and other adaptations for feeding on vegetation.[73]

Undisputed amniotes (reptilomorphs) first appear in the Lower Carboniferous (Brigantian) of East Kirkton, Scotland;[73] there is insufficient evidence to allow them to be placed in trophic relationships, but later (Westphalian) forms were insectivores. Finally, at the beginning of the Upper Permian, the dicynodont therapsids show undisputed adaptations for herbivory.[73]

Thus the early tetrapods, like the earliest arthropods, were predators. Some of them retained extensive adaptations to life in water, and probably fed there, on fish or aquatic invertebrates. The more terrestrial forms were insectivores, so that detritivorous arthropods and potentially herbivorous insects remained, through most of the late Devonian and Carboniferous periods, the mechanism through which primary productivity by plants reached higher trophic levels. Gigantism among myriapods and a few other ground-dwelling arthropod taxa may have been an escalating response to tetrapod predation.[74]

Of the two groups, the insects doubtless had the greater effect on the evolution of plant life. A wider range of relationships was possible because of the diversity of insect feeding strategies[62,63] and their probable early specializaton on fructifications. Even in the late Devonian, closed strobili, hardened seed coats and protected micropyles had evolved to deter insect predation.[75] Most herbivorous vertebrates were and have remained relatively unspecialized as to diet, and consequently have had little evolutionary impact on plant diversification.[76]

Late Paleozoic Diversity and Disparity

An examination of the diversity of paleocommunities has become a standard exercise,[76,77] but often without the required caution as to the taxonomic level at which the diversity is assessed, and without taking into account the varying levels to which the taxonomy of extinct groups has developed.[78,79] In any case, systematic diversity is not synonymous with economical diversity; perhaps a more meaningful concept is the recognition of ecomorphotypes.[80,81] As adaptive groups emerge, they may be compared and contrasted with traditional taxonomic measures of diversity. Gould[82] and Briggs[83] have suggested that the range of body plan types and ecological strategies could be referred to as the "disparity" of the community.

Although such studies are still infrequent in the emerging subdiscipline of Paleozoic terrestrial ecology, what has been done, and what can be inferred from the taxonomic literature, implies strong differences in community organization in the late Paleozoic as compared to extant ecosystems.[42] Carboniferous coal swamp communities have been quantitatively studied through surveys of remains preserved in coal balls. DiMichele, Phillips and colleagues[84–90] have found that, whereas coal swamp communities were characterized by low species-level and ecomorphotype diversity, at high systematic levels an extraordinary range of plants was present. The tree form was assumed by species from every major group of vascular plants—lycopods, seed ferns, sphenopsids, ferns and gymnosperms—in contrast to forest communities of today, where one major taxon (angiosperm seed plants) dominates but shows extraordinary ecomorphotypic diversity.[42,80] Beginning in the Permian, this unusually high taxonomic diversity within ecomorphotypes began its decline,[42] and when angiosperms first appear in the Cretaceous, they rapidly come to dominate the vegetation at nearly every level.[75,91]

Candidates for similar analysis would be the insects and tetrapods of the Carboniferous. In insects, for example, ten extant orders and stem groups of nearly all of the others were present by the early Permian, in addition to several now extinct.[62] An extraordinary variety of amphibian and reptilomorph types has also been documented among the tetrapods.[73]

Other Possible Interpretations

We cannot, however, be complacent about our revised views of Paleozoic communities. For example, the case of two enigmatic plant fossils, *Baragwanathia* and *Longfengshania,* suggest that it may be necessary to reevaluate the accepted view of land-plant evolution, even if Gray's spore evidence from the Ordovician continues to gain wide acceptance. *Baragwanathia* is a large, advanced fossil lycopod, with well-developed leaves, an advanced type of vascular tissue, and sporangia in the leaf axils. Related plants are found abundantly in the Lower Devonian, but *Baragwanathia* occurs anomalously in the Upper Silurian of Australia, well before any other so highly evolved plants appear in the Northern Hemisphere.[22] While there is continuing debate about the dating of the rocks enclosing the *Baragwanathia* flora, it is still possible that vascular plants might have evolved first in Gondwana and reached the present Northern Hemisphere continents by a series of south to north migrations. The controversy over *Baragwanathia* emphasizes the importance of finding more fossil plant sites in Silurian rocks in all parts of the world.

Zhongying Zhang suggested that an abundant but enigmatic 880-Myr-old fossil from the Precambrian, *Longfengshania,* has many features in common with extant mosses and liverworts, as well as generally accepted Silurian and Devonian bryophyte fossils like *Tortilicaulis* and *Sporogonites.*[92] This fossil has previously been interpreted as algal, but Zhang's arguments raise new questions about the tetrads of trilete spores obtained by Volkova from the late Precambrian of Latvia. Gray and Boucot[93] have used Volkova's evidence to suggest an earlier, failed, Precambrian origin and radiation of land plants. New fossils or new interpretations of previously discovered fossils may significantly alter our views of both the timing and geography of terrestrialization.

Similarly, lycopod trees are perhaps the best known of all extinct plants. Yet they were entirely unlike the angiosperm trees of today, having more resemblance to giant weeds.[42] They warn us that generalizations drawn from modern life may give only sketchy or misleading clues about the life of the past, which must be approached on its own terms.

Future Research Directions

The most urgent need in Paleozoic ecological studies today is for more hard data in the form of detailed descriptions of fossils. As with Devonian plants, the repeated hashing over the same body of data has reached a point of diminishing returns. To carry this example further, far more information is required on the Devonian floras of the Southern Hemisphere and ancestral Gondwana,[94] where the presence of the enigmatic *Baragwanathia* hints at the development of advanced floras long before their appearance in the north. Studies are now under way that should aid in defining the elements of early terrestrial arthropod communities, as well as pushing back the earliest dates of occurrence into the middle and early Silurian. As yet, the discovery of few of these communities has been the result of deliberate searching in appropriate deposits; such examination should be rewarding.

As data accumulates, qualitative analyses elucidating adaptive complexes of species, community organization, and the structure of food webs can begin. DiMichele, Phillips and colleagues have pointed to the next step: where possible, quantitative examinations of community composition and the correlation of this information with different geological terrains and time horizons. Important generalizations, such as the impact of abiotic stress on ancient plant communities, have already emerged from their work.[95]

WILLIAM A. SHEAR is in the Department of Biology, Hampden-Sydney College, Hampden-Sydney, Virginia 23943, USA.

References

1. Little, C. *The Colonisation of Land: Origins and Adaptations of Terrestrial Animals* (Cambridge Univ. Press, Cambridge, 1983).
2. Little, C. *The Terrestrial Invasion: An Ecophysiological Approach to the Origins of Land Animals* (Cambridge Univ. Press, Cambridge, 1990).
3. Beerbower, R. in *Geological Factors and the Evolution of Plants* (ed. Tiffney, B. H.) 47–91 (Yale Univ. Press, New Haven, 1985).
4. Kubiena, W. L. *Micromorphological Features of Soil Geography* (Rutgers Univ. Press, New Brunswick, 1970).
5. Schumm, S. A. *Geol. Soc. Am. Bull.* **79,** 1573–1588 (1968).
6. Ratner, M. I. & Walker, J. C. G. *J. atmos. Sci.* **29,** 803–808 (1972).
7. Wright, V. P. *Phil. Trans. R. Soc.* B **309,** 143–145 (1985).
8. Campbell, S. E. *Origins of Life* **9,** 335–348 (1979).
9. Golubic, S. & Campbell, S. E. *Precambr. Res.* **8,** 201–217 (1979).

10. Kobluk, D. R., Pemberton, S. G., Karolyi, M. & Risk, M. J. *Bull. Can. Petrol. Geol.* **25,** 1157–1186 (1977).
11. Sherwood-Pike, M. & Gray, J. *Lethaia* **18,** 1–20 (1985).
12. Retallack, G. & Feakes, C. *Science* **235,** 61–63 (1987).
13. Jones, C. G. & Shachak, M. *Nature* **346,** 839–841 (1990).
14. Wilson, M. J. & Jones, D. *Geol. Soc. Spec. Publ.* **11,** 5–12 (1983).
15. Gray, J. *Spec. Pap. Geol.* **32,** 281–295 (1985).
16. Gray, J. *Phil. Trans. R. Soc.* B **309,** 167–192 (1985).
17. Gray, J. & Boucot, A. *Lethaia* **10,** 145–174 (1977).
18. Gray, J., Massa, D. & Boucot, A. *Geology* **10,** 197–201 (1982).
19. Axelrod, D. I. *Evolution* **13,** 264–275 (1959).
20. Banks, H. P. in *Evolution and Environment* (ed. Drake, E. T.) 73–107 (Yale Univ. Press, New Haven, 1968).
21. Edwards, D. & Fanning, U. *Phil. Trans. R. Soc.* B **309,** 147–165 (1985).
22. Knoll, A. H., Grant, S. W. F. & Tsao, J. W. in *Land Plants* (ed. Gastaldo, R. A.) 45–63 (Univ. Tennessee, Knoxville, 1986).
23. Edwards, D., Fanning, U. & Richardson, J. B. *Nature* **323,** 438–440 (1986).
24. Mishler, B. D. & Churchill, S. P. *Cladistics* **1,** 305–328 (1985).
25. Clarke, G. C. S. *Syst. Assoc. Spec. Vol.* **14,** 231–250 (1979).
26. Taylor, T. N. *Taxon* **37,** 805–833 (1988).
27. Scott, G. A. M. in *Bryophyte Ecology* (ed. Smith, A. J. E.) 105–122 (Chapman & Hall, London, 1982).
28. Wyatt, R. J. *Hattori Bot. Lab.* **52,** 179–198 (1982).
29. Erdtman, G. *An Introduction to Palynology, III* (Almqvist & Wiksell, Stockholm, 1965).
30. Mogensen, G. S. *Bryologist* **84,** 187–207 (1981).
31. Stubblefield, S. & Taylor, T. N. *New Phytol.* **108,** 3–25 (1980).
32. Edwards, D., & Rose, V. *Bot. J. Linn. Soc.* **88,** 35–54 (1984).
33. Edwards, D. *Bot. J. Linn. Soc.* **84,** 223–256 (1982).
34. Strother, P. K. *J. Paleontol.* **62,** 967–982 (1988).
35. Niklas, K. J. *Rev. Palaeobot. Palynol.* **22,** 1–17 (1976).
36. Schmid, R. *Science* **191,** 287–288 (1976).
37. Niklas, K. J. & Pratt, L. M. *Science* **209,** 396–398 (1980).
38. Pratt, L. M., Phillips, T. L. & Dennison, J. M. *Rev. Palaeobot. Palynol.* **25,** 121–149 (1978).
39. Delwiche, C. F., Graham, L. E. & Thomson, N. *Science* **245,** 399–401 (1989).
40. Edwards, D. & Feehan, J. *Nature* **287,** 41–42 (1980).
41. Edwards, D. & Davies, E. C. W. *Nature* **263,** 494–495 (1976).
42. DiMichele, W. A. *et al.* in *Evolutionary Paleoecology of Terrestrial Plants and Animals* (Univ. Chicago Press, in the press).
43. Gensel, P. G. & Andrews, H. N. *Am. Scient.* **75,** 478–489 (1987).
44. Almond, J. *Phil. Trans. R. Soc.* B **309,** 227–237 (1985).
45. Shear, W. A. *et al. Science* **224,** 492–494 (1984).
46. Jeram, A., Selden, P. A. & Edwards, D. *Science* **250,** 658–661 (1990).
47. Scheckler, S. E. in *Land Plants* (ed. Gastaldo, R. A.) 81–96 (Univ. Tennessee, Knoxville, 1986).
48. Hirst, S. *Ann. Mag. Nat. Hist.* **9,** 455–474 (1923).
49. Kethley, J. B., Norton, R. A., Bonamo, P. M. & Shear, W. A. *Micropaleontology* **35,** 367–373 (1989).

50. Norton, R. A., Bonamo, F. M., Grierson, J. G. & Shear, W. A. *J. Paleontol.* **62,** 259–269 (1988).
51. Shear, W. A. & Bonamo, P. M. *Am. Mus. Novit.* **2927,** 1–30 (1988).
52. Shear, W. A., Palmer, J. M., Coddington, J. A. & Bonamo, P. M. *Science* **246,** 479–481 (1989).
53. Shear, W. A., Schawaller, W. & Bonamo, P. M. *Nature* **341,** 527–529 (1989).
54. Shear, W. A., Selden, P. A., Rolfe, W. D. I., Bonamo, P. M. & Grierson, J. D. *Am. Mus. Novit.* **2901,** 1–74 (1984).
55. Weis, A. E. & Berenbaum, M. R. in *Plant-Animal Interactions* (ed. Abrahamson, W. G.) 123–162 (McGraw-Hill, New York, 1989).
56. Lindroth, R. L. in *Plant-Animal Interactions* (ed. Abrahamson, W. G.) 163–206 (McGraw-Hill, New York, 1989).
57. Shear, W. A. & Kukalová-Peck, J. *Can. J. Zool.* **68,** 1807–1834 (1990).
58. Greenslade, P. & Whalley, P. E. S. *2nd Internat. Symp. Apterygota* **1,** 319–323 (1986).
59. Labandeira, C. C., Beall, B. S. & Hueber, F. M. *Science* **242,** 913–916 (1988).
60. Wootton, R. J. *A. Rev. Ent.* **26,** 319–344 (1981).
61. Strong, D. R., Lawton, J. H. & Southwood, R. *Insects on Plants* (Harvard Univ. Press, Cambridge, 1984).
62. Labandeira, C. C. & Beall, B. S. in *Arthropod Paleobiology* (ed. Mikulic, D.) 214–256 (1990).
63. Scott, A. C. & Taylor, T. N. *Bot. Rev.* **49,** 259–307 (1983).
64. Cichan, M. A. & Taylor, T. N. *Palaeogeogr. Palaeoecol.* **39,** 123–127 (1982).
65. Rozefelds, A. C. *Proc. R. Soc. Queensl.* **99,** 77–81 (1988).
66. Kukalová-Peck, J. *Can. J. Zool.* **63,** 933–955 (1985).
67. Crepet, W. L. *BioScience* **29,** 102–108 (1979).
68. Milner, A. R., Smithson, A. C., Milner, M. I., Coates, M. I. & Rolfe, W. D. I. *Mod. Geol.* **10,** 1–28 (1986).
69. Clack, J. A. *Palaeontology* **31,** 699–724 (1988).
70. Panchen, A. L. & Smithson, T. R. in *The Phylogeny and Classification of the Tetrapods:* Vol. 1 *Amphibians, Reptiles, Birds* (ed. Benton, M. J.) 1–32 (Clarendon, Oxford, 1988).
71. Coates, M. I. & Clack, J. A. *Nature* **347,** 66–69 (1990).
72. Bendix-Almgreen, S. E., Clack, J. A. & Olsen, H. *Terra Nova* **2,** 131–137 (1990).
73. Smithson, T. R. *Nature* **342,** 676–678 (1989).
74. Vermeij, G. *Evolution and Escalation* (Princeton Univ. Press, New Haven, 1987).
75. Taylor, T. N. *Paleobotany* (McGraw-Hill, New York, 1981).
76. Wing, S. L. & Tiffney, B. H. *Rev. Palaeobot. Palynol.* **50,** 179–210 (1987).
77. Niklas, K. J. in *Patterns and Processes in the History of Life* (eds. Raup, D. M. & Jablonski, D.) 383–405 (Springer, Berlin, 1986).
78. Niklas, K. J., Tiffney, B. H. & Knoll, A. H. *Evol. Biol.* **12,** 1–89 (1980).
79. Smith, A. B. & Patterson, C. *Evol. Biol.* **23,** 127–216 (1988).
80. Wing, S. L. *et al.* in *Evolutionary Paleoecology of Terrestrial Plants and Animals* (Univ. Chicago Press, in the press).
81. Damuth, J. in *Evolutionary Paleoecology of Terrestrial Plants and Animals* (Univ. Chicago Press, in the press).
82. Gould, S. J. *Wonderful Life* (Norton, New York, 1989).
83. Briggs, D. E. G. in *Arthropod Paleobiology* (ed. Mikulic, D.) 24–43 (Univ. Tennessee Press, Knoxville, 1990).

84. DiMichele, W. A. & Phillips, T. L. *Rev. Palaeobot. Palynol.* **44,** 1–26 (1985).
85. DiMichele, W. A. & DeMaris, P. J. *Palaios* **2,** 146–157 (1987).
86. DiMichele, W. A. & Phillips, T. L. *Rev. Palaeobot. Palynol.* **52,** 115–132 (1988).
87. DiMichele, W. A., Phillips, T. L. & Peppers, R. A. in *Geological Factors and the Evolution of Plants* (ed. Tiffney, B. H.) 223–256 (Yale Univ. Press, New Haven, 1985).
88. Phillips, T. L. *Rev. Palaeobot. Palynol.* **27,** 239–289 (1979).
89. Phillips, T. L., Peppers, R. A. & DiMichele, W. A. *Int. J. Coal Geol.* **5,** 43–109 (1985).
90. Phillips, T. L, Peppers, R. A., Acvin, M. J. & Laugnan, P. J. *Science* **184,** 1367–1369 (1974).
91. Fredricksen, N. D. *Geosci. Man* **4,** 17–28 (1972).
92. Zhang, Z. *28th Internat. Congr. Geol., Abst.* **3,** 441 (1989).
93. Gray, J. & Boucot, A. *Geology* **6,** 489–492 (1978).
94. White, M. E. *Am. Scient.* **78,** 252–262 (1990).
95. DiMichele, W. A., Phillips, T. L. & Olmstead, R. G. *Rev. Palaeobot. Palynol.* **50,** 151–178 (1986).

Acknowledgments

My collaboration with P. Bonamo provided the impetus for this review. I also thank J. Gray, P. Selden. J. Kukalová-Peck, P. Gensel, J. A. Clack and W. D. I. Rolfe for discussion. W. DiMichele, R. Gastaldo, R. Beerbower, J. Boy, N. Hotton, R. Hook, T. Phillips, S. Scheckler and H.-D. Sues participated in the Evolution of Terrestrial Ecosystems Conference at Airlie, Virginia, in 1987, and with me are contributors to chapters in the forthcoming book, *Evolutionary Paleoecology of Terrestrial Plants and Animals.* W. DiMichele permitted citation of this book, where many of the syntheses and ideas used in this review were first developed at length. This paper was written while my research was supported by the NSF and the Power Authority of the State of New York.

Douglas H. Erwin

THE PERMO-TRIASSIC EXTINCTION

The end-Permian mass extinction brought the Paleozoic great experiment in marine life to a close during an interval of intense climatic, tectonic and geochemical change. Improved knowledge of latest Permian faunas, coupled with recent advances in isotopic studies and biostratigraphy, have greatly enhanced our understanding of the events of 250 million years ago and have begun to provide answers to many questions about the causes of extinction.

Killing over 90% of the species in the oceans[1,2] and about 70% of vertebrate families on land[3–5] is remarkably difficult. The end-Permian mass extinction was the closest metazoans have come to being exterminated during the past 600 million years (Myr). The effects of this extinction are with us still, for it changed the structure and composition of marine communities far more than any event since the Cambrian radiation. The end-Permian extinction brought the world of the Paleozoic to a close, permitting the expansion of new marine community types which continue to dominate modern oceans. This event plays a critical role in debates over the nature of mass extinctions and their role in structuring the evolution of life.

By 1840, Phillips[6] recognized that the history of macroscopic life on Earth could be divided into three great eras, which he named the Paleozoic, Mesozoic and Cenozoic; he also recognized two dramatic drops in diversity, each associated with the appearance of new types of organisms. Yet until well into this century, paleontologists paid little attention to these events. Since publication of the impact hypothesis,[7] interest has focused on the Cretaceous/Tertiary mass extinctions and other events with a possible extraterrestrial cause. In part this stems from the widespread, if erroneous, view that the record across the Permo-Triassic boundary is too poor for detailed study, a presumption based on apparent evidence for a widespread marine regression and depositional hiatus at the boundary.

The pattern of disappearances across the Permo-Triassic boundary is complex, with some clades disappearing well below the boundary, others quite diverse right up to the boundary, and still others seemingly oblivious to the extinction.[2,8] Analyzing such patterns is complicated by distortions including the regression, backward smearing of true last occurrences,[9] and many Lazarus taxa. Lazarus taxa[10] disappear from the record during the late Permian, but did not become extinct, for they reappear in middle Triassic rocks. First recognized in gastropods,[11] the phenomenon is widespread among bivalves, brachiopods and other taxa, demonstrating both the extent of sampling problems and the importance of undiscovered refugia in preserving many lineages.

Detailed paleontological, geochemical and sedimentological studies of sections in Italy, Pakistan and South China are revealing a rich and complex pattern of events. New correlations based on conodonts permit more precise and more accurate intercontinental correlations and comparison of extinction patterns among regions. These new data have, in turn, led to the re-examination of older hypotheses about the causes of the extinction and the proposal of additional scenarios.

The cause of the end-Permian mass extinction appears to involve a tangled web rather than a single mechanism. Three phases can be identified. The first began with the onset of the marine regression which dried out many marine basins, reduced habitat area and increased climatic variability. The regression accelerated during the second phase, triggering the release of gas hydrates and the erosion and oxidation of marine carbon. In conjunction with the eruption of the Siberian flood basalts, these carbon sources exacerbated climatic instability; an increase in atmospheric carbon dioxide may have produced oceanic anoxia and global warming. The final phase of the extinction involved the destruction of near-shore terrestrial habitats during the rapid earliest Triassic marine transgression.

Marine Extinctions

The marine fossil record provides the most complete and detailed record of the extinction. Examining the temporal, biogeographical and ecological pattern of extinction allows evaluation of proposed extinction scenarios, all of which ultimately requires high-resolution biostratigraphic correlation between the sections spanning the Permo-Triassic boundary.

Boundary Sections

The Permo-Triassic boundary was traditionally recognized by the first occurrence of the ammonoid *Otoceras woodwardi* and the more easily identifiable bivalve *Claraia.* Correlating between the primary boundary

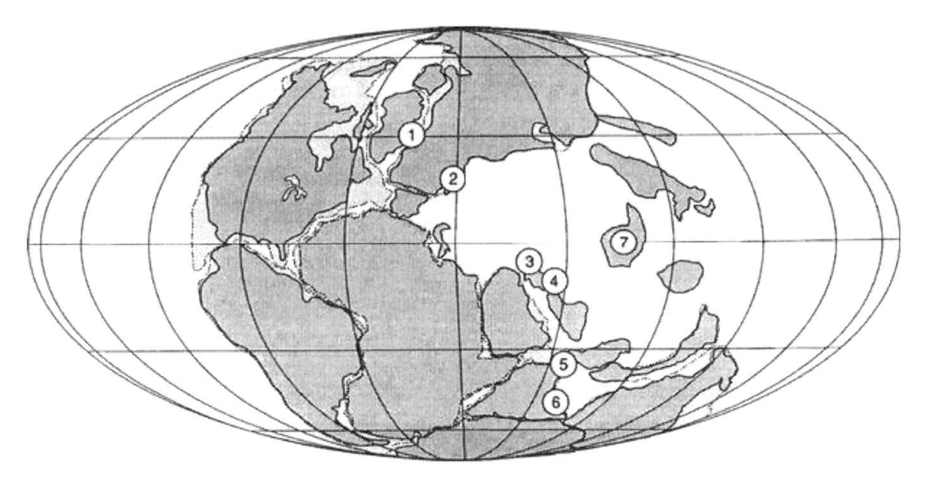

Fig. 1 Continental positions in the late Permian, showing the location of important Permo-Triassic boundary sections. The number and continuity of individual sections in the southern Alps, Iran and South China belies earlier claims for a worldwide period of nondeposition spanning the Permo-Triassic boundary. (1) Greenland; (2) southern Alps; (3) Iran-Armenia border (Kuh-e-Ali Bashi section); (4) central Iran; (5) Salt Range, Pakistan; (6) Guryl Ravine, Kashmir; (7) numerous sections in South China.

sections (Fig. 1) has been difficult, but correlations based on conodonts have resolved these problems, although they have dramatically altered the relationships among boundary sections (reviewed in refs. 2, 8, 12 and 13, but see ref. 14) (Fig. 2). These results suggest *O. woodwardi* first appeared during latest Permian boreal, cold-water faunas.[13] In contrast, *Otoceras* is absent from contemporary warm-water faunas of the Tethyan realm, including the magnificent Changxingian faunas of South China. Moreover, these correlations offer the prospect of the first global, high-resolution analysis of extinction and survival patterns.

Contrary to earlier claims,[15] recent studies of the Permian Bellerophon Formation and the overlying Werfen Formation in the Alps indicate that this classic sequence preserves a detailed paleontological[16] and geochemical[17] record of the extinction interval. The boundary lies within the Tesero Horizon at the base of the Werfen Formation, which contains a fauna transitional between the Permian and Triassic, similar to mixed faunas from South China.[18] A gradual shift in carbon isotopes ($\delta^{13}C$) from +3% to −1% is associated with the extinction horizon.[17] A curious spike in spores of the fungus *Tympanicysta*[19] may indicate widespread collapse and decay of terrestrial ecosystems.

The widespread Permo-Triassic boundary deposits in South China[20–24] contain a diverse fauna. In the latest Permian Changxing Formation, 435 of 476 invertebrate species, or 91%, disappear (ref. 24, cited in ref. 25),

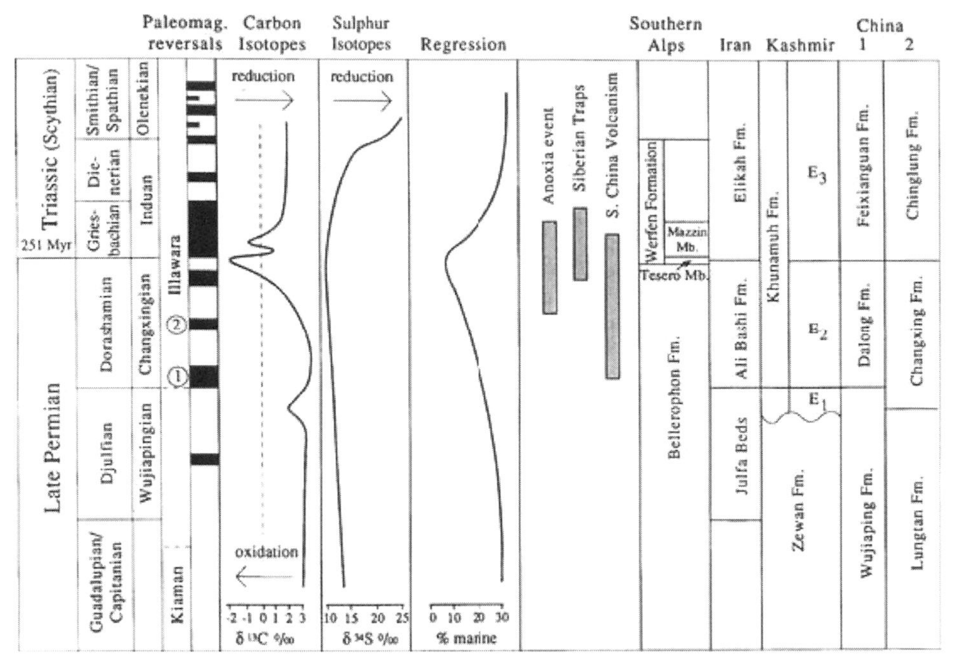

Fig. 2 Permo-Triassic geological events and correlation between major boundary sections. Standard geological stages are shown to left, with two alternative names given for latest Permian. Zircons from the boundary clay at Meishan, China, provide a date for the Permo-Triassic boundary of 251.2 ± 3.4 Myr ago.[72] The end of the Kiaman Reversed Superchron and the onset of the mixed polarity interval at the base of the Illawara Superchron is not well delineated. It has been placed near the base of the Changxingian stage,[73,74] but other work suggests it may have occurred as early as the latest Capitanian (M. Menning, personal communication). *Black* is normal magnetic polarity; *white* is reversed polarity. *1, Neogondollela orientalis* zone; *2,* base of the *N. changxingensis* zone. Shifts in carbon isotopes[17] and sulfur isotopes[50] are schematic. Strontium isotopes show a similar pattern to the carbon isotopes, declining to the lowest value of the Phanerozoic near the boundary, indicating an influx of light strontium from juvenile basalts, generally on mid-ocean ridges. Other significant events include the eruption of the massive Siberian flood basalts, a lengthy (1–3 Myr) anoxia event in deep-water marine rocks in Japan (Y. Isozaki, personal communication), a widespread pyroclastic volcanism in South China. Correlation between boundary sections in the southern Alps; Kuh-e-Ali Bashi, northeastern Iran; Guryl Ravine, Kashmir; and Shangsi *(1)* and Meishan *(2)* in southern China are based on conodont analyses.[12]

including 98% of ammonoid species, 85% of bivalve species and 75% of the shallow-water fusulinid foraminifera. However, taxonomic inconsistencies and lack of Lower Triassic sediments may overstate the magnitude of the extinction. Furthermore, in contrast to the gradual shift in $\delta^{13}C$ in the Alps, sections of South China record a relatively abrupt shift,[26] suggesting that the Chinese sections are condensed relative to those in the Alps. This may account for claims of a catastrophic extinction in the Chinese sections.[27] The youngest known Paleozoic reefs are found in the

Changxing Formation, in Sichuan and Hubei Provinces.[28] Thus a diverse Permian reef fauna disappears before the boundary as evaporites develop, but their presence belies claims of prolonged environmental deterioration before the Permo-Triassic boundary, at least in South China.

Clay layers, apparently of volcanic origin, occur throughout the latest Permian Changxing Formation, and a particularly thick clay marks the basal Triassic. A tuffaceous texture, bipyramidal quartz, volcanic shards and geochemistry all indicate an explosive silica-rich volcanic source for the ash,[29–31] probably associated with a subduction zone. This ash covers $>10^6$ km^2 (ref. 30), representing eruption of 1,000–4,000 km^3 of material.[31] This volcanism has been linked to the extinction,[20,24,29,30] but is no larger than many other pyroclastic eruptions, with no discernable biological effects.[32] Reports of iridium enrichment within boundary clays[25,33,34] cannot be replicated,[30,35,36] arguing against the kind of impact associated with the Cretaceous-Tertiary mass extinction. The lack of an impact signature also raises questions about the nature of the cause of the apparently periodic mass extinctions stretching from the late Permian into the Miocene.[37] Immediately overlying the boundary clay are a series of characteristic transitional beds containing three distinctive mixed faunas of both Permian and Triassic fossils;[23] remixing after deposition has been ruled out as a cause of this assemblage.[38]

Extinction Patterns

Global diversity of durably skeletonized marine families declined from 536 to 267 (49%) between the Capitanian and the end of the Permian; about 72% of corresponding genera disappeared.[37,39–41] In comparison, only 57% of genera disappeared during the end-Ordovician mass extinction, the second largest of the Phanerozoic.[37] Rarefaction analysis suggests that global species extinction was >90% (ref. 1), although the magnitude and pattern varied considerably among different clades (Figs. 3, 4). This included a 79% extinction among families from Sepkoski's Paleozoic evolutionary fauna (predominantly epifaunal, suspension feeders, including Paleozoic corals, articulate brachiopods, stenolaemate bryozoans and stalked echinoderms). In contrast, the modern evolutionary fauna (gastropods, bivalves, echinoids) declined by only 27% (ref. 41).

The reliability of the patterns shown in Fig. 3 is unclear. For example, the decline of the foraminiferid suborder Fusulinina was thought to have begun by mid-Permian times. Yet fusulinids are diverse in the Changxing Formation before disappearing rapidly near the boundary. Non-fusulinid forams, which lived in deeper water, experienced only slight extinction,

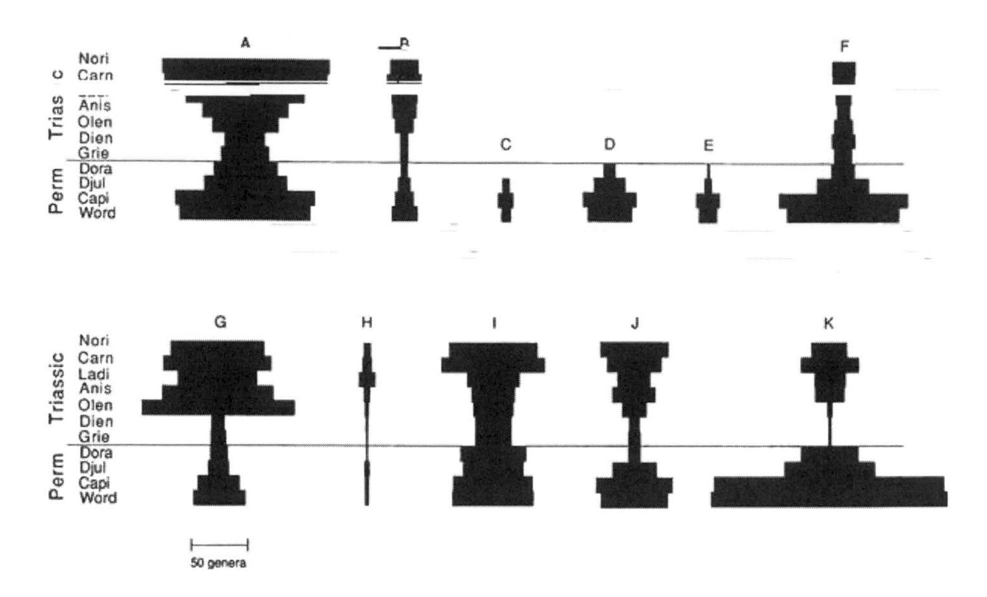

Fig. 3 Generic diversity patterns among marine invertebrates from the mid-Permian through to the Triassic. All stages are of equal width, although they range in duration from 1–2 Myr for the Djulfian, Dorashamian and Griesbachian to ~13 Myr for the Norian. There is great variability in extinction-survival-recovery pattern. Nautiloids *(B)*, sponges *(I)*, non-fusulinid foraminifera, conodonts and some gastropods (none shown) were almost oblivious to the extinction; although some species disappeared, the level of extinction was far less than among other taxa. A second group appears to decline together with the marine regression. These include the tabulate and rugose corals *(C, D)*, the trilobites *(E)*, the bryozoa *(F)*, the articulate brachiopods *(J, K)* and the crinoids and blastoids, although the poor fossil record of the last two during the Permian makes this pattern suspect. This pattern is partly an artefact of the declining quality of the fossil record, however, because many of these taxa, including fusulinid foraminifera, articulate brachiopods, ammonoids *(G)* and several gastropod groups exhibit high diversity in latest Permian rocks, particularly in South China, but decline rapidly near the boundary. *Word,* Wordian; *Capi,* Capitanian; *Djul,* Djulfian; *Dora,* Dorashamian (= Changxingian); *Grie,* Griesbachian; *Olen,* Olenekian; *Anis,* Anisian; *Ladi,* Ladinian; *Carn,* Carnian; *Nori,* Norian. *A,* Bivalvia; *B,* Nautiloidea; *C,* Tabulata; *D,* Rugosa; *E,* Trilobita; *F,* Bryozoa; *G,* Ammonoidea; *H,* Echinoidae; *I,* Porifera; *J,* Rhynchonellida + Terebratulida (brachiopods); *K,* all other articulate brachiopods; mostly spiriferids in the Triassic. These data have not been corrected for the latest conodont correlations and suffer from the deficiencies of such broad-scale compendia. Nonetheless, they provide the best available overview of extinction and survival patterns. The new conodont correlations[12] will permit higher-resolution analysis of the immediate boundary interval. Data provided by J. J. Sepkoski Jr.

although architecturally more complex taxa seem to suffer more than less complex forms across all suborders.[42] Does this mean the apparently gradual extinction simply reflected the loss of shallow marine sediments outside of South China? Perhaps, but South China was an isolated tectonic block during the latest Permian (Fig. 1). It may have served as a refugium for groups already declining on Pangea. Paleontologists do not yet have the high-resolution data on global extinction and survival patterns that

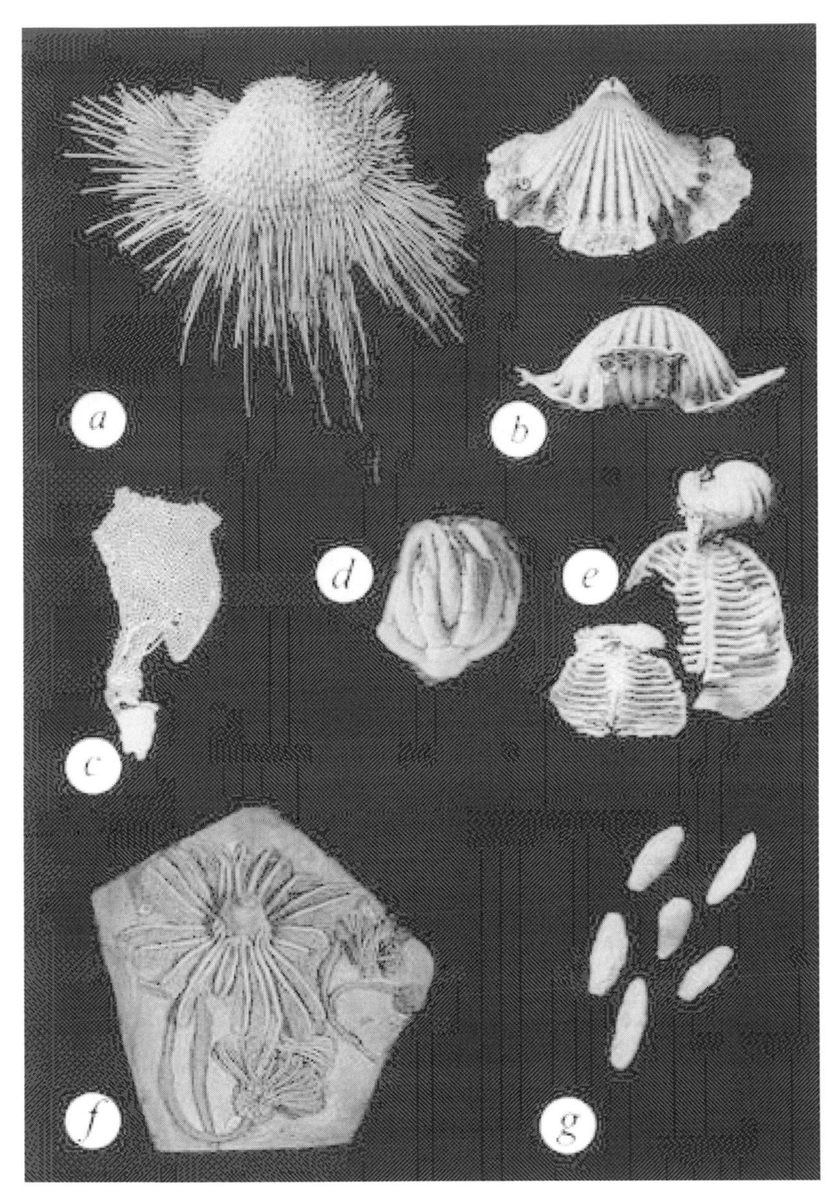

Fig. 4 Representative late Paleozoic marine invertebrates. Brachiopods: *a, Waagenochoncha albichi;* ×1.0. Khisor Range, Pakistan. Late Permian. *b, Stenoscisma venustum;* ×1.0. Texas. Lower Permian. *c,* The bryozoan *Fensetrella* sp.; ×0.9. *d,* An unusual crinoid, *Timorocrinus multicostatus;* ×1.0. Timor. ?Mid-Permian. *e,* An unusual articulate brachiopod, *Leptotus nobilis;* ×0.5. Khisor Range, Pakistan. *f,* The Mississippian crinoid *Platycrinus aggaszi;* ×0.4. *g,* A typical Permian fusulinid foraminifera, *Fusulina* sp.; ×1. Photos *a, b, e* by R. E. Grant. Photos *c, d, f, g* by D.H.E. and E. Valiulis.

BOX 1 Life and times of *Miocidaris*

Although echinoids are ubiquitous and familiar members of modern marine communities, only six echinoid genera are known from the Permian and the group might easily have disappeared like blastoids, or other unusual Paleozoic echinoderm groups. Only a single genus, *Miocidaris* (figure), is known to have survived the extinction (although cladistic analysis suggests that a related, as-yet undiscovered form survived as well and gave rise to the euechinoids[66]). Its persistence ensured the survival of the class, but fortuitously turned echinoid evolution in a new direction. *Miocidaris* has only two columns of interambulacral plates between each of the five files of ambulacral plates (which cover the tube feet), in contrast to the highly variable number (1–8) of interambulacral plates of other Paleozoic echinoids. The survival of *Miocidaris* fixed this relationship for all later echinoids.

Was the survival of *Miocidaris* truly fortuitous? Or did this morphological innovation aid the survival of the genus? Would echinoids with two columns of interambulacral plates dominate modern oceans even if the end-Permian extinction had not occurred? These questions are at the center of one of the great questions in the history of life. To what extent are the long-term patterns in the history of life driven by long-term patterns of adaptation versus the seemingly random patterns of survival during occasional mass extinction events?[43,67,68] Both equilibrium and non-equilibrium diversity models suggest that the Modern evolutionary fauna would have displaced the Paleozoic evolutionary fauna even without help from the Permian mass extinction.[37,40,69] In this view the lower rates of extinction in the clades comprising the Modern fauna ensured their success. Paleoenvironmental analysis suggests that elements of the Modern fauna began to displace the Paleozoic fauna by the late Ordvician.[70,71] My analysis of Paleozoic gastropods (a major component of the modern fauna) indicates the pattern is more complex.[43,68] Paleozoic and post-Paleozoic gastropods are very different, and there is no indication that the post-Paleozoic groups were expanding before the extinction. For gastropods and echinoids at least, the end-Permian mass extinction was a very critical event in their history.

Miocidaris sp.; ×1.0. Texas. A representative of the only clade of echinoids (two species) to survive the end-Permian mass extinction. Note the arrangement of two rows of ambulacral plates separated by two rows of interambulacral plates. This pattern is found only in this lineage in the Permian. The survival of this clade ensured that all subsequent echinoids share this pattern of construction rather than the variable number of rows of interambulacral plates found in other Permian echinoids.

will resolve these questions. There is, however, little support for claims that the mass extinction occurred over 8 Myr,[8] but it is unclear whether the extinction lasted 2 Myr, 1 Myr or even less.

Several generalities seem secure, however. Clades of sessile, epifaunal, filter-feeders generally suffered, but such clades often share other risk factors, including near-shore distribution, or restricted environmental distribution.[43] The apparent increased extinction among marine invertebrates with planktotrophic larval development[44,45] may simply reflect the increased extinction in near-shore environments and in the tropics, both areas where planktotrophs dominate.

The Terrestrial Record

Most paleontologists have emphasized the extensive marine extinctions and hence most descriptions have emphasized marine processes. Data from terrestrial vertebrates and insects reveal widespread extinctions during the late Permian, although the data are not sufficiently detailed to resolve the duration or timing of the extinctions.

Although plagued by taxonomic and sampling problems, 21 of 27 families of reptiles and 6 of 9 amphibian families disappeared during the latest Permian, for an overall 75% drop in diversity.[3] The well-studied vertebrate faunas from the Karoo Basin in South Africa reveal two extinction peaks, one roughly correlative with the end of the Capitanian, the other in the upper Permian, but below the Permo-Triassic boundary.[4] This may reflect nothing more than sampling bias and the difficulties correlating within the Karoo Basin and between marine and terrestrial sections, but does suggest a complex extinction event.

Among insects, 27 orders have been recorded from the Permian, of which 8 disappear during the late Permian, 4 suffer considerable declines in diversity but recover, and 3 straggle into the Triassic with such reduced diversity that they became extinct during the period.[46] The end-Permian mass extinction induced the most profound changes in insect diversity patterns in the history of the class.

The plant record is far more equivocal. The distinctive plant floras of the Carboniferous are replaced by the Mesophytic flora throughout the Permian with the onset of global warming following the Permo-Carboniferous glaciation, but plant fossils show little direct evidence of mass extinction.[47,48] Pollen undergoes a marked change at the Permo-Triassic boundary. In addition to the spike in fungal spores noted earlier, gymnosperm pollen virtually disappears and a new, arid-resistant pollen type appears.[2,19,49] Some of the changes may actually reflect changes in terrestrial communities that accompany the rapid earliest Triassic marine transgression.[2]

Changes in the Physical Environment

The supercontinent of Pangea formed by the early Permian, but the end-Permian regression led to exposure of the continent,[50] widespread evaporite deposition,[50] global warming and increased climatic instability.[51] Other events include eruption of the Siberian flood basalts (1.5 million km^2; the largest flood basalt of the Phanerozoic) in less than 1 Myr,[52] evidence of global warming and marked shifts in carbon, oxygen, sulfur and strontium isotopes[17,49,53] (Fig. 2).

The shifts in carbon isotopes have been linked to the extinction through erosion and oxidation of organic carbon previously sequestered on the continental shelves,[17] or oxidation of deep-sea sapropel-like deposits following overturn of a previously stratified, anoxic ocean.[54–56] Each hypothesis postulates oxidation reducing atmospheric oxygen and increasing atmospheric carbon dioxide, perhaps leading to anoxia and global warming, although there are no data on the extent of likely anoxia or warming. Alternatively, the signal may represent the spread of anoxic bottom waters across the shelves during the earliest Triassic transgression.[14,57,58] The geochemical data present difficulties for each of these hypotheses.[2,59] In particular, they ignore the impact of methane gas hydrates ($\delta^{13}C \approx -65\%$) released during the regression.[2] Regressions release hydrates locked in the outer continental shelf,[60,61] and may modulate ice ages via negative feedback.[61] No negative feedback occurs during non-glacio-eustatic regressions, and larger volumes of methane may be released.

Extinction Mechanisms

Supernovas, declining numbers of marine provinces and salinity changes are among the proposed causes of extinction; these are discussed elsewhere.[2,10,50,62,63] Recent models emphasize volcanism, extraterrestrial impact and global anoxia, but evaluating these is difficult because they often fail to make specific, unique predictions about extinction patterns. Without such predictions, most scenarios are just-so stories: comforting perhaps, but of little use. Given the lack of consistent patterns of extinction or survival which point to a specific cause, geological evidence is crucial in evaluating these scenarios.

The widespread pyroclastic volcanism in South China and the eruption of the Siberian flood basalts have been invoked as causes of the extinction through global cooling,[20,24,29,30,50] although the pyroclastic volcanism is too small to have been effective.[32] Whereas the Siberian traps apparently began erupting near the boundary, most of the flood basalt was emplaced during the earliest Triassic. Moreover, modelling results suggest that the climatic effects of SO_2 are self-limiting: at large volumes the molecules condense into larger particles.[64] Thus it is unclear if this eruption could have caused sufficient cooling to cause the extinction. The claimed iridium enrichment at the boundary in South China[25,33,34] is not supported.[17,35,36,59] It has been suggested that a period of bipolar glaciation triggered the extinction,[65] but without tying it to a specific cause. However, this glaciation was mid-Permian, representing the final pulse of the Permo-Carboniferous glaciation (J. C. Crowell, personal communication).

Ignoring the lack of evidence, is the pattern of extinction consistent with global cooling, either from volcanism, impact or glaciation? Not really. Although filter-feeders suffer severely, carbon isotopes provide no evidence for elimination of primary productivity, unlike the Cretaceous-Tertiary boundary.[59] This also fails to explain the differential extinction of fusulinid over non-fusulinid foraminifera. The apparently heightened extinction among shallow-water taxa is more consistent with global warming or habitat destruction than with global cooling.

Perhaps the various anoxia hypotheses provide the answer.[14,54–58] Suggestive evidence has been developed for the spread of anoxic waters associated with the rapid earliest Triassic transgression.[14,57,58] The geochemical data advanced to support the hypothesis[57] actually provides little support;[2] indeed, the shift in $\delta^{13}C$ permits only a moderate shift in atmospheric oxygen levels, limiting the extent of marine anoxia.[2] The sedimentological data are more promising, but remain subject to alternative interpretations. Additionally, global anoxia is a diversity-independent mode of extinction; thus survival should be enhanced by broad oxygen tolerance and large population size, yet many of the surviving taxa had small population sizes. Additionally, although molluscs contain many groups that are well adapted to dysaerobic environments and had higher-than-average survival, a closer look reveals no association between survival and oxygen tolerance.

Although there is much about this event paleontologists do not understand, our difficulties may stem from our search for a single cause. Few complex events stem from a single cause; more common is a complex web of causality, a web that can be difficult to untangle. My own view is that the cause of the end-Permian extinction lies in such a tangled web. The most plausible explanation would appear to be a three-phase model combining elements of several mechanisms described previously. The extinction began with the loss of habitat area as the regression dried out many marine basins, converting the two-dimensional coastlines of the mid-Permian to more linear coastlines. The increased exposure of Pangea as the regression progressed exacerbated climatic instability. This instability, coupled with the effects of continuing volcanic eruptions and an increase in atmospheric carbon dioxide (with some degree of global warming), led to increasing environmental degradation and ecological collapse. Certainly many of the marine groups that disappeared were those most tightly integrated into the dominant community types. Some degree of oceanic anoxia may have developed also but, as described earlier, it does not appear that either global warming or anoxia were sufficient to cause such a massive extinction. The final phase of the extinction occurred in the earliest Triassic. The rapid transgression destroyed near-shore terrestrial habitats, causing the shifts in spores and pollen and perhaps much of the decline in insects and tetrapods.

Prospects

Recent geochemical, biostratigraphic and paleontological studies have considerably sharpened our understanding of the biodiversity crisis that brought the Permian to a close, and the broad spectrum of environmental perturbations that accompanied it. The development of new correlations brings the promise of high-resolution data on the rate and timing of extinction and survival. Comparative analysis of the paleoecological, geographical and environmental characteristics of these taxa should in turn provide important clues to survival during the greatest biodiversity crisis in the history of life.

DOUGLAS H. ERWIN is at the Department of Paleobiology, National Museum of Natural History, Washington, DC 20560, USA.

References

1. Raup, D. M. *Science* **206,** 217–218 (1979).
2. Erwin, D. H. *The Great Paleozoic Crisis: Life and Death in the Permian* (Columbia, New York, 1993).
3. Maxwell, W. D. *Palaeontology* **35,** 571–584 (1992).
4. King, G. M. *Hist. Biol.* **5,** 239–255 (1991).
5. Benton, M. J. in *Extinction and Survival in the Fossil Record* (ed. Larwood, G. P.) 269–294 (Clarendon, Oxford, 1988).
6. Phillips, J. *Penny Cyclopedia* **17,** 153–154 (1840).
7. Alvarez, L. W., Alvarez, W., Asaro, F. & Mitchel, H. V. *Science* **208,** 1094–1098 (1980).
8. Teichert, C. in *Extinction Events in Earth History* 199–238 (Springer, Berlin, 1990).

9. Signor, P. W. & Lipps, J. H. *Geol. Soc. Amer. Sp. Pap.* **190,** 291–296 (1982).
10. Jablonski, D. in *Dynamics of Extinction* (ed. Elliot, D. K.) 183–299 (Wiley, New York, 1986).
11. Batten, R. L. in *The Permian and Triassic Systems and Their Mutual Boundary* Mem. **2,** 669–677 (Can. Soc. Petrol. Geol., Calgary, 1973).
12. Sweet, W. C. in *Permo-Triassic Boundary Events in the Eastern Tethys* (eds. Sweet, W. C., Yang, Z. Y., Dickins, J. M. & Yin, H. F.) 120–133 (Cambridge Univ. Press, Cambridge, 1992).
13. Kozur, H. in *Occ. Publ. Earth Sci. Res. Inst.* N.S. **9-b,** 139–154 (Columbia, SC, 1993).
14. Wignall, P. B. & Hallam, A. *Palaeogeogr. Palaeoclimatol. Palaoecol.* **102,** 215–237 (1993).
15. Assereto, R., Bosellini, A., Fantini-Sestini, N. & Sweet, W. C. in *The Permian and Triassic Systems and Their Mutual Boundary* Mem. **2,** 176–199 (Can. Soc. Petrol. Geol., Calgary, 1973).
16. Cassinis, G. (ed.) *Mem. Soc. Geol. Ital.* **34,** 1–366 (1988).
17. Holser, W. T., Schönlaub, H. P., Boeckelmann, K. & Magaritz, M. *Abh. Geolog. Bund.* **45,** 213–232 (1991).
18. Broglio Loriga, C., Neri, C., Pasini, M. & Posenato, R. *Mem. Soc. Geol. Ital.* **34,** 5–44 (1988).
19. Visscher, H. & Brugman, W. A. *Mem. Soc. Geol. Ital.* **34,** 121–128 (1988).
20. Yang, Z. Y. *et al. Min. Geol. Min. Res., Geol. Mem., ser. 2* **6,** 1–379 (1987).
21. Yang, Z. Y. & Li, Z. S. in *Permo-Triassic Boundary Events in the Eastern Tethys* (eds. Sweet, W. C., Yang, Z. Y., Dickins, J. M. & Yin, H. F.) 9–20 (Cambridge Univ. Press, Cambridge, 1992).
22. Zhao, J. K. *et al. Bull. Nanjing Inst. Geol. Palaeo. Acad. Sinica* **2,** 1–112 (1981).
23. Sheng, J. Z. *et al. J. Fac. Sci. Hokkaido Univ., ser. 4* **21,** 133–188 (1984).
24. Yin, H. F., Xu, G. R. & Ding, M. H. *Sci. Pap. Geol. Internat. Exch.* (27th Intl Geol. Congr.) 195–204 (1984).
25. Xu, D. Y. *et al. Astrogeological Events in China* (Van Nostrand Reinhold, New York, 1989).
26. Baud, A., Magaritz, M. & Holser, W. T. *Geologische Rundschau* **78,** 649–677 (1989).
27. Xu, G. R. *J. China Univ. Geosci.* **2,** 36–46 (1991).
28. Reinhardt, J. W. *Facies* **18,** 231–288 (1988).
29. Yin, H. F. *et al. Acta geologica sinica* **2,** 417–431 (1989).
30. Yin, H. F. *et al.* in *Permo-Triassic Boundary Events in the Eastern Tethys* (eds. Sweet, W. C., Yang, Z. Y., Dickins, J. M. & Yin, H. F.) 146–157 (Cambridge Univ. Press, Cambridge, 1992).
31. Zhao, L. & Kyte, F. T. *Earth planet. Sci. Lett.* **90,** 411–421 (1988).
32. Erwin, D. H. & Vogel, T. A. *Geophys. Res. Lett.* **19,** 893–896 (1992).
33. Sun, Y. Y. *et al.* in *Developments in Geosciences* (ed. Tu, G.) 235–245 (Science Press, Beijing, 1984).
34. Xu, D. Y. *et al. Nature* **314,** 154–156 (1985).
35. Clark, D. L., Wang, C. Y., Orth, C. J. & Gilmore, J. S. *Science* **233,** 984–986 (1986).
36. Orth, C. J. in *Mass Extinctions: Processes and Evidence* (ed. Donovan, S. K.) 37–72 (Belkhaven, London, 1989).

37. Sepkoski, J. J. Jr. *J. geol. Soc. Lond.* **146,** 7–19 (1989).
38. Yin, H. F. *Newsletter on Strat.* **15,** 13–27 (1985).
39. Sepkoski, J. J. Jr. *Milwaukee Pub. Mus. Contr. Biol. Geol.* **83,** 1–155 (1992).
40. Sepkoski, J. J. Jr. in *Patterns and Processes in the History of Life* (eds. Raup, D. M. & Jablonski, D.) 277–295 (Springer, Berlin, 1986).
41. Sepkoski, J. J. Jr. *Paleobiology* **10,** 246–267 (1984).
42. Brasier, M. D. in *Extinction and Survival in the Fossil Record* (ed. Larwood, G. P.) 37–64 (Clarendon, Oxford, 1988).
43. Erwin, D. H. *Palaios* **4,** 424–438 (1989).
44. Valentine, J. W. *Bull. mar. Sci.* **39,** 607–615 (1986).
45. Valentine, J. W. & Jablonski, D. *Proc. natn. Acad. Sci. USA* **83,** 6912–6914 (1986).
46. Labandeira, C. C. & Sepkoski, J. J. Jr. *Science* **261,** 310–315 (1993).
47. DiMichele, W. A. & Hook, R. W. in *Evolutionary Paleoecology of Terrestrial Plants and Animals* (eds. Behrensmeyer, A. K. *et al.*) 205–325 (Univ. Chicago Press, Chicago, 1992).
48. Knoll, A. H. in *Extinctions* (ed. Nitecki, M. H.) 23–68 (Univ. Chicago Press, Chicago, 1984).
49. Eshet, Y. in *Permo-Triassic Boundary Events in the Eastern Tethys* (eds. Sweet, W. C., Yang, Z. Y., Dickins, J. M. & Yin, H. F.) 134–145 (Cambridge Univ. Press, Cambridge, 1992).
50. Holser, W. T. & Magaritz, M. *Mod. Geol.* **11,** 155–180 (1987).
51. Parrish, J. M., Parrish, J. T. & Ziegler, A. M. in *The Ecology and Biology of Mammal-like Reptiles* (eds. Hotton, N., MacLean, P. D., Roth, J. J. & Roth, E. C.) 109–131 (Smithsonian, Washington, DC, 1986).
52. Campbell, I. H., Czamanski, G. K., Fedorenko, V. A., Hill, R. I. & Stepanov, V. *Science* **258,** 1760–1763 (1992).
53. Magaritz, M., Krishnamurthy, R. V. & Holser, W. T. *Am. J. Sci.* **292,** 727–739 (1992).
54. Gruszczynski, M., Halas, S., Hoffman, A. & Malkowski, K. *Nature* **337,** 64–68 (1989).
55. Hoffman, A., Gruszczynski, M. & Malkowski, K. *Mod. Geol.* **14,** 211–221 (1990).
56. Malkowski, K., Gruszczynski, M., Hoffman, A. & Halas, S. *Hist. Biol.* **2,** 289–309 (1989).
57. Hallam, A. *Hist. Biol.* **5,** 257–262 (1991).
58. Wignall, P. B. & Hallam, A. *Palaeogeogr. Palaeoclimatol. Palaeoecol.* **93,** 21–46 (1992).
59. Holser, W. T. & Magaritz, M. *Geochim. cosmochim. Acta* **56,** 3297–3309 (1992).
60. Kvenvolden, K. A. *Chem. Geol.* **71,** 41–51 (1988).
61. Nisbet, E. G. *Can. J. Earth Sci.* **27,** 148–157 (1990).
62. Erwin, D. H. A. *Rev. Ecol. Syst.* **21,** 69–91 (1990).
63. Maxwell, W. D. in *Mass Extinction: Processes and Evidence* (ed. Donovan, S. K.) 152–173 (Belkhaven, London, 1989).
64. Pinto, J. P., Turco, R. P. & Toon, O. B. *J. geophys. Res.* **94,** 11165–11174 (1989).
65. Stanley, S. M. *Am. J. Sci.* **288,** 334–352 (1988).
66. Smith, A. B. & Hollingsworth, N. T. J. *Proc. York. Geol. Soc.* **48,** 47–60 (1990).
67. Jablonski, D. *Phil. Trans. R. Soc.* B **325,** 357–368 (1989).
68. Erwin, D. H. *Paleobiology* **16,** 187–203 (1990).

69. Kitchell, J. A. & Carr, T. R. in *Phanerozoic Diversity Patterns* (ed. Valentine, J. W.) 277–309 (Princeton Univ. Press, Princeton, NJ, 1985).
70. Sepkoski, J. J. Jr. & Miller, A. I. in *Phanerozoic Diversity Patterns* (ed. Valentine, J. W.) 153–189 (Princeton Univ. Press, Princeton, NJ, 1985).
71. Sepkoski, J. J. Jr. *Paleobiology* **17,** 58–77 (1991).
72. Claoue-Long, J. C., Shang, Z. C., Ma, G. G. & Du, S. H. *Earth planet. Sci. Lett.* **105,** 182–190 (1991).
73. Nikistivskiy, E. A. *Intl. Geol. Rev.* **34,** 1001–1007 (1992).
74. Steiner, M., Ogg, J., Zhang, Z. & Sun, S. *J. geophys. Res.* **94,** 7343–7363 (1989).

Acknowledgments

I thank W. T. Holser and K. Towe for discussion, D. Jablonski and J. J. Sepkoski Jr. for reviews, and E. Valiulis for drafting the figures. This work was supported by the Charles D. Walcott Fund of the Smithsonian Institution.

PART 4

SHAKING THE TREE: CASE HISTORIES IN PHYLOGENY

If this collection had a genesis in a single paper, it was a short piece that passed across my desk in 1991 entitled "Is the Guinea Pig a Rodent?"[1] It was, in part, this paper that inspired Michael Novacek to write the sixth contribution in this section, the review that gives its name to this book.

Novacek's review, on mammalian phylogeny, clearly struck a chord. After its appearance, I was regularly called up by specialists who wished to write a review, in the same style, for other groups of interest. Several of these Novacek-inspired reviews appear here, alongside the archetype. And it was all due to a short paper on rodent phylogeny. Whence the huge influence of the humble guinea pig?

The paper, by Dan Graur and colleagues, came at the right time. By using molecular phylogeny to challenge phylogenies established using anatomical criteria, the paper piqued a debate on the relative importance of molecular and morphological data that resonates still: indeed, it has yet to be resolved. It also came at a time when the potential of cladistics to help us work out the branching pattern of the tree of life was coming to be fully appreciated.

This combination of molecules and cladistics, when added to traditional paleontology, is particularly potent. It forms the core of current thinking in comparative biology. As such, it is represented particularly strongly in

Nature, and reviews based on this sometimes incendiary mixture justifiably form the centerpiece of this book.

Each of the six reviews in this section covers the origin and diversification of a major group. Coverage is not comprehensive, but the reviews included here cover topics that have been controversial, for one reason or another, over the past years. If I had to add a seventh, it would be about the origin and diversification of arthropods—the subject of a recent symposium.[2]

The first two reviews cover two major radiations in the world of plants. The first, by Paul Kenrick and Peter Crane on early land plants, is, in some ways, an update on Shear's review on early terrestrial ecosystems presented in the previous section (see ref. 3 for a book-length treatment by the same authors). The second, again coauthored by Crane, deals with the angiosperms, or flowering plants (see ref. 4 for an earlier, multiauthor discussion of the topic). Both reviews break new ground in their uncompromisingly cladistic treatment of the subject.

Perhaps no area in paleontology has taken cladistics so much to heart as research on fossil fishes. This is hardly a coincidence, as many prominent paleoichthyologists have been instrumental in the development of cladistics as a methodology: one thinks immediately of Gareth Nelson of the American Museum of Natural History in New York, and the late Colin Patterson of the Natural History Museum in London. Two of Patterson's associates, Peter Forey and Philippe Janvier, discuss the origin of gnathostomes (jawed vertebrates) from jawless forms, in the third contribution to this section. (See ref. 5 for an alternative, mechanistic view of the origin of jaws.)

Hagfishes and lampreys are the only two extant agnathans. They have no hard parts, and both are specialized for a predatory or parasitic lifestyle. They present a striking contrast with a wide variety of extinct agnathans, which lived between the Cambrian and the Devonian periods. These were often heavily armored fishes that presumably assumed a largely benthic, detritus- or suspension-feeding existence.

Forey and Janvier's cladistic treatment shows that agnathans are, as a group, paraphyletic: they form a grade of organization rather than a natural, monophyletic group, in that gnathostomes emerge from within their number. But progress in this field has been rapid and dramatic, and has advanced on two fronts.

The first concerns *Sacambambaspis*[6,7] from Bolivia, one of the earliest-known agnathans—indeed, one of the very earliest vertebrates known in any detail. The interpretation of this form catalyzed a new look at agnathan phylogeny. In the phylogeny presented by Forey and Janvier in this volume, hagfishes are regarded as primitive with respect to all other agnathans. They never had hard tissues in their ancestry: indeed, many features of their biology show them to be extremely primitive creatures. Lampreys, in contrast, are seen as a sister group to the anaspids, one of the several extinct groups of agnathans—lampreys, then, had armored ancestors: they are secondarily naked.

But phylogenetic analysis based on new information from *Sacambambaspis* prompted a rethink, in which the hagfish and the lamprey are both regarded as primitively naked.[8–11] This may not be the last word: new insights on the biology of hagfish (see, for example, ref. 12) and the evolution of hard tissues[13] have the potential to force further re-evaluations as new data come in from this lively, exciting field.

The second, explosive advance in the field of the earliest vertebrates comes from the flowering of studies on conodonts, and the admission of conodonts to vertebrate paleontology from the realm of biostratigraphy.

Conodonts are phosphatic microfossils that are locally abundant in Paleozoic rocks. The identity of the conodont-bearing animal remains elusive, but for a very long time this hardly mattered, as conodonts of certain types were often so abundant that stratigraphers could use them as correlative zone-fossils. Their stratigraphic utility was all: apart from a small group of zoologically minded researchers, nobody cared if conodonts were the remains of living things or not. This situation persisted until the discovery of the conodont-bearing fossils of soft-bodied animals[14] that were eventually recognized as vertebrates. Histological, anatomical and phylogenetic studies have forged ahead in the 1990s.[15–20] The majority view is that conodont animals belong among the vertebrates, though precisely where they fit into vertebrate phylogeny remains unclear.[17]

Another field that has witnessed dramatic advances is that of the origin and diversification of tetrapods, addressed here by Per Erik Ahlberg and Andrew Milner. This review falls into two neat parts. The first looks at the initial diversification of tetrapods in the late Devonian; the second at the appearance of essentially "modern" faunas in the early Carboniferous. This approach reveals a "gap" in the lowest Carboniferous in which much interesting evolution presumably took place—evolution about which we know relatively little at present.

For many years, the only well-known Devonian tetrapod was *Ichthyostega,* from East Greenland (see ref. 21 for a long-awaited and somewhat idiosyncratic monograph on the subject from a veteran in the field). Then came other finds from Russia and North America, and perhaps most significantly, a return to East Greenland in 1987 by an Anglo-Danish expedition that produced a large number of informative fossils of the hitherto poorly known tetrapod *Acanthostega gunnari.*[22] The material of *Acanthostega,* collected during this expedition (in which Ahlberg was a participant), has produced a number of now-classic papers by Jennifer Clack of Cambridge and her associates, particularly Michael Coates. Others published since the appearance of Ahlberg and Milner's review are recommended below (see refs. 22 and 23, and papers in ref. 24).

Late Devonian tetrapods already present a diverse assemblage, although a consensus is beginning to emerge on the sequence in which characters were acquired during the earliest stages of tetrapod evolution, namely, the acquisition of digits and the other specializations necessary for land life. *Acanthostega* presents perhaps the most striking challenge to conventional wisdom. Although it had limbs and digits (characteristically polydactylous, as seems to have been the norm for Devonian tetrapods), its internal gills and rigid shoulder girdle suggest that it could hardly have walked on land. Instead, it seems to have been primarily aquatic. Limbs and digits may have initially served for locomotion not on land, but through very shallow water. Indeed, digitlike structures convergent on tetrapod digits appeared in rhizodonts, freshwater lobe-finned fish only distantly related to tetrapods.[25]

The Carboniferous, after the "gap" at the beginning of the period, saw the main lines of tetrapod evolution already well established, although—several years now since Ahlberg and Milner's report—there is still no consensus on the precise pattern of relationships of early tetrapods.[26–28] In particular, an unambiguous picture of the relationships of taxa around the split between amphibians, on the one hand, and amniotes, on the other, has yet to emerge. A proposal that the intriguing Devonian hexadactylous Russian tetrapod *Tulerpeton*[29] might be a stem amniote drives this dichotomy far back into the late Devonian. Of the several groups of tetrapod preserved from Viséan-stage (early Carboniferous) rocks (notably at the *lagerstätte* at East Kirkton in Scotland, ref. 30), some are the so-called anthracosaurs, close to the amniote lineage that eventually bore reptiles, mammals and birds; others are temnospondyls, closer to the extant lissamphibia.

One animal, assigned to a rather shadowy group called the baphetids (formerly loxommatids), does not fit in quite so easily with this simple scheme and is hard to relate either to anthracosaurs or to temnospondyls. This animal, *Eucritta melanolimnetes*[31]—Clack's playfully named "creature from the black lagoon"—is what Ahlberg and Milner describe as the "further . . . new tetrapod from East Kirkton combin[ing] the primitive characters of loxommatids, temnospondyls and anthracosaurs." The "protoanthracosaur" from Delta, Iowa, since described as *Whatcheeria deltae,*[32] and the peculiar *Crassigyrinus,*[33] both seem to be primitive anthracosaurs.

Despite these problems, fossils have continued to come in from far and wide. The geographical range of early (Viséan-Namurian) Carboniferous tetrapods in Euramerica has been extended farther west, to Nevada,[34] and east to eastern Germany,[35] and outside Euramerica to Australia[36]—this last a discovery that should set the 140-year concentration on the Euramerican fauna into its proper context. One or two interesting remains are even beginning to turn up from the frustrating gap between the late Devonian and early Carboniferous faunas. For other recent developments in the study of early tetrapods, see ref. 24, and I recommend ref. 37 for an engaging popular account. Three references (67, 93 and 95) in Ahlberg and Milner's review cited as abstracts or as in press have now appeared. Two are cited below (references 38, 39), and the third, by Milner and Lindsay, is in ref. 24.

The origin and early diversification of birds is perhaps one of the most active topics in paleontology. Luis Chiappe addresses his subject in typically bold and uncompromising style in his contribution here. However, a number of important and often startling discoveries[40–51] have been made in the few short years since Chiappe's review appeared. Readers are encouraged to consult them all and make up their own mind.

Several are worth special mention; *Unenlagia,* a theropod dinosaur from Argentina with forelimb mechanics uncannily similar to those of birds;[48] *Rahona,* a flying bird with the slashing claw seen only in dinosaurs such as *Velociraptor;*[47] *Protarchaeopteryx* and *Caudipteryx,* feathered yet flightless theropod dinosaurs that come off the cladogram below *Archaeopteryx;*[51] and *Shuvuuia,* a flightless animal with a birdlike skull,[50] related to the strange animal *Mononykus*—arguably closer to extant birds than to *Archaeopteryx,* for all that it had tiny, stubby arms quite incapable, one would have thought, for flight. (*Mononykus* is a highlight of Chiappe's review here.)

All reinforce the view, espoused by Thomas Henry Huxley and revived in the 1970s by John H. Ostrom, that birds and dinosaurs are intimately related, and this is repeatedly demonstrated in cladistic analyses. Development in this area is likely to be rapid, as researchers examine the interesting interrelationships around the base of the Aves (= Avialae *sensu* ref. 52), the clade that includes the common ancestor of *Archaeopteryx* and all extant birds. By the time you read this book, the cladogram will have changed, and changed again, as new discoveries are announced and assessed.

Although it is now evident that the origin of birds is only loosely connected with the origins of flight and feathers, respectively, the characters that define the clade Aves occupies a special place in the hearts and minds of most, as it marks the part of the cladogram when dinosaurs took to the air and became "birds" in the sense that we would recognize the term. Some workers continue to maintain that the link between dinosaurs and birds is convergent, citing evidence from a number of sources (see refs. 53, 54). Readers are invited to study this alternative viewpoint; should they do so, they should bear a number of things in mind.

First, the claim of convergence is one that has been repeatedly tested and refuted by cladistics, and can be maintained only by assertion. Workers who deny the link between birds and dinosaurs are not users of cladistics.

Second, the alternative hypothesis, that birds are linked with generalized primitive archosaurs called thecodonts, is prompted by similarities between some thecodonts and birds in a few characters. These similarities may well be convergent, as they are distributed in a sporadic and disconnected way. (For example, the thecodont *Megalancosaurus* is lizardlike in shape but has a beak-shaped snout, and *Longisquama* is an otherwise lizardlike creature but has a dorsal crest of divided structures that look something like feathers, and so on.) Nowhere among thecodonts can one cite the correlated suite of character acquisition that links birds and dinosaurs and that, in itself, acts as a hedge against convergence.

This failure to appreciate that cladistics deals with a number of features together, rather than single, "key" characters, underpins a mind-set that thinks that, if one character is disproved, the rest will soon fall. A study by Burke and Feduccia[53] is a case in point—it looks at the identity of digits in the bird "hand." Some researchers have suggested that, in birds, digits I and V of the amniote pentadactyl limb are lost, giving a characteristic II-III-IV hand. In dinosaurs, though, it is usual to number the three-fingered hand of theropods I-II-III. This leads to a contradiction: if the hands of birds and theropods developed by very different programs, how could they be homologues?

The dilemma is easily resolved. There seems little doubt that theropods had a I-II-III configuration, as some primitive dinosaurs with five-fingered hands show a marked reduction of digits IV and V. The similarities between *Archaeopteryx* and theropods in the form and layout of the bones of the hand, as in much else, make it highly likely that *Archaeopteryx* also had the theropod I-II-III configuration. Yet nobody would doubt the affinities of *Archaeopteryx* with birds: all of which casts doubt on the significance of the development of extant birds to shed light on that of extinct creatures such as theropods and *Archaeopteryx.* A second problem associated with Burke and Feduccia's study is that it leans heavily on the identification of particular elements of the adult skeleton with the condensation of moieties of cartilage in the developing embryo. Cartilage condensations in the developing forelimbs of modern birds can be identified with digits II, III and IV, rather than I, II and III. However, it is likely that the identification of cartilage condensations is open to some interpretation: they certainly don't emerge with their names already embossed for the convenience of comparative anatomists.

For more recent reviews on the origin and evolution of birds, Sereno[55] provides a cladistic treatment with one or two interesting differences from Chiappe's treatment here; comparisons between the two might be fruitful. For other accounts, see refs. 56–60. References 56, 57 and 60 are cladistic; ref. 58 tries to adopt a balanced perspective (wrongly, I think), and ref. 59 is a rather different viewpoint from Chatterjee, a thinker with a unique perspective on all matters connected with birds and dinosaurs.

All of which brings us back to the guinea pig,[1] and Novacek's appraisal of the role of molecular and morphological data in reconstructing the phylogeny of mammals. I had asked Novacek to compare and contrast the different strands of data in the light of Simpson's 1945 benchmark classification of mammals (ref. 10 in Novacek's article). This has also been reconsidered at book length.[61]

Since Novacek's review, discoveries of the fossils of mammals have continued to be made, further shaking the tree. Discoveries of complete skeletons of extremely primitive mammals[62,63] from the Chinese beds that have yielded spectacular fossil birds and feathered dinosaurs[40,41,45,46,51]—and from the Mongolian sequences[64] that have produced enigmatic primitive birds[50]—represent types hitherto known only from fragmentary remains, often just a few teeth. Such finds not only shed light on early mammal lifestyles, but illustrate mosaic evolution in early mammals, with some structures evolving "faster" than others. The discoveries of complete yet primitive mammals with unusual combinations of characters are important because, being so primitive, they have the potential to change the branching order of the phylogeny in fundamental ways, shaking the tree by its roots. Other recent fossil evidence takes the story of ungulates back to the Cretaceous[65] and looks critically at early placental mammal branching order.[66] These studies are just a handful of many paleontological studies, published since Novacek's review, that challenge accepted views of mammalian phylogeny.

The diversification of the placental mammals (Eutheria) and their relationships with marsupials and monotremes has attracted particular interest. The rapid diversification of the Eutheria into recognizably modern orders may be connected with continental breakup at the end of the Cretaceous,[67] although this is highly contentious.

Molecular work has yielded insight (and, it has to be said, cast some dismal fogs) into mammalian phylogeny. References 67–75 are a selection of recent studies, and ref. 76 a comprehensive study on the phylogeny of more than 600 genes, whose results agree very well with paleontological evidence, except for pointing out some surprisingly early Cretaceous divergences in certain mammals—including some subgroups of rodents.

Perhaps no subfield in mammalian phylogenetics has attracted more recent interest than the problem of the origin of whales. Pinpointing the origins of highly specialized, derived groups is always difficult, yet paleontological evidence, especially the discovery of a two-toed, vestigial hindlimb in the Eocene whale *Basilosaurus,* indicated a relationship with the even-toed ungulates, or artiodactyls[77,78] (see again ref. 37 for an engaging popular account). However, a rash of molecular papers made an even bolder claim—that whales evolved from *within* the artiodactyls (even-toed ungulates), so that a hippo is more closely related to any whale than to a camel or a pig.[79–82] Again, this vexed question continues to prompt fruitful research. The interplay between molecules, morphology and phylogeny is discussed in refs. 83–85.

But research never stops. A report on an extremely peculiar 400-million-year-old fossil fish from China, *Psarolepis,*[86] shows that it has features reminiscent of primitive lobe-finned bony fishes, as well as the acanthodians and placoderms, extinct groups of primitive jawed vertebrates. Such a mélange of features from fundamentally different groups, present in a single animal, is likely to shake the tree with hurricane force, rearranging not just a few terminal twigs, but entire boles and trunks.

References

1. D. Graur, W. A. Hide and W.-H. Li. Is the guinea pig a rodent? *Nature* 351 (1991): 649–652.
2. R. A. Fortey and R. Thomas, eds. *Arthropod relationships.* Dordrecht: Kluwer Academic Publishers, 1998.
3. P. Kenrick and P. R. Crane. *The origin and early diversification of land plants: A cladistic study.* Washington, DC: Smithsonian Institution Press, 1997.
4. E. M. Friis, W. G. Chaloner and P. R. Crane. *The origins of angiosperms and their biological consequences.* Cambridge: Cambridge University Press, 1987.
5. J. Mallatt. Ventilation and the origin of jawed vertebrates: A new mouth. *Zoological Journal of the Linnean Society* 117 (1996): 329–404.
6. P.-Y. Gagnier. *Sacabambaspis janvieri,* vertébré Ordovicien de Bolivie: I: Analyse morphologique. *Annales de Paléontologie (Vert.-Invert.)* 79 (1993): 19–51.

7. P.-Y. Gagnier. *Sacabambaspis janvieri,* vertébré Ordovicien de Bolivie: 2: Analyse phylogénétique. *Annales de Paléontologie (Vert.-Invert.)* 79 (1993): 119–166.
8. P. Forey and P. Janvier. Evolution of the early vertebrates. *American Scientist* 82 (1994): 554–565.
9. P. L. Forey. Agnathans recent and fossil, and the origin of jawed vertebrates. *Reviews in Fish Biology and Fisheries* 5 (1995): 267–303.
10. P. Janvier. The dawn of the vertebrates: Characters versus common ascent in the rise of current vertebrate phylogenies. *Paleontology* 39 (1996): 259–287.
11. P. Janvier. *Early vertebrates.* Oxford Monographs in Geology and Geophysics, 33. Oxford: Oxford University Press, 1996.
12. H. Wicht and R. G. Northcutt. Ontogeny of the head of the Pacific hagfish (*Epatretus stouti,* Myxinoidea): Development of the lateral line system. *Philosophical Transactions of the Royal Society* B 349 (1995): 119–134.
13. M. M. Smith, I. J. Sansom and P. Smith. "Teeth" before armour: The earliest vertebrate mineralized tissues. *Modern Geology* 20 (1996): 303–319.
14. D. E. G. Briggs, E. N. K. Clarkson and R. J. Aldridge. The conodont animal. *Lethaia* 16 (1983): 1–14.
15. M. A. Purnell. Microwear on conodont elements and macrophagy in the first vertebrates. *Nature* 374 (1995): 798–800.
16. S. E. Gabbott, R. J. Aldridge and J. N. Theron. A giant conodont with preserved muscle tissue from the Upper Ordovician of South Africa. *Nature* 374 (1995): 800–803.
17. P. Janvier. Conodonts join the club. *Nature* 374 (1995): 761–762.
18. R. J. Aldridge, M. A. Purnell, S. E. Gabbott and J. N. Theron. The apparatus architecture and function of *Promissum pulchrum* Kovács-Endrödy (Conodonta, Upper Ordovician) and the prioniodontid plan. *Philosophical Transactions of the Royal Society* B 347 (1995): 275–291.
19. I. J. Sansom. *Pseudooneotodus:* A histological study of an Ordovician to Devonian vertebrate lineage. *Zoological Journal of the Linnean Society* 118 (1996): 47–57.
20. R. J. Aldridge, D. E. G. Briggs, M. P. Smith, E. N. K. Clarkson and N. D. L. Clark. The anatomy of conodonts. *Philosophical Transactions of the Royal Society* B 340 (1993): 405–421.
21. E. Jarvik. The Devonian tetrapod *Ichthyostega. Fossils and Strata* 40 (1996): 1–213.
22. M. I. Coates. The Devonian tetrapod *Acanthostega gunnari,* Jarvik: Postcranial anatomy, basal tetrapod interrelationships and patterns of skeletal evolution. *Transactions of the Royal Society of Edinburgh: Earth Sciences* 87 (1996): 363–421.
23. P. E. Ahlberg, J. A. Clack and E. Lukševičs. Rapid braincase evolution between *Panderichthys* and the earliest tetrapods. *Nature* 381 (1996): 61–64.
24. D. B. Norman, A. R. Milner and A. C. Milner, eds. A study of fossil vertebrates: Essays in honour of Alec Panchen. *Zoological Journal of the Linnean Society* 122 (1998): 1–384.
25. E. B. Daeschler and N. H. Shubin. A fish with fingers? *Nature* 391 (1998): 133.
26. R. L. Carroll. Problems of the phylogenetic analysis of Paleozoic choanates. *Bulletin du Muséum National d'Histoire Naturelle, Paris* 17C (1995): 389–445.
27. M. Laurin and R. R. Reisz. A new perspective on tetrapod phylogeny. In S. S. Sumida and K. L. M. Martin, eds., *Amniote origins: Completing the transition to land,* 9–59. San Diego: Academic, 1997.

28. M. Laurin. The importance of global parsimony and historical bias in understanding tetrapod evolution: Part 1: Systematics, middle-ear evolution and jaw suspension. *Annales des Sciences Naturelles* 1 (1998): 1–42.
29. O. A. Lebedev and M. I. Coates. The postcranial skeleton of the Devonian tetrapod *Tulerpeton curtum* Lebedev. *Zoological Journal of the Linnean Society* 114 (1995): 307–348.
30. E. N. K. Clarkson, A. L. Panchen and W. D. I. Rolfe, eds. Volcanism and early terrestrial biotas. *Transactions of the Royal Society Edinburgh: Earth Science* 84 (1994): 175–464.
31. J. A. Clack. A new early Carboniferous tetrapod with a mélange of crown-group characters. *Nature* 394 (1998): 66–69.
32. R. E. Lombard and J. R. Bolt. A new primitive tetrapod, *Whatcheeria deltae,* from the Lower Carboniferous of Iowa. *Palaeontology* 38 (1995): 471–494.
33. J. A. Clack. The Scottish Carboniferous tetrapod *Crassigyrinus scoticus* (Lydekker) cranial anatomy and relationships. *Transactions of the Royal Society Edinburgh: Earth Science* 88 (1998): 127–142.
34. K. S. Thomson, N. Shubin and F. G. Poole. A problematic early tetrapod from the Mississippian of Nevada. *Journal of Vertebrate Paleontology* 18 (1998): 315–320.
35. R. Werneburg. Temnospondyle amphibien aus dem Karbon Mitteldeutschlands. *Veröffentlichungen Naturhistorisches Museum Schleusingen* 11 (1996): 23–64.
36. T. Thulborn, A. Warren, S. Turner and T. Hamley. Early Carboniferous tetrapods in Australia. *Nature* 381 (1996): 777–780.
37. C. Zimmer. *At the water's edge.* New York: Free Press, 1998.
38. H.-P. Schultze and J. R. Bolt. The lungfish *Tranodis* and the tetrapod fauna from the Upper Mississippian of North America. In A. R. Milner, ed., *Studies on Carboniferous and Permian vertebrates,* Special Papers in Palaeontology 52, 31–54. London: Palaeontological Association, 1996.
39. J. A. Clack. The palate of *Crassigyrinus scoticus,* a primitive tetrapod from the Lower Carboniferous of Scotland. In A. R. Milner, ed., *Studies on Carboniferous and Permian Vertebrates,* Special Papers in Palaeontology 52, 55–64. London: Palaeontological Association, 1996.
40. L. Hou, Z. Zhou, L. D. Martin and A. Feduccia. A beaked bird from the Jurassic of China. *Nature* 377 (1995): 616–618.
41. L. Hou, L. D. Martin, Z. Zhou and A. Feduccia. Early adaptive radiation of birds: Evidence from fossils from northeastern China. *Science* 274 (1996): 1164–1167.
42. J. L. Sanz, L. M. Chiappe, B. P. Pérez-Moreno, A. D. Buscalioni, J. J. Moratalla, F. Ortega and F. J. Poyato-Ariza. An early Cretaceous bird from Spain and its implications for the evolution of avian flight. *Nature* 382 (1996): 442–445.
43. K. Padian. Early bird in slow motion. *Nature* 382 (1996): 400–401.
44. J. L. Sanz, L. M. Chiappe, P. Pérez-Moreno, J. J. Moratalla, F. Hernández-Carrasquila, A. D. Buscalioni, F. Ortega, F. J. Poyato-Ariza, D. Rasskin-Gutman and X. Martínez-Delclòs. A nestling bird from the Lower Cretaceous of Spain: Implications for avian skull and neck evolution. *Science* 276 (1997): 1543–1546.
45. P. J. Chen, Z.-M. Dong and S.-N. Zhen. An exceptionally well-preserved theropod dinosaur from the Yixian Formation of China. *Nature* 391 (1998): 147–152.
46. D. M. Unwin. Feathers, filaments and theropod dinosaurs. *Nature* 391 (1998): 119–120.

47. C. A. Forster, S. D. Sampson, L. M. Chiappe and D. W. Krause. The theropod ancestry of birds: New evidence from the late Cretaceous of Madagascar. *Science* 279 (1998): 1915–1919.
48. F. E. Novas and P. F. Puerta. New evidence concerning avian origins from the late Cretaceous of Patagonia. *Nature* 387 (1997): 390–392.
49. L. M. Witmer. A new missing link. *Nature* 387 (1997): 349–350.
50. L. M. Chiappe, M. A. Norell and J. M. Clark. The skull of a relative of the stem-group bird *Mononykus*. *Nature* 392 (1998): 275–278.
51. Q. Ji, P. J. Currie, M. A. Norell and S. A. Ji. Two feathered dinosaurs from northeastern China. *Nature* 393 (1998): 753–761.
52. J. A. Gauthier. Saurischian monophyly and the origin of birds. *Mem. California Academy of Sciences* 8 (1986): 1–55.
53. C. A. Burke and A. Feduccia. Developmental patterns and the identification of homologies in the avian hand. *Science* 278 (1997): 666–668.
54. A. Feduccia. *The origin and evolution of birds.* New Haven: Yale University Press, 1996.
55. P. Sereno. The origin and evolution of dinosaurs. *Annual Reviews of Earth and Planetary Sciences* 25 (1997): 435–489.
56. K. Padian and L. Chiappe. The origin of birds and their flight. *Scientific American,* February 1998, 38–47.
57. K. Padian and L. M. Chiappe. The origin and early evolution of birds. *Biological Reviews* 73 (1998): 1–42.
58. P. Shipman. *Taking wing: Archaeopteryx and the evolution of bird flight.* New York: Simon and Schuster, 1998.
59. S. Chatterjee. *The rise of birds: 225 million years of evolution.* Baltimore: Johns Hopkins University Press, 1997.
60. L. Dingus and T. Rowe. *The mistaken extinction: Dinosaur evolution and the origin of birds.* New York: W. H. Freeman and Co., 1997.
61. M. C. McKenna and S. K. Bell. *Classification of mammals above the species level.* New York: Columbia University Press, 1997.
62. Q. Ji, Z. Luo and S. A. Ji. A Chinese triconodont mammal and the mosaic evolution of the mammalian skeleton. *Nature* 398 (1999): 326–330.
63. Y. Hu, Y. Wang, Z. Luo and C. Li. A new symmetrodont mammal from China and its implications for mammalian evolution. *Nature* 390 (1997): 137–142.
64. M. J. Novacek, G. Rougier, J. R. Wible, D. Dashzeveg, M. C. McKenna and I. Horovitz. Epipubic bones in eutherian mammals from the late Cretaceous of Mongolia. *Nature* 389 (1997): 483–485.
65. J. D. Archibald. Fossil evidence for a late Cretaceous origin of "hoofed" mammals. *Science* 272 (1996): 1150–1153.
66. T. J. Gaudin, J. R. Wible, J. A. Hopson and W. D. Turnbull. Reexamination of the morphological evidence for the cohort Epitheria (Mammalia, Eutheria). *Journal of Mammalian Evolution* 3 (1) (1996): 31–79.
67. S. B. Hedges, P. H. Parker, C. G. Sibley and S. Kumar. Continental breakup and the ordinal diversification of birds and mammals. *Nature* 381 (1996): 226–229.
68. A. Janke, X. Xu and U. Arnason. The complete mitochondrial genome of the wallaroo *(Macropus robustus)* and the phylogenetic relationships among Monotremata, Marsupialia, and Eutheria. *Proceedings of the National Academy of Sciences USA* 94 (1997): 1276–1281.

69. U. Arnason, A. Gullberg and A. Janke. Phylogenetic analyses of mitochondrial DNA suggest a sister groups relationship between Xenarthra (Edentata) and Ferungulates. *Molecular Biology and Evolution* 14 (7) (1997): 762–768.
70. M. A. Springer and J. A. W. Kirsch. A molecular perspective on the phylogeny of placental mammals based on mitochondrial 12S rDNA sequences, with special reference to the problem of Paenungulata. *Journal of Mammalian Evolution* 1 (1993): 149–166.
71. M. S. Springer, G. C. Cleven, O. Madsen, W. W. De Jong, V. G. Waddell, H. M. Amrine and M. J. Stanhope. Endemic African mammals shake the phylogenetic tree. *Nature* 388 (1997): 61–64.
72. A. M. D'Erchia, C. Gissi, G. Pesole, C. Saccone and U. Arnason. The guinea pig is not a rodent. *Nature* 381 (1996): 597–600.
73. S. Easteal, C. Collett and D. Betty. *The mammalian molecular clock.* Austin, TX: R. G. Landes, 1995.
74. Y. Cao, J. Adachi, A. Janke, S. Pääbo and M. Hasegawa. Phylogenetic relationships among eutherian orders estimated from inferred sequences of mitochondrial proteins: Instability of a tree based on a single gene. *Journal of Molecular Evolution* 39 (1994): 519–527.
75. A. Janke, G. Feldmaier-Fuchs, W. K. Thomas, A. von Haeseler and S. Pääbo. The marsupial mitochondrial genome and the evolution of placental mammals. *Genetics* 137 (1994): 243–256.
76. S. Kumar and S. B. Hedges. A molecular timescale for vertebrate evolution. *Nature* 392 (1998): 917–920.
77. P. D. Gingerich, B. H. Smith and E. L. Simons. Hindlimbs of Eocene *Basilosaurus:* Evidence of feet in whales. *Science* 249 (1990): 154–157.
78. M. C. Milinkovitch and J. G. M. Thewissen. Even-toed fingerprints on whale ancestry. *Nature* 388 (1997): 622–623.
79. M. Hasegawa, J. Adachi and M. C. Milinkovitch. Novel phylogeny of whales supported by total molecular evidence. *Journal of Molecular Evolution* 44, suppl. 1 (1997): S117–S120.
80. M. Shimamura, H. Yasue, K. Ohshima, H. Abe, H. Kato, T. Kishiro, M. Goto, I. Munechika and N. Okada. Molecular evidence from retroposons that whales form a clade with even-toed ungulates. *Nature* 388 (1997): 666–670.
81. J. Gatesey, C. Hayashi, M. A. Cronin and P. Arctander. Evidence from milk casein genes that cetaceans are the close relatives of hippopotamid artiodactyls. *Molecular Biology and Evolution* 13 (1996): 954–963.
82. D. Graur and D. G. Higgins. Molecular evidence for the inclusion of cetaceans within the order Artiodactyla. *Molecular Biology and Evolution* 11 (1994): 357–364.
83. M. J. Novacek. Morphological and molecular inroads to phylogeny. In L. Grande and O. Rieppel, eds., *Interpreting the hierarchy of nature: From systematic patterns to evolutionary process theories,* 85–131. New York: Academic, 1994.
84. M. M. Miyamoto. A congruence study of molecular and morphological data for eutherian mammals. *Molecular Phylogeny and Evolution* 6 (3) (1996): 373–390.
85. M. A. Norell and M. J. Novacek. The fossil record and evolution: Comparing cladistic and paleontologic evidence for vertebrate history. *Science* 255 (1992): 1690–1693.
86. M. Zhu, X. Yu and P. Janvier. A primitive fossil fish sheds light on the origin of bony fishes. *Nature* 397 (1999): 607–610.

Paul Kenrick and Peter R. Crane

THE ORIGIN AND EARLY EVOLUTION OF PLANTS ON LAND

The origin and early evolution of land plants in the mid-Paleozoic era, between about 480 and 360 million years ago, was an important event in the history of life, with far-reaching consequences for the evolution of terrestrial organisms and global environments. A recent surge of interest, catalyzed by paleobotanical discoveries and advances in the systematics of living plants, provides a revised perspective on the evolution of early land plants and suggests new directions for future research.

The origin and early diversification of land plants marks an interval of unparalleled innovation in the history of plant life. From a simple plant body consisting of only a few cells, land plants (liverworts, hornworts, mosses and vascular plants) evolved an elaborate two-phase life cycle and an extraordinary array of complex organs and tissue systems. Specialized sexual organs (gametangia), stems with an intricate fluid transport mechanism (vascular tissue), structural tissues (such as wood), epidermal structures for respiratory gas exchange (stomates), leaves and roots of various kinds, diverse spore-bearing organs (sporangia), seeds and the tree habit had all evolved by the end of the Devonian period. These and other innovations led to the initial assembly of plant-dominated terrestrial ecosystems, and had a great effect on the global environment.

Early ideas on the origin of land plants were based on living groups, but since the discovery of exceptionally well preserved fossil plants in the early Devonian Rhynie Chert, research has focused almost exclusively on the fossil record of vascular plants.[1,2] During the 1970s, syntheses of paleobotanical and stratigraphic data emphasized the late Silurian and Devonian periods as the critical interval during which the initial diversification of vascular plants occurred,[1,2] and identified a group of simple fossils (rhyniophytes, such as *Cooksonia* and *Rhynia*) as the likely ancestral forms.[2] They also supported earlier hypotheses of two main lines of evolution: one comprising clubmosses (Fig. 1f) and extinct relatives, the other including all other living vascular plants (ferns, horsetails and seed plants;

Fig. 1 Morphological diversity among basal living land plants and potential land-plant sister groups. *a, Coleochaete orbicularis* (Charophyceae) gametophyte; ×75 (photograph courtesy of L. E. Graham). *b, Chara* (Charophyceae) gametophyte; ×1.5 (photograph courtesy of M. Feist). *c, Riccia* (liverwort) gametophyte showing sporangia *(black)* embedded in the thallus; ×5 (photograph courtesy of A. N. Drinnan). *d, Anthoceros* (hornwort) gametophyte showing unbranched sporophytes; ×2.5 (photograph courtesy of A. N. Drinnan). *e, Mnium* (moss) gametophyte showing unbranched sporophytes with terminal sporangia (capsule); ×4.5 (photograph courtesy of W. Burger). *f, Huperzia* (clubmoss) sporophyte with leaves showing sessile yellow sporangia; ×0.8. *g, Dicranopteris* (fern) sporophyte showing leaves with circinate vernation; ×0.08. *h, Psilotum* (whisk fern) sporophyte with reduced leaves and spherical synangia (three fused sporangia); ×0.4. *i, Equisetum* (horsetail) sporophyte with whorled branches, reduced leaves and a terminal cone; ×0.4. *j, Cycas* (seed plant) sporophyte showing leaves and terminal cone with seeds; ×0.05 (photograph courtesy of W. Burger). (A color version of this figure is available on-line at <http://www.press.uchicago.edu/books/gee>.)

Fig. 1g–j) and related fossils.[1,2] During the past two decades, the discovery of fossil spores from as far back as the mid-Ordovician period,[3] improved knowledge of living green algae,[4,5] renewed interest in the phylogenetic position of other relevant groups such as mosses and liverworts,[5] and advances in molecular systematics,[5–14] together with unexpected new data on the structure and biology of Silurian and Devonian fossils,[15–25] have provided a broader perspective on the origin of a land flora.[26] These new data indicate that the early diversification of land plants substantially predates the late Silurian to early Devonian, and suggest that the main basal lineages originated over a period of more than 100 million years (Myr).

Patterns in the Early Fossil Record

Evidence on the origin and diversification of land plants has come mainly from dispersed spores and megafossils. Gray recognized three new plant-based epochs (Eoembryophytic, Eotracheophytic and Eutracheophytic) spanning the origin and early establishment of land plants: each is characterized by the relative abundance of spore types and megafossils.[3] This synthesis highlights diversification and floral change in the Ordovician and Silurian,[3,27,28] and emphasizes a major discrepancy between evidence from spores and megafossils: unequivocal land-plant megafossils are first recognized in the fossil record roughly 50 Myr after the appearance of land-plant spores.

Eoembryophytic (Mid-Ordovician [Early Llanvirn: ~476 Myr] to Early Silurian [Late Llandovery: ~432 Myr])[3]

Spore tetrads (comprising four membrane-bound spores; Fig. 2d) appear over a broad geographic area in the mid-Ordovician and provide the first good evidence of land plants.[3,26,29] The combination of a decay-resistant wall (implying the presence of sporopollenin) and tetrahedral configuration (implying haploid meiotic products) is diagnostic of land plants. The precise relationships of the spore producers within land plants are controversial, but evidence of tetrads and other spore types (such as dyads) in late Silurian and Devonian megafossils,[16,30] as well as data on spore wall ultrastructure[25] and the structure of fossil cuticles,[31] support previous suggestions of a land flora of liverwort-like plants (Fig. 1c).[3] Some early spores and cuticles may also represent extinct transitional lineages between charophycean algae (Fig. 1a, b) and liverworts (Box 1), but precise understanding of their affinities is hindered by the dearth of associated megafossils.

Eotracheophytic (Early Silurian [Latest Llandovery: ~432 Myr] to Early Devonian [Mid-Lochkovian: ~402 Myr])[3]

The early Silurian (latest Llandovery) marks the beginning of a decline in diversity of tetrads and a rise to dominance of individually dispersed, simple spores, which are found in several basal land-plant groups (such

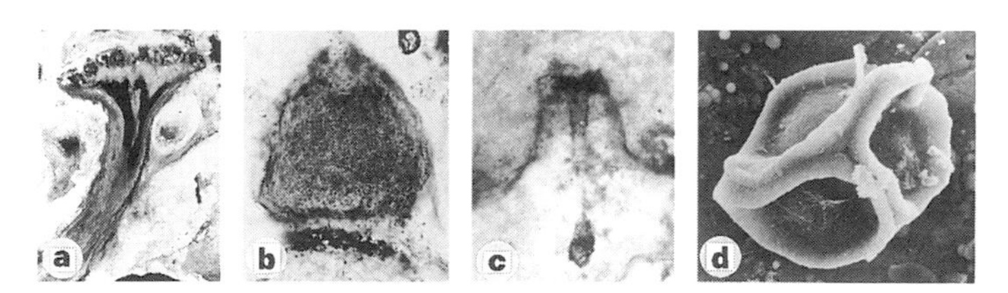

Fig. 2 *a,* Longitudinal section of part of a silicified early fossil gametophyte (*Kidstonophyton discoides* from the Rhynie Chert). Antheridia (male sexual organs) are located on the upper surface of the branch; ×3.4. *b,* Longitudinal section of antheridium of *Lyonophyton rhyniensis* from the Rhynie Chert; ×40. *c,* Longitudinal section of archegonium (female sexual organ) of *Langiophyton mackiei* from the Rhynie Chert; ×80. *a–c* are from the Remy Collection (slides 200, 90 and 330), Abteilung Paläobotanik, Westfälische Wilhelms-Universität, Münster, Germany (photographs courtesy of H. Hass and H. Kerp). *d,* Scanning electron micrograph of *Tetrahedraletes medinensis* showing a spore tetrad of possible liverwort affinity from the late Ordovician (photograph courtesy of W. A. Taylor); ×670.

BOX 1 Relationships among land plants

Land plants (embryophytes) are most closely related to the Charophyceae, a small group of predominantly freshwater green algae, within which either Coleochaetales (~15 living species; Fig. 1a) or Charales (~400 living species; Fig. 1b), or a group containing both, is sister group to land plants.[4,5,10,12,74]

Land-plant monophyly is supported by comparative morphology[4,5,26,75] and gene sequences (18S rRNA, mitochondrial DNA: *cox III*).[12,14] Relationships among the major basal living groups are uncertain,[4,5,26,76,77] but the best-supported hypothesis resolves liverworts (Fig. 1c) as basal and either mosses (Fig. 1e) or hornworts (Fig. 1d) as the living sister group to vascular plants (tracheophytes).[4,5,13,14,26,75] Less parsimonious hypotheses recognize bryophyte monophyly and either a sister-group relationship with vascular plants[26] or an origin from within basal vascular plants.[14,76,78]

Among vascular plants, living ferns (Fig. 1g), horsetails (Fig. 1i) and seed plants (Fig. 1j) (euphyllophytes) are the sister group to clubmosses (Fig. 1f).[13,14,26,75,79] Euphyllophyte monophyly is strongly supported by comparative morphology[26] and a unique 30-kb inversion in the chloroplast genome,[8] as well as sequence data from 18S rRNA[13] and mitochondrial DNA *(cox III)*.[14] These data also provide evidence that the enigmatic Psilotaceae (Fig. 1h) (a group of simple plants once thought to be living relicts of the earliest vascular plants) are more closely related to the fern–seed-plant lineage than to basal vascular plants (clubmosses or the extinct rhyniophytes). Within vascular plants, molecular and morphological assessments of phylogeny at the level of orders and below give similar results,[11] but at deeper levels (for example, the divergence of major groups of ferns, horsetails and seed plants) phylogenetic resolution is poor. These difficulties highlight the problems of approaches based solely on living species.[7,80] Combined analyses of molecular sequences from multiple loci, and large-scale structural characteristics of the genome (such as introns and inversions), may be more useful in assessing deep phylogenetic patterns.

Megafossils fill some of the substantial morphological "gaps" among living groups. Phylogenetic analyses[19,26] interpolate two early Devonian Rhynie Chert plants, *Aglaophyton* and *Horneophyton*, between bryophytes and basal vascular plants as they possess some features unique to vascular plants (a branched, nutritionally independent sporophyte) but also retain bryophyte-like characteristics (terminal sporangia, columella in *Horneophyton*, and the absence of leaves, roots and tracheids with well-defined thickenings). The discovery of previously unrecognized diversity in extinct *Cooksonia* and

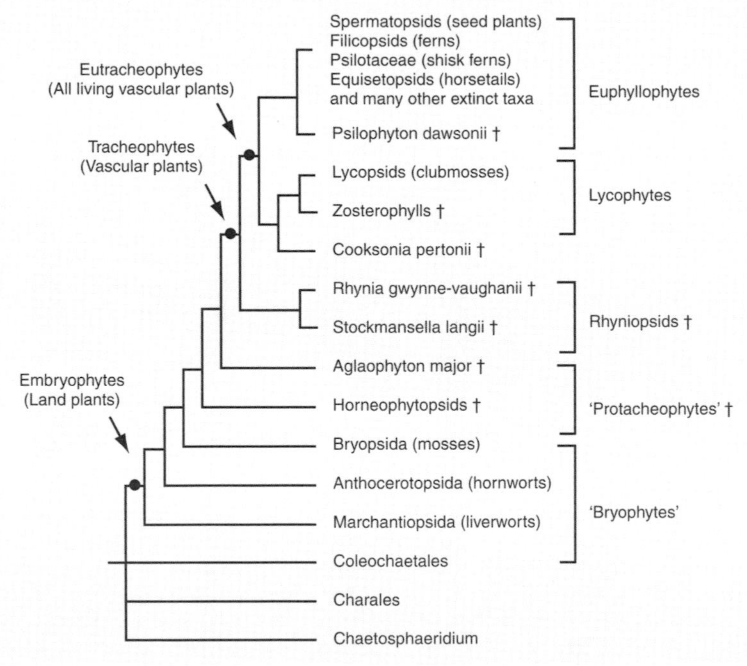

similar early fossils (such as *Tortilicaulis, Uskiella, Caia*[15–17,81]) (Fig. 3) suggests that simple early land plants (once grouped as rhyniophytes[1]) are an unnatural assemblage.[26] Some *Cooksonia* species may be among the precursors to vascular plants (protracheophytes), whereas others are vascular plants apparently allied to the clubmoss lineage.[26]

Clubmosses emerge from a poorly resolved grade of extinct *Zosterophyllum*-like plants (Fig. 4), although most zosterophylls form a monophyletic group.[26] Within clubmosses, early leafy herbaceous fossils such as *Baragwanathia* and *Asteroxylon* are basal,[26,82] and living Lycopodiaceae are resolved as sister group to a clade that comprises the extinct herbaceous Protolepidodendrales, living *Selaginella* and the predominantly arborescent carboniferous lepidodendrids, including living *Isoetes*[26,82] (Fig. 4).

Euphyllophytes make up more than 99% of living vascular plants and exhibit much greater diversity than lycophytes. Relationships among basal euphyllophytes are still poorly understood.[26] Further progress requires a better understanding of the relationships of several fossil groups of uncertain status (such as Trimerophytina, Cladoxylales and Zygopteridales).[26,79]

as hornworts, some mosses, and early vascular plants).[3] Although tetrads remain dominant in some early Devonian localities from northwestern Europe,[32] the elaboration of simple spores and turnover of spore "species"[3] provide evidence of increasing land-plant diversity and vegetational change. Although spores have been observed in Silurian megafossils, the affinities of most dispersed forms remain unknown, indicating that substantial land-plant diversity is currently undetected in the megafossil record.[30]

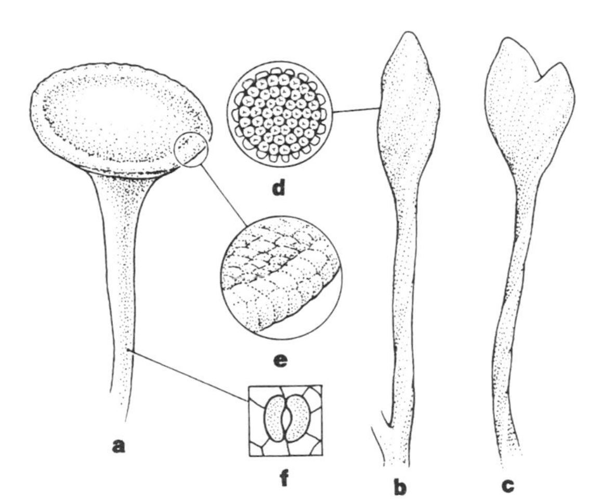

Fig. 3 Sporophyte diversity in early Devonian rhyniophyte fossils. *a, Cooksonia pertonii apiculispora:* sporophyte (incomplete proximally) with terminal sporangium;[15] ×15. *b, Tortilicaulis offaeus:* sporophyte (incomplete proximally) with terminal sporangium;[81] ×40. *c, Tortilicaulis offaeus:* sporophyte (incomplete proximally) with terminal bifurcating sporangium;[81] ×30. *d,* Transverse section of sporangium showing thick wall and central spore mass; ×70. *e,* Details of epidermis at rim of sporangium; ×45. *f,* Stomate with two reniform guard cells *(stippled);* ×120.

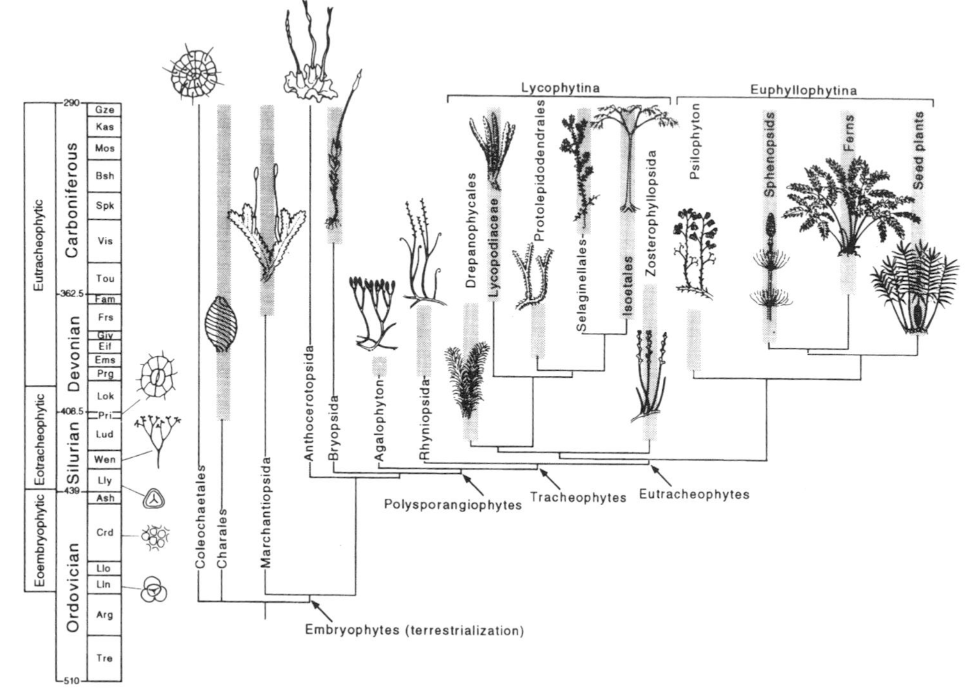

Fig. 4 Simplified phylogenetic tree showing the minimum stratigraphic ranges of selected groups based on megafossils *(thick bars)* and their minimum implied range extensions *(thin lines)*. Also illustrated alongside timescale are minimum age estimates for the appearance of certain important land-plant features (from the bottom: spore tetrads, cuticles, single trilete spores, megafossils and stomates). The first unequivocal record of charophycean algae is based on calcified charalean oogonia (female sexual organs) from the late Silurian (Pridoli, ~410 Myr)[83] and distinctive gametophytes from the early Devonian Rhynie Chert.[44] Proposed similarities between living *Coleochaete* and early Devonian *Parka* remain to be confirmed.[44] Note that megafossil evidence for vascular plants precedes megafossil evidence of bryophytes and charophycean algae. Confirmation that the early Devonian *Sporogonites* is a plant at the bryophyte grade could help to reduce this discrepancy. *Tre,* Tremadoc; *Arg,* Arenig; *Lln,* Llanvirn; *Llo,* Llandeilo; *Crd,* Caradoc; *Ash,* Ashgill; *Lly,* Llandovery; *Wen,* Wenlock; *Lud,* Ludlow; *Pri,* Pridoli; *Lok,* Lochkovian (Gedinnian); *Prg,* Pragian (Siegenian); *Ems,* Emsian; *Eif,* Eifelian; *Giv,* Givetian; *Frs,* Frasnian; *Fam,* Famennian; *Tou,* Tournaisian; *Vis,* Visean; *Spk,* Serpukhovian; *Bsh,* Bashkirian; *Mos,* Moscovian; *Kas,* Kasimovian; *Gze,* Gzelian.

The earliest unequivocal land-plant megafossils are from the mid-Silurian of northern Europe,[33] and lowermost Upper Silurian of Bolivia[34] and Australia,[35] and the uppermost Silurian of northwestern China.[36] Early assemblages include clubmosses (such as *Baragwanathia*) and related early fossils (such as zosterophylls, some species of *Cooksonia*), and various other plants of uncertain affinity (such as *Salopella* and *Hedeia;* Fig. 3). These data document an influx into land-plant communities of diverse but generally small (usually less than 10 cm tall) organisms related to vascular plants (Fig. 3). Exceptions to the generally small size include the clubmoss *Baragwanathia*[37] and the large and much-branched *Pinnatiramosus* from the early Silurian of China.[38] The habit and branching of *Pinnatiramosus* is similar to that of green algae in the Caulerpales, but the presence of tracheid-like tubes is inconsistent with this interpretation.[39] Additional details, including conclusive data on reproductive structures, are needed to clarify the relationships of this enigmatic plant.

Data from northern Europe, Siberia, Podolia (southwestern Ukraine), Libya, Vietnam, Bolivia, Australia and Xinjiang and Yunnan (China) document increasing land-plant diversity into the base of the Devonian.[33–36,40] These fossils, together with the relative chronology implicit in current hypotheses of relationship, imply a minimum mid-Silurian origin for several important vascular plant groups (Box 1; Fig. 4).

Eutracheophytic (Early Devonian [Late Lochkovian: ~398 Myr] to Mid-Permian [~256 Myr])[3]

In the early Devonian (late Lochkovian) the diversity of spores and megafossils increased dramatically.[29,40–42] Early assemblages include the classic floras from the Rhynie Chert,[20,43,44] the Gaspé Peninsula of eastern Canada,[43,44] New York State,[43,44] the Rhine Valley of Germany,[45] Belgium,[46] Australia[35] and Yunnan Province (China),[33] which document a substantial increase in vascular plant diversity, including the appearance and early diversification of many important living groups.

Building a Land Plant

Phylogenetic studies favor a single origin of land plants from charophycean green algae (Box 1). Based on the ecology of living species, a freshwater origin of land plants seems likely, but direct evidence from the fossil record is inconclusive as mid-Paleozoic charophytes are found in both freshwater and, more commonly, marine facies.[47] Living charophycean algae (Fig. 1a, b) possess several biosynthetic attributes that are expressed more fully among land plants, including the capacity to produce sporopollenin, cutin, phenolic compounds and the glycolate oxidase pathway.[4,48] However, the absence of well-developed sporophytes, gametophytes with sexual organs of land-plant type, cuticle and non-motile, airborne, sporopollenin-walled spores suggests that these innovations evolved during the transition to the land.[4,18] In contrast to animal groups, the entire multicellular diploid phase of the plant life cycle probably evolved in a terrestrial setting.

The transition from an aqueous to a gaseous medium exposed plants to new physical conditions that resulted in key physiological and structural changes. Important metabolic pathways leading to lignins, flavenoids, cutins and plant hormones in vascular plants probably arose from pre-existing elements of primary metabolism in charophycean algae and bryophytes.[4] Although the evolution of these pathways is poorly understood, possible phenolic precursors have been detected in charophycean algae,[4,31] and elements of auxin metabolism have been recognized in mosses and hornworts.[49]

Phylogenetic studies predict that early land plants were small and morphologically simple, and this hypothesis is borne out by fossil evidence (Fig. 3). Early fossils bear a strong resemblance to the simple spore-producing phase of living mosses and liverworts (Fig. 1d, e),[16,26,50,51] and these similarities extend to the anatomical details of the spore-bearing organs and the vascular system (Fig. 5).[19] The fossil record also documents significant differences from living groups, particularly in life cycles and the early evolution of the sexual phase (Box 2).

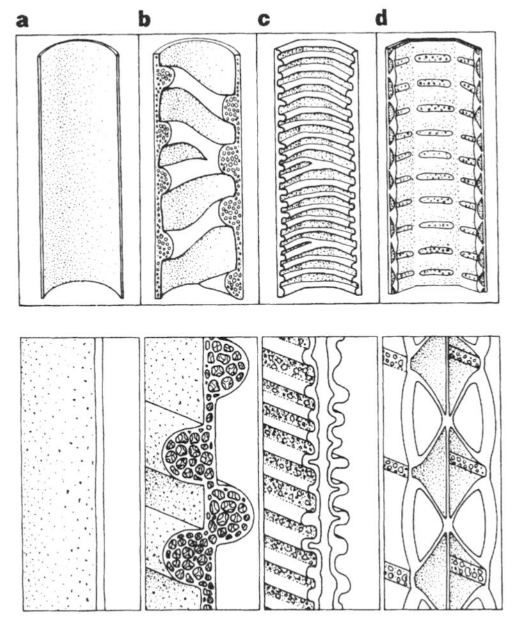

Fig. 5 Diversity of water-conducting cells (tracheids) in early land plants (median longitudinal section through cells, basal and proximal end walls not shown; cells are 20–40 μm diameter). *a, Top,* bryophyte hydroid; *bottom,* details of hydroid wall showing distribution of plasmodesmata-derived micropores (10–50 nm diameter; *stipple*).[84] *b, Top,* S-type tracheid (fossil) of Rhyniopsida; *bottom,* details of S-type cell wall showing distribution of plasmodesmata-derived micropores *(stipple)* and "spongy" interior to thickenings.[19] *c, Top,* G-type tracheid (fossil) of basal extinct eutracheophytes, which closely resembles the tracheids of some living vascular plants; *bottom,* details of G-type cell wall showing pores distributed between thickenings.[19] *d, Top,* scalariform pitted P-type tracheid (fossil) typical of trimerophyte-grade plants (euphyllophytes); *bottom,* details of P-type cell wall showing pit chambers and sheet with pores that extends over pit apertures.[26]

BOX 2 Early evolution of the land-plant life cycle

Land-plant life cycles are characterized by alternating multicellular sexual (haploid gametophyte, n) and asexual phases (diploid sporophyte, $2n$). Phylogenetic studies indicate that land plants inherited a multicellular gametophyte from their algal ancestors but that the sporophyte evolved during the transition to the land. Most megafossils are sporophytes, and until recently there was no direct early fossil evidence for the gametophyte phase. Recent discoveries of gametophytes in the Rhynie Chert (early Devonian, 408–380 Myr) have shed new light on the evolution of land-plant life cycles.[18,20]

Early gametophytes (*a* in figure) are more complex than in living plants and have branched stems bearing sexual organs on terminal cup- or shield-shaped structures (Fig. 2a). Archegonia (female gametangia) are flask-shaped with a neck canal and egg chamber, and are sunken as in hornworts and most vascular plants (Fig. 2c). Antheridia (male gametangia) are roughly spherical, sessile or with a poorly defined stalk, and superficial (Fig. 2b). Gametophytes are very similar to associated sporophytes, and shared anatomical features (water-conducting tissues, epidermal patterns and stomates) have been used to link corresponding elements of the life cycle.[18,20] Our provisional reconstruction of the life cycle of an early vascular plant is based on information from anatomically preserved plants and contemporaneous compression fossils.

The similarities between gametophyte and sporophyte in early fossil vascular plants contrast strongly with the marked dissimilarities typical of living land plants (*b* in figure). The phylogenetic position of fossils suggests that, after the development of a simple, unbranched, "parasitic" sporophyte among early land colonizers at the bryophyte grade (such as mosses), there was elaboration of both gametophyte and sporophyte in vascular plants. The implications for interpreting life cycles in living vascular plants[18,26] are shown. The small, simple, often subterranean and saprophytic gametophytes of living clubmosses (such as Lycopodiaceae) and ferns (such as Psilotaceae, Stromatopteridaceae, Ophioglossaceae) result from morphological loss. Phylogenetic evidence indicates that gametophyte reduction was independent in clubmosses and the fern–seed-plant lineage. These data provide a new interpretation of the gametophyte morphology of living clubmosses (Lycopodiaceae).[18]

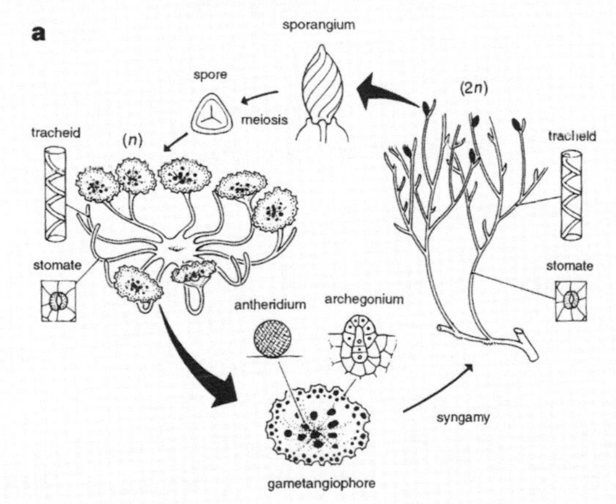

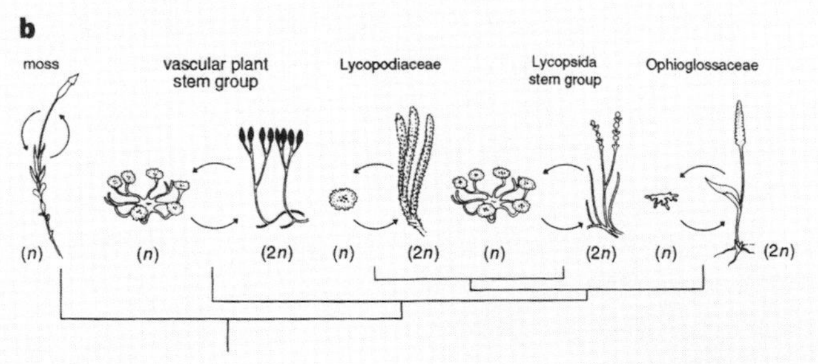

In common with some animal groups, internalization of vital functions and organs (such as gas exchange surfaces and sexual organs), combined with the development of impermeable exterior surfaces, seems to have been a primary response to life on land. Together, these changes resulted in more highly differentiated plants with stomates, multicellular sexual and spore-bearing organs, water-conducting and other tissue systems.[52–54] Morphological differentiation occurred in both phases of the life cycle (gametophyte and sporophyte), but there was subsequently a dramatic reduction in the gametophyte and a great increase in sporophyte complexity among vascular plants (Box 2). Apical growth and branching coupled with delayed initiation of spore-bearing organs were important innovations of vascular plants that led to a more complex architectural framework on which subsequent morphological diversification was based. The fossil record clearly shows that many vascular-plant organs can be interpreted in terms of modification (especially duplication and sterilization) of basic structural units such as the spore-bearing tissues and the stem.[26,54] In ferns and seed plants, much morphological diversity is clearly attributable to modifications of branching systems into a variety of leaf-like organs, whereas the relatively conservative clubmoss bauplan has a dearth of organ systems that can be interpreted as modified branches. In both lineages, however, meristem dormancy and abortion were early innovations, providing evidence of hormonal control and substantial phenotypic flexibility.[21,26]

Early Terrestrial Ecosystems

The advent of land plants had important consequences for energy and nutrient fluxes among terrestrial and freshwater ecosystems[29,55] and hence for the evolution of animal groups that live in these habitats. The vegetational changes of the Silurian and Devonian also had a major impact on the atmosphere and other aspects of the global environment. The evolution of roots is thought to have been an important factor in the reduction of atmospheric CO_2 concentrations through increased weathering of Ca-Mg silicate minerals brought about by mechanical disruption and soil acidification.[56,57] Accelerated weathering has also been linked to the formation of Devonian and early Carboniferous marine black shales,[58] but this requires further investigation in view of similar deposits earlier and later in the geological record. Root-like impressions have been recognized in late Silurian paleosols,[59] but the earliest unequivocal evidence comes from early Devonian vascular plants,[26] which have modified prostrate stems bearing rhizoids resembling those of living bryophytes. More substantial roots capable of anchoring large trees evolved independently in several groups during the middle to late Devonian.

A further series of innovations in vascular plants, including the biosynthesis of lignin and the origin of lateral meristems (cambium), were critical to the development of large plants, and these developments may have been stimulated by competition for light. Trees evolved independently in several major groups, resulting in stratified forest communities by the end of the middle Devonian and the production of large amounts of highly decay-resistant organic material (in the form of lignified wood). The early evolution of lignin-decomposing fungi (some Ascomycetes, and Basidiomycetes) is still poorly understood,[24] but these groups would have been essential for recycling much of the organic carbon.

The earliest land plants probably encountered terrestrial ecosystems that had been occupied by bacteria and protists,[60,61] algae,[4] lichens[23,62] and fungi[24] since the late Proterozoic. A variety of enigmatic plants (such as *Protosalvinia*[44,63]) were also present, and some of the largest elements (*Prototaxites* "trunks" >69 cm in diameter) may have been fungi.[64] Such organisms, or perhaps some rhyniophytes,[16] may be the source of the microscopic tubular fragments commonly extracted from Silurian and early Devonian sediments.[28] These tubes are often associated with cellular cuticular fragments *(Nematothallus* and *Cosmochlaina)* that may represent fragmented cuticular material from bryophyte-like plants.[31] The discovery of fungal arbusculae in early Devonian megafossils[22] confirms earlier suggestions that endomycorrhizal associations were an important innovation in the colonization of the land.[65]

In contrast to megascopic plants, which appear to have colonized the land only once, many animal groups made the transition to terrestrial existence independently and overcame the problems of water relations in different ways.[52,66,67] Early evidence for terrestrial animals is sparse,[29,67–69] but by the early Devonian exquisitely preserved arthropod faunas are known from several localities in North America, Germany and the United Kingdom.[29,66,67] These faunas document the appearance of diverse arthropod communities including centipedes, millipedes, trigonotarbids and their living relatives spiders, pseudoscorpions, mites (orbatids and endeostigmatids), arthropleurids (extinct arthropods), archaeognathans (primitive wingless insects), collembolans and possibly bristletails. Available evidence indicates that these animals were mainly predators and detritivores and, until the appearance of vertebrate herbivores in the latest Paleozoic, most energy flow into animal components of early terrestrial ecosystems was probably through the decomposer pathway rather than direct herbivory.[29] Indirect evidence for herbivory comes from wound responses in the tissues of some fossil plants,[70,71] and perhaps also from fossil fecal pellets containing abundant spores.[70,72]

Future Directions

The fossil record of spores, combined with phylogenetic studies, indicates that groups related to living bryophytes were early colonizers of the land, and suggests that several major lineages of vascular plant had already evolved by the mid-Silurian. Megafossils of land plants, however, appear much later, and in these assemblages there is a conspicuous bias toward the recognition and perhaps representation of vascular plants. The most important source of data on early megafossils has been the northern European (Laurussian) region, but the appearance of megafossils in this area coincides with facies changes driven by a widespread marine regression,[28,73] and all Silurian land-plant megafossils are from marine sediments.[33] It seems likely that the onset of continental conditions in the Devonian of northern Europe allowed megafossils to be preserved at a time when vascular plants were well established but still diversifying. The rapid appearance of vascular plants in this region[40–42] owes as much to changing geological conditions as to rapid biological diversification.[27,28] Intensified sampling in areas that are remote from these regional events is therefore a high priority.

Paleobotanical evidence shows that the major groups of living land plants are relicts, even though much modern species diversity within these groups may have evolved more recently. Data from the fossil record are therefore especially important for clarifying homologies among major organ systems which may otherwise be difficult to detect as a result of morphological divergence and extinction. Such combined studies of living and fossil plants provide an improved basis for comparative studies of plant development. They indicate, for example, that the ontogeny of leaves and spore-bearing organs in clubmosses are likely to share substantial similarities, but are unlikely to exhibit common features with leaves in seed plants, ferns and horsetails. They also suggest that fundamental features of land plants, such as the spore-bearing organs, stems, stomates and sexual organs, are each under the same kind of developmental control in all main groups. To explore these issues further, data are needed on the molecular basis of plant development from a broader selection of land plants than are currently under study. In the context of a more complete understanding of plant diversity than that provided by living plants alone, such data should be expected to confirm the underlying unity and relative simplicity of developmental processes in land plants.

PAUL KENRICK is at the Swedish Museum of Natural History, Box 50007, S104 05, Stockholm, Sweden; Peter R. Crane is at the Field Museum, Roosevelt Road at Lake Shore Drive, Chicago, Illinois 60605, and the Department of the Geophysical Sciences, University of Chicago, USA.

References

1. Banks, H. P. Reclassification of Psilophyta. *Taxon* **24,** 401–413 (1975).
2. Chaloner, W. G. & Sheerin, A. in *The Devonian System* (eds. House, M. R., Scrutton, C. T. & Bassett, M. G.) 145–161 (The Palaeontological Association, London, 1979).
3. Gray, J. Major Paleozoic land plant evolutionary bio-events. *Palaeogeog. Palaeoclimatol. Palaeocol.* **104,** 153–169 (1993).
4. Graham, L. E. *Origin of Land Plants* (Wiley, New York, 1993).
5. Mishler, B. D. *et al.* Phylogenetic relationships of the "green algae" and "bryophytes." *Ann. MO Bot. Gard.* **81,** 451–483 (1994).
6. Manhart, J. R. & Palmer, J. G. The gain of two chloroplast tRNA introns marks the green algal ancestors of land plants. *Nature* **345,** 268–270 (1990).
7. Manhart, J. R. Phylogenetic analysis of green plant *rbc*L sequences. *Mol. Phylogenet. Evol.* **3,** 114–127 (1994).
8. Raubeson, L. A. & Jansen, R. K. Chloroplast DNA evidence on the ancient evolutionary split in vascular land plants. *Science* **255,** 1697–1699 (1992).
9. Chapman, R. L. & Buchheim, M. A. Ribosomal RNA gene sequences: Analysis and significance in the phylogeny and taxonomy of green algae. *Crit. Rev. Plant Sci.* **10,** 343–368 (1991).
10. McCourt, R. M., Karol, K. G., Guerlesquin, M. & Feist, M. Phylogeny of extant genera in the family Characeae (Charales, Charophyceae) based on *rbc*L sequences and morphology. *Am. J. Bot.* **83,** 125–131 (1996).
11. Pryer, K. M., Smith, A. R. & Skog, J. E. Phylogenetic relationships of extant ferns based on evidence from morphology and *rbc*L sequences. *Am. Fern J.* **85,** 205–282 (1995).
12. Kranz, H. D. *et al.* The origin of land plants: Phylogenetic relationships among charophytes, bryophytes, and vascular plants inferred from complete small-subunit ribosomal RNA gene sequences. *J. Mol. Evol.* **41,** 74–84 (1995).
13. Kranz, H. D. & Huss, V. A. R. Molecular evolution of pteridophytes and their relationships to seed plants: Evidence from complete 18S rRNA gene sequences. *Plant Syst. Evol.* **202,** 1–11 (1996).
14. Hiesel, R., von Haeseler, A. & Brennicke, A. Plant mitochondrial nucleic acid sequences as a tool for phylogenetic analysis. *Proc. Natl. Acad. Sci. USA* **91,** 634–638 (1994).
15. Edwards, D., Davies, K. L. & Axe, L. A vascular conducting strand in the early land plant *Cooksonia. Nature* **357,** 683–685 (1992).
16. Edwards, D., Duckett, J. G. & Richardson, J. B. Hepatic characters in the earliest land plants. *Nature* **374,** 635–636 (1995).
17. Fanning, U., Edwards, D. & Richardson, J. B. A diverse assemblage of early land plants from the Lower Devonian of the Welsh Borderland. *Bot. J. Linn. Soc.* **109,** 161–188 (1992).
18. Kenrick, P. Alternation of generations in land plants: New phylogenetic and morphological evidence. *Biol. Rev.* **69,** 293–330 (1994).
19. Kenrick, P. & Crane, P. R. Water-conducting cells in early fossil land plants: Implications for the early evolution of tracheophytes. *Bot. Gaz.* **152,** 335–356 (1991).
20. Remy, W. Gensel, P. G. & Hass, H. The gametophyte generation of some early Devonian land plants. *Int. J. Plant Sci.* **154,** 35–58 (1993).
21. Remy, W. & Hass, H. New information on gametophytes and sporophytes of *Aglaophyton major* and inferences about possible environmental adaptations. *Rev. Palaeobot. Palynol.* **90,** 175–194 (1996).

22. Remy, W., Taylor, T. N., Hass, H. & Kerp, H. Four-hundred-million-year-old vesicular arbuscular mycorrhizae. *Proc. Natl. Acad. Sci. USA* **91,** 11841–11843 (1994).
23. Stein, W. E., Harmon, G. D. & Hueber, F. M. in *International Workshop on the Biology and Evolutionary Implications on Early Devonian Plants* (Westfälische Wilhelms-Universität, Münster, 1994).
24. Taylor, T. N. & Osborne, J. M. The importance of fungi in shaping the paleoecosystem. *Rev. Palaeobot. Palynol.* **90,** 249–262 (1996).
25. Taylor, W. A. Ultrastructure of lower Paleozoic dyads from southern Ohio. *Rev. Palaeobot. Palynol.* **92,** 269–280 (1996).
26. Kenrick, P. & Crane, P. R. *The Origin and Early Diversification of Land Plants: A Cladistic Study* (Smithsonian Institution Press, Washington, DC, 1997).
27. Gray, J. The microfossil record of early land plants: Advances in understanding of early terrestrialization, 1970–1984. *Phil. Trans. R. Soc.* B **309,** 167–195 (1985).
28. Gray, J. & Boucot, A. J. Early vascular land plants: Proof and conjecture. *Lethaia* **10,** 145–174 (1977).
29. DiMichele, W. A. *et al.* in *Terrestrial Ecosystems through Time: Evolutionary Paleoecology of Terrestrial Plants and Animals* (ed. Behrensmeyer, A. K.) 205–325 (Univ. Chicago Press, 1992).
30. Fanning, U., Richardson, J. B. & Edwards, D. in *Pollen and Spores* (eds. Blackmore, S. & Barnes, S. H.) 25–47 (Clarendon, Oxford, 1991).
31. Kroken, S. B., Graham, L. E. & Cook, M. E. Occurrence and evolutionary significance of resistant cell walls in charophytes and bryophytes. *Am. J. Bot.* **83,** 1241–1254 (1996).
32. Wellman, C. H. & Richardson, J. B. Sporomorph assemblages from the "Lower Old Red Sandstone" of Lorne, Scotland. *Spec. Pap. Palaeontol.* **55,** 41–101 (1996).
33. Edwards, D. in *Palaeozoic Palaeogeography and Biogeography* (eds. McKerrow, W. S. & Scotese, C. R.) 233–242 (Geological Society, London, 1990).
34. Morel, E., Edwards, D. & Iñiquez Rodriguez, M. The first record of *Cooksonia* from South America in the Silurian rocks of Bolivia. *Geol. Mag.* **132,** 449–452 (1995).
35. Tims, J. D. & Chambers, T. C. Rhyniophytina and Trimerophytina from the early land flora of Victoria, Australia. *Palaeontology* **27,** 265–279 (1984).
36. Cai, C.-Y., Dou, Y.-W. & Edwards, D. New observations on a Pridoli plant assemblage from north Xinjiang, northwest China, with comments on its evolutionary and palaeographical significance. *Geol. Mag.* **130,** 155–170 (1993).
37. Hueber, F. M. Thoughts on the early lycopsids and zosterophylls. *Ann. MO Bot. Gard.* **79,** 474–499 (1992).
38. Cai, C. *et al.* An early Silurian vascular plant. *Nature* **379,** 592 (1996).
39. Geng, B.-Y. Anatomy and morphology of *Pinnatiramosus,* a new plant from the middle Silurian (Wenlockian) of China. *Acta Bot. Sin.* **28,** 664–670 (1986).
40. Raymond, A. & Metz, C. Laurussian land-plant diversity during the Silurian and Devonian: Mass extinction, sampling bias, or both? *Paleobiology* **21,** 74–91 (1995).
41. Edwards, D. & Davies, M. S. in *Major Evolutionary Radiations* (eds. Taylor, P. D. & Larwood, G. P.) 351–376 (Clarendon, Oxford, 1990).
42. Knoll, A. H., Niklas, K. J., Gensel, P. G. & Tiffney, B. H. Character diversification and patterns of evolution in early vascular plants. *Paleobiology* **10,** 34–47 (1984).
43. Gensel, P. G. & Andrews, H. N. *Plant Life in the Devonian* (Praeger, New York, 1984).
44. Taylor, T. N. & Taylor, E. L. *The Biology and Evolution of Fossil Plants* (Prentice Hall, Englewood Cliffs, NJ, 1993).

45. Schweitzer, H.-J. Die Unterdevonflora des Rheinlandes. *Palaeontographica* B **189,** 1–138 (1983).
46. Gerrienne, P. Inventaire des végétaux éodévoniens de Belgique. *Ann. Soc. Géol. Belg.* **116,** 105–117 (1993).
47. Tappan, H. N. *The Paleobiology of Plant Protists* (Freeman, San Francisco, 1980).
48. Raven, J. Plant responses to high O_2 concentrations: Relevance to previous high O_2 episodes. *Palaeogeog. Palaeoclimatol. Palaeocol.* **97,** 19–38 (1991).
49. Sztein, A. E., Cohen, J. D., Slovin, J. P. & Cooke, T. J. Auxin metabolism in representative land plants. *Am. J. Bot.* **82,** 1514–1521 (1995).
50. Edwards, D. New insights into early land ecosystems: A glimpse of a Lilliputian world. *Rev. Palaeobot. Palynol.* **90,** 159–174 (1996).
51. Edwards, D., Fanning, U. & Richardson, J. B. Stomata and sterome in early land plants. *Nature* **323,** 438–440 (1986).
52. Raven, J. A. Comparative physiology of plant and arthropod land adaptation. *Phil. Trans. R. Soc.* B **309,** 273–288 (1985).
53. Raven, J. A. The evolution of vascular plants in relation to quantitative functioning of dead water-conducting cells and stomata. *Biol. Rev.* **68,** 337–363 (1993).
54. Niklas, K. J. *Plant Allometry: The Scaling of Form and Process.* (Univ. Chicago Press, 1994).
55. Beerbower, R. in *Geological Factors and the Evolution of Plants* (ed. Tiffney, B. H.) 47–92 (Yale Univ. Press, New Haven, 1985).
56. Berner, R. A. GEOCARB II: A revised model of atmospheric CO_2 over Phanerozoic time. *Am. J. Sci.* **294,** 56–91 (1994).
57. Mora, C. I., Driese, S. G. & Colarusso, L. A. Middle to late Paleozoic atmospheric CO_2 levels from soil carbonate and organic matter. *Science* **271,** 1105–1107 (1996).
58. Algeo, T. J., Berner, R., Maynard, J. B. & Scheckler, S. E. Late Devonian oceanic anoxic events and biotic crises: "Rooted" in the evolution of vascular land plants? *GSA Today* **5,** 45, 64–66 (1995).
59. Retallack, G. J. in *Paleosols: Their Recognition and Interpretation* (ed. Wright, V. P.) (Blackwell, Oxford, 1986).
60. Knoll, A. H. The early evolution of eukaryotes: A geological perspective. *Science* **256,** 622–627 (1992).
61. Bengtson, S. (ed.) *Early Life on Earth.* (Columbia Univ. Press, New York, 1994).
62. Taylor, T. N., Hass, H., Remy, W. & Kerp, H. The oldest fossil lichen. *Nature* **378,** 244 (1995).
63. Hemsley, A. R. in *Ultrastructure of Fossil Spores and Pollen* (eds. Kurmann, M. H. & Doyle, J. A.) 1–21 (Royal Botanic Gardens, Kew, 1994).
64. Hueber, F. M. in *International Workshop on the Biology and Evolutionary Implications of Early Devonian Plants* (Westfälische Wilhelms-Universität, Münster, 1994).
65. Simon, L., Bousquet, J., Léveque, C. & Lalonde, M. Origin and diversification of endomycorrhizal fungi with vascular plants. *Nature* **363,** 67–69 (1993).
66. Selden, P. A. & Edwards, D. in *Evolution and the Fossil Record* (eds. Allen, K. C. & Briggs, D. E. G.) 122–152 (Belhaven, London, 1989).
67. Gray, J. & Shear, W. Early life on land. *Am. Sci.* **80,** 444–456 (1992).
68. Gray, J. & Boucot, A. J. Early Silurian nonmarine animal remains and the nature of the early continental ecosystem. *Acta Palaeontol. Pol.* **38,** 303–328 (1994).
69. Retallack, G. J. & Feakes, C. R. Trace fossil evidence for late Ordovician animals on land. *Science* **235,** 61–63 (1987).

70. Scott, A. C., Stephenson, J. & Chaloner, W. G. Interaction and coevolution of plants and arthropods during the Palaeozoic and Mesozoic. *Phil. Trans. R. Soc.* B **336,** 129–165 (1992).
71. Banks, H. P. & Colthart, B. J. Plant-animal-fungal interactions in early Devonian trimerophytes from Gaspé, Canada. *Am. J. Bot.* **80,** 992–1001 (1993).
72. Edwards, D., Seldon, P. A., Richardson, J. B. & Axe, L. Coprolites as evidence for plant-animal interaction in Siluro-Devonian terrestrial ecosystems. *Nature* **377,** 329–331 (1995).
73. Allen, J. R. L. Marine to fresh water: The sedimentology of the interrupted environmental transition (Ludlow-Siegenian) in the Anglo-Welsh region. *Phil. Trans. R. Soc.* B **309,** 85–104 (1985).
74. Melkonian, M. & Surek, B. Phylogeny of the Chlorophyta: Congruence between ultrastructural and molecular evidence. *Bull. Soc. Zool. Fr.* **120,** 191–208 (1995).
75. Bremer, K., Humphries, C. J., Mishler, B. D. & Churchill, S. P. On cladistic relationships in green plants. *Taxon* **36,** 339–349 (1987).
76. Garbary, D. J., Renzaglia, K. S. & Duckett, J. G. The phylogeny of land plants: A cladistic analysis based on male gametogenesis. *Plant Syst. Evol.* **188,** 237–269 (1993).
77. Capesius, I. A molecular phylogeny of bryophytes based on the nuclear encoded 18S rRNA genes. *J. Plant Physiol.* **146,** 59–63 (1995).
78. Taylor, T. N. The origin of land plants: Some answers, more questions. *Taxon* **37,** 805–833 (1988).
79. Rothwell, G. W. in *Pteridiology in Perspective* (eds. Camus, J. M., Gibby, M. & Johns, R. J.) (Royal Botanic Gardens, Kew, in the press).
80. Albert, V. A. *et al.* Functional constraints and *rbc*L evidence for land plant phylogeny. *Ann. MO Bot. Gard.* **81,** 534–567 (1994).
81. Edwards, D., Fanning, U. & Richardson, J. B. Lower Devonian coalified sporangia from Shropshire: *Salopella* Edwards & Richardson and *Tortilicaulis* Edwards. *Bot. J. Linn. Soc.* **116,** 89–110 (1994).
82. Bateman, R. M., DiMichele, W. A. & Willard, D. A. Experimental cladistic analysis of anatomically preserved lycopsids from the Carboniferous of Euramerica: An essay on paleobotanical phylogenetics. *Ann. MO Bot. Gard.* **79,** 500–559 (1992).
83. Feist, M. & Grambast-Fessard, N. in *Calcareous Algae and Stromatolites* (ed. Riding, R.) 189–203 (Springer, Berlin, 1991).
84. Hébant, C. in *Bryophyte Systematics* (eds. Clarke, G. C. S. & Duckett, J. G.) 365–383 (Academic, London, 1979).

Acknowledgments

We thank W. G. Chaloner, D. Edwards, J. A. Raven, P. S. Herendeen, E. M. Friis, S. Bengtson and especially J. Gray for criticisms of earlier drafts of this manuscript; W. Burger, J. Cattel, A. N. Drinnan, M. Feist, L. E. Graham, H. Haas, H. Kerp, W. A. Taylor and P. Lidmark for assistance with illustrations. This work was supported in part by the Swedish Natural Science Research Council (NFR) and the National Science Foundation.

Peter R. Crane, Else Marie Friis and Kaj Raunsgaard Pedersen

THE ORIGIN AND EARLY DIVERSIFICATION OF ANGIOSPERMS

The major diversification of flowering plants (angiosperms) in the early Cretaceous, between about 130 and 90 million years ago, initiated fundamental changes in terrestrial ecosystems and set in motion processes that generated most of the extant plant diversity. New paleobotanical discoveries, combined with recent phylogenetic analyses of morphological and molecular data, have clarified the initial phases of this radiation and changed our perspective on early angiosperm evolution, though important issues remain unresolved.

Angiosperms dominate the vegetation of most terrestrial ecosystems and consist of roughly 250,000–300,000 extant species, more than all other groups of land plants combined. For almost 150 years, understanding the origin and early diversification of these dominant land plants was hindered by what was thought to be an uninformative fossil record, uncertain relationships among extant angiosperms, and apparently insuperable morphological "gaps" between angiosperms and other seed plants (gymnosperms). The 1960s and 1970s saw notable progress toward breaking this impasse. Syntheses of data from extant angiosperms demonstrated that the subclass Magnoliidae is a phylogenetically basal assemblage,[1–3] and studies of fossil pollen, leaves and reproductive structures documented a major diversification of angiosperms in the early Cretaceous between about 130 and 90 million years (Myr) before present.[4–9]

Recently there has been renewed interest in the patterns and processes of early angiosperm evolution, and explicit phylogenetic analyses, based both on morphological and on molecular data, have clarified and revitalized many of the old debates.[10–14] In the fossil record, bulk sieving techniques have yielded diverse and exquisitely preserved Cretaceous flowers (both mummified and charcoalified) that have resolved the systematic relationships of many early angiosperms and identified the source of important dispersed Cretaceous pollen.[15–19] New information on floral form and reproductive biology in putatively basal extant groups has also catalyzed comparative studies and enhanced the interpretation of fossil material.[20–22]

Here we review how these developments have sharpened arguments over the relationships among basal angiosperms and their seed-plant relatives, and have changed our perspective on early angiosperm evolution.

Origin of Angiosperms and Their Flowers

Ideas on the origin of angiosperms have sometimes invoked polyphyly, have treated the pteridosperms (seed ferns) as a natural group, and have implicated almost all groups of fossil and living gymnosperms as potential angiosperm ancestors.[9,23] More recent work has emphasized cladistic discrimination of relationships among major clades of extant and extinct seed plants[10] (Box 1). Parsimony analyses, based on morphological data, have shown that angiosperms are one of the most strongly supported monophyletic groups in the plant kingdom (see Box 1) and that the pteridosperms (as traditionally defined) are a highly unnatural group.[10,11] However, the most striking result from recent phylogenetic analyses is the support for earlier ideas that identified Bennettitales (extinct) and Gnetales (extant; Fig. 5) as the seed plants that are most closely related to angiosperms (see Box 1). The resulting group (Bennettitales, Gnetales and angiosperms, plus *Pentoxylon* in some analyses) has been termed the "anthophytes," to emphasize their shared possession of flower-like reproductive structures.[11] Among extant taxa, angiosperm monophyly and a close relationship between Gnetales and angiosperms (to the exclusion of *Ginkgo,* conifers and cycads), is also supported by parsimony analysis of partial 18S and 26S ribosomal RNA sequence data[14,24] as well as analyses of combined morphological and molecular (*rbc*L, 18S, 26S) data.[14,25]

Although all of the explicit cladistic studies conducted to date[10,11,26–28] broadly support the anthophyte concept, they differ markedly in their resolution of relationships within the clade, and in their identification of the closest relatives of anthophytes (see Box 1). In addition, the long-standing question of whether angiosperm flowers are derived from a single branch (euanthial) or derived from multiple branches (pseudanthial) is still unresolved (Fig. 1). The origin of angiosperm stamens and carpels, and their homologies with structures in other seed plants also remain problematic.[29] Angiosperm stamens are both more regular and more simple than the pollen organs of the Bennettitales, Gnetales and most Mesozoic pteridosperms. In contrast, angiosperm carpels and bitegmic ovules are relatively complex compared to the ovulate structures of Gnetales and Bennettitales. Particular concerns are how the carpel and the outer layer of the typical angiosperm ovule should be compared to structures in related groups.[30,31] Current explanations are all inadequate, and lean heavily on critical inter-

BOX 1 Hypotheses of relationship among angiosperms and related plants

Two contrasting hypotheses of phylogenetic relationships among anthophytes. Morphological characters supporting the anthophyte group include: the presence of an additional integumentary envelope, granular-columellate structure of the pollen wall and at least some stomates syndetocheilic.[10,11,27] Other potential defining characters include scalariform pitting in the secondary xylem, whorled or opposite arrangement of microsporophylls, apical meristems with tunica-corpus organization, wood with syringal lignin subunits (Mäule reaction) and (secondarily) non-saccate pollen.[10,11,27] New information on fertilization in *Ephedra* and *Gnetum*[88,89] also raises the possibility that double fertilization may be a general feature of anthophytes. Monophyly of the Bennettitales is supported by the distinctive interseminal scales, and the aggregation of pollen sacs into bivalved synangia (present in all Bennettitales except a few Triassic taxa).[10,72] Monophyly of angiosperms is supported by sieve tubes and companion cells derived from the same initials, stamens with two pairs of pollen sacs, anthers with a hypodermal endothecium, microgametophyte of only three nuclei, carpel with stigmatic pollen germination, carpel enclosing a (probably) bitegmic ovule, megaspore wall lacking sporopollenin, megagametophyte with only eight nuclei and the formation of a triploid endosperm.[10,11]

Uncertainty concerning relationships among anthophytes is introduced by: (1) possible secondary loss or transformation of critical morphological features (such as loss of syndetocheilic stomates in *Ephedra* and many angiosperms; loss of granular exine in most angiosperms); (2) the homologous or convergent origin of certain angiosperm-like features in Gnetales (especially in *Welwitschia* and *Gnetum,* Fig. 5) and Bennettitales (such as bisexual flowers); (3) the ambiguous homology of the second integument in the three anthophyte groups and the "cupule" of potentially related taxa (such as *Caytonia,* corystosperms); and (4) uncertainty regarding the basic condition of many characters among basal angiosperms (such as orthotropous versus anatropous ovules, one-seeded versus many-seeded carpel, few-parted versus many-parted flowers).

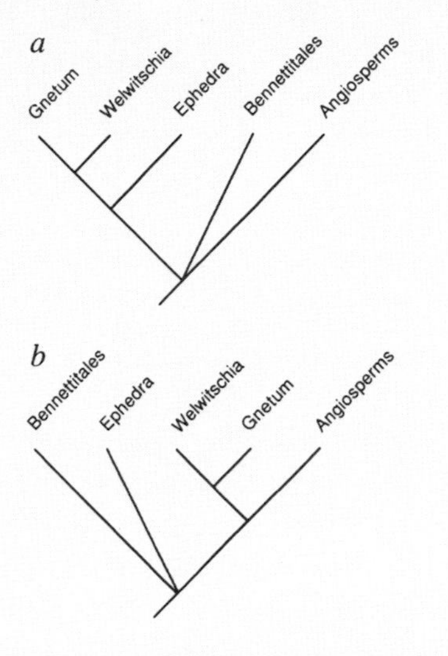

a, Hypothesis A.[10,11,90] Gnetales are monophyletic and the derived similarities of *Gnetum/Welwitschia* and angiosperms are interpreted as convergence (such as reticulate-veined leaves, reduction of male gametophyte, tetrasporic female gametophyte lacking archegonia and with free nuclei serving as eggs, cellular early embryogeny). This hypothesis is consistent with parsimony analyses of current molecular data (*rbc*L,[13] rRNA[14,24]), as well as analyses of combined morphological and molecular data (morphology plus *rbc*L[25] and morphology plus rRNA[14]).

b, Hypothesis B.[27] Gnetales are paraphyletic, with *Gnetum* and *Welwitschia* the sister group to angiosperms. This hypothesis interprets the derived similarities of *Welwitschia, Gnetum* and angiosperms as homologous.

pretations of fossil specimens that are not well understood (such as *Caytonia,* corystosperms).

Such uncertainties highlight the need to clarify structural and developmental homologies among the reproductive structures of angiosperms and related groups; in many respects, the morphological gap between angio-

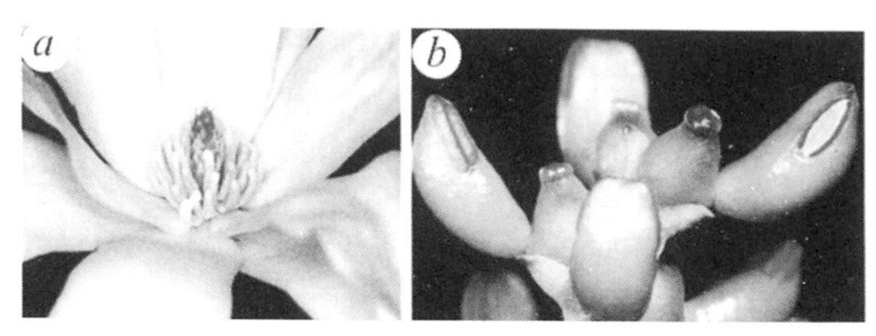

Fig. 1 Extremes of floral diversity in extant basal (magnoliid) angiosperms. *a,* Large multiparted flower of extant *Magnolia stellata* (Magnoliaceae), showing nine tepals and numerous stamens and carpels; ×1. *b,* Small few-parted flowers of *Sarcandra glabra* (Chloranthaceae); each flower consisting of a single bract, carpel and stamen (photograph courtesy of P. K. Endress), ×3.8. Interpretations based on recent phylogenetic results[14] suggest that flowers of *Magnolia* and *Sarcandra* are, respectively, elaborated and reduced compared to the basic angiosperm condition. More radical interpretations of floral evolution suggest that all angiosperm flowers may not be homologous, and that the simple flowers of some (such as *Sarcandra*) are equivalent only to parts of flowers in others (such as *Magnolia*).[44,86] (A color version of this figure is available on-line at <http://www.press.uchicago.edu/books/gee>.)

sperms and other seed plants remains as wide as ever. This gap can be closed most directly through renewed morphological documentation and phylogenetic analysis of well-preserved fossil gymnosperm material from Mesozoic fossil floras.[32,33] Progress may also be possible by understanding how the genes controlling angiosperm stamen and carpel differentiation are expressed in the reproductive organs of extant gymnosperms (such as cycads, conifers, Gnetales).[29]

Angiosperm Phylogeny and Floral Evolution

Current hypotheses of angiosperm evolution recognize two large clades (monocotyledons and eudicots) embedded within a poorly defined basal assemblage (grade) of magnoliid dicots (Magnoliidae[1]). The monocotyledons are defined as monophyletic by their single cotyledon and other features.[34] Eudicots are circumscribed by the production of triaperturate or triaperturate-derived pollen (convergent in Illiciaceae, Schisandraceae[35]).

Generalizations developed around the turn of the century emphasized large multiparted flowers like those of extant *Magnolia* (Figs. 1a, 2A) as a starting point for angiosperm floral evolution.[36,37] Recent ideas accept that plants in the subclass Magnoliidae retain a broad array of probable unspecialized angiosperm features[20,22] (such as parts generally free, lack of differentiation within the perianth), and have also documented several previously unrecognized floral features that are general among extant magnoliids (such as valvate anther dihiscence,[38] ascidiate carpel development[39]). Modern studies have emphasized the great diversity of floral form, biology and structure among magnoliids.[22,40] Variation in the number and arrangement of floral parts is extreme (Fig. 2), and both large, multiparted bisexual flowers and small, simple, frequently unisexual flowers are widespread[22,40] (Fig. 1).

Identifying the likely basic condition of angiosperm flowers is therefore intimately connected with the recognition of phylogenetic patterns among magnoliids (Box 2). Unfortunately, the substantial morphological differences that separate angiosperms from all other seed plants, combined with uncertain relationships among the anthophytes, makes it difficult to "polarize" many of the crucial characters. This, in turn, complicates attempts to "root" the angiosperm tree and contributes to the instability of current phylogenetic results.[12,14,41] Among the studies currently available, there is emerging agreement that undue attention has focused on extant Magnoliaceae and its allies, and combined morphological and molecular results tend to favor phylogenetic models in which taxa with small, trimerous or even more simple flowers are basal in angiosperms[14,42–45] (see Box 2).

Fossil Evidence of Early Magnoliid Diversity

The variety of leaves and pollen in the early Cretaceous[6,7] implies that magnoliids were diverse early in angiosperm evolution, and this is supported by the surprisingly rich, emerging record of fossil flowers (Fig. 3 and Box 3). In the extensive assemblages of early Cretaceous angiosperm reproductive structures from Portugal and eastern North America, all of the fossils described so far are small (frequently less than 2 mm in length), and because they occur with larger fossils of other plants (such as conifers) it seems unlikely that this is a result of depositional bias. The flowers are generally few-parted and often with an undifferentiated perianth. Stamens generally have small pollen sacs with valvate dehiscence, and a relatively

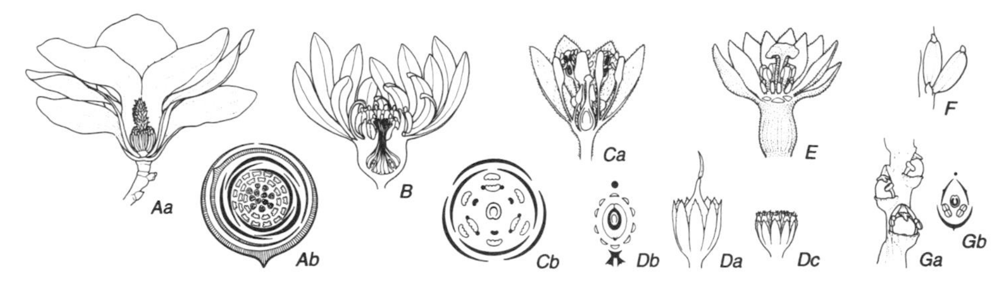

Fig. 2 Floral diversity in extant basal (magnoliid) angiosperms, illustrating variation in the number of carpels, stamens and perianth parts per flower. Generalized floral diagrams are shown in *Ab, Cb, Db* and *Gb*. *A, Magnolia* (Magnoliaceae); *B, Calycanthus* (Calycanthaceae); *C, Cinnamomum* (Lauraceae); *D, Ceratophyllum* (Ceratophyllaceae) (*a,* pistillate flowers; *c,* staminate flowers); *E, Hazomalania* (Hernandiaceae); *F, Ascarina* (Chloranthaceae), showing two staminate flowers; *G, Piper* (Piperaceae), showing three flowers.

BOX 2 Hypotheses of relationships among basal angiosperms

Two of several published hypotheses of relationships among basal angiosperms. Taxa placed in a basal position by different recent studies include *Ceratophyllum*,[13,42] Idiospermaceae-Calycantha-ceae,[43] Chloranthaceae/Piperales[44,45] and Nymphaeales.[14]

a, Hypothesis A.[14] Phylogenetic analysis of combined morphological and rRNA data showing relationships among extant Gnetales and basal angiosperms. Under this interpretation angiosperms are "rooted" close to Nymphaeales, monocots are interpreted as an early branch in angiosperm evolution, and basal magnoliids are various more or less herbaceous taxa with basically simple flowers (secondarily complex in derived Nymphaeaceae). Magnoliales (including Annonaceae, Degeneriaceae, Magnoliaceae, Myristicaceae) are placed in a relatively derived position. Eudicots are monophyletic and diverge from the basal magnoliid assemblage below Laurales/Magnoliales. Note that an alternative "rooting" in the vicinity of Magnoliales would imply a relatively late divergence of monocotyledons from magnoliids.

b, Hypothesis B.[25] Phylogenetic analysis of combined morphological and *rbc*L data showing relationships among extant Gnetales and basal angiosperms. Under this interpretation *Ceratophyllum* is the basal angiosperm lineage (not considered in hypothesis A) and eudicots are monophyletic; most other relationships are unresolved. Aristolochiaceae are not considered.

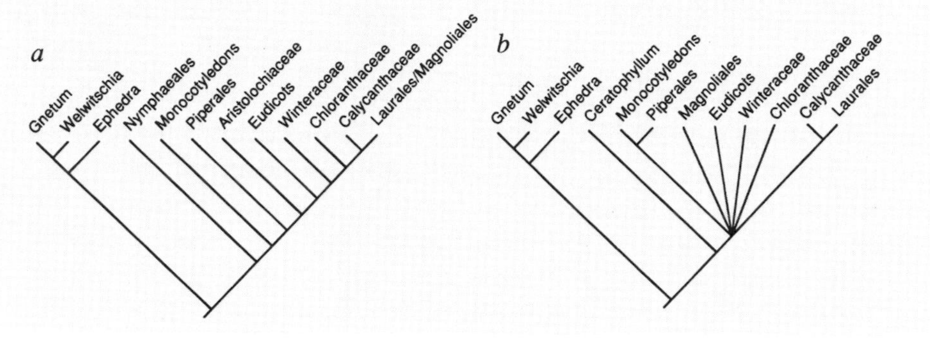

large apically expanded connective.[46] Carpels generally have a poorly differentiated stigmatic surface. Among magnoliids, the fossil record indicates that several extant families were already differentiated by about the Turonian (Box 3).

Fossil Evidence of Early Monocot Diversity

Among extant plants monocotyledons comprise only about 22% of angiosperm species. Over half of this diversity is accounted for by four families (Orchidaceae, orchids; Poaceae, grasses; Cyperaceae, sedges; Arecaceae, palms). The Cretaceous fossil record of monocotyledons is depauperate compared to that of contemporary magnoliids and eudicots, and perhaps reflects a variety of biases working against both their preservation and recognition. During the mid-Cretaceous, monocots are probably represented by several leaves,[47] and more equivocally by dispersed pollen.[48,49] However, evidence of rapid monocot diversification is provided later in

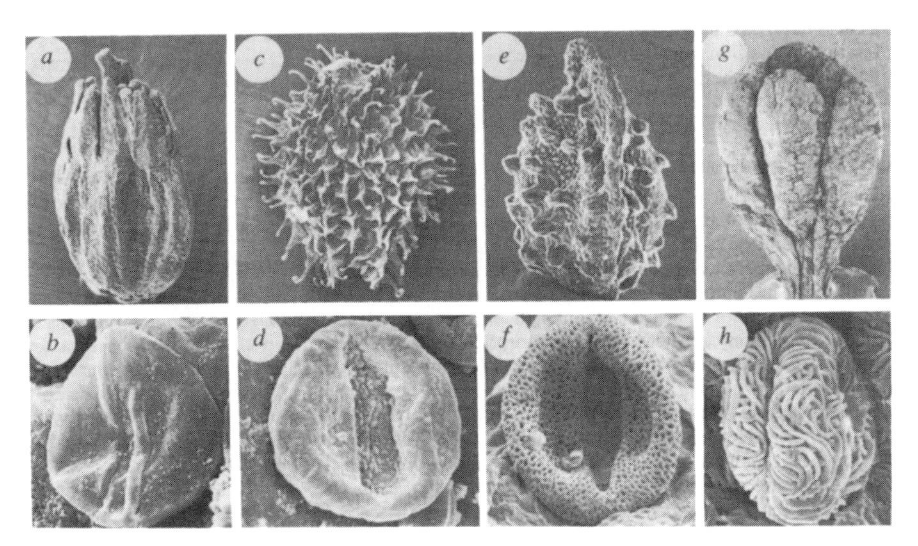

Fig. 3 Scanning electron micrographs of fossil early angiosperm flowers and carpels with their corresponding pollen. *a,* Epigynous flower showing tepals and protruding stamens, western Portugal, early Cretaceous (Barremian or Aptian); ×20. *b,* Monocolpate pollen with striate exine ornamentation from stamens of above type of flower (see *a*); ×2,200. *c,* Unilocular fruiting unit covered with hooked unicellular hairs, and containing a single, orthotropous, pendulous seed; Potomac Group, Virginia, USA, early Cretaceous (early-middle Albian); ×22. *d,* Monocolpate *Transitoripollis/Tucanopollis*-like pollen from the stigmatic surface of above type of fruiting unit (see *c*); ×1,680. *e, Couperites mauldinensis,* unilocular fruiting unit, showing protruding "resin bodies," and containing a single pendulous, anatropous seed; Potomac Group, Virginia, USA, late Cretaceous (early Cenomanian); ×44. *f,* Monocolpate *Clavatipollenites*-type pollen from the stigmatic surface of *Couperites mauldinensis* (see *e*); ×1,200. *g, Spanomera mauldinensis,* staminate flower with five tepals and five stamens; Potomac Group, Virginia, USA, late Cretaceous (early Cenomanian); ×47. *h,* Tricolpate *Retitricolpites*-type pollen from stamens of *Spanomera mauldinensis* (see *g*); ×1,825.

the Cretaceous by fruits of Zingiberales (gingers and their allies), and leaves and stems of palms.[34] By the early Tertiary many monocot groups had differentiated[34,50] (Fig. 4).

Fossil Evidence of Early Eudicot Diversity

Eudicots comprise about 75% of extant angiosperm species, and are recognized as a monophyletic group in phylogenetic analyses of combined morphological and molecular data.[14,25] Triaperturate pollen, which is diagnostic of the group, is first recorded in dispersed palynological assemblages around the Barremian-Aptian boundary (about 125 Myr) or perhaps slightly earlier.

Basal groups in the eudicot clade include the ranunculids (Ranunculidae, sometimes included in the Magnoliidae) and also the "lower" hama-

BOX 3 Fossil evidence of floral diversity among early angiosperms

The most extensive information on early and mid-Cretaceous angiosperm reproductive structures is from western Portugal (1–14)[19] and the Atlantic Coastal Plain of eastern North America (15–32),[18] which together provide a sequence of floras ranging in age from Valanginian(?)-Hauterivian(?) to earliest Cenomanian.

Valanginian(?)-Aptian (1–14). Although only partially described, the angiosperm component of the earliest floras (Aptian and older, for example, 1–14) is dominated by small, simple flowers and flower parts. Epigynous flowers of possible lauralean affinity (such as 2, 3, 6, 7) and unilocular, single-seeded fruits or fruitlets with monocolpate pollen (such as 8, 9, 11–14) are especially common and diverse.[19] *Clavatipollenites*-type pollen, for which a chloranthoid affinity has been inferred,[91] occurs on the stigmatic surfaces of some of these fruits. *Clavatipollenites*-type pollen is also known *in situ* within distinctive angiosperm stamens and is widespread in dispersed pollen floras from the early Cretaceous.[49] Triaperturate pollen diagnostic of eudicots is also present.[19] Other putative angiosperm macrofossils from Aptian and older rocks include probable magnoliid leaves[92] and horned (?*Ceratophyllum*-like) fruits of uncertain system-

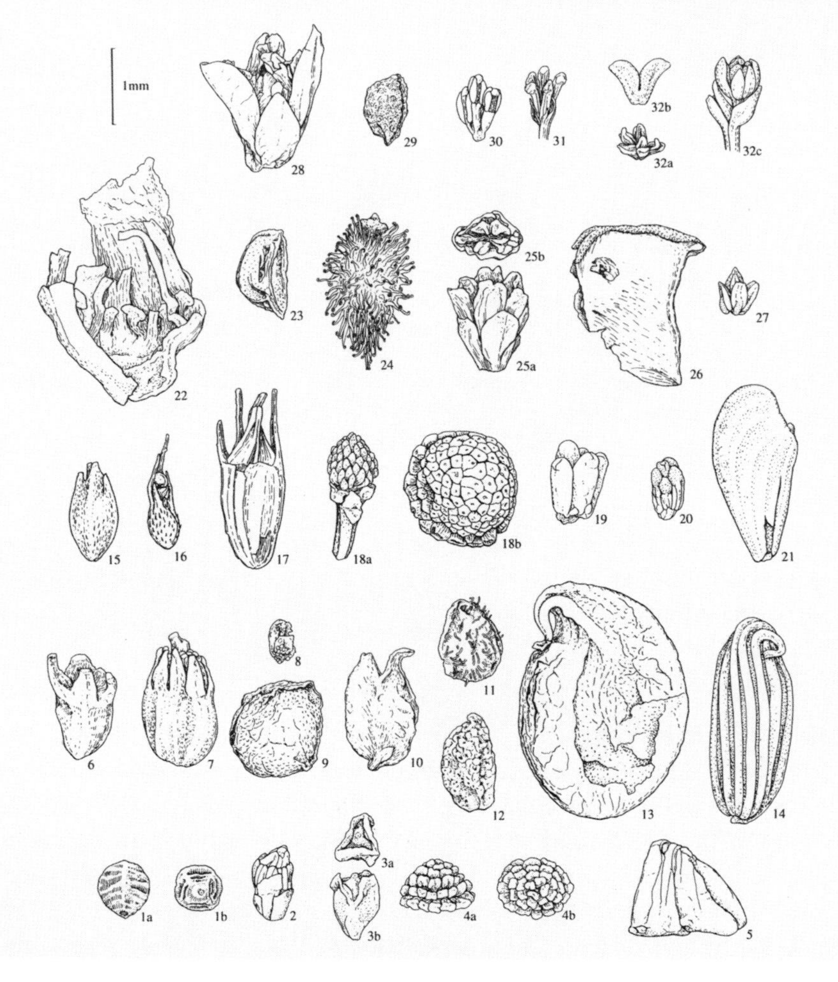

atic affinity.[93] One small specimen with attached leaves and reproductive structures, which is reported to show a mosaic of chloranthoid and piperalean features,[94] is too poorly preserved for detailed study of the flowers, fruits, stamens or pollen. In addition, dispersed pollen grains from Aptian and older rocks include forms perhaps related to extant Winteraceae[35] and Magnoliales.[95]

Albian (15–27). In Albian floras (such as 15–27) small epigynous flowers (such as 15–17) and unilocular single-seeded fruits or fruitlets (such as 21, 23, 24) remain common. One of the most distinctive of these (24) appears to combine characters of Piperales, Chloranthaceae and perhaps Circaeasteraceae (Ranunculidae).[18] A variety of other taxa are also present, including probable flowers of Lauraceae (22)[18] and Calycanthaceae,[96] possible magnolialean fruits (such as 18) and stamens,[18] and possible fruits of Ceratophyllaceae.[93] Flowers of basal eudicots become a distinctive component of Cretaceous floras from the mid-Albian onwards and several lineages of "lower" hamamelidids can be recognized during the mid-Cretaceous phase of angiosperm diversification, including Platanaceae[16,52] (such as 25, 31) and perhaps trochodendroids[6,7,51,52] and buxoids[52,97] (such as 32) (see Fig. 3g, h).

Earliest Cenomanian (28–32). By the early part of the late Cretaceous the magnoliid groups include chloranthoids[98] and Lauraceae (such as 28),[99] both of which were apparently geographically widespread. Probable Magnoliaceae[100] were also present. Eudicots of this age include a variety of platanoids,[16,53] rosiids[17] and other taxa.[20]

Fossils illustrated are from the Valanginian(?)-Hauterivian(?) of Torres Vedras, western Portugal (1–5); the Barremian or Aptian of Catefica, western Portugal (6, 12), Famalicão, western Portugal (13, 14), and Vale de Agua, western Portugal (7–11); the early or mid-Albian of Puddledock, Virginia (15–22, 24), Bank near Brooke, Virginia (25), and Kenilworth, Maryland, (23); the late Albian of West Brothers, Maryland (26, 27, 30, 31); and the early Cenomanian of Mauldin Mountain, Maryland (28, 29, 32). *1,* Four-angled fruit or seed in lateral *(a)* and apical *(b)* view. *2,* Epigynous (?) flower with monocolpate, reticulate pollen. *3,* Epigynous *Hedyosmum*-like flower in apical *(a)* and lateral *(b)* view. *4,* Floral structure with several whorls of stamens (?unistaminate flowers) in lateral *(a)* and apical *(b)* view. *5,* Group of seeds. *6,* Four-angled, epigyrous structure.[18,19] *7,* Epigynous flower with monocolpate, finely striate *Cabomba*-like pollen (see Fig. 3a, b).[19] *8, 9, 11,* Unilocular single-seeded fruiting units with strengthening tissue in seed wall; *8* and *11* have adhering *Clavatipollenites*-type pollen. *10,* Bicarpellate fruit. *12–14,* Unilocular single-seeded fruiting units with strengthening tissue in fruit wall. *15–17,* Epigynous floral structures. *18,* Multiparted floral structure; smaller specimen in lateral view *(a),* larger specimen in apical view *(b).* *19,* Apocarpous pistillate? flower. *20,* Staminate flower. *21,* Unicarpellate fruit with strengthening layer in fruit wall. *22,* Fragment of lauralean flower. *23,* Unilocular single-seeded fruiting unit with strengthening tissue in the seed wall and with monocolpate pollen. *24,* Spiny, unilocular and singleseeded fruiting unit (see Fig. 3c, d). *25,* Pistillate flowers of *Platanocarpus brookensis* in lateral *(a)* and apical *(b)* view.[53] *26, 27,* Fruitlet *(26)* and staminate flower *(27)* of *Spanomera marylandensis.*[98] *28,* Trimerous flower of *Mauldinia mirabilis.*[99] *29,* Fruit of *Couperites mauldinensis* with adhering *Clavatipollenites*-type pollen (Fig. 3e, f).[91] *30,* Chloranthoid androecium.[93] *31,* Pistillate flower of *Platanocarpus marylandensis.*[16] *32,* Staminate *(a)* and pistillate *(b, c)* flowers of *Spanomera mauldinensis* (Fig. 3g, h).[97]

melidids.[51,52] Embedded within this basal eudicot assemblage are the diverse groups that comprise the bulk of extant angiosperms, including the "higher" hamamelidids along with the subclasses Rosidae, Caryophyllidae, Dilleniidae and Asteridae, which contain many highly diverse extant families (such as Asteraceae, the sunflower family). Among the eudicots that can be recognized during the mid-Cretaceous diversification are several "lower" hamamelidid lineages and a diversity of generalized rosiid types (Fig. 4). By around the Campanian, eudicots were very diverse.[20,50,52]

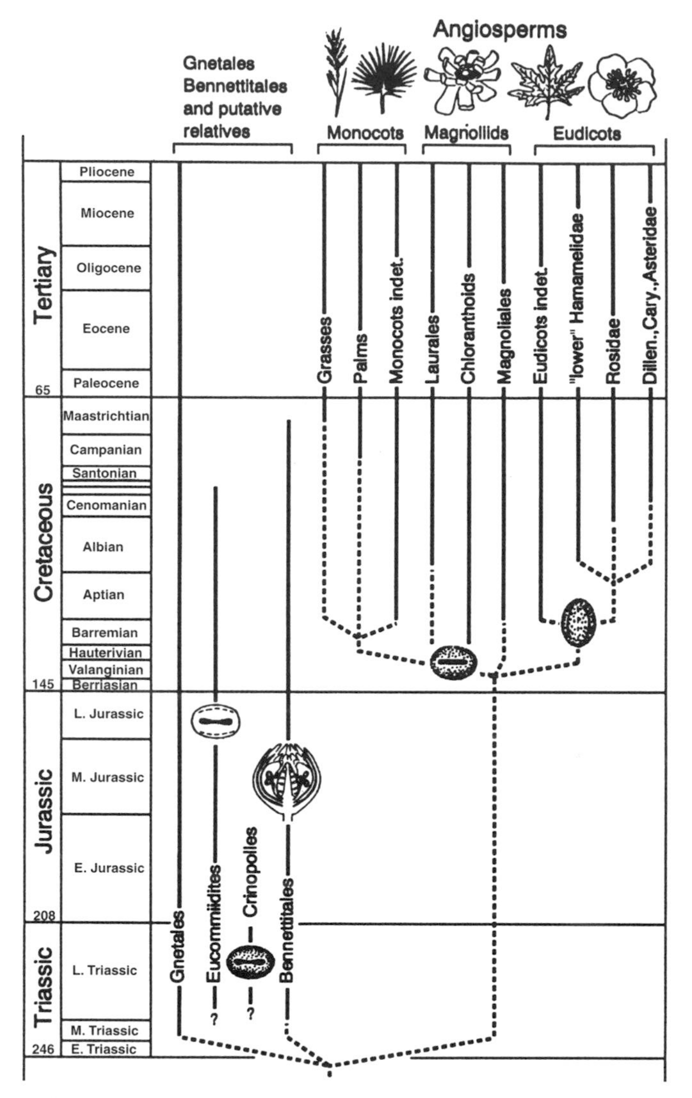

Fig. 4 Simplified phylogenetic tree showing the relative stratigraphic ranges of selected angiosperm groups and putative close gymnosperm relatives. The fossil record of Gnetales is based primarily on diagnostic dispersed ribbed pollen, with supporting evidence from rare Triassic[72,87] and Cretaceous[72] macrofossils. The fossil record of *Eucommiidites* is based primarily on dispersed pollen, with seeds and pollen organs known from the Jurassic and Cretaceous.[32] The combined evidence suggests that *Eucommiidites*-producing plants are closely related to the anthophyte group.[32] Crinopolles pollen is reported from the late Triassic[57] but other organs of the plant are not known. Bennettitales have an extensive fossil record based on leaves and reproductive structures.[10,72] The bennettitalean flower-like reproductive structure illustrated is *Williamsoniella.* Monocolpate pollen is the basic condition in angiosperms that is retained by monocots and magnoliid dicots. Triaperturate pollen is diagnostic of eudicots. *Dillen.,* Dilleniidae; *Cary.,* Caryophyllidae.

Pre-Cretaceous Angiosperms?

Current studies of the fossil record show an orderly sequence of appearance of angiosperm remains beginning with putative magnoliid pollen in the Valanginian, and triaperturate pollen of eudicots around the Barremian-Aptian boundary (or slightly earlier). By the early Cenomanian there are diverse magnoliids, a variety of hamamelidids and a few rosiids, which show clearly that several angiosperm families had already differentiated. By the Campanian, Maastrichtian and Paleocene, many extant angiosperm families in all subclasses had already differentiated.[34,50–54]

Claims of pre-Cretaceous angiosperms need to confront this orderly sequence, and although the literature on angiosperm origins is replete with putative pre-Cretaceous angiosperms, most have been shown to be stratigraphically misplaced, unrelated to angiosperms, lacking in diagnostic characters, or too poorly preserved for reliable determination.[55] Recently, however, there has been renewed discussion of the possibility of pre-Cretaceous angiosperms stimulated by the anthophyte hypothesis (which implies that the lineage leading to angiosperms diverged from other known groups of seed plants before the late Triassic);[10,11,41] the description of new and potentially relevant fossils from Jurassic and Triassic rocks;[56,57] and calculations of divergence times based on molecular-clock assumptions.[58–60] These developments have highlighted the importance of distinguishing clearly between the timing of angiosperm divergence (splitting of the stem lineage from its sister groups) and the timing of the angiosperm diversification (splitting of the crown group into extant clades).[41]

In our view, the absence of distinctive triaperturate pollen grains in numerous, rich, Triassic and Jurassic palynofloras from both hemispheres precludes the long cryptic period of evolution implied by some estimates of rates of molecular evolution, at least for eudicots.[61] The systematic affinities of recently described putative Jurassic and Triassic angiosperms[56,57,62,63] are either clearly with other groups (such as dipteridaceous ferns[64]) or remain equivocal (such as Triassic *Sanmiguelia*-like plants and Crinopolles-type pollen grains[64,65]). Although some of these fossils may be attributable to the anthophyte clade, they highlight the difficulties of accurate systematic determination based on inadequate material. Most defining features of angiosperms (Box 1) are unlikely to be preserved in the fossil record. The most useful diagnostic morphological features of angiosperms are stamens with two pairs of pollen sacs, and a carpel enclosing the ovule. Both can be observed only in well-preserved fossil material.

Coevolutionary Interactions

Angiosperm diversification has often been linked to the diversification of pollen- and nectar-collecting insects on the supposition that insect pollination provides new possibilities for reproductive isolation and therefore elevation of speciation rates.[66–68] Compared to wind pollination, which is widespread in other seed plants, insect pollination may also permit effective outcrossing at lower population densities and in a greater range of environments, perhaps thereby reducing extinction rates.[66–68] These factors may explain the diversity of some of the most speciose angiosperm families (such as orchids), but the extent to which they provide a general explanation for the massive mid-Cretaceous increase in angiosperm diversity is uncertain.[69,70] In the Bennettitales, flower-like reproductive structures provide indirect evidence of insect pollination from the late Triassic to late Cretaceous,[71,72] and insect pollination may also have been established in some Mesozoic pteridosperms.[33] The size and morphology of diverse mid-Cretaceous gnetalean pollen, combined with evidence from extant taxa,[73] is also suggestive of insect pollination in the Gnetales. Early angiosperms may therefore have co-opted pollinators from previously established relationships with other groups of seed plants.

Whatever the effect of insect pollination on the initial angiosperm diversification, discoveries of early and mid-Cretaceous flowers leave no doubt that early members of the group were insect pollinated. Stamens in fossil flowers have small anthers with low pollen production, anther dehiscence is valvate, pollen grains are often covered with a pollenkitt-like material, stigmatic surfaces are generally unelaborated and pollen grains are often smaller than the most effective size for wind dispersal.[18,19] Comparison with modern relatives suggests that these flowers were probably pollinated by pollen-collecting or pollen-eating insects. Flowers pollinated by nectar-collecting Hymenoptera and Lepidoptera occur in more derived groups of angiosperms and appear later in the fossil record.[17]

Although early Cretaceous angiosperms may have been very similar to their living relatives in pollination syndrome, modes of fruit and seed dispersal were probably substantially different.[22] Cretaceous angiosperm fruits and seeds are generally very small compared to their modern relatives, and there is no evidence of specialized mammal or bird dispersal. Among basal groups, as in angiosperms as a whole, the evolution of fleshy fruits, arillate seeds and other apparent adaptations for animal dispersal, seems to be correlated with the evolution of frugivorous/granivorous birds and mammals, perhaps during the latest Cretaceous, but most strikingly during the early Tertiary. The only suggestion of animal dispersal of angiosperms in the early Cretaceous is provided by a single species of small fruits covered with hooked spines (Box 3, Fig. 3c).

Rise to Ecological Dominance

The mid-Cretaceous diversification of angiosperms marks the transition from Mesozoic ecosystems dominated by ferns, conifers, cycads and Bennittitales to more modern late Cretaceous and Tertiary ecosystems dominated by angiosperms.[74,75] In the paleobotanical record this change is much more profound than that occurring at the Cretaceous-Tertiary boundary and is also reflected in the transition from sauropod-dominated to ornithopod-dominated dinosaur faunas.[76] Data on the floristic composition of both macrofloras and microfloras show that angiosperms had attained diversity levels of about 50–80% by the end of the Cretaceous,[74] but there is evidence that angiosperm abundance still remained subordinate to gymnosperms and ferns in some habitats, perhaps over large geographical areas.[77,78] The extent to which this is a general phenomenon is currently uncertain, but it is supported in part by the palynological data and by the prominence of certain open-habitat ferns in some late Cretaceous floras. These considerations emphasize the value of complementary macrofossil-based and pollen/spore-based assessments of diversity and abundance in evaluating the rate and magnitude of the angiosperm radiation. They also raise the possibility that disturbance by herbivorous dinosaurs may have been a significant ecological factor in the first half of angiosperm history.[76]

Problems of large-scale stratigraphic resolution make it difficult to resolve geographic patterns in the angiosperm radiation, but compilations of palynological data, especially from the Northern Hemisphere, show that the initial increase in angiosperm diversity occurred in low paleolatitude areas.[75,77] This result is of biogeographic interest for understanding the current distribution of relictual magnoliid families, and also has significant ecological implications. During the early Cretaceous, low latitude areas experienced semiarid or seasonally arid conditions that may have promoted a weedy life history with precocious reproduction (progenesis).[1,11,41] Associated effects may have included simplification and aggregation of sporophylls to form a flower, enclosure of ovules in a carpel, truncation of the gametophyte phase of the life cycle, and major reorganization of leaf and stem anatomy. In turn, several of these features may have contributed to angiosperm "success" through elevated speciation rates[11,41,79] and/or more rapid and more flexible vegetative growth.[80,81] It remains uncertain, however, whether such effects were manifested at the base of the angiosperm clade or within one or more angiosperm subgroups.[82,83]

Fig. 5 Gnetales consist of only three extant genera: *Ephedra* (*a,* about 40 species in arid and semiarid areas); *Welwitschia mirabilis* (*b,* one species restricted to the Namibian Desert); and *Gnetum* (*c,* about 30 species of tropical rainforest lianas and small trees). Despite the obvious morphological differences the group is widely regarded as monophyletic (Box 1) and is united by multiple axillary buds, opposite/decussate phyllotaxis (also reflected in the reproductive structures), circular bordered pits in the protoxylem, vessels (?derived independently to those in angiosperms) and a feeder in the embryo.[10,11] In addition, *Welwitschia* and *Gnetum* share several derived characters that are thought to be convergent with similar features in angiosperms[10,11] (see Box 1). (A color version of this figure is available on-line at <http://www.press.uchicago.edu/books/gee>.)

Interestingly, the Gnetales, which show many remarkable structural and biological convergences to angiosperms (Fig. 5), also diversified during the mid-Cretaceous in low paleolatitude areas, although they never became significant at middle and high paleolatitudes and experienced rapid decline during the earliest late Cretaceous.[75,77] The temporal, paleogeographic, and perhaps ecological, parallels between this radiation of Gnetales and angiosperms implies a common response to changing environmental conditions. Explanations of angiosperm diversification may therefore have underemphasized the effects of environmental changes during the early and mid-Cretaceous, which included high rates of sea-floor spreading, high sea-level stands and probably high global temperatures.[84]

Future Directions

The five living groups of seed plants are a poor sample of the total historical diversity of the seed-plant clade and thus additional molecular phylogenetic analyses are likely to lead only to limited progress in evaluating the phylogenetic relationships of angiosperms. However, reducing the gap between angiosperms and their gymnosperm relatives is important because of its implications for rooting the angiosperm tree. Renewed paleobotanical efforts with the early members of the Gnetales, Bennettitales, angiosperms and other potentially closely related groups are therefore a high priority. Recognition of "stem group" taxa with some but not all of the defining features of the "crown group" has been possible in other studies of land-plant phylogeny, and would help to resolve several outstanding problems.

Current data from the fossil record and combined morphological/molecular phylogenies of extant taxa challenge the view that earliest members of the angiosperm clade were large, woody plants with *Magnolia*-like flowers. Instead they suggest a very different concept of early angiosperms as perhaps herbaceous plants of small stature. Flowers would have been small, simple (perhaps unisexual) and probably lacking clear differentiation into sepals and petals (suggesting that molecular genetic models of flora morphogenesis developed for the eudicots *Arabidopsis* and *Antirrhinum*[85] may require modification among magnoliids). Stamens would have had a poorly developed filament and a well-developed anther with valvate dehiscence. Pollen would have been small, monocolpate, and tectate with a weakly developed end exine in non-aperturate areas. The gynoecium would have been composed of one or more unilocular carpels containing one or two ovules. The stigmatic surface would have been unelaborated.

Ecologically, the early diversity of angiosperms at low paleolatitudes, and the parallel Aptian-Cenomanian radiation of angiosperm and gnetalean pollen in these areas suggest that both groups responded in similar ways to the same environmental cues. Possible linkages between mid-Cretaceous climatic, tectonic and other environmental changes, and their possible impact on terrestrial ecosystems need to be explored. Early Cretaceous fossil plants from low paleolatitudes are likely to be particularly informative.

PETER R. CRANE is at the Field Museum, Roosevelt Road at Lake Shore Drive, Chicago, Illinois 60605 and the Department of the Geophysical Sciences, University of Chicago, USA; Else Marie Friis is at the Swedish Museum of Natural History, Box 50007, S104 05, Stockholm, Sweden; Kaj Raunsgaard Pedersen is at the Department of Geology, University of Aarhus, Universitetsparken DK-8000, Aarhus, Denmark.

References

1. Takhtajan, A. *Flowering Plants: Origin and Dispersal* (Oliver and Boyd, Edinburgh, 1969).
2. Cronquist, A. *An Integrated System of Classification of Flowering Plants* (Columbia Univ. Press, New York, 1981).

3. Walker, J. W. in *Origin and Early Evolution of Angiosperms* (ed. Beck, C. B.) 241–291 (Columbia Univ. Press, New York, 1976).
4. Brenner, G. *Maryland Dept. Geol. Mines Water Res. Bull.* **27,** 1–215 (1963).
5. Dilcher, D. L. *Rev. Palaeobot. Palynol.* **27,** 291–328 (1979).
6. Doyle, J. A. & Hickey, L. J. in *Origin and Early Evolution of Angiosperms* (ed. Beck, C. B.) 139–206 (Columbia Univ. Press, New York, 1976).
7. Hickey, L. J. & Doyle, J. A. *Bot. Rev.* **43,** 3–104 (1977).
8. Wolfe, J. A., Doyle J. A. & Page, V. M. *Ann. Missouri bot. Gard.* **62,** 801–824 (1975).
9. Hughes, N. F. *Palaeobiology of Angiosperm Origins* (Cambridge Univ. Press, Cambridge, 1976).
10. Crane, P. R. *Ann. Missouri bot. Gard.* **72,** 716–793 (1985).
11. Doyle, J. A. & Donaghue, M. J. *Bot. Rev.* **52,** 321–431 (1986).
12. Donoghue, M. J. & Doyle, J. A. in *Evolution, Systematics and History of the Hamamelidae* (eds. Crane, P. R. & Blackmore, S.) 17–45 (Clarendon, Oxford, 1989).
13. Chase, M. W. *et al. Ann. Missouri bot. Gard.* **80,** 526–580 (1993).
14. Doyle, J. A., Donoghue, M. J. & Zimmer, E. A. *Ann. Missouri bot. Gard.* **81,** 419–450 (1994).
15. Friis, E. M. *Ann. Missouri bot. Gard* **71,** 403–418 (1984).
16. Friis, E. M., Crane, P. R. & Pedersen, K. R. *Biol. Skr.* **31,** 1–55 (1988).
17. Friis, E. M & Crepet, W. L. in *Origins of Angiosperms and Their Biological Consequences* (eds. Friis, E. M., Chaloner, W. G. & Crane, P. R.) 145–179 (Cambridge Univ. Press, Cambridge, 1987).
18. Crane, P. R., Friis. E. M. & Pedersen. K. R. *Pl. Syst. Evol.* **8** (suppl.), 51–72 (1994).
19. Friis, E. M., Pedersen K. R. & Crane, P. R. *Pl. Syst. Evol.* **8** (suppl.), 31–49 (1994).
20. Friis, E. M. & Endress, P. K. *Adv. Bot. Res.* **17,** 99–162 (1990).
21. Endress, P. K. *Bot. Jahrb. Syst.* **109,** 153–226 (1987).
22. Endress, P. K. *Mem. NY Bot. Gard.* **55,** 5–34 (1989).
23. Hughes, N. F. *The Enigma of Angiosperm Origins* (Cambridge Univ. Press, Cambridge, 1994).
24. Hamby, R. K. & Zimmer, E. A. in *Molecular Systematics of Plants* (eds. Soltis, P. S., Soltis, D. E. & Doyle, J. J.) 50–91 (Chapman & Hall, New York, 1992).
25. Albert, V. A. *et al. Ann. Missouri bot. Gard.* **81** 534–567 (1994).
26. Loconte, H. & Stevenson, D. W. *Brittonia* **42,** 197–211 (1990).
27. Nixon, K. C., Crepet, W. L., Stevenson, D. W. & Friis, E. M. *Ann. Missouri bot. Gard.* **81,** 484–533 (1994).
28. Rothwell, G. W. & Serbert, R. *Syst. Bot.* **19,** 443–482 (1994).
29. Doyle, J. A. *Pl. Syst. Evol.* **8** (suppl.), 7–29 (1994).
30. Stebbins, G. L. *Flowering Plants: Evolution above the Species Level* (Harvard Univ. Press, Cambridge, MA, 1974.)
31. Doyle, J. A. *A. Rev. Ecol. Syst.* **9,** 365–392 (1978).
32. Pedersen, K. R., Crane, P. R., & Friis, E. M. *Grana* **28,** 279–294 (1989).
33. Pedersen, K. R., Friis, E. M. & Crane, P. R. *Grana* **32,** 273–289 (1993).
34. Herendeen, P. S. & Crane, P. R. in *Monocotyledons: Systematics and Evolution* (eds. Rudall, P. J., Cribb, P. J., Cutler, D. F. & Humphries, C. J.) (Royal Botanic Gardens, Kew, in the press).
35. Doyle, H. A., Hotton, C. L. & Ward, J. V. *Am. J. Bot.* **77,** 1558–1568 (1990).

36. Arber, E. A. N. & Parkin, J. *Bot. J. Linn. Soc.* **38,** 29–80 (1907).
37. Bessey, C. E. *Ann. Missouri bot. Gard.* **2,** 109–164 (1915).
38. Endress, P. K. & Hufford, L. D. *Bot. J. Linn. Soc.* **100,** 45–85 (1989).
39. Van Heel, W. A. *Blumea* **27,** 499–522 (1981).
40. Endress, P. K. *Biol. J. Linn. Soc.* **39,** 153–175 (1991).
41. Doyle, J. A. & Donoghue, M. J. *Paleobiology* **19,** 141–167 (1993).
42. Les, D. H., Garvin, D. K. & Wimpee, C. F. *Proc. natn. Acad. Sci. USA* **88,** 10119–10123 (1991).
43. Loconte, H. & Stevenson, D. W. *Cladistics* **7,** 267–296 (1991).
44. Burger, W. C. *Bot. Rev.* **43,** 345–393 (1977).
45. Taylor, D. W. & Hickey, L. J. *Pl. Syst. Evol.* **180,** 137–156 (1992).
46. Friis, E. M., Crane, P. R. & Pedersen, K. R. in *Pollen and Spores: Patterns of Diversification* (eds. Blackmore, S. & Barnes, S. H.) 197–224 (Clarendon, Oxford, 1991).
47. Doyle, J. A. *Q. Rev. Biol.* **48,** 399–413 (1973).
48. Walker, J. W. & Walker, A. G. *Ann. Missouri bot. Gard.* **71,** 464–521 (1984).
49. Walker, J. W. & Walker, A. G. in *Pollen and Spores: Form and Function* (eds. Blackmore, S. & Ferguson, I. K.) 203–217 (Academic, London, 1986).
50. Collinson, M. E., Boulter, M. C. & Holmes, P. L. in *The Fossil Record 2* (ed. Benton, M. J.) 809–841 (Chapman & Hall, London, 1993).
51. Crane, P. R. *Pl. Syst. Evol.* **162,** 165–191 (1989).
52. Drinnan, A. N., Crane, P. R. & Hoot, S. *Pl. Syst. Evol.* **8** (suppl.), 93–122 (1994).
53. Crane, P. R., Pedersen, K. R., Friis, E. M. & Drinnan, A. N. *Syst. Bot.* **18,** 328–344 (1993).
54. Muller, J. *Bot. Rev.* **47,** 1–142 (1981).
55. Scott, R. A., Barghoorn, E. S. & Leopold, E. B. *Am. J. Sci.* **258** A, 284–299 (1960).
56. Cornet, B. *Evol. Theory* **7,** 231–309 (1986).
57. Cornet, B. *Palaeontographica* **213** B, 37–87 (1989).
58. Martin, W., Gieri, A. & Saedler, H. *Nature* **339,** 46–48 (1989).
59. Martin, W. *et al. Molec. Biol. Evol.* **10,** 140–162 (1993).
60. Wolfe, K. H., Gouy, M., Yang, Y. W., Sharp, P. & Li, W.-H. *Proc. natn. Acad. Sci. USA* **86,** 6201–6205 (1989).
61. Crane, P. R., Donoghue, M. J., Doyle, J. A. & Friis, E. M. *Nature* **342,** 131–132 (1989).
62. Cornet, B. *Mod. Geol.* **19,** 81–99 (1993).
63. Taylor, D. W. *Am. J. Bot.* **81,** 103 abstract (1994).
64. Crane, P. R. *Nature* **366,** 631–632 (1993).
65. Doyle, J. A. & Hotton, C. L. in *Pollen and Spores: Patterns of Diversification* (eds. Blackmore, S. & Barnes, S. H.) 169–195 (Clarendon, Oxford, 1991).
66. Regal, P. J. *Science* **196,** 622–629 (1977).
67. Crepet, W. L. *Bioscience* **29,** 102–108 (1979).
68. Burger, W. C. *Bioscience* **31,** 572; 577–581 (1981).
69. Pellmyr, O. *Trends Ecol. Evol.* **7,** 46–49 (1992).
70. Labandeira, C. C. & Sepkoski, J. J. *Science* **262,** 310–315 (1993).
71. Crepet, W. L., Friis, E. M. & Nixon, K. C. *Phil. Trans. R. Soc.* B **333,** 187–195 (1991).
72. Crane, P. R. in *Origin and Evolution of Gymnosperms* (ed. Beck, C. B.) 218–272 (Columbia Univ. Press, New York, 1988).

73. Kato, M. & Inoue, T. *Nature* **368,** 195 (1994).
74. Lidgard, S. & Crane, P. R. *Paleobiology* **16,** 77–93 (1990).
75. Crane, P. R. & Lidgard, S. *Science* **246,** 675–678 (1989).
76. Wing, S. L. & Tiffney, B. H. *Rev. Palaeobot. Palynol.* **50,** 179–210 (1987).
77. Crane, P. R. & Lidgard, S. in *Major Evolutionary Radiations* (eds. Taylor, P. D. & Larwood, G. P.) 377–407 (Clarendon, Oxford, 1990).
78. Wing, S. L., Hickey, L. J. & Swisher, C. C. *Nature* **363,** 342–344 (1993).
79. Mulcahy, D. L. *Science* **206,** 20–23 (1979).
80. Stebbins, G. L. *Bioscience* **31,** 573–577 (1981).
81. Bond, W. J. *Biol. J. Linn. Soc.* **36,** 227–249 (1989).
82. Sanderson, M. J. & Donoghue, M. J. *Science* **264,** 1590–1593 (1994).
83. Nee, S. & Harvey, P. H. *Science* **264,** 1549–1550 (1994).
84. Larson, R. L. *Geology* **19,** 963–966 (1991).
85. Coen, E. S. & Meyerowitz, E. M. *Nature* **353,** 31–37 (1991).
86. Leroy, J.-F. *Origine et évolution des plantes à fleurs* (Masson, Paris, 1993).
87. Van Konijnenburg-Van Cittert, J. H. A. *Rev. Palaeobot. Palynol.* **71,** 239–254 (1992).
88. Friedman, W. E. *Am. J. Bot.* **81,** 1468–1486 (1990).
89. Carmichael, J. S. & Friedman, W. E. *Am. J. Bot.* **81,** 20 abstract (1994).
90. Doyle, J. A. & Donoghue, M. J. *Brittonia* **44,** 89–106 (1992).
91. Pedersen, K. R., Crane, P. R., Drinnan, A. N. & Friis, E. M. *Grana* **30,** 577–590 (1991).
92. Upchurch, G. R. *Ann. Missouri bot. Gard.* **71,** 522–550 (1984).
93. Dilcher, D. L. *Am. J. bot.* **76,** 162 abstract (1989).
94. Taylor, D. W. & Hickey, L. J. *Science* **247,** 702–704 (1991).
95. Ward, J. V., Doyle, J. A. & Hotton, C. L. *Pollen Spores* **33,** 101–120 (1989).
96. Friis, E. M., Eklund, H., Pedersen, K. R. & Crane, P. R. *Int. J. Pl. Sci.* **155,** 772–785 (1994).
97. Drinnan, A. N., Crane, P. R., Pedersen, K. R. & Friis, E. M. *Am. J. Bot.* **78,** 153–176 (1991).
98. Crane, P. R., Friis, E. M. & Pedersen, K. R. *Pl. Syst. Evol.* **165,** 211–226 (1989).
99. Drinnan, A. N., Crane, P. R., Friis, E. M. & Pedersen, K. R. *Bot. Gaz.* **151,** 370–384 (1990).
100. Dilcher, D. L. & Crane, P. R. *Ann. Missouri bot. Gard.* **71,** 351–383 (1984).

Acknowledgments

We thank P. K. Endress for helpful discussion and S. Wing for his review of the manuscript. This work was supported by the National Science Foundation, the Swedish Natural Science Research Council, the Carlsberg Foundation and the Field Museum Bass Fellowship Fund.

Peter Forey and Philippe Janvier

AGNATHANS AND THE ORIGIN OF JAWED VERTEBRATES

The origins of jawed vertebrates (gnathostomes) lie somewhere within the ranks of long-extinct jawless fishes, represented today as the lampreys and hagfishes. Recent discoveries of hitherto unknown kinds of jawless fishes (agnathans), together with re-examination of known agnathans and advances in systematic methods, have revitalized debates about the relationships of ancient fishes and given fresh insights into early vertebrate history.

For many years fossil agnathans were known primarily by cephalaspids (Osteostraci) (Fig. 1d) and pterapsids (Heterostraci) (Fig. 1c), accompanied by a small group of anaspids (Anaspida) (Fig. 1f) and the very poorly known thelodonts (Fig. 1i). All of these fishes occur in Silurian and Devonian deposits and were most numerous in the late Silurian/early Devonian (about 420–390 million years before present [Myr BP]). These extinct fishes were unlike their modern counterparts in possessing a dermal skeleton of minute scales and/or bony plates. They are therefore sometimes known collectively as ostracoderms (bony-skinned) and it is thought that one or other ostracoderm groups gave rise to the gnathostomes.

During the past 15 years, important discoveries of new kinds of extinct agnathans have been made worldwide which considerably increase our knowledge of the variety of agnathans and, at the same time, question some of the traditional ideas about the interrelationships of these early fishes and the history of vertebrate specializations. Some of the important finds are listed.

New Fossil Finds

1. In 1977 *Arandaspis* was described from the Lower Ordovician of Australia[1] and in 1986 a very similar form, *Sacambambaspis,* was described[2] from the Upper Ordovician of Bolivia (Fig. 1h). Both of these are early representatives of articulated heterostracan fishes with well-developed dermal skeletons and they may offer insights into the anatomy of the earliest vertebrate armor.

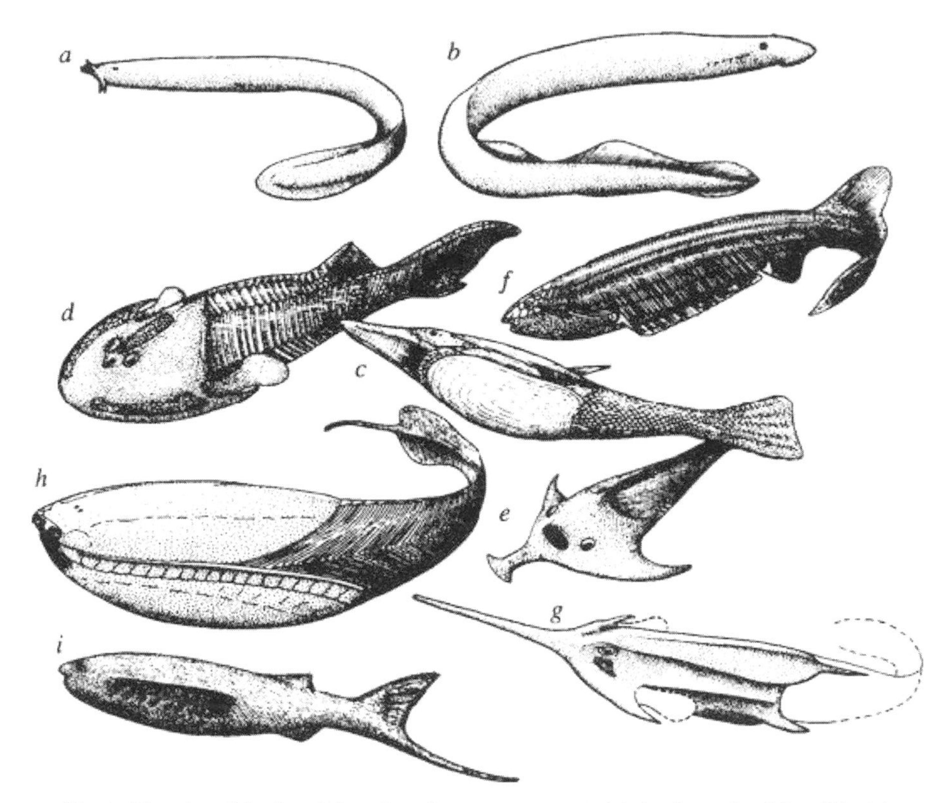

Fig. 1 Diversity of jawless fishes. Agnathans are represented today by *a,* hagfishes (Myxiniformes) and *b,* lampreys (Petromyzontiformes). Both are naked, eel-shaped and without paired fins. They are collectively known as cyclostomes (round-mouths). In the Paleozoic there were several different groups of agnathans showing a wide variety of morphology of bony dermal armor. These are collectively known as ostracoderms. *c,* Pteraspids (Heterostraci) and *d,* cephalaspids (Osteostraci) were very common and their solid head shields are often preserved in Upper Silurian and Lower Devonian rocks of North America, Europe and Siberia. The pteraspids (*Pteraspis* illustrated) were the dominant group of heterostracans. Their armor consisted of large dorsal and ventral oval shields surrounded by a series of smaller plates. The eyes were placed laterally and the mouth surrounded by tiny rod-shaped plates. In cephalaspids (*Cephalaspis* illustrated) the eyes were placed dorsally with a single nasohypophysial opening, and the flattened head shield was fenestrated by enlarged sensory fields connected to the inner ear. They were enhanced parts of the lateral-line system and probably used to detect water vibrations. Cephalaspids had pectoral fins. *e,* Galeaspids (Galeaspida) were very unusual agnathans, also with solid head shields, and restricted to eastern Asia. Like cephalaspids, the galeaspid head shield was flattened and in some species may have an unusual shape (*Sanchaspis* illustrated here). The eyes were placed more laterally and there was a large median opening leading to the nasal sacs and the mouth. *f,* Anaspids (Anaspida) were a small, poorly known group of fusiform fishes known from North America, Europe and China. They had a dorsal nasohypohysial opening like cephalaspids and lampreys. The gill openings were arranged in an oblique line as in lampreys. Some species, such as *Pharyngolepis,* illustrated here, had long paired fin folds running along the ventral sides of the body. *g,* Pituriaspids from the Devonian of Australia have only recently been discovered. The shield of *Pituriaspis* was long, sutureless and pierced by triangular openings, of unknown function, on either side of the eyes. It is probable that paired pectoral fins were present as in cephalaspids.

2. From the Silurian and Devonian of China and northern Vietnam we now have evidence[3–5] of a highly distinctive and varied group of agnathans known as galeaspids (Galeaspida) (Fig. 1e). The internal structures of some of these galeaspids are beautifully preserved[6] such that it is possible to make detailed anatomical comparisons with cephalaspids and gnathostomes.

3. Very recent discoveries[7] of articulated thelodonts from the Silurian and Devonian of northwestern Canada have revealed a surprising new body shape among agnathans and traces of a stomach, as well as providing evidence of the affinities of this enigmatic group.

4. Another new agnathan body form was revealed with the recent discovery of pituriaspids[8] from the middle Devonian of Australia (Fig. 1g). Superficially, they resemble the galeaspids but as yet very little is known about these animals and any statement about their affinities must be speculative.

5. Last, among the recent fossil finds there has been the discovery of a superficially hagfish-like animal which carries a feeding apparatus consisting of conodonts.[9–11] Conodonts are tiny phosphatic tooth-like structures, commonly occurring in marine rocks ranging from the Cambrian to the Triassic, and of great biostratigraphic importance for correlating rock sequences. Their biological affinities, however, have never been satisfactorily explained. The discovery of this conodont-bearing animal suggests that at least some conodonts may be among the most primitive of vertebrates.

Alongside the discoveries of these new animals there has also been a considerable increase in our knowledge of the anatomy of heterostracans[12] and osteostracans.[13,14] Many new detailed comparisons may be made between the fossils and Recent representatives, leading to new theories about the relationships between agnathan groups and new theories about how jawed vertebrates acquired their characters.

Lampreys, Hagfishes and Relatives

Lampreys (Petromyzontiformes) and hagfishes (Myxiniformes) are eel-like, naked animals with no trace of dermal armor or paired fins. Both

---

h, The earliest known vertebrates (Arandaspida) show that by early Ordovician times vertebrate life had already developed complex body armor. *Sacambambaspis,* illustrated here, shows large dorsal and ventral head shields separated by tiny branchial plates and a body covered with elongated scales. *i,* The thelodonts (Thelodontida) are poorly known, usually found as minute isolated scales. *Thelodus* appears to have been flattened, bearing several gill openings beneath pectoral flaps. Some newly discovered thelodonts suggest that at least some were markedly compressed laterally.

have a single median nostril and circular mouth (accompanied by a sucker in lampreys), a complex toothed tongue operated by circular and longitudinal muscles, and pouch-like gills opening through circular openings. The feeding habits of both are rather specialized. Most lampreys are filter-feeders as young and ectoparasitic blood-suckers as adults, whereas hagfishes are predators or they rasp away the flesh of dead or moribund fishes. Because of these common anatomical and behavioral features lampreys and hagfishes were thought to be most closely related to one another among the modern fauna and were recognized as cyclostomes (Fig. 2A, a).

This theory carries with it the implication that the cyclostome characters mentioned above were not part of the history of gnathostomes but were instead specializations restricted to lampreys and hagfishes.

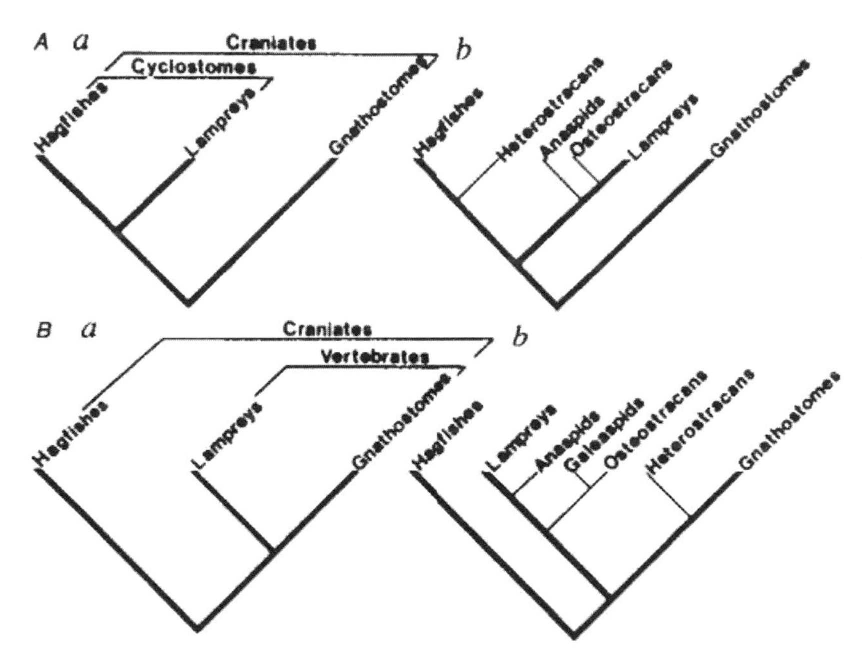

Fig. 2 Theories of craniate relationships. On the left are two theories of the relationships among modern animals. On the right are two schemes including fossil agnathan groups congruent with the classifications to the left. *A, a,* A traditional classification which suggests that lampreys and hagfishes are most closely related to one another. *A, b,* This classification[44] assumes that the fossil groups are related to one or other of the Recent agnathans (galeaspids were not known at this time). *B, a,* This theory suggests that lampreys are more closely related to gnathostomes than to hagfishes (agnathans are considered paraphyletic). *B, b,* This classification[36,39] is one of several suggesting that one or other of the fossil groups are more closely related to jawed vertebrates than they are to other agnathans. In this case heterostracans are considered the closest relatives of gnathostomes; this is based on the speculation that heterostracans had paired nasal sacs and paired nostrils as seen in jawed vertebrates. Thelodonts are not included in these classifications; they are too poorly known.

Lampreys and hagfishes have long, separate histories. Fossil representatives of lampreys and hagfishes may be found in the Upper Carboniferous (230 Myr BP) where, in most preserved details, they are similar to their modern counterparts.[15–18]

Fossil agnathan groups, such as the heterostracans, anaspids and osteostracans were considered related to one or other of the modern types (Fig. 2A, b). Osteostracans (cephalaspids) have long been considered close relatives of lampreys despite a total difference in overall body shape. Like lampreys, cephalaspids show a single nostril joined with a single hypophysial opening which together form a keyhole-shaped opening on top of the head. This is a highly unusual feature, and for many years paleontologists have interpreted cephalapsids using the lamprey as the model.[19,20] Anaspids also show a dorsally placed nasohypophysial opening and are therefore considered to be closely related to lampreys and cephalaspids. The heterostracans were associated with the hagfishes, chiefly because they lacked the specializations of the nasohypophysial opening and because they have a single branchial opening (see below).

This theory, relating armored forms to each of the living groups, implies that any feature such as armor, scales or modified fins was independently lost during the history of hagfishes and lampreys.

Another theory concerning the relationships of the Recent taxa suggests that lampreys are more closely related to gnathostomes than either is to hagfishes, implying that hagfishes are the most primitive of craniate animals (Fig. 2B, a). Lampreys and gnathostomes share a great many features which are, at the same time, absent from hagfishes,[21,22] such as well-developed neural and hemal arches (the beginnings of a true backbone) and radial muscles in the fins (which allow voluntary flexing of the fins), true neuromast organs distributed along a lateral line, extrinsic eye muscles, nervous regulation of the heart and hyperosmoregulation.

This theory raises the possibility that cyclostome characters such as the median nostril, complex tongue and pouch-like gills are either primitive craniate characters and are truly part of the history of gnathostomes, or are convergences, that is, accidental similarities, developed independently in lampreys and in hagfishes.

There have been several suggestions as to how fossil agnathans may fit into a scheme such as this. Usually, one or other fossil group is considered more closely related to gnathostomes than to any other agnathan, living or fossil. One theory is shown in Fig. 2B, b, another in Fig. 4. With these different theories come different conclusions about the history of the skeleton, fins, nasal sacs and so on.

Choice between these competing theories of relationship among the

Recent forms has not received consensus (compare refs. 23, 24 to refs. 21, 22, 25). Unfortunately, molecular evidence, which has proved itself useful in other areas of disagreement, has yet to prove itself here. Amino-acid data for a variety of proteins has accumulated throughout the 1980s. Sequences for globin molecules are available for lampreys, hagfishes and a variety of gnathostomes, and comparisons of these sequences[26] imply that lampreys and hagfishes are sister groups. But this may be misleading because agnathan globin is homologous with both gnathostome myoglobin and hemoglobin. It is probable that the agnathan globin sequences simply represent the primitive craniate sequences from which other gnathostome sequences were derived, perhaps by gene duplication.

Another line of molecular evidence comes from nucleotide sequences of 18S ribosomal RNA.[27] This analysis used a bootstrapping technique and examined 1,631 base pairs. When cephalochordates or cephalochordates plus tunicates were used as the outgroup, the analysis suggested that lampreys and hagfishes are sister groups. But when tunicates alone were used to root the analysis, the grouping of lampreys and gnathostomes was supported.

A somewhat similar result was obtained using 28S rRNA. The results of analyses are equivocal, and these appear to depend on which outgroup (cephalochordate or tunicate) is chosen to root the analysis. Lecointre (unpublished memoir, DEA, University of Paris VII [1989]) has examined sequences of 361 nucleotides of the 5′ extremity of 28S rRNA and found that there are only two base-pair positions, out of 361, common to lampreys and hagfishes, and there is only one position uniquely shared by lampreys and gnathostomes. This is hardly conclusive evidence for one theory or another. Therefore, the molecular data, while adding a new dimension, also requires additional explanation and does not clearly arbitrate between the competing theories of relationship. It is clear that one of the challenges of molecular systematics is to deal with sequence data of ancient and possibly highly specialized lineages.

Significance of the New Fossil Finds

Heterostracans

One of the chief reasons for believing that heterostracans were fossil relatives of hagfishes (Fig. 2A, b) centered on the fact that in the modern hagfish, *Myxine,* and heterostracans there is a single external branchial opening emptying from several internal gill pouches. This is undoubtedly a specialization which developed from the condition where there are many separate branchial openings as seen, for instance, in lampreys, anaspids, cephalaspids, galeaspids and thelodonts as well as in cephalochordates (one likely sister group of craniates). Although *Myxine* shows a single paired branchial opening, another Recent hagfish *(Eptatretus)* has a series of separate openings and this has cast doubt on reasons for associating hagfishes and heterostracans.

With the discovery of *Arandaspis* and *Sacambambaspis* this doubt is reinforced. These earliest armored vertebrates are clearly related to heterostracans: they have large dorsal and ventral shields encasing the head and a similar pattern of lateral lines penetrating the shields. At the same time they show a series of separate gill openings wedged between these shields and each associated with a tiny branchial plate. A similar series of separate openings has recently been found in *Astraspis*[28] which is another early heterostracan but without the large shield-like armor. Thus, these early heterostracans show a primitive vertebrate condition and suggest that the single branchial opening was developed independently within the heterostracan lineage and within the hagfish lineage.

The discovery of articulated Ordovician heterostracans has other more general implications for our ideas of the evolution of the vertebrate skeleton and tail structure.

For nearly a century it was assumed that the primitive vertebrate armor consisted of scales and small polygonal bony plates. These had been found in the fragmentary vertebrates such as *Astraspis* and *Eriptychius,* long known from Middle Ordovician rocks of North America, and they were often regarded as showing primitive vertebrate conditions from which the large shields of more typical pteraspids developed. *Arandaspis* and *Sacambambaspis* both show large continuous head shields and a trunk covered with elongated scales (Fig. 1h). This suggests that considerable diversification of the vertebrate skeleton had already taken place by Lower Ordovician times and, furthermore, that it is not possible on the basis of antiquity alone to predict whether a micromeric or macromeric skeleton is the more primitive vertebrate condition.

The detailed histology of the bone of *Arandaspis* and *Sacambambaspis* is, as yet, unknown. But the Ordovician agnathans *Astraspis* and *Eriptychius* were originally associated with heterostracans because the skeleton was made of aspidin (acellular bone), a condition regarded by some as primitive and by others as derived. The early occurrence of aspidin has always strengthened the opinion that it is the more primitive condition. Recently, however, investigation[29] has confirmed earlier suspicions[30] that the presence of cellular bone is of equal antiquity. Furthermore the particular type of cellular bone found in the same deposits as *Eriptychius* suggests that cephalaspids (osteostracans) may be equally as old as heterostracans.

Traditional ideas about the primitive vertebrate tail were derived from studies on pteraspid heterostracans. These ideas[31,32] proposed that the tail was asymmetrical, with a larger lower lobe and, when in use, tended to drive the head up to compensate for a heavy head shield and a lack of paired fins. *Sacambambaspis* and several other recently discovered well-preserved heterostracans from the Silurian of the Northwest Territories[33] and the Canadian Arctic[34] all show perfectly symmetrical tails. The tail was deep with enlarged ridge scales along the upper and lower edges for stiffening, and the fan was strengthened by special radiating rows of enlarged scales (Fig. 1e). This type of tail drove the fish directly forward.

Cephalaspids

During the last decade or so cephalaspids have been re-examined in great detail.[13,14] Leaving aside the undoubted similarities in the nasohypophysial opening between cephalaspids and lampreys (see above), these recent studies show that there is a striking resemblance between the cephalaspid skull and the skull of an early jawed vertebrate such as the placoderm (Fig. 3). Externally the cephalaspid body, like that of primitive gnathostomes, shows a heterocercal tail and paired fins (only pectoral fins are developed in cephalaspids). These similarities, together with several other less obvious features, suggest that cephalapsids are more closely related to gnathostomes than either is to lampreys. This would mean that the single nasohypophysial opening is either a convergent feature or a feature primitive for craniates.

Galeaspids

The galeaspids are a group of about 40 species which superficially resemble cephalaspids in having a large, flattened and solid head skeleton which encloses a complex brain and inner ear (Fig. 3). They have been considered as directly related to cephalaspids.[35,36] But unlike cephalaspids they lack paired fins and, instead of a dorsal nasohypophysial opening, they have a large median dorsal opening which communicates with paired nasal cavities and, farther below, with the pharynx. In some respects this arrangement is similar to that seen in Recent hagfishes where a single nostril opens into a duct which passes the single nasal sac before opening into the roof of the mouth.

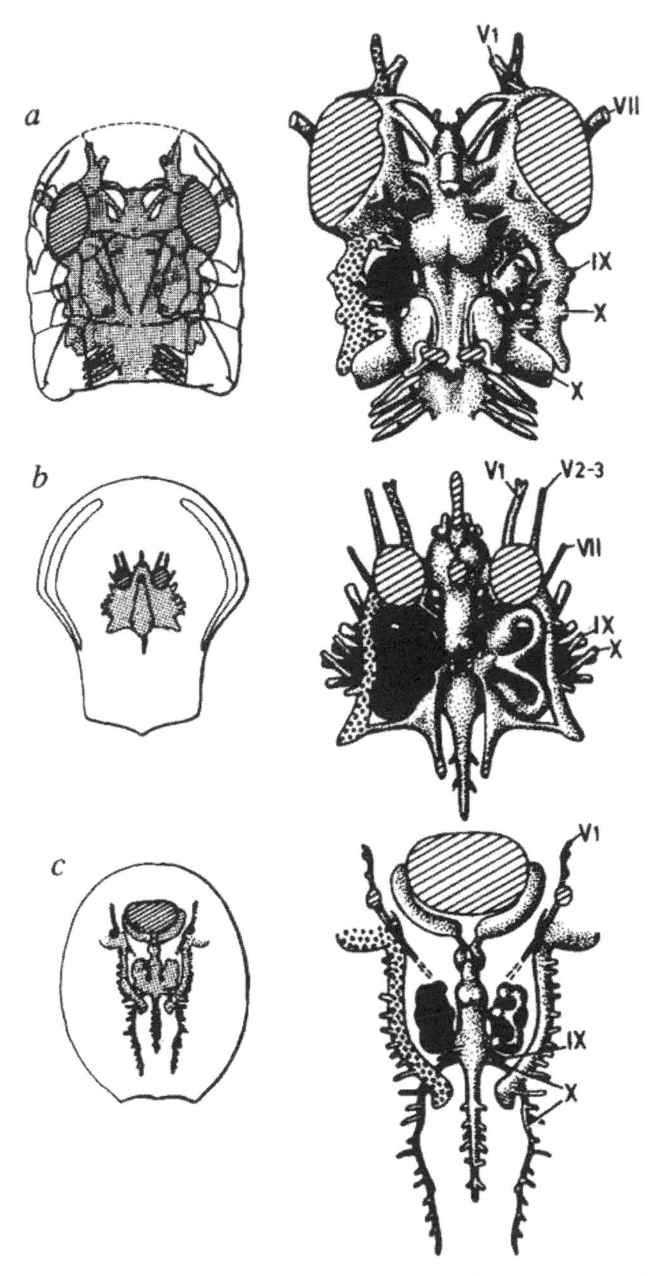

Fig. 3 Internal anatomy. In recent years we have learned much about the internal anatomy of cephalaspids and galeaspids which had a bony skeleton surrounding the brain and gills. There is a remarkable similarity between these fishes and primitive gnathostomes in the structure of the orbit, the labyrinth of the ear and the position of the head vein. *a, Brindabellaspis,* a placoderm gnathostome (from ref. 45); *b, Norselapsis,* an osteostracan (from ref. 13); *c, Duyunolepis,* a galeaspid (ref. 46). The outline of the head *(left)* shows the position of the internal cavities *(right).* The head vein is dotted, the labyrinth of the inner ear is black. Some of the cranial nerves are indicated by roman numerals.

Thelodonts

Thelodonts are usually represented in the fossil record as isolated scales which, because of their abundance, are useful for stratigraphic correlation. But we know very little about the animals which bore these scales and although a few articulated remains have been known for several years, they show very little anatomical detail and it is not easy to identify immediate relatives. Hitherto, known thelodonts appear to be flattened dorsoventrally and to have paired flaps on either side of the head.

Very recently, discoveries of some new and very well preserved thelodonts from the Silurian of British Columbia were announced.[7] These new thelodonts show an unexpected body shape in which the body is deep and compressed from side to side and the tail is perfectly symmetrical, rather like that of early heterostracans. They also show that several very different types of scale could be borne by the same animal, including some complex scales immediately surrounding the gill openings. Interestingly, the gill openings are arranged in an oblique line, a situation very reminiscent of the gill opening of anaspids and lampreys. This may suggest that at least some thelodonts are related to these two groups.

Another interesting feature about these new thelodonts is that there is evidence that a well-developed stomach was present. Lampreys and hagfishes lack a differentiated stomach, and it is generally assumed that all agnathans were similarly microphagous with no need for a stomach. Jaws, macrophagy and stomachs were thought to go together. However, in these thelodonts we have evidence that stomachs preceded jaws in vertebrate history.

Conodonts

Conodonts have traditionally been used as form taxa, useful for stratigraphic correlation within Paleozoic rocks. During the last few years there have been a number of discoveries suggesting that conodonts may be vertebrates.

Briggs and colleagues[9,10] discovered fossils of 4 cm–long eel-like animals from the early Carboniferous of Scotland, each of which carried a set of conodont elements at the anterior end. Each of these animals, one of which has been named *Clydagnathus,* shows a laterally flattened body, caudal fin and "V-shaped" muscle blocks, together with the faint outline of a notochord and large paired eyes. These are features typically found in chordates, although it is difficult to make more precise comparisons.

More recently the histological detail of the minute conodont elements themselves has been re-examined. Most conodonts, or euconodonts, are tooth-like phosphatic structures with a shiny outer layer and an inner core. Recent research[37] advocates that the minute spaces within the substance of the core, which have long been known to exist, are true bone cell or osteocyte cell spaces. The basal tissue of the conodont has also been compared to globular calcified cartilage as seen associated with the earliest authenticated bone of *Eriptychius*. But the surface layer is far more problematic. It appears to be composed of centrifugally deposited lamellar tissue, superficially like vertebrate enamel. But in the conodonts examined so far the crystalline structure within this layer is very variable. In two species examined the crystals lie parallel to the surface and in another two they lie at a steep angle. Neither type corresponds precisely to that seen in vertebrate enamel, and the extreme variation in crystal orientation is puzzling. Furthermore, there appears to be a total absence of dentine, which is unexpected if conodonts are vertebrates. Dentine is universal in vertebrates and is thought to be the most primitive of vertebrate hard tissues.[29]

Clearly, much more research needs to be done on a wide variety of conodonts before the histological structure may be used as evidence of vertebrate affinities. If that supportive evidence is forthcoming then the implication is that vertebrates existed in the late Cambrian, 510 Myr BP, and some 40 Myr before the appearance of vertebrate armor. Conodonts are not considered further here.

Relationships and Ideas of Vertebrate History

The new features that have been discovered by re-examining Recent taxa alongside the many new fossil finds during the past 15 years lead to classifications that suggest that lampreys are more closely related to gnathostomes than either is to hagfishes. In other words, lampreys and hagfishes do not form a monophyletic group. One of the classifications that inserts the fossil groups is illustrated in Fig. 4. This classification is based on a cladistic analysis of morphological characters using the PAUP parsimony program. Details of this classification will be published elsewhere. Interpreting this as a phylogeny, we can use this classification to trace the history of the transformation of some characters leading to the condition in gnathostomes.

One of the most debated areas of anatomy is the early history of the vertebrate nasal and hypophysial regions already mentioned above. It is not clear whether paired or single nasal sacs and nostrils are primitive for craniates. The phylogeny given here predicts that the condition in hagfishes is primitive. That is, the earliest craniate had a single median in-

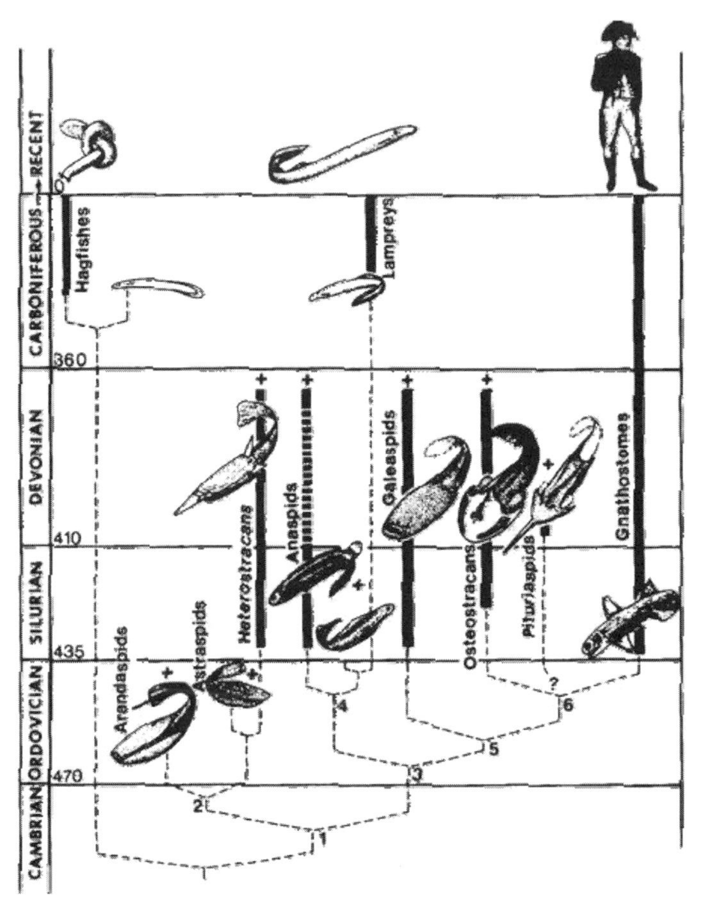

Fig. 4 Our favored phylogenetic tree of principal agnathan groups with stratigraphic ranges superimposed *(bold black lines).* This tree is simplified from a PAUP analysis, the details of which will be published elsewhere. The position of pituriaspids and thelodonts is not clear. Some of the skeletal characters supporting the nodes of this classification are as follows: *1,* Lateral-line sensory canal, two semicircular canals in the ear, bone (lost in lampreys). *2,* Two large plates (dorsal and ventral) covering the head, separate branchial plates. *3,* True dorsal and anal fins with fin rays, asymmetrical tail shape. *4,* Branchial openings arranged in a slanting line (suggesting a separation of pharynx from esophagus as in modern lampreys). *5,* Perichondrial bone, large head vein placed dorsally, large orbits. *6,* Dermal bone with cells, epicercal tail with modified scale cover (hinge line), paired pectoral fins, open endolymphatic duct, sclerotic ossification.

halent duct conveying water over the unpaired nasal sac on its way to the pharynx and gills. The gnathostome condition, with paired nasal sacs not associated with a through duct, would therefore be a derived condition.

The nature of the nasal sacs in heterostracans has been the subject of some debate. Stensiö[38] reconstructed the snout after the hagfish model. Others[36,39] have suggested that there were paired nasal sacs opening separately and gave rise to the theory that heterostracans might be ancestral to jawed vertebrates (Fig. 2B, b). As long as the endoskeleton of the heterostracans remains unknown there must always be doubt about the pattern. Some faint impressions on the inside of the dermal skeleton suggest that at least paired nasal sacs may have been present[40] (unlike in modern hagfishes). But the inhalent duct leading to the outside may well have been single. We now know that this was the condition in galeaspids where paired nasal sacs opened into a median duct and the duct itself communicated with the pharynx. This may represent an example of a second stage of transformation. Subsequent modification of this pattern would be the confluence of both nasal sacs and the failure of the inhalent duct to communicate with the mouth in lampreys and cephalaspids. A different modification would have taken place in gnathostomes where the separate nasal sacs remained large but the common inhalent duct was lost.

The phylogeny suggested here implies that hagfishes are primitively naked, whereas lampreys are secondarily so. For the history of the dermal skeleton the recent finds have tended to complicate the issue because it is no longer clear from the stratigraphic record whether cellularity or acellularity is primitive, or whether a solid shield or polygonal plates are primitive.

The evolution of fins may also be traced across this phylogeny. Hagfishes have a weakly developed symmetrical tail and the web is supported by poorly developed cartilage rods without associated muscles. This was probably very similar in the heterostracans except that here the tail lobe was scale covered, perhaps with some special scale rows aligned along the length of fin (see above). From here anaspids, lampreys and perhaps some thelodonts developed a downturned asymmetrical tail (hypocercal), and in the first two there were definite endoskeletal fin rays with, in at least lampreys, associated musculature. In cephalaspids the tail is turned upwards (epicercal) as in gnathostomes and here too there were probably muscularized endoskeletal fin rays. The tail of galeaspids remains poorly known.

Paired pectoral fins of gnathostome-type were only developed in cephalaspids (and perhaps pituriaspids) where there is evidence of an endoskeleton and associated muscles. Some anaspids have specialized elongate paired fins which may have been developed in and restricted to this group (Fig. 1f).

The outline above suggests that agnathans are a paraphyletic group (that is, some are more closely related to jawed vertebrates than to other agnathans) and therefore the presence of circular mouths and pouch-like gills was part of gnathostome history. The acquisition of many gnathostome characters occurred through several transformation series which can be traced across the phylogeny of agnathans. But some characters appear suddenly, and this applies to the feature which characterizes the gnathostomes: the development of jaws.

The way in which jaws developed remains unknown. There is a long-held belief that jaws developed from a true mandibular gillbearing arch. Yet no animal showing this has ever been found. The idea stems from the belief that in gnathostomes the mandibular arch is a serial homologue of the hyoid arch and the more posterior gill arches. Agnathans have gill arches but no jaws. Hence jaws might be a transformation of the anteriormost gill skeleton. But we know that the gill lamellae in a lamprey develop medially to the supporting skeleton, whereas the gills of a gnathostome develop laterally to the skeleton. For this reason some workers have denied any homology.[41,42] One possible alternative may be to suggest that jaws correspond to the velum in a larval lamprey, a special pumping device located at the entrance to the pharynx. The velum develops from the same embryonic segment as the mandibular arch, and part of its skeleton is located more medially than that of subsequent arches.

Summary

The discovery of many new facts and new types of agnathans with hitherto unknown anatomical features has questioned some of our traditional ideas. At the same time, it has permitted the formation of a more precise theory of relationships from which we may deduce evolutionary pathways.

We may ask if we would have different views on the interrelationships of modern craniates if we had no fossils. In this case we doubt it. But if we asked about the history of the appearance of characters then the answer is "yes." Fossils do throw light on the history of the lateral line and tail. In these cases the fossils show that these features have developed through an ordered series of transformations leading to conditions in the jawed vertebrates. The fossils show that paired fins constructed like those in gnathostomes are a relatively late development but that there may have been an experiment with elongated paired fins as seen in anaspids. Fossils also show that the structure of the nasohypophysial region is considerably variable and more complicated than the Recent animals might suggest. At the same time, the great variety of skeletal anatomy has complicated our theories of the early evolution of the vertebrate dermal skeleton.

Interesting future directions of study might be attempts to understand the varied locomotory adaptations of agnathans which show different development of paired and median fins. Detailed anatomical studies need to be done alongside modern hydrodynamic experiments.[43] There is also considerable scope for informed speculation about the feeding mechanisms among the extinct agnathans. The wealth of new information is a reminder that there is still plenty of scope for new studies in old fishes.

PETER FOREY is at the Natural History Museum, Cromwell Road, London SW7 5BD, UK; Philippe Janvier is at the Muséum National d'Histoire Naturelle, 8 rue Buffon, 75005 Paris, France.

References

1. Ritchie, A. & Gilbert-Tomlinson, J. *Alcheringa* **1,** 351–368 (1977).
2. Gagnier, P.-Y., Blieck, A. R. M. & Rodrigo, S. *Geobios* **19,** 629–634 (1986).
3. Pan Jiang, K. *Proc. Linn. Soc. N.S.W.* **107,** 303–319 (1984).
4. Janvier, P. *J. vert. Paleont.* **4,** 344–358 (1984).
5. Tong-Dzuy, T. & Janvier, P. *Bull. Mus. nat. Hist. nat. Paris* **12,** 143–223 (1990).
6. Halstead, L. B. *Nature* **282,** 833–836 (1979).
7. Wilson, M. V. H. & Caldwell, M. W. *Nature* (in the press).
8. Young, G. C. in *Early Vertebrates and Related Problems of Evolutionary Biology* (eds. Chang, M. M., Liu, Y. H. & Zhang, G. R.) 67–85 (Science Press, Beijing 1991).
9. Briggs, D. E. G., Clarkson, E. N. K. & Aldridge, R. J. *Lethaia* **16,** 1–14 (1983).
10. Aldridge, R. J., Briggs, D. E. G., Clarkson, E. N. K. & Smith, M. P. *Lethaia* **19,** 279–291 (1986).
11. Briggs, D. E. G. *Bull. Field Mus. nat. Hist.* **55,** 11–18 (1984).
12. Blieck, A. *Les Hétérostracés (Vertébrés, Hétérostracés) de l'horizon Vogti (Groupe de Red Bay, Dévonien inférieur de Spitsberg)* (CNRS, Paris, 1985).
13. Janvier, P. *Palaeovertebrata* **11,** 19–130 (1981).
14. Janvier, P. *Les Cephalaspides du Spitsberg* (CNRS, Paris, 1985).
15. Bardack, D. & Zangerl, R. *Science* **1962,** 1265–1267 (1962).
16. Janvier, P. & Lund, R. *Geobios* **19,** 647–652 (1986).
17. Bardack, D. *Bull. Soc. Hist. nat. Autun (France)* **116,** 97–99 (1985).
18. Bardack, D. *Science* **254,** 701–703 (1991).
19. Stensiö, E. A. *Skrifter øm Svalbard og Ishavet* **12,** 1–391 (1927).
20. Stensiö, E. A. *The Cephalaspids of Great Britain* (British Museum [Natural History], London, 1932).
21. Hardisty, M. W. in *The Biology of Lampreys* (eds. Hardisty, M. W. & Potter, I. C.) 165–259 (Academic, London, 1982).
22. Janvier, P. *J. vert. Paleont.* **1,** 121–159 (1981).
23. Schaeffer, B. & Thomson, K. S. in *Aspects of Vertebrate History* (ed. Jacobs, L. L.) 19–33 (Museum of Northern Arizona Press, Flagstaff, 1980).
24. Yalden, D. W. *Zool. J. Linn. Soc.* **84,** 291–300 (1985).
25. Løvtrup, S. *The Phylogeny of the Vertebrata* (Wiley, London, 1977).
26. Goodman, M., Weiss, M. L. & Czelusniak, J. *Syst. Zool.* **31,** 376–399 (1982).
27. Stock, D. W. & Whitt, G. S. *Science* **257,** 787–789 (1992).
28. Elliott, D. K. *Science* **237,** 190–192 (1987).

29. Smith, M. M. & Hall, B. K. *Biol. Rev.* **65,** 277–373 (1990).
30. Denison, R. H. *Fieldiana Geol.* **16,** 131–192 (1967).
31. Stensiö, E. in *Traité de Paléontologie* Vol. 4 (ed. Piveteau, J.) (Masson, Paris, 1961).
32. Kermack, K. A. *J. exp. Biol.* **20,** 23–27 (1947).
33. Soehn, K. & Wilson, M. V. H. *J. vert. Paleont.* **10,** 405–419 (1990).
34. Dineley, D. L. & Loeffler, E. J. *Spec. Pap. Palaeont.* **18,** 1–214 (1976).
35. Janvier, P. *Bull. Mus. nat. Hist. nat. Paris* **278,** 1–16 (1975).
36. Halstead, L. B. in *Problems of Phylogenetic Reconstruction* (eds. Joysey, K. A. & Friday, A. E.) 159–196 (Academic. London, 1982).
37. Sansom, I. J., Smith, M. P., Armstrong, H. A. & Smith, M. M. *Science* **256,** 1308–1311 (1992).
38. Stensiö, E. A. *Nobel Symposium,* 13–71 (Almqvist & Wiksell, Stockholm, 1968).
39. Novitskaya, L. *Trudy Palaeont, Inst. Akad. Nauk SSSR* **196,** 1–178 (1983).
40. Janvier, P. & Blieck, A. *Zool. Scripta* **8,** 287–296 (1979).
41. Jarvik, E. *Basic Structure and Evolution of the Vertebrates* (Academic, London, 1980).
42. Moy-Thomas, J. A. & Miles, R. S. *Palaeozoic Fishes* 3rd edn. (Chapman & Hall, London, 1971).
43. Bunker, S. & Machin, K. E. in *Biomechanics in Evolution* (eds. Rayner, J. M. V. & Wootton, R. J.) 113–129 (Cambridge Univ, Press, Cambridge, 1991).
44. Jarvik, E. *Annls Soc. r. Zool. Belg.* **94,** 11–95 (1965).
45. Young, G. C. *Palaeontographica* **167,** 10–76 (1980).
46. P'an, J. & Wang, S.-T. in *Early Vertebrates and Related Problems of Evolutionary Biology* (eds. Chang, M. M., Liu, Y. H. & Zhang, G. R.) 299–333 (Science Press, Beijing 1991).

Acknowledgments

We thank P. Ahlberg (Oxford University) for reading this manuscript and the numerous discoverers and researchers for making this review desirable.

Per E. Ahlberg and Andrew R. Milner

THE ORIGIN AND EARLY DIVERSIFICATION OF TETRAPODS

A series of new fossil discoveries, coupled with cladistic analysis of old and new data, is beginning to resolve the origin of tetrapods into a documented sequence of character acquisition. Devonian tetrapods were more fish-like than believed previously, whereas Lower Carboniferous tetrapod faunas contain early representatives of the amphibian and amniote lineages. These very different assemblages are separated by a 20-million-year "Tournaisian gap" which has yielded very few tetrapod fossils.

The late Paleozoic invasion of the land by vertebrates was a key episode in the history of life, and has had a profound impact on the development of all terrestrial ecosystems. However, this event is still poorly understood. Although it has been known for a century that the terrestrial vertebrates (tetrapods) evolved from the lobe-finned fishes (sarcopterygians) during the Devonian period, only very recently has there been more direct evidence of this transition.[1,2]

In the late nineteenth and early twentieth centuries, the earliest known fossil tetrapods were from Lower Carboniferous strata in Scotland, and were assumed to have originated during the Carboniferous, in the lowland permanent freshwater swamps that gave rise to these deposits. Barrell's theory,[3] of a shift onto land triggered by the periodic aridity exemplified by the Devonian red-beds, was not supported by "hard" paleontological data. In 1932, the tetrapod record was pushed back in the Upper Devonian by the description of the skull of *Ichthyostega*[4] from East Greenland, and this discovery was interpreted as verification of Barrell's theory. The 363-million-year-old, meter-long *Ichthyostega* rapidly acquired symbolic status as "the earliest land vertebrate," but it possessed a bizarre mosaic of characters, combined obvious relict fish-like structures, such as fin rays, preopercular bones and a notochord entering the braincase, with unique features of skull and ankle construction which seemed to bar it from ancestry to later tetrapods.[5–7] This has resulted in a tendency to put *Ichthyostega* to one side and instead to look at tetrapod origins by comparing Devonian sarcopterygians with Carboniferous and Permian tetrapods.

For much of the middle of this century, there was general agreement that most of the Tetrapoda had arisen from osteolepiform lobefins (Fig. 2a–c). The principal debate centered on whether the entire Tetrapoda was monophyletic, or whether the Urodela (salamanders) had a separate origin from the others. Jarvik[5,8–11] held to the latter view, arguing that urodeles were derived from porolepiform lobefins. Other paleoichthyologists and paleoherpetologists argued for monophyly,[12–16] and by 1981, most workers believed that all tetrapods were derived monophyletically from within or near the osteolepiforms (Fig. 1). However, while osteolepiforms are anatomically tetrapod-like in several respects, they are morphologically rather undistinguished fishes, with no unambiguous structural adaptations to an amphibious mode of life. There was thus little hard evidence to suggest how, or in what sequence, most of the key tetrapod characters had arisen. In accordance with the "process-based" evolutionary thinking of the day, many scenarios were constructed to "explain" the environmental background and selection pressures behind the move onto land. Most incorporated the implicit assumption that tetrapod characteristics had evolved to facilitate terrestrial life. One class of scenarios centered on the concept of vertebrates surviving drought by moving from one water body to another,[17,18] another that vertebrates emerged from the water either to disperse[19,20] or to exploit the developing terrestrial ecosystem.[21,22]

Although cladistic methodology had already been applied to problems of the interrelationships of sarcopterygian fishes[16,23] and to tetrapod monophyly,[24] a conceptual turning-point in the study of tetrapod origins occurred with the publication in 1981 of a large-scale cladistic review of sarcopterygian and tetrapod interrelationships,[25] which used a wide-ranging examination of character distributions and polarities as the basis for a direct attack on the existing phylogenetic schemes and the reasoning underpinning them. Rosen *et al.*'s morphological arguments centered on the distribution of the choanae (internal nostrils) among sarcopterygian fish and lower tetrapods, but their main point was methodological. They stated categorically that evolutionary scenarios, such as those that purported to reconstruct the evolution of tetrapods from osteolepiforms, were unfalsifiable "stories" and not testable science. They concluded that the lungfishes (Dipnoi) are the closest relatives (living and fossil) of the tetrapods, and dismissed the osteolepiforms altogether as a paraphyletic assemblage of basal sarcopterygians defined on primitive characters. Although many have strongly criticized this bold article,[26–32] and its sys-

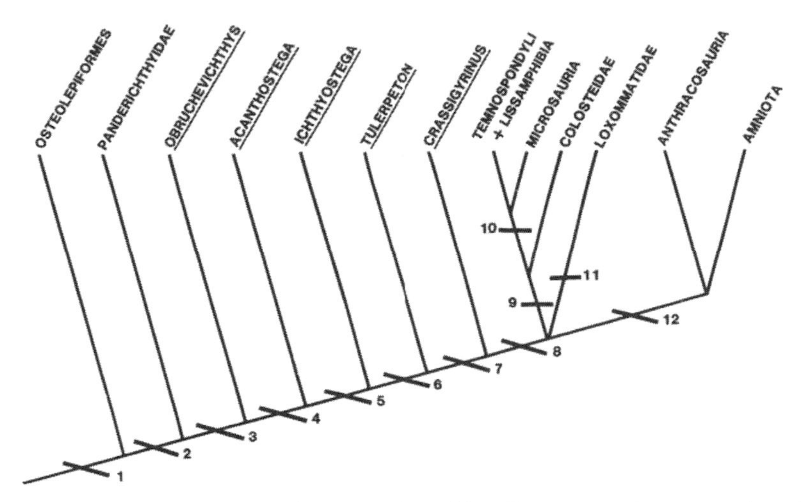

Fig. 1 A cladogram spanning the fish-tetrapod transition. The taxa have been grouped on the basis of the shared derived characters that they possess; the principal characters supporting each node are listed below. Inferences can be drawn about the nature of the transition from the sequence in which the derived characters (which can also be considered as evolutionary novelties) appear. The characters shared by panderichthyids and tetrapods appear to be morphological adaptions to a shallow-water habitat; Devonian tetrapods had acquired limbs and a large pelvis, but the known forms must have been semi-aquatic as they retained prominent lateral-line systems and gills; only post-Devonian tetrapods possess indisputably terrestrial adaptations, such as a strengthened vertebral column, and occipital condyles instead of a notochordal support for the skull. The move onto land could thus be interpreted as a very gradual process, with the earliest tetrapods using their limbs to move in shallow water rather than on land. This cladogram was produced iteratively rather than through computer analysis, and excludes the less completely known Devonian forms and several minor Carboniferous groups of enigmatic relationship. As the *Obruchevichthys*-grade tetrapods are poorly known, some of the characters listed at node 4 may eventually prove to be present at node 3. Significant characters defining the numbered nodes are as follows: *1,* Choana present; single pair of external nares. *2,* Flattened head with relatively elongate preorbital region; dorsally positioned orbits with narrow, transversely concave, interorbital region; frontal bones present; intracranial joint immobilized; body dorsoventrally flattened; absence of separate median fins; enlarged ribs. *3,* Meckelian bone not exposed dorsal to the prearticular; humerus with thin, flat entepicondyle continuous with humerus body, and narrow tall ectepicondyle; tibia with articulation surfaces for intermedium and tibiale. *4,* Single pair of nasals meeting in midline; compact otico-occipital region to skull; cheek with broad jugal-quadratojugal contact; absence of coronoid fangs; absence of operculars, median gular and submandibulars; presence of pre- and postzygapophyses; well-developed, ventrally directed ribs; large ornamented interclavicle; carpus, tarsus and dactyly of up to eight digits; iliac blade of pelvis extending dorsally and attached to vertebral column by sacral rib; ischia contribute to pubo-ischiadic symphysis; femur with adductor crest. *5,* Presence of olecranon process on ulna. *6,* Open lateral-line system on most or all dermal bones; elongate scapula blade with distinct cleithrum; dactyly of six digits or fewer. *7,* Absence of anocleithrum. *8,* Occipital condyles present; loss of preopercular; loss of anterior tectal; notochord excluded from braincase in adults; loss of ectepicondylar foramen in humerus; loss of "d" canal in humerus; palatal dentition reduced to paired fangs (reversed in some later groups). *9,* Exoccipital-postparietal contact on occiput; loss of cranial kinesis; small/incipient interpterygoid vacuities; dactyly of manus reduced to four digits or fewer. *10,* Long cultriform process of the parasphenoid contacting vomers; ceratobranchial ossicles compact with only five or six denticles; narrow-waisted humerus, not L-shaped. *11,* Antorbital vacuities at least as large as orbits. *12,* Premaxillaries less than half of skull width, vomers taper anteriorly, maxillaries reach level of anterior ends of vomers; tabular-parietal contact; oblique ridge of femur reduced to ventral ridge system; pes with phalangeal formula of 2.3.4.5.4-5.

tematic conclusions have not been widely accepted, some of Rosen *et al.*'s methodological criticisms were generally perceived as justified and have consequently shifted the basis on which the topic is discussed. Nearly all subsequent papers on sarcopterygian systematics and tetrapod origins have been framed within a cladistic methodology,[16,29–33] and most are computer-based. This has focused attention on the sequence of acquisition of tetrapod characters, which provides a more robust framework for reconstructing the base of the tetrapod tree. It also permits the construction of paleobiological models that are more constrained by character distributions and are hence partly testable.

Despite extensive debate, no single consensus about sarcopterygian interrelationships has emerged since 1981, but some precladistic theories have proved to be relatively robust. Virtually all later authors have rejected the lungfish-tetrapod link proposed by Rosen *et al.,* whereas both the monophyly of the osteolepiforms and the relationship of this group to the tetrapods have been reasserted.[16,29,32–34] Increasingly, however, the debate about tetrapod origins has focused on a previously obscure group of sarcopterygians, the family Panderichthyidae.[15,35–37]

Panderichthyids in the Lower Frasnian

The panderichthyids, comprising the genera *Panderichthys* (Fig. 2d–f) and *Elpistostege,* are a group of early Frasnian lobe-finned fishes that in many respects resemble osteolepiforms, and indeed were originally included within that group. However, new material from Latvia and Canada has shown the Panderichthyidae to be substantially more tetrapod-like than previously realized. With some dissent,[29,30] an increasing number of workers perceive them as the closest known relatives of tetrapods.[15,35–37] Indeed, one panderichthyid fragment, the holotype skull roof of *Elpistostege,* was initially described as a tetrapod,[38] while two other supposed panderichthyids have recently proved to be Devonian tetrapods.[37,39] The fundamental importance of panderichthyids lies in the combination of characters they possess. Unlike osteolepiforms, panderichthyids actually look like early tetrapods with paired fins: they have the same superficially crocodile-like skulls with dorsally placed orbits, straight tails and slightly flattened bodies without dorsal or anal fins (Figs. 1, 2d). Like tetrapods, but unlike all other fishes, they also have frontal bones in the skull roof (Fig. 2f).[35] In evolutionary terms, this simply means that a tetrapod-like body morphology appeared before the limbs. Functionally, however, it may indicate that panderichthyids had adopted a shallow-water predatory

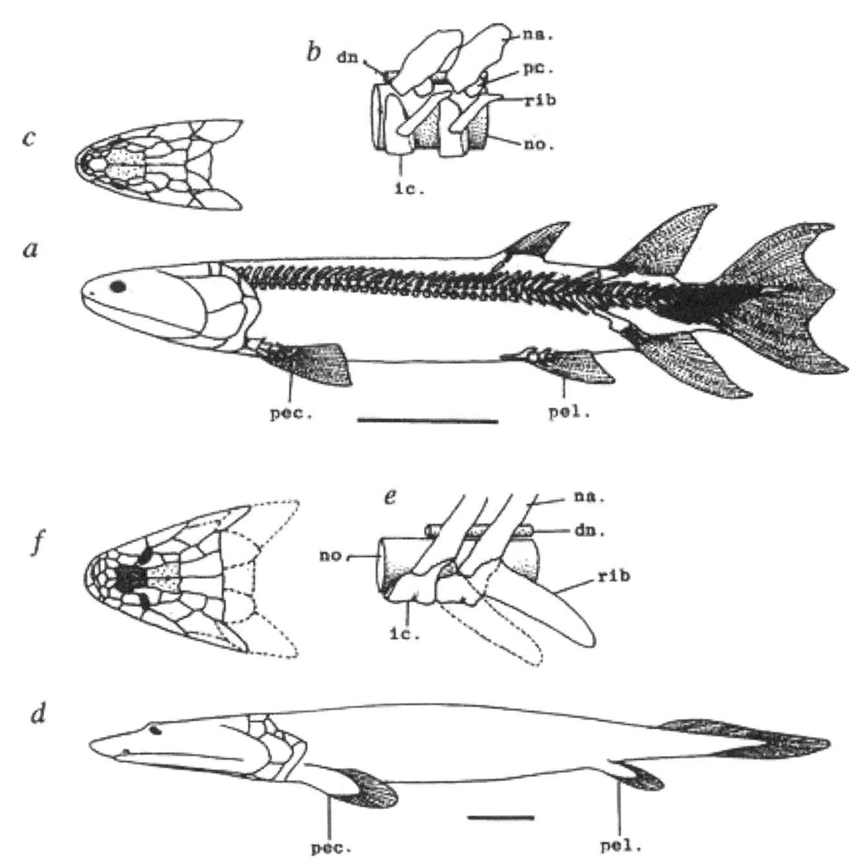

Fig. 2 Osteolepiforms and panderichthyids. *a–c,* The Lower Frasnian osteolepiform *Eusthenopteron. a,* Lateral view showing axial skeleton and fin supports (modified from ref. 11). Scale bar, 100 mm. *Eusthenopteron* is a fusiform predatory fish with no obvious terrestrial adaptations; the paired appendages are pectoral *(pec.)* and pelvic *(pel.)* fins. As in tetrapods, there is only one pair of external nostrils. *b,* Section of vertebral column (modified from ref. 54); the notochord *(no.)* and dorsal nerve cord *(dn.)* have been reconstructed. The intercentra *(ic.)* and pleurocentra *(pc.)* are comparable to those of early tetrapods, but there are no articulations between the neural arches *(na.),* and the short ribs were probably directed dorsally. *c,* Skull roof; the region anterior to the parietals *(light shading)* is occupied by a mosaic of small bones, as in many other sarcopterygians (from ref. 11). *d–f,* The Lower Frasnian panderichthyid *Panderichthys rhombolepis. d,* Lateral view (modified from ref. 36). Scale bar, 100 mm. *Panderichthys* is dorsoventrally flattened, lacks separate dorsal and anal fins, and has dorsally placed eyes; these are derived characters shared with tetrapods. The axial skeleton and girdles are not completely known. *e,* Section of vertebral column (modified from ref. 36); there are no ossified pleurocentra, but the ribs are larger than in osteolepiforms. *f,* Skull roof (modified from ref. 15); panderichthyids and tetrapods both have paired frontal bones *(dark shading)* anterior to the parietals.

niche. The dorsally placed orbits of panderichthyids are particularly noteworthy, as this configuration is associated with aerial vision in some shallow-water teleosts, such as *Anableps* the four-eyed fish and *Periophthalmus* the mud-skipper, as well as in crocodiles. It has also recently been suggested that *Panderichthys* was capable of terrestrial locomotion similar to that in the extant catfish *Clarias.*[40]

In addition to their tetrapod-like features, panderichthyids possess some unique characters (for example, the vertebral construction [Fig. 2e]) which indicate that they form a small clade rather than a paraphyletic segment of the tetrapod stem lineage. Nevertheless, they appear to provide a Lower Frasnian "window" on the earliest recognizable stages of tetrapod evolution. The immediately following fossil record is, however, highly "patchy" and much of our understanding of the diversification of tetrapods is based on three subsequent "windows" at widely spaced intervals (Fig. 3).

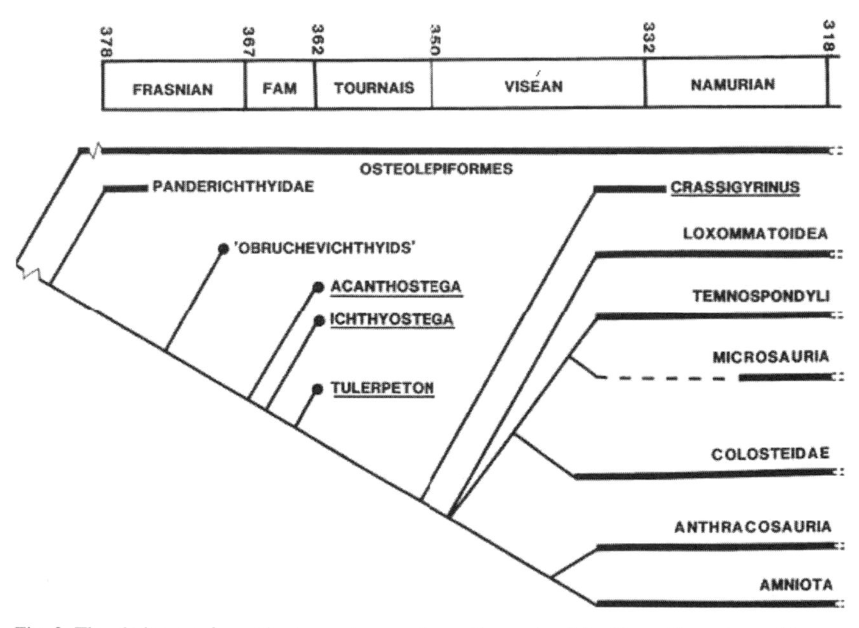

Fig. 3 The cladogram from Fig. 1 superimposed on a timescale of the Upper Devonian and Lower Carboniferous. The *numbers* at the top are recent datings. The *black dots* represent taxa known from a single locality and hence point in time; *thicker bars* represent the known stratigraphic range of taxa known from several records. The osteolepiforms originated before the beginning of the middle Devonian; hence the *zigzag lines* in the lowermost part of the cladogram indicate a time span of at least 15 Myr. Seven of the groups extend on beyond the Namurian. The diagram emphasizes the large chronological gaps between the known assemblages through this period of time. *FAM,* Famennian; *TOURNAIS,* Tournaisian.

Basal Stem Tetrapods

The next body of evidence consists of a range of fragments and some footprints apparently derived from true tetrapods. These date from the Upper Frasnian and Lower Famennian (middle Upper Devonian), 5–10 million years later than the panderichthyids, and have been found in Latvia *(Obruchevichthys),*[15,37] Scotland[37] and Australia *(Metaxygnathus).*[37,41] Most of the osteological evidence is either recently collected or recently recognized, and little can yet be said about these animals. Many of the specimens are jaw fragments, but the Scat Craig locality in Scotland has produced the earliest limb bones, from both fore- and hindlimbs.[37] A humerus from this locality resembles those of tetrapods in having a thin flat entepicondyle continuous with the flat body of the humerus, and a narrow tall ectepicondyle, but in other respects it is like that of *Eusthenopteron* (Fig. 4f). A tibia from Scat Craig bears articulation surfaces for the intermedium and tibiale, suggesting the presence of a tarsus (Fig. 4h). One of the contemporaneous Australian finds is a pair of tetrapod trackways with clear footprints,[42] confirming that tetrapods were present at this time.

Primitive Stem Tetrapods

After a further interval comprising most of the Famennian, a more substantial sample of early tetrapod material, including complete skeletons, appears in late Famennian horizons, representing the latest Devonian (Figs. 4a–e, g, 5). Much of this is new material, collected during the past decade. The *Ichthyostega*-bearing sediments of East Greenland have recently yielded many fossils of the previously poorly known genus *Acanthostega* which is proving to be a very different form (Fig. 4d, e).[43–49] A third genus, *Tulerpeton,* apparently more advanced in having relatively gracile limbs, has been collected recently in European Russia,[50–52] while more fragmentary tetrapod fossils from the late Famennian of Latvia are under description.[39] *Ichthyostega, Acanthostega* and *Tulerpeton* seem to be morphologically disjunct representatives of a diverse basal tetrapod fauna. The latter genera relieve *Ichthyostega* of its isolated position at the base of the tetrapod tree, and give us a clearer and broader perspective of the morphological range of the most primitive tetrapods. Although they have not yet been fully described, we know that they were carnivores about 50 to 120 cm long. They differ considerably from one another, but all three genera are strikingly archaic compared with all later tetrapods except for the genus *Crassigyrinus* (see below).

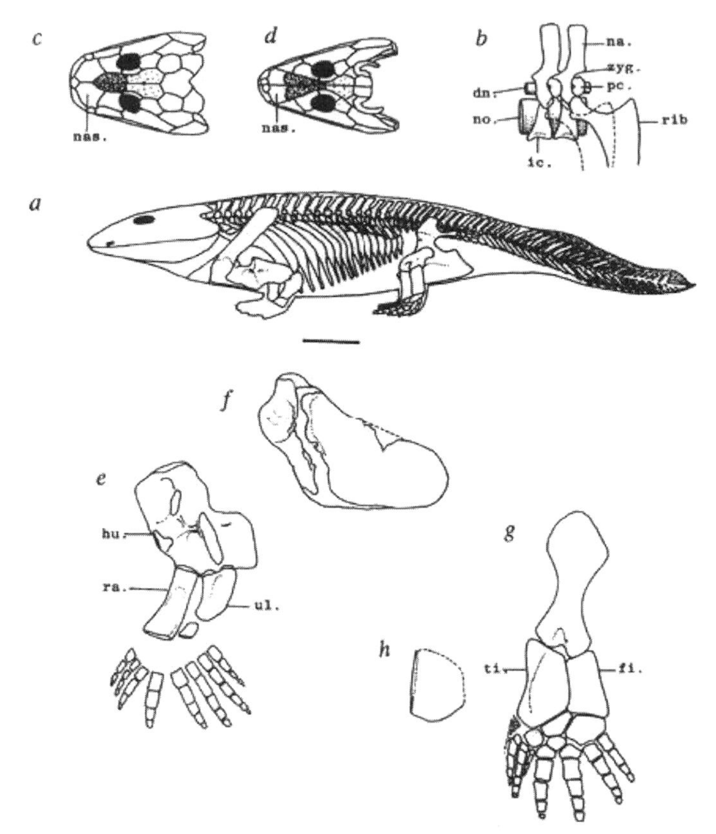

Fig. 4 Devonian tetrapods. *a,* Reconstruction of the Upper Famennian genus *Ichthyostega* in lateral view (modified from ref. 11). This is the only whole-body reconstruction yet published for a Devonian tetrapod; note the large pelvic girdle, the rib cage composed of broad overlapping ribs, and the loss of the dorsal series of bones connecting the pectoral girdle with the head. The forelimb has a permanently flexed elbow, but the structure of the manus is unknown. The tail carries bony fin rays similar to those in *Panderichthys.* Scale bar, 100 mm. *b,* Section of vertebral column (modified from ref. 11); the general structure is similar to that of *Eusthenopteron,* but the neural arches *(na.)* interlock by means of zygapophyses *(zyg.),* and the ribs are much larger. *c, d,* Skulls of *Ichthyostega* (from ref. 11) and *Acanthostega* (from refs. 43, 44) in dorsal view; all known Devonian tetrapods possessed frontals *(dark shading)* and paired nasals *(nas.). e,* Forelimb of the Upper Famennian tetrapod *Acanthostega* in dorsal view (from ref. 48); unlike *Ichthyostega,* this animal had a straight elbow with limited mobility. There are eight digits, and the proportions of the radius *(ra.)* and ulna *(ul.)* are very fish-like, although the humerus *(hu.)* resembles those of other early tetrapods. Coates and Clack[48] regard this as the most primitive known tetrapod forelimb. *f,* Upper Frasnian humerus from Scat Craig, Scotland (from ref. 37); this humerus, about 7 million years older than *Acanthostega,* has more fish-like proportions. *g,* Hindlimb of *Ichthyostega* in dorsal view (after ref. 48); the foot has seven digits preceded by a poorly ossified mass, and the tibia *(ti.)* and fibula *(fi.)* are strikingly broad. *h,* Upper Frasnian tibia from Scat Craig (from ref. 37); this incomplete tibia is closely comparable to that of *Ichthyostega,* but seems to be proportionately shorter.

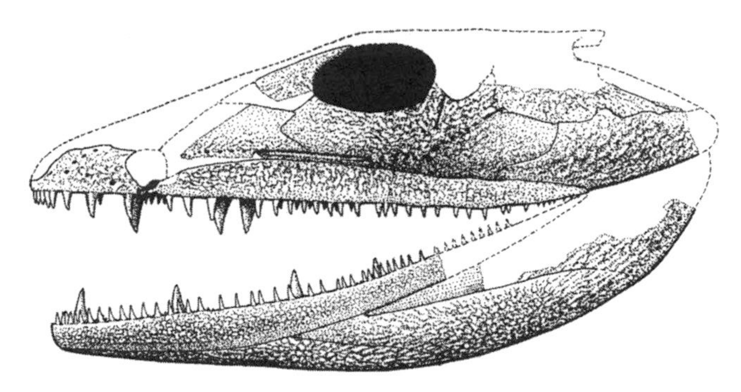

Fig. 5 Skull reconstruction of *Ventastega curonica,* a recently discovered Devonian (Upper Famennian) tetrapod from Latvia (after ref. 39). Scale bar, 10 mm.

The most striking feature shared by these late Famennian forms is polydactyly: *Acanthostega* had eight digits on fore- (Fig. 4e) and hindlimbs, *Ichthyostega* had seven on the hindlimb (Fig. 4g) (forelimb unknown) (refs. 47, 48 and M. I. Coates, personal communication), while *Tulerpeton* appears to have had six digits on both.[50] As Coates and Clack point out,[48] the discovery of these configurations is of profound importance in constraining theories of the origin and development of limbs (Fig. 6). Virtually all previous investigations have assumed that the primitive tetrapod limb pattern was pentadactyl (five-toed) and incorporated a specific arrangement of wrist and ankle bones.[53] After comparison of this "canonical" pattern with the skeletons of the paired fins of sarcopterygians, most workers concluded that the limb skeleton has the same basic structure as the internal skeleton of the corresponding osteolepiform paired fin (Fig. 6b).[11,54] Wrists, ankles and jointed digits were argued to have arisen by fragmentation and slight realignment of the distal elements of the fin skeleton. However, it now appears that neither the pentadactyl condition nor the canonical wrist pattern is primitive for tetrapods, which invalidates such earlier theories substantially if not completely. Furthermore, recently described embryological evidence[55] indicates that the digits cannot be equated with any known part of a sarcopterygian fin skeleton.

The pectoral or pelvic fin skeleton of a lobe-finned fish consists of a jointed axis carrying preaxial, and sometimes postaxial, elements (see Fig. 6a). In the living Australian lungfish *Neoceratodus,* the axial and preaxial elements arise from a single branching process during development, whereas postaxial elements develop separately as condensations of skeletal tissue (Fig. 6c, d).[55] In embryonic tetrapod limbs,[55] the humerus-radius-ulna complex and the femur-tibia-fibula complex arise by a very similar branching process, the radius and tibia being preaxial branches

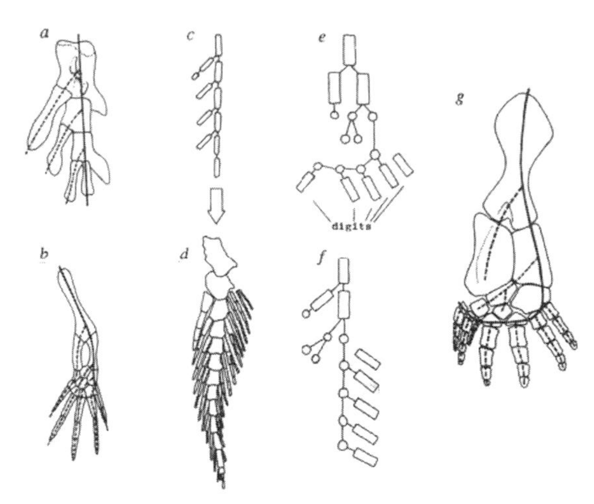

Fig. 6 Paired fins and limbs. *a,* Pectoral fin skeleton of the osteolepiform *Eusthenopteron,* a relatively short structure consisting of an axis *(solid line)* carrying preaxial radials *(dashed lines).* Phylogenetic evidence suggests that the tetrapod limb is derived from this general type of fin skeleton. *b,* Gegenbaur's[97] plan of an idealized limb, showing inferred position of the axis and radials. In this interpretation (and those of most subsequent authors), the tetrapod digits correspond to preaxial radials, and the overall structure closely resembles the fin skeleton of *Eusthenopteron* (*a* and *b* after ref. 48). *c, d,* Pectoral fin skeleton of the Recent lungfish *Neoceratodus. c,* Diagrammatic representation of early development (after ref. 55). The axis and preaxial radials are laid down by a branching process; the fin skeleton of *Eusthenopteron* was probably formed in the same way. *d,* Adult fin. The numerous postaxial radials form as condensations of tissue, not through branching. *e,* Forelimb skeleton of mouse, *Mus musculus;* diagrammatic representation of early development (after ref. 55). The proximal part of the limb skeleton consists of an axis with preaxial branches (radials), as in *Eusthenopteron (a)* or embryonic *Neoceratodus (c),* but the digits are formed as postaxial branches. This pattern is unknown in fishes. *f,* Limb skeleton of *Mus musculus* straightened out to facilitate comparison with fin skeletons. *g,* Hindlimb of the Devonian tetrapod *Ichthyostega,* showing inferred position of axis *(solid line)* and radials *(dashed lines),* following ref. 55 (after ref. 48). Other known Devonian tetrapod limbs have eight *(Acanthostega)* or six *(Tulerpeton)* digits. All limbs/fins shown with anterior edge to left.

(Fig. 6e, f). The proximal elements of tetrapod limbs thus correspond closely to those of sarcopterygian paired fins. However, the process then "switches sides" to produce the digits as a series of postaxial branches (Fig. 6e–g). No process resembling this is known to occur or to have occurred in lobe-finned fishes. It appears that hands and feet are "new" structures, produced by a major developmental change affecting the distal parts of the paired appendages.[47,48]

A puzzling aspect of limb evolution is that sarcopterygian pectoral fins tend to be larger then the pelvic fins, whereas tetrapods are "rear-wheel drive" animals with larger hindlimbs than forelimbs. The earliest known tetrapods have not yet clarified the nature of this transition. In *Ichthyostega,* the forelimb is a strong prop with a permanently bent elbow (Fig.

4a),[11] but the forelimb of *Acanthostega* is rather feeble and much more fish-like than the hindlimb (Fig. 4e).[47,48] The Frasnian limb bones from Scat Craig, Scotland, include a tibia (Fig. 4h) resembling those of *Ichthyostega* and *Acanthostega,* and a humerus (Fig. 4f) roughly intermediate between the osteolepiform and early tetrapod patterns.[37] If these bones belong to the same genus—which is at present unprovable but seems likely—this suggests a degree of limb disparity even greater than in *Acanthostega.* This may indicate that the predominance of the hindlimb is a very ancient tetrapod feature, and possibly that the hindlimb was first to appear, although more evidence is needed before any firm conclusions can be drawn. Interestingly, the *hox* genes expressed in the tetrapod forelimb are interpreted by some as copies of those expressed in the hindlimb,[56,57] but others dispute this interpretation.[58,59]

Although the Famennian tetrapods have dactylous limbs and more heavily constructed vertebral columns than lobe-finned fishes, they still appear to have been largely aquatic. Both *Ichthyostega* and *Acanthostega* retain true tail fins with fin rays. *Acanthostega* also has a large and well-ossified branchial skeleton similar to that of modern lungfishes,[49] and a large supportive stapes that does not seem to have been associated with a tympanum.[45,46] Until recently, it was widely assumed that the origin of tetrapods was synonymous with the invasion of the land, a view epitomized by countless popular illustrations of lobe-finned fishes hauling themselves laboriously onto Devonian pond margins and riverbanks. However, the new discoveries have led Coates and Clack[48,49] to suggest that tetrapods evolved initially to exploit a shallow-water environment. The Famennian forms bear a broad morphological resemblance to the panderichthyids, despite the presence of limbs and longer tails, and may have filled similar niches, evolving limbs and digits to facilitate movement in weed-choked shallows rather than on land. The Sargassum frogfish *(Histrio)* provides a living example of "underwater walking."[48] The revised view that *Acanthostega* and *Ichthyostega* were in themselves aquatic will probably not prove controversial. Whether they were primarily aquatic (that is, their ancestors had never left the water) or secondarily so (descended from more terrestrial precursors) is a more contentious issue. If *Acanthostega* was primarily aquatic, then not only did the dactylous tetrapod limb evolve underwater, but also the large tetrapod pelvic girdle and sacral rib, together with the reduction of opercular and gular ossifications.

By the end of the Devonian, tetrapods had been in existence for at least 7 million years and had begun to diversify. Their fossils have been found in fluviatile or deltaic "Old Red Sandstone" sediments and in lagoonal limestones, in association with fish assemblages dominated by antiarchs, osteolepiforms and lungfishes. At the Devonian-Carboniferous boundary, these assemblages cease to appear in the record, although this may represent a phenomenon of taphonomic overlay rather than sudden extinction. The vertebrate record for the ensuing 30 million years through the Tournaisian to the middle Viséan is extremely poor (Fig. 3). A few isolated bones have been collected from the Tournaisian of Nova Scotia,[60] but have yet to be described. When, following this interval, the middle Viséan–Namurian "window" is reached, the tetrapods have diversified substantially and, except for one relict, have moved substantially beyond the grade of structural organization seen in the Famennian.

Late Lower Carboniferous Tetrapod Diversity

The Middle Viséan–Lower Namurian tetrapod fauna is known from Scotland,[61] Nova Scotia,[60] West Virginia,[62–65] Iowa,[66] Illinois[67] and Utah.[68] Most assemblages represent communities in lowland swamps or lakes on coastal plains, and all are situated in the continent of Euramerica, close to the paleoequator of the time.[69] Forms representing 15 families (effectively distinct lineages) have been or are being described and the presence of several other lineages may be deduced by phylogenetic inference. With the exception of the enigmatic *Crassigyrinus,* all the described forms are clearly of a more derived grade of organization than the Famennian tetrapods. The anthracosaurs *Eoherpeton* and *Proterogyrinus* are widely accepted to be early relatives of the amniotes,[63,70,71] whereas the colosteids *Pholidogaster* and *Greererpeton* are probably aberrant early relatives of the modern amphibians[65,72] (see Fig. 1 for a hypothesis of relationships of these forms). The remaining forms, the loxommatids *Loxomma*[73] and *Spathicephalus,* the aïstopod *Lethiscus,*[74] the adelogyrinids[75] and the enigmatic *Caerorhachis,*[76] are still of uncertain position.

For most of the last century, the Viséan-Namurian tetrapod fauna was known largely from the specialized aquatic forms described above. It was clear that much of the Lower Carboniferous fauna, including the more terrestrial elements, was missing through taphonomic bias. The discovery in the 1980s of three new tetrapod assemblages is filling this gap. The richest of these assemblages is that discovered at East Kirkton Quarry near Bathgate in Midlothian, Scotland.[77–86] The tetrapods were found in a local sequence of freshwater limestones, black shales and tuffs, and are associated principally with scorpions, the eurypterid *Hibbertopterus,* millepedes, a harvestman spider and a wide range of plant material. The only elements of the assemblage that seem to have been completely aquatic are abundant ostracodes. The tetrapods themselves seem to have been terrestrial forms and several of the most readily characterized forms can be assigned to groups already known from the Upper Carboniferous. These include temnospondyls, an aïstopod and three members of the anthracosaur-amniote radiation.[80–84] The common temnospondyl appears to be more advanced in skull construction than several Upper Carboniferous members of the group, and this position within the Temnospondyli implies that several other, more primitive, lineages of temnospondyl were already present in the Viséan.[80] The aïstopod shows that this group of snake-convergent limbless tetrapods had already differentiated in the Viséan.[81] An amphibious anthracosaur,[82] a terrestrial anthracosaur-like form of uncertain position[83] and the small, long-bodied stem amniote *Westlothiana*[84–86] show that the anthracosaur-amniote radiation was well differentiated on land at this time. As well as these forms, a further and as yet unpublished new tetrapod from East Kirkton combines the primitive characters of loxommatids, temnospondyls and anthracosaurs and may be close to the amphibian-amniote branching point.

The second recently discovered assemblage is that from the Viséan St. Louis Formation of Delta, Iowa.[66,87] The commonest tetrapod here is a completely new form having many of the features of anthracosaurs, but retaining a primitive tetrapod skull roof construction. It has been preliminary described as a "protoanthracosaur," but may represent a lower grade of organization. It is represented by several large complete skeletons, somewhat crushed, and should provide the head-to-tail osteology of a tetrapod close to the amphibian-amniote dichotomy. It is accompanied by a colosteid amphibian and fishes.

Third, the basal Namurian Manning Canyon Shale near Utah Lake in Utah has produced two specimens of a microsaur, together with some of the earliest known winged insects.[68] The microsaur, *Utaherpeton,* can be attributed to the *Microbrachomorpha,* one of the subgroups recognized in later microsaurs. Thus, like the East Kirkton material, it takes the diversification of an Upper Carboniferous group well back in the Lower Carboniferous.

The Systematics of Early Tetrapods

For many years, the basal tetrapods were conveniently placed in two large groups, the Labyrinthodontia characterized by labyrinthine teeth and multipartite vertebral centra, and the Lepospondyli characterized by simple teeth and unipartite centra. The advent of cladistic methodology has left the Labyrinthodontia exposed as a paraphyletic group based solely on primitive characters,[70,88] while the Lepospondyli is more controversially seen as a polyphyletic assemblage of dwarf early tetrapods.[1,70,88,89] Consequently, the use of these terms is declining but no consensus has yet emerged on the groupings that may replace Labyrinthodontia and Lepospondyli. For many workers the problem can be seen in terms of the living Tetrapoda. There is widespread (though not uniform) agreement that the Lissamphibia (the living amphibians) and the Amniota (the reptiles, birds and mammals) are each monophyletic groups. If this is accepted, it follows that early tetrapods could be representatives of, or sidelines off, the stem of the Lissamphibia, the stem of the Amniota, or the common tetrapod stem. To place early tetrapods into one of these three categories would be considerable progress, not only for the systematist, but also for the paleobiologist wishing to understand the sequence of character acquisition in early tetrapods.

Figure 1 represents a set of relationships for the tetrapod taxa discussed above and below. No consensus exists, but this proposal is based largely on published arguments.[65,78,88] *Acanthostega* and *Ichthyostega* are perceived as the most adequately known primitive stem tetrapods, with *Crassigyrinus* (discussed below) farther up the stem. The loxommatids are ambiguous in their position and could be either stem tetrapods or stem amniotes. Other early tetrapods fall within the amphibian/lissamphibian clade (the microsaurs, colosteids and temnospondyls), or the amniote clade (the anthracosaurs and true amniotes). The discovery of all these groups in the Viséan-Namurian sequences shows that the radiation of the "crown-group" tetrapods was well under way.

Crassigyrinus

Crassigyrinus, from the Viséan and Namurian of Scotland, is a bizarre and systematically challenging form which has recently generated much discussion.[90,91] It was a basically eel-like form with tiny limbs but was at least 130 cm long (Fig. 7). The short-snouted skull had deep massive cheeks, prominent labyrinthodont teeth and fangs; *Crassigyrinus* was clearly a large aquatic predator. There are two conflicting interpretations

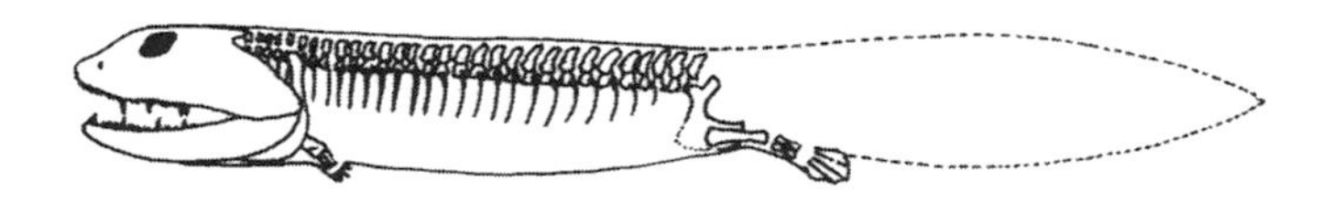

Fig. 7 The primitive Carboniferous (Viséan and Namurian) tetrapod *Crassigyrinus.* This large predator had tiny forelimbs and must have been wholly aquatic (modified from refs. 90 and 91). Scale bar, 100 mm.

of its systematic position. Panchen[90,92] has argued that, although an extremely primitive tetrapod, *Crassigyrinus* shares significant characteristics with the anthracosaurs, which in turn are generally accepted as primitive amphibious relatives of the amniotes (reptiles, birds and mammals). In this view, *Crassigyrinus* would be the most primitive representative of the line leading to amniotes and would show that the division between the two lineages giving rise to the modern amphibians and to the amniotes had occurred at a very basal, possibly pre-terrestrial, stage in tetrapod evolution. The alternative view[36,65,78,93] is that *Crassigyrinus* lacks so many characteristics shared by other Carboniferous and later tetrapods that it must be an early offshoot of the tetrapod stem, close to *Acanthostega* and *Ichthyostega* but convergently resembling the anthracosaurs.

The new interpretations of *Acanthostega* and *Ichthyostega,* as primary aquatic forms that had not yet become terrestrial, raises the possibility that *Crassigyrinus* is not only a structural relict but also an ecological relict. Instead of being the first tetrapod to "return to the water," it may be the last survivor of the primitive tetrapods that never left the water.

Basal Crown-Group Tetrapods

The loxommatids, a group of early tetrapods known since the last century, are uniquely characterized by curious keyhole-shaped orbits.[73] The loxommatids clearly form a compact group, but their extrinsic relationships are enigmatic. They were previously considered to be temnospondyls and hence a side branch on the lineage leading to modern amphibians. Recently it has been shown that their similarity to temnospondyls is based entirely on primitive tetrapod features, and some derived similarities to anthracosaurs—and hence amniotes—have been proposed.[70,88,94] Others have suggested that loxommatids belong on the tetrapod stem above *Crassigyrinus.*[78] The limited information available from loxommatid skulls seems insufficient to place them precisely, but they appear to be very close to the amphibian-amniote dichotomy. A new loxommatid specimen from the Upper Carboniferous of Lancashire, UK, has the first partial postcranial skeleton associated with this family and should generate considerable data about their position.[95]

Another form that may belong to this grade of organization is *Caerorhachis.* Described as a very primitive temnospondyl with anthracosaur-like vertebrae, this incomplete specimen could equally be a juvenile loxommatid, a basal temnospondyl or a basal anthracosaur.[80] Two of the most recent discoveries that also fall in this category are the Delta "protoanthracosaur" and the Kirkton archaic tetrapod. Neither is a loxommatid, but both are almost universally primitive in the characteristics they share with temnospondyls, loxommatids and anthracosaurs. This cluster of forms, when fully described, will give us a basic morphotype for the tetrapod crown group. Whereas the Devonian tetrapods and *Crassigyrinus* clearly had only a partial set of tetrapod characteristics, these forms appear to have most, if not all, of them and are close to the point at which the modern amphibians and amniotes diverged.

Stem Amphibians

The true amphibians (that is, all those fossil forms that are more closely related to living amphibians than to living amniotes) are widely perceived as a group that were successful in the Carboniferous but later were outcompeted by the amniotes for all but a restricted range of terrestrial niches. This is a somewhat exaggerated view, not only because of the immense diversity of modern amphibians (more than 4,000 species), but also because the Carboniferous record largely derives from ponds, abandoned channels and ox-bow lakes, in which amphibians might be expected to be better represented than amniotes. Nevertheless, early members of the amphibian clade, such as the large aquatic colosteids, the small terrestrial, burrowing and aquatic microsaurs and the crocodile- and salamander-like temnospondyls, were certainly diverse in the Upper Carboniferous. Recent discoveries are tending to push this diversity down into the Lower Carboniferous. The Viséan colosteid *Pholidogaster* was described in the last century, but recognized as a colosteid only in 1975.[96] The first namurian microsaur from the Manning Canyon Shale of Utah was described in 1990.[68] This form, *Utaherpeton,* can be assigned to one of the subgroups of microsaurs, the *Microbrachomorpha,* and suggests some diversification by that time. An undescribed assemblage from the Viséan of Illinois also includes what appear to be microsaur vertebrae.[67] The common temnospondyl at East Kirkton is an unexpectedly advanced form and implies that several temnospondyl families previously known only from later horizons must also have had antecedents in the Viséan.[80]

Stem Amniotes

The anthracosaurs are widely accepted to be stem amniotes, and were known to exist in the Viséan and Namurian, based on such forms as *Eoherpeton* from Scotland and *Proterogyrinus* from Scotland and West Virginia, USA. The East Kirkton fauna appears to include a variety of stem amniotes, one of which can be readily characterized as a true anthracosaur,[82] but other forms are also present. The enigmatic *Westlothiana,* initially reported as the earliest reptile,[84–86] may prove to be slightly farther down the amniote stem, but still appears to be a post-anthracosaurian offshoot. Yet another Kirkton tetrapod also seems to be a terrestrial stem amniote, but of less certain position.[83] Again, the principal significance of the new material is, at this stage, that it shows that much of the diversity previously associated with Upper Carboniferous faunas was already present in the Viséan.

A Tournaisian Radiation?

As new information from these four sets of assemblages, from panderichthyids to late Lower Carboniferous tetrapod faunas, emerges, it appears increasingly that there was an immense radiation of terrestrial tetrapods in the Tournaisian. On the one hand, Viséan and early Namurian faunas are proving to be more advanced than previously realized, with temnospondyls, anthracosaurs, aïstopods, microsaurs and possible amniotes, and thus resembling later Carboniferous faunas in their taxonomic composition. On the other hand, the sequence of Lower Frasnian panderichthyids, Upper Frasnian *Obruchevichthys*-grade forms and Famennian tetrapods looks increasingly like a series of "state-of-the-art faunas," and suggests that basal tetrapod morphology was being acquired during the Upper Devonian, but that the latest Devonian forms may not yet have been fully terrestrial. Thus, more advanced tetrapod morphologies are being described from the Viséan, while Famennian tetrapods prove to be more primitive than previously suggested. The Tournaisian and early Viséan may thus be the critical time both for the functional water-land transition and the ensuing diversification of the truly terrestrial tetrapods.

The alternative hypothesis is that tetrapod diversification did occur gradually during the Famennian, and that forms such as *Acanthostega* and *Ichthyostega* are already relicts when we see them in the late Famennian. Such a theory permits a more generous time for the diversification of the tetrapods present by the Upper Viséan, but the only direct evidence to support it is Famennian postcranial material of *Tulerpeton*[50] which is not yet fully described. This is clearly more derived than that of the other Devonian tetrapods, and resembles that of later terrestrial forms in certain respects.

Future Research Directions

We can expect a substantial increase in information about each of the three tetrapod assemblages reviewed above. New material of the Upper Frasnian *Obruchevichthys*-grade from Scotland will be described and further collecting at a productive locality is under way. For the Famennian tetrapods, detailed descriptions of *Acanthostega, Ichthyostega* and a new Latvian genus (Fig. 5) are in press or preparation, and a collaborative Anglo-Russian study of *Tulerpeton* in conjunction with the Greenland material is being undertaken, details of the cranial component having just been published.[52] Most of the nineteenth-century material from the classical Viséan-Namurian localities has been further prepared and redescribed over the past two decades, but description of articulated tetrapod skeletons in the East Kirkton, Delta and Manning Canyon Shale assemblages will add substantially to the known diversity of amphibious-terrestrial tetrapods for this interval. The most interesting forms are proving to be those nearest the base of the tetrapod crown group, and the description of the Delta "protoanthracosaur," the East Kirkton archaic tetrapod and the later Wigan loxommatid postcranium will together provide us with a view of tetrapod construction at, or close to, the point when the ancestors of modern amphibians and the amniotes diverged. These forms are probably relicts of the "state-of-the-art" in the Tournaisian and the nearest we can come to understanding tetrapod evolution at that time. The next challenge will be to find such Tournaisian tetrapod assemblages.

PER E. AHLBERG is at the Department of Palaeontology, Natural History Museum, Cromwell Road, London SW7 5BD, UK and Andrew R. Milner is at the Department of Biology, Birkbeck College, Malet Street, London WC1E 7HX, UK.

References

1. Carroll, R. L. A. *Rev. Earth planet Sci.* **20,** 45–84 (1992).
2. Thomson, K. S. *Am. J. Sci.* **293** A, 33–62 (1993).
3. Barrell, J. *Bull. geol. Soc. Am.* **27,** 387–436 (1916).
4. Säve-Söderbergh, G. *Meddr. Grønland* **94,** 1–107 (1932).
5. Jarvik, E. *Zool. Bidr. Upps.* **21,** 235–675 (1942).
6. Jarvik, E. *Ark. Zool.* **41** A, No. 13, 1–8 (1948).
7. Jarvik, E. *Meddr. Grønland* **114,** 1–90 (1952).
8. Jarvik, E. *Scient. Mon.* **80,** 141–154 (1955).
9. Jarvik, E. *Colloq. int. Cent. natn. Rech. Scient.* **104,** 87–101 (1962).
10. Jarvik, E. *Kung. Svensk. Vetens. Handl.* (4) **9,** 1–74 (1963).
11. Jarvik, E. *Basic Structure and Evolution of Vertebrates* Vol. 1 (Academic, London, 1980).
12. Parsons, T. S. & Williams, E. E. *Q. Rev. Biol.* **38,** 26–53 (1963).
13. Thomson, K. S. *Nobel Symp.* **4,** 285–305 (1968).
14. Jurgens, J. D. *Ann. Univ. Stellenbosch A* **46,** 1–146 (1971).
15. Vorobyeva, E. I. *Trudy paleont. Inst.* **163,** 1–239 (1977).

16. Schultze, H.-P. *Paläont. Z.* **55,** 71–86 (1981).
17. Romer, A. S. *Vertebrate Paleontology* 3rd edn. (Univ. Chicago Press, Chicago, 1966).
18. Schmalhausen, I. I. *The Origin of Terrestrial Vertebrates* (Academic, New York, 1968).
19. Ewer, D. W. *Science* **122,** 467–468 (1955).
20. Feduccia, J. A. *Texas J. Sci.* **22,** 255–263 (1971).
21. Gunter, G. *Science* **123,** 495–496 (1956).
22. Cox, C. B. *Proc. Linn. Soc. Lond.* **178,** 37–47 (1967).
23. Andrews, S. M. in *Interrelationships of Fishes* (eds. Greenwood, P. H., Miles, R. S. & Patterson, C.) 137–177 (Academic, London, 1973).
24. Gaffney, E. S. *Bull Carnegie Mus. nat. Hist.* **3,** 92–105 (1979).
25. Rosen, D. E., Forey, P. L., Gardiner, B. G. & Patterson, C. *Bull. Am. Mus. nat. Hist.* **167,** 159–276 (1981).
26. Jarvik, E. *Syst. Zool.* **30,** 378–384 (1981).
27. Holmes, E. B. *Biol. J. Linn. Soc.* **25,** 379–397 (1985).
28. Schultze, H.-P. in *The Biology and Evolution of Lungfishes* (eds. Bemis, W. E., Burggren, W. W. & Kemp, N. E.) 39–74 (Liss, New York, 1987).
29. Panchen, A. L. & Smithson, T. R. *Biol. Rev.* **62,** 341–438 (1987).
30. Chang, M.-M. in *Origins of the Higher Groups of Tetrapods: Controversy and Consensus* (eds. Schultze, H.-P. & Trueb, L.) 3–28 (Cornell, Ithaca, NY, 1991).
31. Ahlberg, P. E. *Zool. J. Linn. Soc.* **96,** 119–166 (1989).
32. Ahlberg, P. E. *Zool. J. Linn. Soc.* **103,** 241–287 (1991).
33. Schultze, H.-P. *Verh. dt. zool. Ges.* **84,** 135–151 (1991).
34. Long, J. A. *J. Vert. Paleont.* **9,** 1–17 (1989).
35. Schultze, H.-P. & Arsenault, M. *Palaeontology* **28,** 293–309 (1985).
36. Vorobyeva, E. & Schultze, H.-P. in *Origins of the Higher Groups of Tetrapods: Controversy and Consensus* (eds. Schultze, H.-P. & Trueb, L.) 68–109 (Cornell, Ithaca, NY, 1991).
37. Ahlberg, P. E. *Nature* **354,** 298–301 (1991).
38. Westoll, T. S. *Nature* **141,** 127 (1938).
39. Ahlberg, P. E., Lukševičs, E. & Lebedev, O. *Phil Trans. R. Soc.* B **343,** 303–328 (1994).
40. Vorobyeva, E. & Kuznetsov, A. in *Fossil Fishes as Living Animals* (ed. Mark-Kurik, E.) 131–140 (Acad. Sci. Estonia, Tallinn, 1992).
41. Campbell, K. S. W. & Bell, M. W. *Alcheringa* **1,** 369–381 (1977).
42. Warren, J. W. & Wakefield, N. A. *Nature* **238,** 469–470 (1972).
43. Clack, J. A. *Palaeontology* **31,** 699–724 (1988).
44. Clack, J. A. *Geol. Today* **4,** 192–194 (1988).
45. Clack, J. A. *Nature* **342,** 425–430 (1989).
46. Clack, J. A. in *The Evolutionary Biology of Hearing* (eds. Webster, D. B., Fay, R. R. & Popper, A. N.) 405–420 (Springer, New York, 1992).
47. Coates, M. I. in *Developmental Patterning of the Vertebrate Limb* (eds. Hinchliffe, J. R., Hurle, J. M. & Summerbell, D.) 325–337 (Plenum, New York, 1991).
48. Coates, M. I. & Clack, J. A. *Nature* **347,** 66–69 (1990).
49. Coates, M. I. & Clack, J. A. *Nature* **352,** 234–235 (1991).
50. Lebedev, O. A. *Dokl. Akad. Nauk SSSR* **278,** 1470–1473 (1984).
51. Lebedev, O. A. *Priroda* **1985,** 26–36 (1985).

52. Lebedev, O. A. & Clack, J. A. *Palaeontology* **36,** 721–734 (1993).
53. Watson, D. M. S. *Anat. Anz.* **44,** 24–27 (1913).
54. Andrews, S. M. & Westoll, T. S. *Trans. R. Soc. Edinb.* **68,** 207–329 (1970).
55. Shubin, N. H. & Alberch, P. *Evol. Biol.* **20,** 319–387 (1986).
56. Tabin, C. J. *Development* **116,** 289–296 (1992).
57. Tabin, C. & Laufer, E. *Nature* **361,** 692–693 (1993).
58. Coates, M. *Nature* **364,** 195–196 (1993).
59. Thorogood, P. & Ferretti, P. *Nature* **364,** 196 (1993).
60. Carroll, R. L., Belt, D., Dineley, D. L., Baird, D. & McGregor, D. C. *24th int. Geol. Congr., Montreal, Field Excursion A59: Guidebook* 1–113 (McAra, Calgary, Canada, 1972).
61. Smithson, T. R. *Scott. J. Geol.* **21,** 123–142 (1985).
62. Smithson, T. R. *Zool. J. Linn. Soc.* **76,** 29–90 (1982).
63. Holmes, R. *Phil. Trans. R. Soc.* B **306,** 431–527 (1984).
64. Godfrey, S. J. *Kirtlandia* **43,** 27–36 (1988).
65. Godfrey, S. J. *Phil. Trans. R. Soc.* B **323,** 75–133 (1989).
66. Bolt, J. R. *Nat. Geogr. Res.* **6,** 339–354 (1990).
67. Schultze, H.-P. & Bolt, J. R. *Spec. Pap. Palaeontology* (in the press).
68. Carroll, R. L., Bybee, P. & Tidwell, W. D. *J. Paleont.* **65,** 314–322 (1991).
69. Milner, A. R. in *Palaeozoic Vertebrate Biostratigraphy and Biogeography* (ed. Long, J. A.) 324–353 (Belhaven, London, 1993).
70. Smithson, T. R. *Zool. J. Linn. Soc.* **85,** 317–410 (1985).
71. Smithson, T. R. *Palaeontology* **29,** 603–628 (1986).
72. Milner, A. R. *Herpetol. Monogr.* **7,** 8–27 (1993).
73. Beaumont, E. H. *Phil. Trans. R. Soc.* B **280,** 29–101 (1977).
74. Wellstead, C. F. *Palaeontology* **25,** 193–208 (1982).
75. Andrews, S. M. & Carroll, R. L. *Trans. R. Soc. Edinb., Earth Sci.* **82,** 239–275 (1991).
76. Holmes, R. & Carroll, R. L. *Bull. Mus. comp. Zool. Harv.* **147,** 489–511 (1977).
77. Wood, S. P., Panchen, A. L. & Smithson, T. R. *Nature* **314,** 355–356 (1985).
78. Milner, A. R., Smithson, T. R., Milner, A. C., Coates, M. I. & Rolfe, W. D. I. *Mod. Geol.* **10,** 1–28 (1986).
79. Rolfe, W. D. I. *et al. Geol. soc. Am. Spec. Pap.* **244,** 13–24 (1990).
80. Milner, A. R. & Sequeira, S. E. K. *Trans R. Soc. Edinb., Earth Sci.* **84,** 331–361 (1994).
81. Milner, A. C. *Trans R. Soc. Edinb., Earth Sci.* **84,** 363–368 (1994).
82. Clack, J. A. *Trans R. Soc. Edinb., Earth Sci.* **84,** 369–376 (1994).
83. Smithson, T. R. *Trans R. Soc. Edinb., Earth Sci.* **84,** 377–382 (1994).
84. Smithson, T. R., Carroll, R. L., Panchen, A. L. & Andrews, S. M. *Trans R. Soc. Edinb., Earth Sci.* **84,** 383–412 (1994).
85. Smithson, T. R. *Nature* **342,** 676–678 (1989).
86. Smithson, T. R. & Rolfe, W. D. I. *Scott. J. Geol.* **26,** 137–138 (1990).
87. Bolt, J. R., McKay, R. M., Witzke, B. J. & McAdams, M. P. *Nature* **333,** 768–770 (1988).
88. Panchen, A. L. & Smithson, T. R. in *The Phylogeny and Classification of the Tetrapods:* Vol. 1 *Amphibians, Reptiles, Birds* (ed. Benton, M. J.) 1–32 (Clarendon, Oxford 1988).
89. Thomson, K. S. & Bossy, K. H. *Forma functio* **3,** 7–31 (1970).

90. Panchen, A. L. *Phil. Trans. R. Soc.* B **309,** 505–568 (1985).
91. Panchen, A. L. & Smithson, T. R. *Trans. R. Soc. Edinb., Earth Sci.* **81,** 31–44 (1990).
92. Panchen, A. L. in *Origins of the Higher Groups of Tetrapods: Controversy and Consensus* (eds. Schultze, H.-P. & Trueb, L.) 110–144 (Cornell, Ithaca, NY, 1991).
93. Clack, J. A. in *Spec. Pap. Palaeontology* (in the press).
94. Panchen, A. L. in *The Terrestrial Environment and the Origin of Land Vertebrates* (ed. Panchen, A. L.) 319–350 (Academic, London, 1980).
95. Milner, A. C. & Lindsay, W. *First World Congress of Herpetology,* Abstr. 201 (1989).
96. Panchen, A. L. *Phil. Trans. R. Soc.* B **259,** 581–640 (1975).
97. Gegenbaur, C. *Morph. Jb.* **2,** 396–420 (1876).

Luis M. Chiappe

THE FIRST 85 MILLION YEARS OF AVIAN EVOLUTION

More than half of the evolutionary history of birds is played out in the Mesozoic. A recent burst of fossil discoveries has documented a tremendous diversity of early avians. Clarification of phylogenetic structure of this diversity has provided clues for a better understanding of the evolution of functional, developmental and physiological characteristics of modern birds. Yet their long Mesozoic history is only beginning to be deciphered.

The spectacular specimens of the late Jurassic *Archaeopteryx,* and to a lesser extent the more derived and much younger hesperornithiforms and ichthyornithiforms, have formed the core of our understanding of avian origins and early history for more than a century.[1] Today, numerous new findings of Mesozoic birds—the number of taxa of which have doubled in the past five years—have shed light on the large chronological and phylogenetic gap between *Archaeopteryx* on the one hand and hesperornithiforms and ichthyornithiforms on the other, and research on early avian evolution has flourished with new ideas.

The theropodan origin of birds, namely the hypothesis that birds are most closely related to bipedal, carnivorous dinosaurs, is strongly supported by the known evidence.[2–4] Yet, a few skeptics still regard crocodilomorphs,[5,6] basal archosaurs[7] or even a variety of non-archosaurian diapsids[7,8] (such as *Megalancosaurus, Longisquama*) as the closest avian relatives. These arguments have consistently relied on excessive weights for a few similarities[5–7] (such as inner ear, periotic pneumaticity)—found to be of broader distribution among archosaurs (including dinosaurs)—or on pointing at the presence of "bird-like" structures in one or other taxon[8] (such as feather-like scales in *Longisquama,* beak-like snout in *Megalancosaurus*). None of these interpretations, however, has furnished more than an *a priori* argument of convergence to explain the enormous body of morphological evidence supporting the notion that birds are nested within theropod dinosaurs.

A wealth of new Mesozoic birds has provided data for elucidating the series of branching events and structural transformations from non-avian dinosaurs to their extant descendants (Fig. 1), reinforcing the idea that modern birds are short-tailed, feathered dinosaurs. Coupled with a surge in physiological, functional and developmental studies on non-avian theropods and birds, these phylogenetic developments have illuminated new frontiers in early avian evolution. Here I will review these advances and discuss their implications for understanding the evolution of the biology of modern avians.

Archaeopteryx and the Earliest Bird

The known history of birds starts in the late Jurassic (Fig. 1). Although presumably this history pre-dates the end of this period, attempts to unravel its earlier phases have failed to provide reliable evidence. In recent years, two sets of fossils have been suggested to substantiate the record of birds before the late Jurassic: *Protoavis* from the late Triassic of Texas[9] and an ensemble of traces from the Triassic-Jurassic of Africa[10,11] and North America.[11]

Chatterjee[9] not only considered *Protoavis* a bird but also regarded it as much more closely related to modern, neornithine birds than is *Archaeopteryx*. Except for a few elements, the available material of *Protoavis* is extremely fragmentary. Chatterjee's interpretations of certain bones (such as furcula, sternum) are questionable, and even the association of elements into specimens and then into a single taxon seems difficult to support. Despite detailed reconstructions of the skull,[9] only portions of the braincase, quadrate and orbital roof are reasonably preserved. Nevertheless the anterior cervical vertebrae approach a heterocoelic condition and dorsal vertebrae have large vertebral canals, both avian synapomorphies within theropods.[12] Despite this, I concur with others[13] that until better specimens are recovered and support for association of these into a single taxon is provided, *Protoavis* should not be considered relevant to avian evolution.

In the early 1970s, Ellenberger[10] regarded as avian a variety of late Triassic–early Jurassic footprints and traces from South Africa. More recently, Lockley *et al.*[11] reported on other early Jurassic bird-like footprints from northern Africa and North America, suggesting again that these traces may represent evidence of birds before the late Jurassic. Although these footprints deserve serious consideration, it is well known that the association of a footprint with a particular tracemaker always involves some degree of uncertainty, because several factors influence its shape and therefore its taxonomic identification.[14] Given the lack of reliable oste-

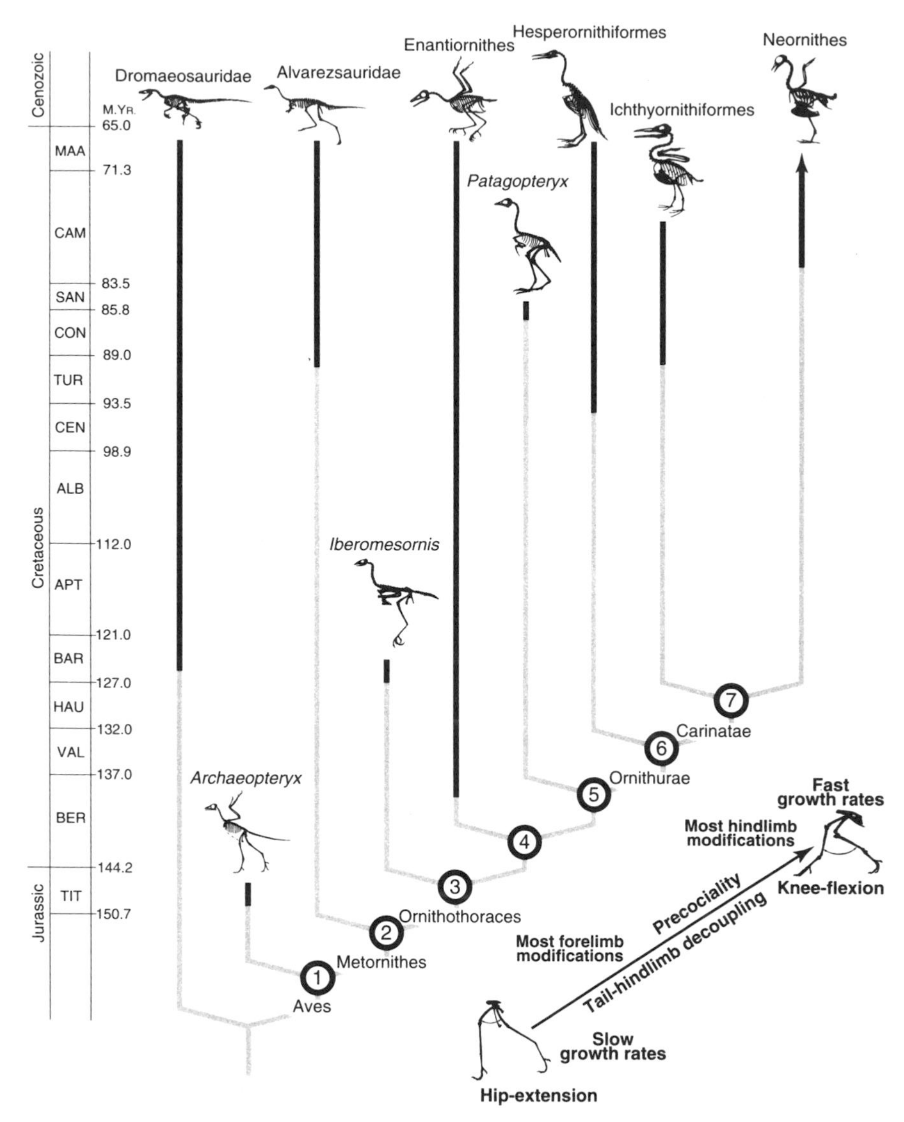

Fig. 1 Phylogenetic relationships of the best-known taxa of mesozoic birds (data from Chiappe and Calvo;[12] Chiappe *et al.*;[60] 146 steps, consistency index = 0.75, retention index = 0.81). *Solid bars* indicate the actual record of each lineage. Phylogenetic calibration of this record implies that most lineages must have had their origin at the beginning of the Cretaceous or even earlier. For example, Turonian ichthyornithiforms imply that, even though reliable evidence of neornithines is Campanian, we should expect to find earlier members of this clade in the Turonian. Indeed, the early Cretaceous ornithurines *Enaliornis* and *Ambiortus* (here not included because their relationships are not clear) indicate that carinate birds, as ornithurines, must have been differentiated at least in the early Cretaceous. Nodes 1 to 7 are diagnosed by the following unequivocal synapomorphies among others: *Node 1* (Aves = Aviale *sensu* Gauthier[3]): fewer than 26 caudals, caudals with short prezygapophyses, teeth with unserrated crowns and crown-base constrictions, completely

ological evidence before the late Jurassic, caution is advised in identifying these footprints as avian.

Thus, the presence of birds before the end of the Jurassic is far from documented. Yet, the singular position of *Archaeopteryx* as the oldest bird has been challenged by recent reports of alleged late Jurassic avians. These new specimens have only been preliminarily studied or reported informally. A feathered specimen from North Korea was dubbed the "North Korean *Archaeopteryx*."[15] Differences in its wing proportions suggest that this specimen is not *Archaeopteryx,* and its stratigraphic provenance has yet to be announced. *Confuciusornis,* from the Yixian Formation of northeastern China, is another purported late Jurassic bird.[16,17] The holotype is known from a skull associated with a wing. Other specimens from the same locality have been referred to this taxon,[16,17] but lack of overlapping elements between these and the holotype make this assignment conjectural. Furthermore, no consensus exists on the age of the Yixian Formation, and palynological data support a younger, early Cretaceous, age.[18] Despite the importance of these birds, further chronostratigraphic studies are necessary before placing confidence in their alleged age.

The spectacular specimens of *Archaeopteryx*[1] are still the most enlightening testimony of late Jurassic birds. However, aspects of its biology remain controversial. Important anatomical complexes such as the braincase, orbit and palate are uncertain, and even whether these specimens belong to one or more closely related species is still controversial. In recent years, the burden of the debate has been concentrated on its aerodynamic ability and mode of life. Some opinions regard *Archaeopteryx* as

---

reverted hallux; *Node 2* (Metornithes,[29] the common ancestor of *Mononykus* and Neornithes plus all its descendants): prominent ventral processes on cervico-dorsals, carinate and longitudinally rectangular sternum, carpometacarpus; *Node 3* (Ornithothoraces,[12] *Iberomesornis,* Neornithes and all descendants of their common ancester): pygostyle, strut-like coracoid, sharp caudal end of scapula, radial shaft/ulnar shaft ratio smaller than 0.7; *Node 4:* synsacrum with more than eight vertebrae, heterocoelous cervicals, complete calcaneo-astragalar-tibial fusion, distal tarsals completely fused to metatarsals; *Node 5:* quadrate with three distal condyles forming a triangle, caudal prezygapophyses absent or extremely reduced, absence of pubic foot, tarso-metatarsal vascular distal foramen; *Node 6* (Ornithurae,[5,42,69] all birds derived from the common ancestor of Hesperornithiformes and Neornithes): sharp and pointed quadrate orbital process, fewer than 11 dorsals, procoracoid process, small acetabulum, pubis parallel to ischium and ilium, femur with prominent patellar groove; *Node 7* (Carinatae,[5,42] Ichthyornithiformes, Neornithes, and all descendants from their common ancestor): globe-shaped, cranio-caudally convex humeral head. *Tit,* Tithonian; *Ber,* Berriasian; *Val,* Valanginian; *Hau,* Hauterivian; *Bar,* Barremian; *Apt,* Aptian; *Alb,* Albian; *Cen,* Cenomanian; *Tur,* Turonian; *Con,* Coniacian; *San,* Santonian; *Cam,* Campanian; *Maa,* Maastrichtian (boundaries between stages follow Gradstein *et al.*[85]).

primarily a terrestrial animal, with little or no capability for flight and a morphology and ecology similar to non-avian theropods such as *Deinonychus.*[3,19,20] Others see it as an arboreal bird, with a design for flight approaching the condition of modern birds.[6,21,22] Despite this broad range of opinions, most authors would agree that the *Archaeopteryx* was capable of some sort of flight.[1,23–26] The popular notion of an essentially arboreal *Archaeopteryx*—so frequently used in adaptational models of the origins of flight[21,22]—has encountered resistance from several detailed morphological studies.[19,24,27] Moreover, this idea has consistently relied on a paleoenvironmental misconception, for the Solnhofen lagoons were not surrounded by forests but by sparse small plants which at the most reached 3 m in height.[28] Phylogenetic reconstruction does not provide any direct evidence about the specific aerodynamic ability or mode of life of *Archaeopteryx.* Nevertheless, it strongly supports that *Archaeopteryx* was morphologically very different from its living relatives (Fig. 1), and that the ancestral mode of life for birds was predominately terrestrial as it must have been for their closest theropod outgroups and other basal avians (such as *Mononykus*[29]).

The Cretaceous Diversity

Cretaceous birds have often been regarded as early members of extant lineages.[30–32] Only recently have these fossils been interpreted as remotely connected to modern groups and the enormous Cretaceous diversity fully appreciated.

The most informative specimens of early Cretaceous birds come from Spain, China and Mongolia (Fig. 2). Although several of these early Cretaceous birds can be placed within the Enantiornithes—a diverse, worldwide group of Cretaceous volant birds—a number of them represent lineages so far known only by a single species.

The lowermost Cretaceous limestones of Montsec (northern Spain) have yielded the remains of the finch-sized *Noguerornis.*[33] Although the single known species is fragmentary, *Noguerornis* documents important structural novelties correlated with enhanced flying ability, such as interlocking of manual elements into a rigid structure, elongation of the distal portions of the wing, and enlargement of the wing surface, only a few million years after *Archaeopteryx.* Assessment of its phylogenetic relationships is complicated because it is poorly preserved. Nevertheless, retention of several plesiomorphies (such as ischiadic symphysis, subequal widths of radius and ulna) suggest a sister-taxon relationship to all Ornithothoraces.[12]

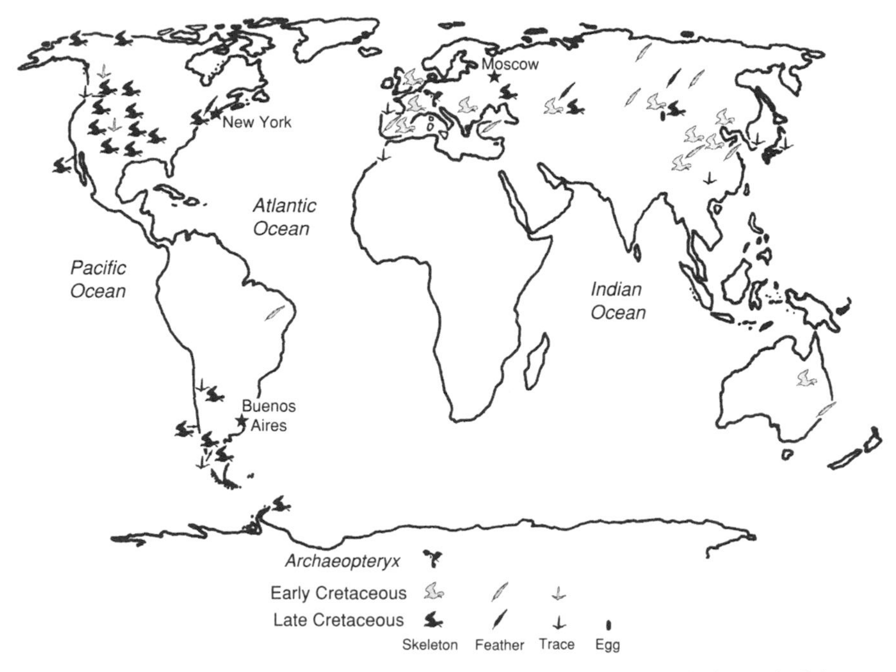

Fig. 2 Geographical distribution of Mesozoic birds. Note the relatively limited record of the southern continents, in particular Africa. *Archaeopteryx* is here regarded as the only indisputable evidence of pre-Cretaceous birds (see text for further discussion).

More complete specimens of other taxa were found in slightly younger rocks from central Spain. The Barremian limestones from Las Hoyas have yielded the basal *Iberomesornis*[34] and the enantiornithine *Concornis*[35] (Fig. 3). Sparrow-sized *Iberomesornis* is the most basal Ornithothoraces (Fig. 1). Like *Noguerornis,* and in contrast to *Archaeopteryx, Iberomesornis* shows characters (such as strut-like coracoid, ulna longer than humerus) suggesting an improved flying capability.[34] It also shows a fully specialized perching foot, indicating that arboreality and perching capacity arose very early in avian history.

Remains of early Cretaceous birds have been found in several Chinese localities (Fig. 2). Most informative specimens come from the lacustrine deposits of the Valanginian Jiufotang Formation at Liaoning Province,[36–38] including the sparrow-sized *Sinornis*[36] and *Cathayornis,*[37] which are unquestionably enantiornithines. Other less complete specimens, such as the larger *Chaoyangia,*[38] are more difficult to place in a particular clade. Despite the retention of plesiomorphies such as unfused pelvic elements and pubic symphysis, *Chaoyangia* stands out among early avians in possess-

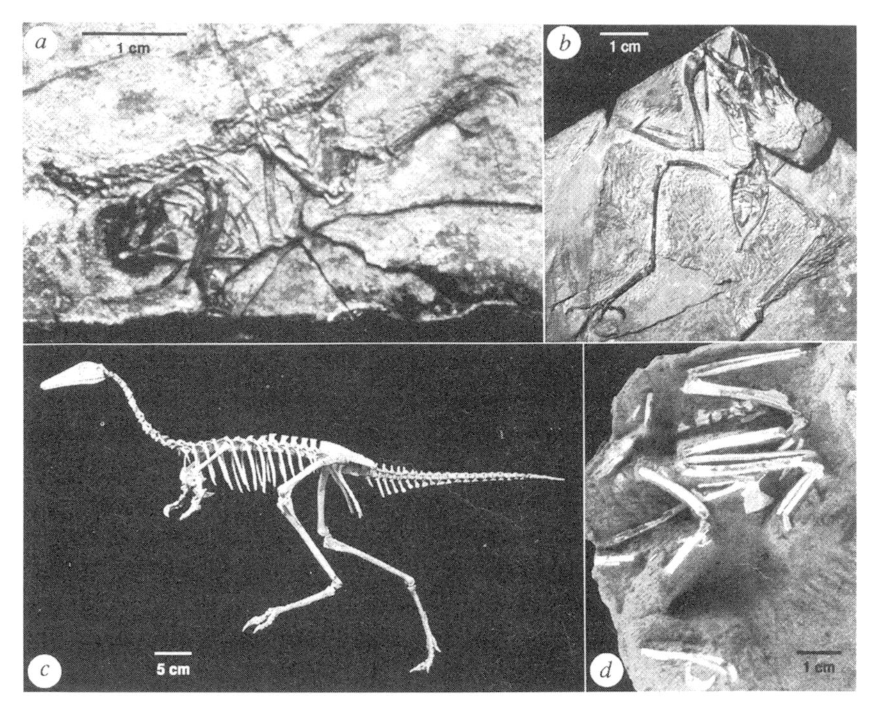

Fig. 3 *a, b, Iberomesornis* and the enantiornithine *Concornis,* respectively, from the early Cretaceous (Barremian) of Las Hoyas, central Spain. *c,* Skeletal mounting of *Mononykus* from the late Cretaceous (Campanian-Maastrichtian) of southern Mongolia. *d,* Enantiornithine *Neuquenornis* from the late Cretaceous (Coniacian-Santonian) of northwestern Patagonia, Argentina.

ing large, ossified uncinate processes otherwise unknown in the ribs of any non-ornithurine bird[12] (Fig. 1).

Known early Cretaceous birds are relatively small and most of them have perching, arboreal specializations (such as *Iberomesornis, Sinornis, Cathayornis, Concornis*). At the end of the early Cretaceous and during the late Cretaceous, however, the known fossil specimens range widely in size and lifestyle, from flightless cursores and foot-propelled divers to waders and tree dwellers. Several of these lineages belong to the Enantiornithes, but many of them are members of the Ornithurae (Fig. 1).

Two remarkable, flightless, ground-dwelling lineages have their first records in the late Cretaceous: the central Asian, stout-forelimbed *Mononykus*[29] and its South American relatives (see below), and the Patagonian *Patagopteryx.*[39] Hen-sized *Patagopteryx* was originally regarded as a primitive ratite,[39] though the absence of many derived features of ornithurine birds (Fig. 1) indicates that it is neither a ratite nor a member of the Ornithurae; *Patagopteryx* is the sister group of all ornithurine birds.[12,40] *Patagopteryx* acquired flightlessness independent from all other birds, and stands out among birds in having a fused quadrate-pterygoid complex, which suggests a unique kind (if any) of cranial kinesis.

Reconstructing the early history of the Ornithurae is hampered by the fragmentary nature of their putative earliest members. The early Cretaceous *Ambiortus*[41] from Mongolia is probably the oldest ornithurine. Nevertheless, its position within the Ornithurae is not yet clear;[25,42] the presence of amphicoelus vertebrae might suggest a close relationship with ichthyornithiforms. Sereno and Rao,[36] however, have placed *Ambiortus* outside Ornithurae, but they have not presented evidence (character distribution) supporting their claim.

The best-known Mesozoic ornithurines are the hesperornithiforms: toothed, foot-propelled, flightless divers known primarily from marine late Cretaceous rocks of North American Western Interior.[5,25,43] In recent years, these birds have also been recorded in rocks of the late Cretaceous Turgai strait,[44] between eastern Europe and western Asia. Remains of these birds have been found in continental[45] and estuarine[46] deposits, indicating that they also inhabited non-marine environments. Whether *Enaliornis*[47] from the early Cretaceous of the United Kingdom is a primitive member of this group is not yet clear, for despite similarities in pedal morphology,[5] the braincase appears to resemble modern birds more than *Hesperornis*.[47] Hesperornithiforms have often been regarded as primitive grebes and loons,[30,31,48] or as aquatic palaeognaths.[32] Recent studies, however, have consistently regarded them as basal ornithurines,[5,12,42] and their flightlessness interpreted as a secondary specialization (Fig. 1). Elzanowski,[49] however, has suggested a closer connection to neognaths, namely as the sister group of all modern birds except palaeognaths (ratites and tinamous). Nevertheless, the latter hypothesis requires explanation of additional homoplasies, such as the independent loss of teeth in both palaeognaths and neognaths.

Also from the late Cretaceous of the North American Western Interior[5,25,43] are the flying, toothed ichthyornithiforms. Claims reporting the occurrence of these birds in the late Cretaceous of Asia[44] have been recently refuted.[45] Though known since the last century, few papers have dealt with them after Marsh's 1880 monograph.[43] Earlier authors saw them as primitive charadriiforms, although most recent phylogenetic studies have placed them outside modern birds[5,42] (Fig. 1).

There is compelling evidence documenting the presence of several lineages of modern birds at the end of the Cretaceous. Taxa closely related to extant anseriforms,[50] gaviiforms,[51] charadriiforms[52,53] and procellariiforms[44,45,53] have been found in Campanian-Maastrichtian rocks, and phylogenetic inferences predict that these and other lineages (such as palaeo-

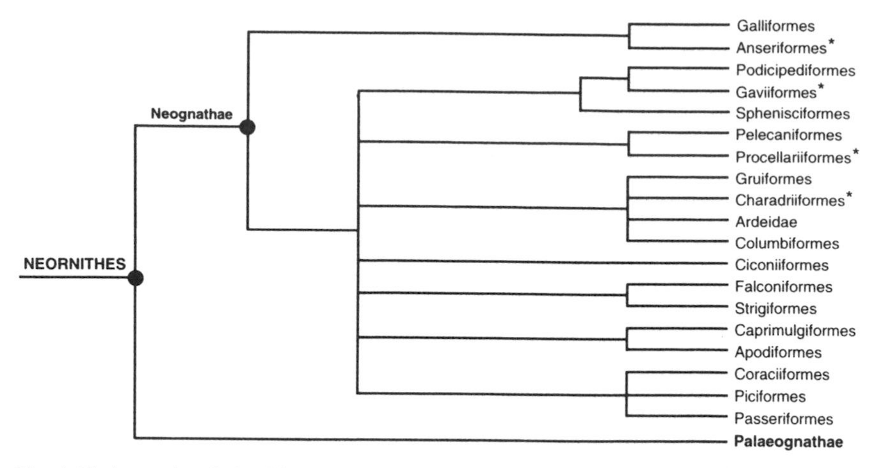

Fig. 4 Phylogenetic relationships among modern, neornithine birds (after Cracraft[86]). Asterisked taxa are known to occur in late Cretaceous deposits. Note that the phylogenetic pattern among these taxa suggests that several other lineages (such as palaeognaths) must have been differentiated before the end of the Cretaceous. Neornithes includes the common ancestor of living birds plus all its descendants; this clade is subdivided in Palaeognathae (ratites and tinamous) and Neognathae (all other extant birds).

gnaths) must have differentiated even earlier (Fig. 4). The documented earlier history of Neornithes, however, is obscured again by the fragmentary preservation of their putative oldest representatives. Early Cretaceous *Gansus*[54] from China and *Paleocursornis*[55] from Romania have been related to modern charadriiforms and ratites, respectively, but the evidence presented so far is inconclusive.

The *Mononykus* Problem

Perhaps the most startling recent finding is the flightless, stout-forelimbed, turkey-sized *Mononykus* from the late Cretaceous of central Asia[29,56] (Fig. 3). Close relatives of *Mononykus* lived in the late Cretaceous of Argentina[57] and probably western North America.[58] Novas[57] has reinterpreted the Patagonian *Alvarezsaurus*—formerly thought to be a non-avian theropod[59]—as a bird, and placed *Mononykus* and a new Argentinean taxon within the monophyletic Alvarezsauridae (Fig. 1).

Placement of these bizarre creatures within Aves has created a great deal of controversy. Nevertheless, aside from its grotesque forelimbs, *Mononykus* and its relatives exhibit several avian synapomorphies and other derived characters only known for birds more derived than *Archaeopteryx*[29,56,60] (Fig. 1). Perhaps its most manifest avian characteristic has been documented by a recently found complete skull, in which the orbit is connected to the infratemporal fenestra, a condition exclusively known for birds among archosaurs. The placement of *Mononykus* outside Aves requires the less parsimonious conclusion that all these characters evolved independently in these two taxa. This fact has not avoided criticism in which *Mononykus* is *a priori* disregarded as a bird.[61–64] This criticism seems to stem from the conjecture that *Mononykus* does not fit the "stereotype" of a basal bird.[65] For example, Wellnhofer[61] finds it "very difficult to imagine how a primitive bird wing, such as that of *Archaeopteryx,* could have evolved into a forelimb like that of *Mononykus,*" and Feduccia[62] claims that "the keeled breastbone doesn't resemble that of birds . . . and it is probably associated with a digging function, not any birdlike actions." However, assumptions about evolutionary processes or adaptational scenarios, such as these, are misleading when identifying historical relationships. Phylogenetic reconstruction should be based exclusively on the hierarchical distribution of homologies among taxa, and a given phylogenetic hypothesis can be rejected only by providing an alternative hypothesis for which supporting evidence outweighs that of the original hypothesis. Clearly, such a procedure has not been implemented by these that reject out-of-hand the avian relationship of *Mononykus.*[65]

The Enantiornithes: An Unexpected Radiation

Distributed worldwide, ranging from the Valanginian to the Maastrichtian, and known by over a dozen distinct species, the Enantiornithes document a major cladogenetic event of Mesozoic volant birds (Fig. 1). Quite surprisingly, this clade remained unrecognized until 1981.[66] Enantiornithine fossils had been found earlier—even a century ago—but they remained misidentified as non-avian theropods (such as *Ornithomimus*) or were included within modern groups (such as *Gobipteryx*[67]).

Except for a few occurrences in near-shore deposits, most enantiornithines are known from non-marine rocks. Most likely, this bias in the fossil records reflects their preponderance in Cretaceous terrestrial avifaunas. Despite the fact that all enantiornithines show morphological features suggesting an enhanced flying capacity,[12] they exhibit a broad morphological range. For example, although early forms such as *Cathayornis* and *Sinornis* were tiny and toothed, late Mesozoic taxa such as *Enantiornis*[66] reached a wingspan of over 1 m, and some like *Gobipteryx*[67] lacked teeth. Interpretation of *Sinornis* as an enantiornithine modifies Sereno and Rao's hypothesis regarding this taxon as the most basal bird apart from *Archaeopteryx.*[36] Although these authors did not provide support for their hypothesis, *Sinornis* shows several derived characters that connect it to the Enantiornithes (such as reduced metatarsal IV, dorsal process on ischium, parapophyses of dorsal vertebrae centrally located). One of the most striking structural characteristics of the Enantiornithes has been revealed by histological studies of their bones which have shown a unique pattern when compared with those of other birds.[68]

The relationship of the Enantiornithes to other Mesozoic avian lineages has been the subject of intense debate.[40,69] This clade has been regarded variously as closely related to *Archaeopteryx,*[5,6] included within the Ornithurae,[42] or placed in an intermediate position between *Archaeopteryx* and ornithurine birds.[12,40,66,69] The latter hypothesis is clearly the one that best summarizes the available evidence (Fig. 1). The sister-group relationship between *Archaeopteryx* and Enantiornithes has been consistently argued by Martin,[5,6] who has envisioned a basal avian dichotomy between the "Sauriurae" (Enantiornithes and *Archaeopteryx*) and its alleged sister taxon the Ornithurae. The "Sauriurae," however, is a paraphyletic group supported by characters that are erroneous or, at best, plesiomorphic.[40] This unjustified dichotomy turns the otherwise monophyletic Enantiornithes into a wastebasket for everything that is neither *Archaeopteryx* nor Ornithurae. This arrangement has become even more complex because of indiscriminate use of paraphyletic taxa as practiced by Feduccia[70] who, though endorsing the "Sauriurae," represents the Enantiornithes as more closely related to Neornithes than to *Archaeopteryx,* and regards the Ornithurae as a terminal taxon not including modern, neornithine birds.

Growth, Development Modes and Physiology

Comparative studies on bone microstructure have been extensively used to infer growth rates, and in some instances, physiological aspects of extinct vertebrates. The bone histology of Mesozoic birds is known only for a few taxa. In *Hesperornis,* the bone tissue is like that of its extant counterparts,[71] with uninterrupted, fast-deposited fibro-lamellar bone. This type of bone is consistent with rapid growth; modern birds grow rapidly, reaching adult size within the first year.[72] A different pattern, however, has been found in Enantiornithes and *Patagopteryx.* Microstructural studies of these birds have documented punctuations interrupting bone deposition (lines of arrested growth or LAGs), suggesting cyclical pauses during postnatal growth.[68,73] Such a pattern contrasts with the uninterrupted growth pattern of extant birds. Furthermore, although *Patagopteryx* shares the modern condition of a highly vascularized, fibro-lamellar bone, the enantiornithines exhibit a lamellar, slowly deposited bone tissue, with virtually no vascularization,[68,73] a pattern remarkably different from that of modern birds and their non-avian theropod relatives.

Most probably, LAGs of Enantiornithes and *Patagopteryx* formed annually, as occurs in extant non-avian reptiles.[74] This pattern of discontinuous growth, and the slowly deposited bone tissue of Enantiornithes, suggest that these birds had slower rates of growth than their modern relatives. Four and five LAGs observed in two enantiornithine specimens indicate that these individuals had at least four and five years, respectively, of post-hatching growth.[68,73] Moreover, the fact that six specimens of *Archaeopteryx* were found to lie on a linear size trajectory[75] also hints at slow rates of postnatal growth for basal birds, because it would be very unlikely that all these *Archaeopteryx* individuals died differing by only a few months of age.

The inferred growth rates for these birds shed light on their developmental modes. Hatchlings of extant birds form a broad spectrum of appearance and conduct, ranging from downy, precocial species capable of independent locomotion to the usually naked and blind altricial taxa that are incapable of locomotion.[72,76] Precociality has been traditionally regarded as primitive for it is typical of basal neornithines such as palaeognaths and galliforms.[72] Evidence of precociality in basal Mesozoic birds was suggested by the discovery of embryos of the enantiornithine *Gobipteryx* showing a remarkable degree of ossification,[77] a condition also found in non-avian theropod embryos.[78] The slow growth rates inferred for basal birds are consistent with this hypothesis, for living precocial taxa have slower rates of growth than do altricial birds.[72]

Observed interruptions during bone deposition of Enantiornithes and *Patagopteryx* also hint at other physiological differences with respect to modern birds. Growth, as is the case for other activity levels, is strongly correlated with body temperature.[79] Hence inference of cyclical growth in basal birds suggests that they may not have been endothermic homeotherms.[68,73] The fact that these basal birds may not have been endotherms does not necessarily imply that they were ectotherms, for they conceivably may have had intermediate levels of activity as it has been suggested for non-avian dinosaurs.[74] If this is correct, however, characterizations of *Archaeopteryx* and the enantiornithine *Sinornis* as endotherms should be re-evaluated.[68,73,80]

Interpretation of these primitive birds as non-endothermic conflicts with ideas envisioning the origin of feathers in conjunction with endothermic metabolism. For example, Ostrom[19] maintained that endothermy "must have been achieved before there was any extensive feather covering" and that it "must have preceded flight," but the Enantiornithes were covered by feathers and their anatomy suggests an enhanced flying ability.[12] Furthermore, not all feathered creatures are endothermic (downy hatchlings are unable to generate internal heat[76]) nor does any kind of flight appear to be necessarily related to endothermy.[80,81] Although some of these aspects may be conjectural, the histological differences documented for early birds suggest that important physiological changes may have occurred late in avian history. The notion that classic, modern endothermy may have evolved after the development of feathers and an enhanced flying ability should be seriously considered.[68,73,80]

Terrestrial Locomotion and Flight Apparatus

Birds are unique among extant vertebrates in that fore- and hindlimbs are part of two independent locomotory systems. Functional limb decoupling certainly took place long before the origin of birds, for bipedalism is ancestral even for dinosaurs. The modern avian pattern of hindlimb kinematics, however, differs from the ancestral theropod one. As Gatesy[82] has shown, in non-avian theropods—as in lizards and crocodiles—hip extension was the primary mechanism of hindlimb kinematics, namely a pattern of extensive femoral retraction during each stride. In contrast, a knee-flexion mechanism is characteristic of modern birds, in which the stride is coupled by a large tibiotarsal displacement.[82] Most probably, reduction of the tail and caudal musculature throughout theropod evolution is the principal agent implicated in the development of the latter mechanism. Correlated with this transformation are the functional decoupling of the tail from the primitive pattern of hindlimb kinematics and its linkage with the flight apparatus, and the forward migration of the center of gravity and development of a modern avian stance.[82] Interestingly, these transformations were not fully achieved until late in avian evolution. The relatively long tails of *Archaeopteryx* and *Mononykus* suggest that these basal birds retained a pattern of hindlimb kinematics similar to that of non-avian theropods. Furthermore, although the pygostyle and relatively short tails of early ornithothoracines (such as *Iberomesornis*) indicate tail decoupling, retention of several primitive pelvic characters suggests that in these birds the transformation was not yet completed. For example, pubic and ischiadic symphyses are present in *Concornis* and *Noguerornis,* respectively. *Patagopteryx* has an ample brevis fossa (origin of the M. caudofemoralis brevis) and a fairly long tail. In fact, it is only in the Ornithurae that the pelvis and hindlimb acquired a fully modern appearance. Thus, though the tail of several basal birds is poorly known and specific morphofunctional analyses are still pending, the modern pattern of hindlimb kinematics was probably not fully developed until the differentiation of the Ornithurae.[40,69]

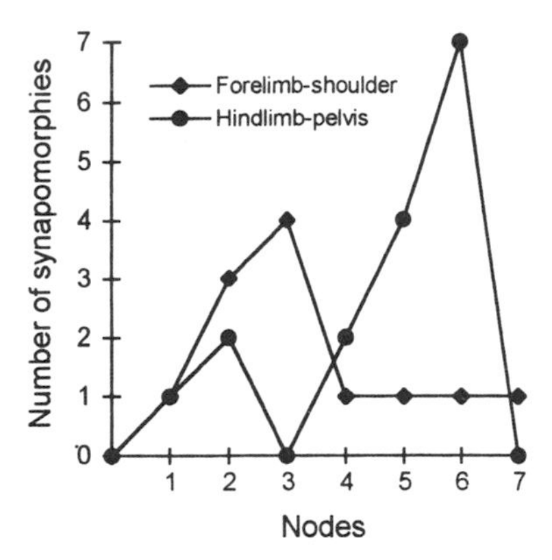

Fig. 5 Sequence of transformations of the forelimb and shoulder girdle (including sternum), and the hindlimb and pelvis during early avian phylogeny. The chart shows the number of synapomorphies (only unequivocally optimized) for each node of the cladogram in Fig. 1. The fact that most synapomorphies of the forelimb and shoulder girdle precede those of the hindlimb and pelvis suggests that, during the early history of birds, the structures correlated to an enhanced flying ability were acquired earlier than those correlated to a modern kind of terrestrial locomotion.[40] Character data derived from Chiappe *et al.*[60]

Conversely, a condition approaching a modern design of wing and shoulder girdle was acquired very early in avian history. Although the aerodynamic capabilities of *Archaeopteryx* are still a matter of controversy, the anatomy of the slightly younger *Noguerornis* and *Iberomesornis* documents an enhanced flying ability. Thus, the evolution of the modern flight apparatus appears to have preceded the modern mechanism of hindlimb kinematics, a conclusion supported by the fact that the sequence of structural modifications leading to the former appear earlier (that is, are hierarchically more inclusive)[40,69] (Fig. 5).

Temporal Branching and Taxonomic Dynamics

Even among the most complete sequences, the fossil record can hardly ever be taken at face value.[83] Despite the large number of recent discoveries, large segments of the early history of birds are still missing (Figs. 1, 6). The available fossils, however, can provide a clearer account of temporal patterns if they are calibrated using a phylogenetic hypothesis.[84] The minimal age for a lineage and its sister taxon can be retrieved from their oldest member, for sister taxa must have a common origin. Calibrating the phylogeny of Mesozoic groups (Fig. 1) indicates that most, if not all, the Cretaceous lineages were present very early in this period.[42] Unfortunately this conclusion tells us nothing about the amount of time between the ancestor of all birds and those of each branching event. It shows,

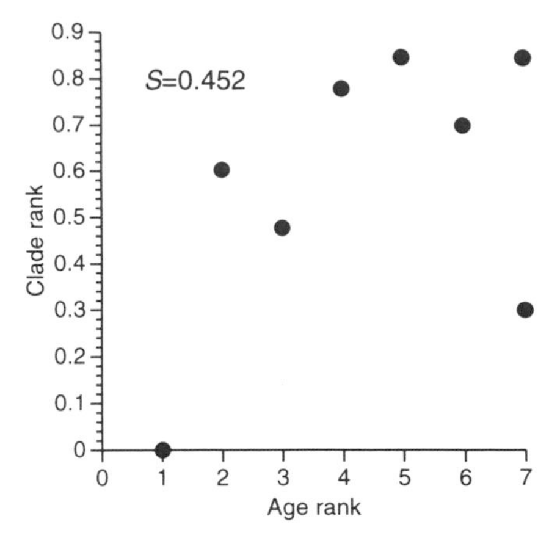

Fig. 6 Congruence between stratigraphic record and branching pattern for the cladogram of Fig. 1. Plots show age ranks (derived from the oldest record of terminal taxa) as a function of clade ranks (assigned to each branching event; for example, *Archaeopteryx* is 1, Alvarezsauridae is 2 and so on). Chart constructed by using the method described by Norell and Novacek.[87] Clade ranks are rescaled from 0 to 1. The correlation is not even significant at $P < 0.05$. This low correlation indicates that, despite the numerous new findings, the fossil record of basal birds is still incomplete and great portions of their evolutionary history are yet missing. *S* is the Spearman rank correlation coefficient.

however, that recent claims proposing an explosive cladogenesis of modern birds at the base of the Tertiary[70] are inaccurate. Feduccia's new model[70] not only illustrates the misleading effects of using the fossil record at face value[84] but also ignores evidence documenting several modern lineages in late Cretaceous rocks[50–53] (Fig. 4). His interpretation of massive avian extinctions at the Cretaceous-Tertiary boundary[70] is also paradoxical, for in this case the information available in the fossil record has been disregarded (Fig. 1). No non-neornithine Mesozoic lineages are known for the Tertiary, and most probably became extinct before the end of Cretaceous. There is no positive evidence, however, that these extinctions occurred at the Cretaceous-Tertiary boundary, for the last record of several lineages is before the Maastrichtian[25] (Fig. 1).

Prospects and Conclusion

Birds flourished during the Mesozoic, acquiring specializations for a variety of different habitats. This was a period of extensive diversification, though one in which extinction was also a common phenomenon. The many new findings have provided additional support for the hypothesis that birds evolved from cursorial, bipedal carnivorous dinosaurs. Furthermore, these fossils have documented that several important morphological, functional and physiological systems characteristic of modern birds were acquired only late in avian history. Clearly, the common practice of extrapolating modern avian characteristics to these basal avian members or farther down to non-avian dinosaurs is at best a naive procedure.

Although the new fossil evidence has opened a realm of scientific inquiry, it has also exposed our relative ignorance about the first 85 million years of avian history. Further research is needed before a more comprehensive, finely tuned understanding of the interrelationships among the basal taxa can be developed and the importance of these fossils in clarifying the still controversial historical relationships of their extant relatives assessed.

LUIS M. CHIAPPE is in the Department of Vertebrate Paleontology and the Department of Ornithology, American Museum of Natural History, Central Park West at 79th Street, New York, New York 10024, USA.

References

1. Hecht, M. K., Ostrom, J. H., Viohl, G. & Wellnhofer, P. (eds.) *The Beginnings of Birds* (Jura Museum, Eichstätt, 1985).
2. Ostrom, J. H. *Biol. J. Linn. Soc.* **8,** 91–182 (1976).
3. Gauthier, J. *Mem. Calif. Acad. Sci.* **8,** 1–55 (1986).
4. Witmer, L. in *Origin of the Higher Groups of Tetrapods* (eds. Schultze, H.-D. & Trueb, L.) 427–466 (Comstock, Ithaca, NY, 1991).
5. Martin, L. D. in *Perspectives in Ornithology* (eds. Bush, A. H. & Clark, G. A. Jr.) 291–338 (Cambridge Univ. Press, Cambridge, 1983).
6. Martin, L. D. in *Origin of the Higher Groups of Tetrapods* (eds. Schultze, H.-D. & Trueb, L.) 485–540 (Comstock, Ithaca, NY, 1991).
7. Tarsitano, S. in *Origin of the Higher Groups of Tetrapods* (eds. Schultze, H.-D. & Trueb, L.) 541–576 (Comstock, Ithaca, NY, 1991).
8. Feduccia, A. & Wild, R. *Naturwissenschaften* **80,** 564–566 (1993).
9. Chatterjee, S. *Phil. Trans. R. Soc. Lond.* B **332,** 277–346 (1991).
10. Ellenberger, P. *Palaeovertebrata (Mem. Extraor.)* 1–141 (1974).
11. Lockley, M. G., Yang, S. Y., Matsukawa, M., Fleming, F. & Lim, S. K. *Phil. Trans. R. Soc.* B **336,** 113–134 (1992).
12. Chiappe, L. M. & Calvo, J. O. *J. Vert. Paleont.* **14,** 230–246 (1994).
13. Ostrom, J. H. *Nature* **353,** 212 (1991).
14. Padian, K. & Olsen, P. E. *J. Paleont.* **58,** 178–184 (1984).
15. *Korean Pictorial* **2,** 25 (1994) (in Japanese).
16. Hou, L. in *Short Papers, IV Symp. Mes. Terrestr. Ecosyst.* (eds. Sun, A. & Wang, Y.) 193–201 (China Ocean, Beijing, 1995).
17. Hou, L., Zhou, Z., Martin, L. D. & Feduccia, A. *Nature* **377,** 616–618 (1995).
18. Li, W. & Liu, Z. *Cretaceous Res.* **15,** 333–365 (1994).
19. Ostrom, J. H. *Q. Rev. Biol.* **49,** 27–47 (1974).
20. Vazquez, R. J. *Res. Explor.* **8,** 387–388 (1992).

21. Feduccia, A. & Tordoff, H. B. *Science* **203,** 1021–1022 (1979).
22. Feduccia, A. *Science* **259,** 790–793 (1993).
23. Olson, S. L. & Feduccia, A. *Nature* **278,** 247–248 (1979).
24. Speakman, J. R. *Evolution* **47,** 336–340 (1993).
25. Olson, S. L. in *Avian Biology* Vol. 8 (eds. Farner, D. S., King, J. & Parkes, K. C.) 79–238 (Academic, New York, 1985).
26. Dodson, P. *Vert. Paleont.* **5,** 177–179 (1985).
27. Peters, D. S. & Görgner, E. *Sci. Ser. nat. Hist. Mus. Los Angeles Co.* **36,** 29–37 (1992).
28. Barthel, K. W., Swinburne, N. H. M. & Morris, S. C. *Solnhofen: A Study in Mesozoic Paleontology* (Cambridge Univ. Press, Cambridge, 1990).
29. Perle, A., Norell, M. A., Chiappe, L. M. & Clark, J. M. *Nature* **362,** 623–626 (1993).
30. Howard, H. *Ibis* **92,** 1–21 (1950).
31. Brodkorb, P. in *Avian Biology* (eds. Farner, D. S. & King, J. R.) 19–55 (Academic, New York, 1971).
32. Simpson, G. G. *Contrib. Sci. nat. Hist. Mus. Los Angeles Co.* **330,** 3–8 (1980).
33. Lacasa, A. *Estud. Geol.* **45,** 417–425 (1989).
34. Sanz, J. L. & Bonaparte, J. F. *Sci. Ser. nat. Hist. Mus. Los Angeles Co.* **36,** 39–49 (1992).
35. Sanz, J. L., Chiappe, L. M. & Buscalioni, A. D. *Am. Mus. Nov.* **3133,** 1–23 (1995).
36. Sereno, P. & Rao, C. *Science* **255,** 845–848 (1992).
37. Zhou, Z., Jin, F. & Zhang, J. *Chinese Sci. Bull.* **37,** 1365–1368 (1992).
38. Hou, L. & Zhang, J. *Vert. PalAs.* **31,** 217–224 (1993) (in Chinese).
39. Alvarenga, H. M. F. & Bonaparte, J. F. *Sci. Ser. nat. Hist. Mus. Los Angeles Co.* **36,** 51–64 (1992).
40. Chiappe, L. M. *Courier Forschungsinstitut Senckenberg* **181,** 55–63 (1995).
41. Kurochkin, E. N. *Cretaceous Res.* **6,** 271–278 (1985).
42. Cracraft, J. *Paleobiology* **12,** 383–399 (1986).
43. Marsh, O. C. *United States Geological Exploration of the 40th Parallel* 1–201 (Government Printing Office, Washington, DC, 1880).
44. Nessov, L. A. *Russ. J. Ornithol.* **1,** 7–50 (1993) (in Russian).
45. Kurochkin, E. N. in *Short Papers, IV Symp. Mes. Terrestr. Ecosyst.* (eds. Sun, A. & Wang, Y.) 203–208 (China Ocean, Beijing, 1995).
46. Fox, R. C. *Can. J. Earth Sci.* **11,** 1335–1338 (1974).
47. Elzanowski, A. & Galton, P. M. *J. Vert. Paleont.* **11,** 90–107 (1991).
48. Cracraft, J. *Syst. Zool.* **31,** 35–56 (1982).
49. Elzanowski, A. *Postilla* **207,** 1–20 (1991).
50. Noriega, J. I. & Tambussi, C. P. *Ameghiniana* **32,** 57–61 (1995).
51. Olson, S. L. *Vert. Paleontol.* **12,** 122–124 (1992).
52. Brodkorb, P. *Acta. XIII Congr. Int. Ornithol.* 55–70 (1963).
53. Olson, S. L. & Parris, D. *Smith. Contr. Paleobiol.* **63,** 1–22 (1987).
54. Hou, L. & Liu, Z. *Sci. Sin.* **27,** 1296–1302 (1984).
55. Kessler, E. & Jurcsák, T. *Trav. Mus. d'Hist. Nat. Grigore Antipa* **28,** 289–295 (1986).
56. Perle, A., Chiappe, L. M., Barsbold, R., Clark, J. M. & Norell, M. A. *Am. Mus. Nov.* **3105,** 1–29 (1994).
57. Novas, F. *Internatnl symp. Gond. Dinos. Mem. Queens. Mus.* (Trelew, 1994)

58. Holtz, T. R. Jr. *J. Vert. Paleont.* **14,** 480–519 (1994).
59. Bonaparte, J. F. *Rev. Mus. Arg. Cien. Alat. "Bernardino Rivadavia" (Paleont.)* **4,** 17–123 (1991) (in Spanish).
60. Chiappe, L. M., Norell, M. A. & Clark, J. M. *Internatnl symp. Gond. Dinos. Mem. Queens. Mus.* (Trelew, 1994).
61. Wellnhofer, P. *C. R. Acad. Sci. Paris* **319,** 299–308 (1994).
62. Feduccia, A. *Living Bird* **13,** 28–33 (1994).
63. Ostrom, J. H. in *Major Features in Vertebrate Evolution* (convs. Prothero, D. R. & Schoch, R. M.) *Short Course in Paleontology,* Vol. 7, 160–177 (1994).
64. Martin, L. D. & Rinaldi, C. *Maps Digest* **17,** 190–196 (1994).
65. Chiappe, L. M., Norell, M. A. & Clark, J. M. *C. R. Acad. Sci. Paris* **320,** 1031–1032 (1995).
66. Walker, C. A. *Nature* **292,** 51–53 (1981).
67. Elzanowski, A. *Palaeont. Pol.* **37,** 153–165 (1977).
68. Chinsamy, A., Chiappe, L. M. & Dodson, P. *Nature* **368,** 196–197 (1994).
69. Chiappe, L. M. *Alcheringa* **15,** 333–338 (1991).
70. Feduccia, A. *Science* **267,** 637–638 (1995).
71. Houde, P. *Auk* **103,** 125–129 (1987).
72. Starck, J. M. *Curr. Ornithol.* **10,** 275–366 (1993).
73. Chinsamy, A., Chiappe, L. M. & Dodson, P. *Paleobiology* **21** (in the press).
74. Chinsamy, A. & Dodson, P. *Am. Sci.* **83,** 174–180 (1995).
75. Houck, M. A., Gauthier, J. A. & Strauss, R. E. *Science* **247,** 195–198 (1990).
76. Campbell, B. & Lack, E. (eds.) *A Dictionary of Birds* (Buteo, Vermillion, 1985).
77. Elzanowski, A. *Acta XVIII Congr. Int. Ornithol.* **1,** 178–183 (1985).
78. Norell, M. A. *et al. Science* **266,** 779–782 (1994).
79. Prosser, C. L. (ed.) *Comparative Animal Physiology* (Saunders, New York, 1973).
80. Randolph, S. E. *Zool. J. Linn. Soc.* **112,** 389–397 (1994).
81. Ruben, J. *Evolution* **45,** 1–17 (1991).
82. Gatesy, S. M. in *Functional Morphology in Vertebrate Paleontology* (ed. Thomason, J.) 219–234 (Cambridge Univ. Press, Cambridge, 1995).
83. Holland, S. M. *Paleobiology* **21,** 92–109 (1995).
84. Norell, M. A. in *Extinction and Phylogeny* (eds. Novacek, M. J. & Wheeler, Q. D.) 89–118 (Columbia Univ. Press, New York, 1992).
85. Gradstein, F. M. *et al. Geophys. Res.* **99,** 24051–24074 (1994).
86. Cracraft, J. in *The Phylogeny and Classification of the Tetrapods:* Vol. 1 *Amphibians, Reptiles, Birds* (ed. Benton, M. J.) 339–361 (Clarendon, Oxford, 1988).
87. Norell, M. A. & Novacek, M. J. *Science* **255,** 1690–1693 (1992).

Acknowledgments

I thank A. Chinsamy, P. Dandonoli, D. Frost, M. Norell, K. Padian, J. Wible and L. Witmer for comments and discussions. P. Conversano and M. Ellison prepared the illustrations. Research supported by the Frick Fund of the American Museum of Natural History, the Philip M. McKenna Foundation and the National Science Foundation (DEB-9407999).

Michael J. Novacek

MAMMALIAN PHYLOGENY: SHAKING THE TREE

Recent paleontological discoveries and the correspondence between molecular and morphological results provide fresh insight on the deep structure of mammalian phylogeny. This new wave of research, however, has yet to resolve some important issues.

Although several major lineages of fossil and extant mammals are recognized, all of the more than 3,000 living species are members of the monotremes (comprising only the duck-billed platypus and the echidna), the metatherian (marsupials) or the eutherian (placental) mammals. This basic tripartite division leaves unclarified the relationships among a bewildering array of marsupial and eutherian lineages. Recently, this problem has been investigated using new systematic methods,[1,2] a growing molecular database[3,4,5] and continuing paleontological discoveries.[6–8] Current work mirrors a long history of forays and setbacks in untangling higher mammalian relationships,[9] culminating in George Gaylord Simpson's classification of the Mammalia[10] in 1945 (Box 1). The influence of this classification not only reflected Simpson's authority in the field but also his role as one of the architects of the modern synthesis,[11] the melding of Darwinian evolutionary theory with genetics and paleontology that emerged in the middle of the twentieth century.

Refinement of Simpson's arrangement largely applied to generic and familial level taxa. An exponential increase in knowledge of fossil mammals,[12] as well as developing molecular techniques, laid the groundwork for the current attention to the broader outlines of mammalian phylogeny. Molecular approaches were rooted in studies[13] contemporaneous with Simpson's classification and eventually were related to phylogenetic reconstruction.[14] Techniques evolved through immunological comparisons,[15,16] protein sequencing[17,18] and, most recently, direct comparisons of DNA sequences in selected genes.[5,19–21] New systematic methods, such as cladistics[2,22,23] and powerful computer programs,[24,25] have allowed for analyses of diverse anatomical characters and nucleotide sequences under the same logical framework.

BOX 1 Simpson's 1945 classification of mammals

Simpson's classification had two primary objectives: to outline the author's version of the principles of systematics and to illustrate the application of such principles with an exhaustive catalogue, down to the generic level, of the living and fossil mammals. Genera, families and orders were arranged under several cohorts and superorders. Some of these higher categories were drawn from earlier work of W. K. Gregory[9] and others. A few of these concepts, such as the grouping of carnivore and ungulate orders within the Ferungulata, were audacious and original. Moreover, they were heavily influenced by information on early fossil lineages. As Simpson remarked (ref. 10, pp. 173–174), "To students of recent mammals, the association of carnivores and ungulates in a single cohort must appear thoroughly unnatural . . . Students of early mammals have for some time been increasingly aware that the earliest carnivores and ungulates *(sensu lato)* are more closely similar to each other than either group is to the contemporaneous insectivores and their special allies." Despite these convictions, Simpson did not take great pains to defend various mammalian superordinal categories nor did he make much reference to them in his numerous publications postdating the classification.

The following higher categories were recognized in Simpson's classification (extinct groups are indicated by *):

Class Mammalia
Subclass Prototheria
  Order Monotremata (duck-billed platypus, echidna)
Subclass *Allotheria
  Order *Multituberculata
Subclass uncertain
  Order *Triconodonta
Subclass Theria
  Infraclass *Pantotheria
  Infraclass Metatheria
    Order Marsupialia (opossums, kangaroos and so on)
  Infraclass Eutheria
    Cohort Unguiculata
      Order Insectivora (hedgehogs, shrews, moles and so on)
      Order Dermoptera (flying lemurs)
      Order Primates
      Order Chiroptera (bats)
      Orders *Tillodontia and *Taeniodonta
      Order Edentata (sloths, anteaters, armadillos and so on)
      Order Pholidota (pangolins)
    Cohort Glires
      Order Lagomorpha (rabbits, pikas)
      Order Rodentia (rats, mice, porcupines and so on)
    Cohort Mutica
      Order Cetacea
    Cohort Ferungulata
      Superorder Ferae
        Order Carnivora (including *Creodonta, Fissipeda, Pinnipedia)
      Superorder Protungulata
        Orders *Condylarthra, *Litopterna, *Notoungulata, *Astrapotheria
        Order Tubulidentata (aardvarks)
      Superorder Paenungulata
        Orders *Pantodonta, *Dinocerata, *Pyrotheria
        Order Proboscidea (elephants, mastodons and so on)
        Order *Embrithopoda
        Order Hyracoidea (hyraxes)
        Order Sirenia (sea cows, manatees)
      Superorder Mesaxonia
        Order Perissodactyla (horses, rhinos and so on)
      Superorder Paraxonia
        Order Artiodactyla (even-toed ungulates: pigs, hippos, camels, deer, cows and so on)

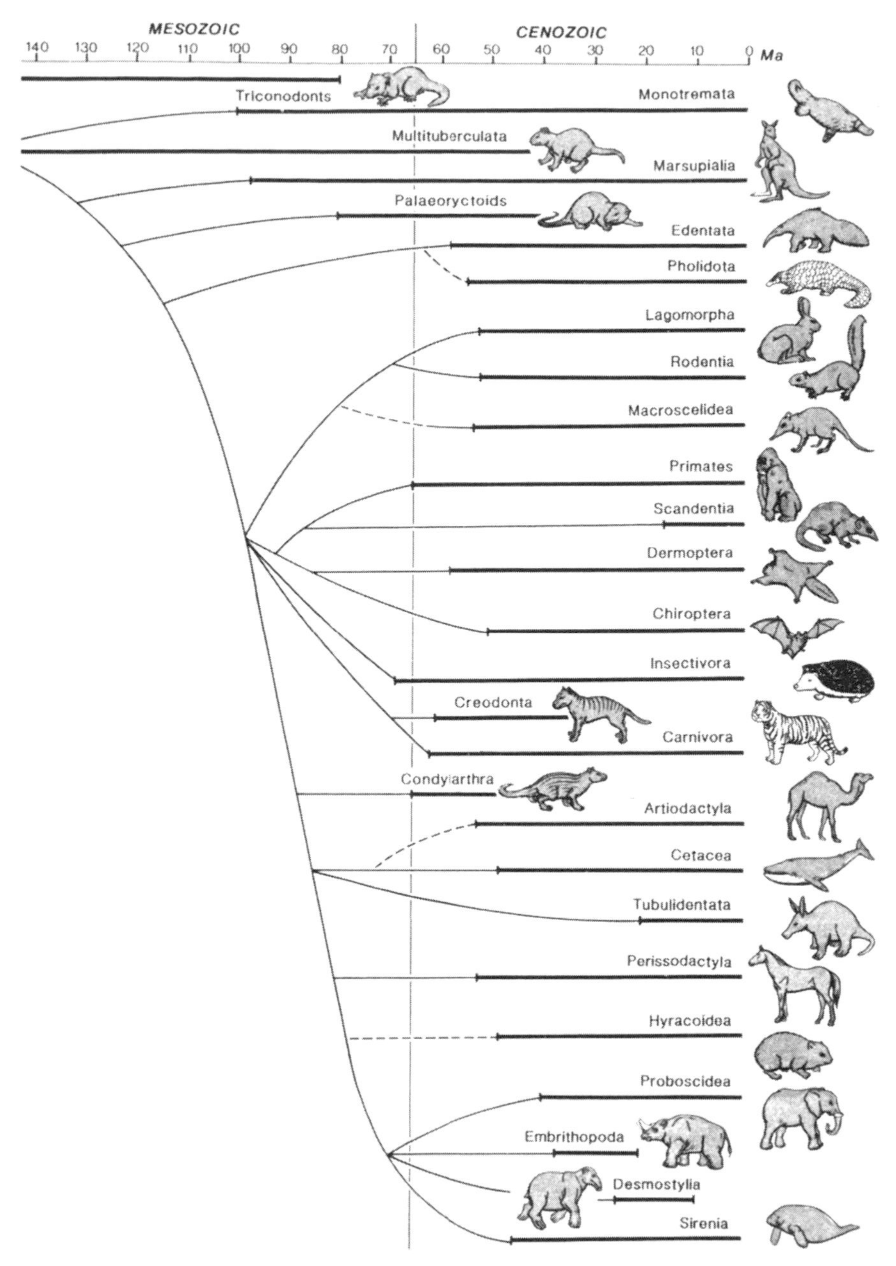

Fig. 1 A phylogenetic tree showing relationships among the major mammalian clades. The *solid horizontal bars* indicate the age range of the clade on the basis of dated first appearance in the fossil record. *Solid lines* indicate the branching sequence, although the date of the actual splitting event can only be inferred from the relationships of the clades and their known ages. *Dashed lines* indicate relatively more ambiguous relationships.

Morphological and Molecular Discrepancies

This conformity in method prompts an essential question: to what extent do phylogenies based on morphological traits either conflict or comply with molecular results? Given the limitations both in molecular[19,26,27] and in morphological[23,28] approaches, it is not surprising that higher mammalian phylogeny is still unclear (Fig. 1). For example, there is only limited congruence among various data sets.[29] Some morphological traits[29] and protein sequence data[18,30] offer strong support for Simpson's Paenungulata, the superorder which includes the living hyraxes, elephants and sirenians (sea cows and dugongs).[31] Nonetheless, a group of morphologists[32] dispute the association of hyraxes with other alleged paenungulates and favor a pre-Simpsonian idea[33] that hyraxes and perissodactyls (the order including modern horses, rhinos and tapirs) are very close relatives. This alternative scheme, however, finds almost no support from protein sequence studies[18,30] that clearly favor the Paenungulata (Fig. 2).

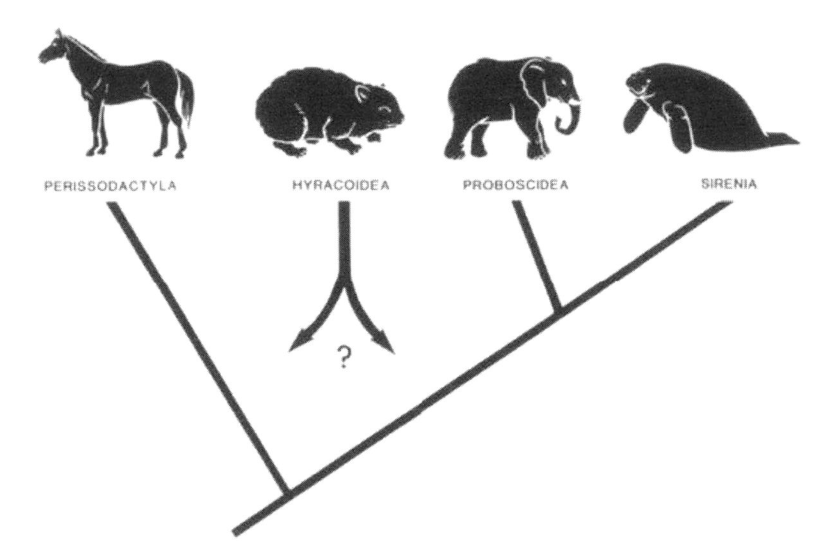

Fig. 2 The debate over hyracoid relationships has been spawned by a curious mismatch of anatomical evidence. Although the Paenungulata (Hyracoidea, Sirenia and Proboscidea) is affirmed by specializations involving the serial arrangement of the carpal elements and the reduction of certain basicranial bones and other features,[29] at least some living perissodactyls and hyracoids share notable and rather bizarre specializations including an expanded eustachian sac in the middle-ear region.[32] Nonetheless, molecular results, primarily amino-acid sequence data from lens α-crystallin and other proteins,[18,30] clearly support the paenungulate grouping, as does a paleontological accounting of early fossil horses.[48] New features of ear morphology in fossil proboscideans[8] have significant bearing on the origin of this group and their relationships with sirenians, hyraxes and other ungulates.

There are also some marked discrepancies between molecular and morphological results. Perhaps most notable among these concerns the cohort Glires, which includes lagomorphs and rodents. Glires was promoted in early studies,[9] preserved in Simpson's classification[10] and strongly supported in many modern investigations of skull structure, architecture of the ankle joint, fetal membranes and tooth development (for review, see refs. 29, 31). Such results starkly contradict immunological comparisons[34] and fail to find clear affirmation in protein sequence[18] as well as gene sequence[35] studies. Finally, neither molecular nor morphological probes have been successful in resolving major sectors of the mammalian tree.

Morphological and Molecular Concordance

Yet this somewhat bumpy effort to scan data from complex anatomy to nucleotides has produced some illumination. Both molecular and morphological results seem to converge on the idea that New World edentates (sloths, armadillos and anteaters) represent an early (perhaps even the earliest) branch in the placental mammal tree[18,22,23,31] (Fig. 1). Although this case is rather tenuous[31] from an anatomical standpoint, the phylogenetic isolation of edentates is consistent with their long historical isolation in South America.[22] Protein sequences[18] also suggest a very remote divergence point for edentates. Much greater correspondence is seen in the case of the ungulate branching sequence. For example, results from a cytochrome *b* analysis[19] are compatible with the morphological perspective[29] that cetaceans and artiodactyls are segregated from perissodactyls and proboscideans.

But perhaps the most encouraging result to date is the agreement among comprehensive studies of molecules and morphology in certain orders. Such correspondence has been known for some time in primates, where a comparatively large amount of protein sequence data[17,18] is available. In addition to the cytochrome *b* work,[19] mitochondrial (mt) DNA sequences for a sampling of artiodactyls[5] shows a high compliance with the basic outlines of artiodactyl phylogeny supported by morphological data.[36] Rapid cladogenesis and early splitting events are best detected with long sequences of highly conserved genes (for example, 18S and 28S ribosomal RNA genes).[5,37] Transversions (changes from purine to pyrimidines or vice versa) for these genes seem to have a linear relationship with time for at least 75 million years before present.[5] By contrast, transitions (purine-purine or pyrimidine-pyrimidine changes) in mitochondrial genes are rapid, probably more susceptible to parallelisms, and therefore less reliable as indicators of higher level (superordinal) relationships.[5,37] Mitochondrial studies on artiodactyls, though confined to a single order of mammals, are comprehensive enough to suggest that deeper phylogenetic structure might be attained when conservative mitochondrial and nuclear genes are more widely sampled.

Origins of Mammalian Orders

The current multidisciplinary flavor of mammalian systematics has also taken researchers down long-neglected pathways. The very integrity of certain mammalian orders has been subject to assault. New findings on brain-visual organization have inspired a claim that not all bats (order Chiroptera) share a unique common ancestor.[38] The bat diphyly argument (which would probably have caused G. G. Simpson [ref. 10, p. 179] much agitation), has fomented a passionate debate,[39,40] conducted mainly on the grounds of conflicting anatomical evidence. Recent molecular and some new morphological data are clearly more consistent with the traditional view of a single origin for bats (Fig. 3).

Another unexpected attack has been made on traditional concepts of rodent origins. Central to theories concerning rodent phylogeny[12] is the notion that the separation of the hystricognaths from other rodents represents a very early divergence event. Recently, this distinction for hystricognaths has been carried to a new extreme. It has been proposed[41] that the removal of the hystricognath *Cavia porcellus* (the guinea pig) from the Rodentia resolves some of the paradoxical qualities of gene evolution in this species, including rampant parallelisms with molecular changes observed in birds and drastically accelerated rates of gene evolution. The idea is provocative, but it minimizes the impressive morphological evidence for the monophyly of Rodentia inclusive of hystricognaths.[9,10,42] A recent analysis of 12S rRNA genes[43] showed that transversions applied to an unrooted network of four eutherian mammal taxa support rodent monophyly. In the same study,[43] a network rooted with a nonmammal outgroup (chicken) disassembled Rodentia. But this solution required only one mutation less than the tree opting for rodent monophyly. The higher statistical support for rodent monophyly in the unrooted analysis suggests that a bird taxon is too remotely related to be a reference outgroup. An alternative (perhaps a marsupial or an edentate) would provide a better test. In any case, the molecular sampling is as yet too sketchy to resolve the issue.

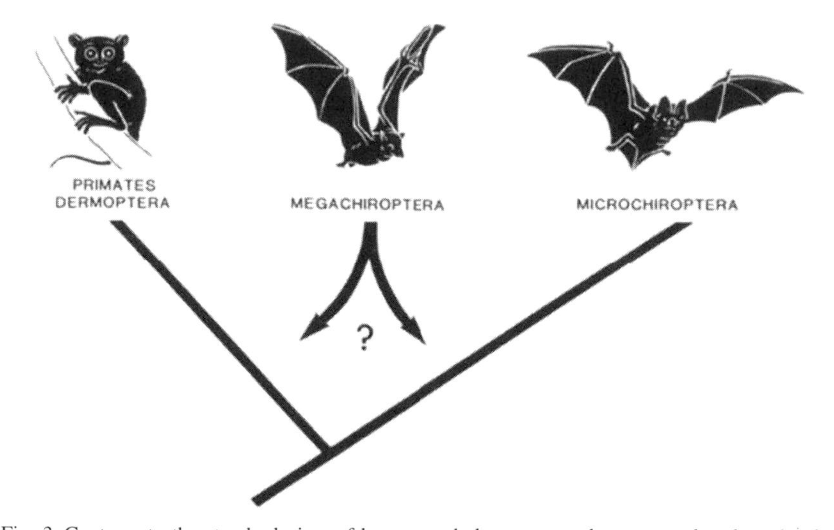

Fig. 3 Contrary to the standard view of bat monophyly, some workers argue that the suborder Megachiroptera (which includes the large Old World fruit bats) is more closely related, by virtue of specialized visual neural pathways in the brain, to primates (and possibly also to flying lemurs and tree-shrews) than to the echolocating Microchiroptera.[38] Under this view, myriad similarities in wing structure and the unique capabilities for powered flight evolved independently in the two bat groups, whereas the brain-visual traits, being less obviously adaptive, are clearly indicative of a true genealogical relationship.[39] Opponents stress that this argument has been tied to a naive assumption linking the obvious adaptation of a feature with a high likelihood that it is not good evidence for monophyly.[40] Such an assumption, they argue, would effectively eliminate many essential characters (including the notochord in chordates, or mammary glands in mammals) from use in biological classifications. From the molecular side, restriction-map analysis of the ribosomal DNA gene complex[20] failed to clearly indicate either bat monophyly or diphyly. But amino-acid sequences relating to three globin genes,[86] and DNA sequence changes in 12S rRNA,[87,88] cytochrome oxidase subunit I (COI)[87] and cytochrome oxidase subunit II (COII)[44] all strongly support a single origin for bats. New morphological evidence on the intricate and unique nature of innervation of appendicular and wing musculature[89] further strengthens the case for chiropteran monophyly. Bats are generally associated with primates, flying lemurs and tree-shrews in the superorder Archonta.[9,29,90] Finally, recent neurological work[91] suggests that brain-visual traits may be misleading evidence for bat diphyly.

New Fossil Evidence

Given the energies invested in molecular research, it might be assumed that the repository of more traditional (that is, anatomical and paleontological) evidence has been depleted. This is not the case. Recent and extraordinary discoveries of fossils continue to shake the tree. Prominent among these is the disclosure that a group of early primate-like taxa, once known primarily from dentitions, have skeletons that show bizarre adaptations similar to the gliding dermopterans.[6] The discovery has reopened the whole question of primate affinities with dermopterans and other putatively related orders. For some time,[9] the monophyly of a superordinal category, the Archonta, has been recognized to include the primates, dermopterans, tree-shrews (order Scandentia) and bats. One concept favors the close affinity between dermopterans and bats on one hand and tree-shrews and primates on the other.[29–31] Nucleotide sequences of cytochrome oxidase (COII) genes[44] in a sampling of archontans suggests that tree-shrews are no more closely related to primates than are flying lemurs. Interestingly, the COII study weakens the case for the Archonta by isolating the monophyletic Chiroptera from the other archontan groups.

Another remarkable advance on the paleontological front is the discovery of evidence in Eocene whales[7] that elucidates the long-suspected connection between cetaceans and early artiodactyl relatives. Vestigial hindlimbs in the Eocene *Basilosaurus* distally show a paraxonic arrangement,[7] a condition where digits symmetrically extend about a plane between digits III and IV (digit II is essentially lacking in *Basilosaurus*). This paraxonic arrangement bears striking resemblance to that in an archaic group of ungulates, the mesonychid condylarths, as well as the even-toed (paraxonic) artiodactyls.[7,45] Mesonychid condylarths are, moreover, very like primitive whales (archaeocetes) in skull and dental structure. The picture resolves to a branching sequence wherein whales and artiodactyls are closely related by virtue of their close affinities with different groups of mesonychids. Immunodiffusion, protein sequence[46] and cytochrome *b* sequence[19] results suggest an even more intimate derivation of cetaceans somewhere within Artiodactyla. Nonetheless, in a general sense, the cetacean-artiodactyl association so well illuminated by the fossil record is supported by the new molecular data. The impact of paleontology here and in other cases should not be underestimated. Fossils will continue to reveal crucial branching points in the deep structure of mammalian phylogeny, patterns that might never be retrieved from the molecular and anatomical structure of extant taxa alone.[47,48]

The Insectivore Problem

The importance of the paleontological database is also demonstrated by the reorganization and clarification of the Insectivora, a group central to the question of higher relationships among placental mammals. As originally conceived, Insectivora comprised hedgehogs, moles, shrews, tenrecs, solenodons, golden moles and a diverse suite of fossils in the suborder Lipotyphla along with tree-shrews (Scandentia) and elephant-shrews (Macroscelidea) in the suborder Menotyphla. For some time, however, Menotyphla has been rendered obsolescent[9] by systematic fission. As noted above, tree-shrews have been drawn in with primates, bats and flying lemurs.[9,29] Elephant-shrews have been moved toward the Glires complex,[29] but recent molecular and morphological studies spread them over the mammalian map, providing contradictory evidence for affinities with various archontans[49] or primitive ungulates[50] or for a remote branching position at the base of the eutherian tree.[51]

This distillation notwithstanding, Insectivora still presented many problems, largely created by the association of lipotyphlans with palaeoryctids, leptictids and other fossil taxa of primitive skeletal structure. As the fossil evidence accumulated, Insectivora simply became a receptacle for any taxon that defied reasonable allocation elsewhere.[52] A reassessment of well-preserved fossils[53–56] has etched a clearer distinction between archaic insectivorans and those of a more modern stamp. There is now some case for the monophyly of a more restricted Lipotyphla.[53,56] The fossil leptictids are clearly excluded from this assemblage, although their skull structure suggests they may be the nearest relatives of the lipotyphlans.[56] Cretaceous palaeoryctids are of extremely primitive morphology[55] and probably diverged much earlier in the eutherian branching sequence[56] (Fig. 1).

Carnivores and the Origins of Pinnipeds

Problems of insectivore phylogeny readily merge with questions concerning carnivore origins. The sharp, high-crowned teeth of early fossil carnivores evoke comparisons with archaic insectivorans, particularly palaeoryctids.[56] Unfortunately, the relations of carnivores with other extant orders remain enigmatic. This uncertainty also applies to Simpson's proposal for a carnivore-ungulate connection in the cohort Ferungulata (Box 1). A recent study of gene sequences[35] in a few mammalian species suggests a closer affinity between carnivores and artiodactyls than between the former and either primates or rodents. Nonetheless, taxonomic coverage for gene sequence comparisons relevant to this issue must be greatly expanded.

Within Carnivora, the picture is more satisfactorily resolved both from morphological[12,57] and from molecular[58] perspectives, and it bears on an important problem in mammalian phylogeny. Although there is resounding consensus that pinnipeds (seals, sea lions and walruses) are closely related to terrestrial carnivores, there is much controversy over the nature of this linkage. Throughout the twentieth century, pinnipeds were most consistently viewed as monophyletic,[59] but a diphyletic origin for pinnipeds has also been proposed.[60–62] The diphyletic argument calls for the alliance of odobenids (walruses) and otariids (sea lions) within an arctoid clade somewhere near ursids (bears) and a separate connection between the phocids (seals) and mustelids (weasels, skunks and kin). But recent morphological work[63] revives the monophyletic argument with an added refinement suggesting a close odobenid-phocid affinity. Molecular studies emphatically support a case for pinniped monophyly.[63–66]

Marsupials and Monotremes

The problems noted so far apply to the diverse orders assigned to the infraorder Eutheria, the placental mammals. The other major radiation of living mammals, represented by the Australian and New World marsupials, has also attracted much attention, although developments on the molecular front are less dramatic. Recent proposals[67–69] favor basic divisions that roughly parallel the pattern of biogeographic separation between Australian and New World marsupials. Intriguingly, some of these classifications[67,68] converge on the notion that the New World microbiotheres (whose sole living member is *Dromiciops,* the monitos del monte) represents the first twig of the branch leading to the Australian radiation. Indeed, the discovery of fossil marsupials in Antarctica[70] conforms with the long-standing view that New World and Australian marsupials diverged from a Mesozoic group distributed broadly over then connected landmasses of South America, Antarctica and Australia.

So far, the molecular input on questions of marsupial relationships primarily concern immunological[71,72] and DNA-DNA hybridization[73–75] studies. Some immunological comparisons allegedly support the close affinity between Australian marsupials and microbiotheres (see comment in ref. 67, p. 457), but this is not clearly indicated by DNA-DNA hybridization results.[75] Protein and gene sequence studies[51,58] merely corroborate the expected aggregation of a few representative marsupials (for example, opossums and kangaroos). The nascent area of gene sequence studies in marsupials includes findings[76] on 12S rRNA and cytochrome *b* which link the extinct *Thylacinus* (the marsupial wolf) with the Australian *Dasyurus* (tiger cat) and *Sarcophilus* (Tasmanian devil), an association compatible with some current classifications.[67]

Until recently, the fossil record of the presumably oldest but least diverse of the major mammalian clades, the monotremes, was limited to scattered and partial remains in the late Cenozoic of Australia. That the fossil record still offers many surprises is fully demonstrated by findings in Australia of an exquisite Miocene platypus skull[77] and a Cretaceous jaw and dentition also claimed to represent an early platypus.[78] These specimens not only provide new data on the evolution of monotreme characters, but also indicate a closer relationship than was traditionally recognized[78] between monotremes and the other extant mammalian clades relative to various Mesozoic mammals.

Search for Congruence

The above cases show a mixed picture of an attack, with a diverse arsenal, on a very tough problem. The goal here is, however, important. A better map of mammalian history provides a critical framework for wrestling with the problems of genic and taxic rates of change, biogeographic deployment and the connection between phylogeny and character and gene transformation.[31] It is, of course, difficult to measure progress in fashioning this framework. For one thing, agreement among different data sets does not guarantee that we are on the right track. Moreover, the failure to achieve congruence may simply mean that not all approaches are equally effective in revealing phylogeny. Critics[23,53] note that an abiding fascination with tooth structure seems responsible for some very confusing and inefficient descriptions of higher mammal relationships. Dental traits seem prone to rapid and parallel change. Likewise, certain genes and proteins seem ineffective in retrieving ancient splitting events[79] because of their tendency for multiple changes at a single locus after the split.[19] Indeed, a recent analysis[28] suggests that amino-acid replacements in certain proteins produced trees that were in fact less efficient in describing the variation than trees produced from a random mix of characters and taxa.

These shortcomings, and the conflicting results they produce, make one look more carefully at the nature of the basic evidence. Nevertheless, in the absence of knowledge of the true phylogeny, a measure of congruence among data sets must be heeded. This is why congruent patterns produced from molecular data, morphological traits of various systems, and the fossil record are greeted with some feeling of comfort.[37] If a signal for phylogeny is there, it should be seen from several different directions.

Prospects

At present, the study of higher mammalian phylogeny is so carried by its own momentum that it seems presumptuous to consider where and how new insights will arrive. There is an obvious, auspicious direction to the study of gene sequences, with merely the threshold attained for survey of the more conserved, and possibly more illuminating, genes.[80] Another promising route here concerns the experimental disclosure of genic influence on mammalian development and morphology,[81] as a basis for further insights on character transformation in phylogeny. Likewise, the fossil record of mammals seems capable of unabated offerings of new evidence.

Where do the sort of "middle of the road" studies of morphology in living taxa fit? There is a tendency to observe that such approaches, despite their long cultivation, have not yielded their share of productive results.[82] But if this is true, why is there startling new phylogenetic information on brain-visual systems,[38] fetal development[83] and minutiae of the middle ear?[29] In reality, the traditional study of morphology has little to do with the new cladistic methods for seeking homology, coding data, and searching for trees that today bring together morphological and molecular research. One can recast morphological studies, as well as molecular systematics,[82] as a new age of exploration.

Will all this interaction eventually produce a precise and elegant picture of higher mammal phylogeny? There is no reason to expect that all the basic lines among, for instance, the placental mammal orders will be teased apart. And there may be some truth in Simpson's remarks[31,84] that the great burst of radiating mammalian orders more than 65 million years ago will not completely yield to these probes. Simpson neither encouraged further research on the subject nor did he seem very enchanted[85] with one of the methods (namely, cladistics) now adopted to carry it out. Yet enough insight and collaboration has come from this venture, that its momentum seems justified. It is hard to imagine that even Simpson himself wouldn't have agreed.

MICHAEL J. NOVACEK is at the American Museum of Natural History, New York, New York, 10024, USA.

References

1. Hennig, W. *Phylogenetic Systematics* (Univ. Illinois Press, Urbana, 1966).
2. Nelson, G. & Platnick, N. I. *Systematics and Biogeography: Cladistics and Vicariance* (Columbia Univ. Press, New York, 1981).
3. Hillis, D. M. & Dixon, M. T. in *The Hierarchy of Life* (eds. Fernholm, B., Bremer, K. & Jörnvall, H.) 355–367 (Elsevier, Amsterdam, 1989).
4. Miyamoto, M. M. *et al. Proc. natn. Acad. Sci. USA* **85,** 7627–7631 (1988).
5. Miyamoto, M. M. & Boyle, S. M. in *The Hierarchy of Life* (eds. Fernholm, B., Bremer, K. & Jörnvall, H.) 437–450 (Elsevier, Amsterdam, 1989).
6. Beard, K. C. *Nature* **345,** 340–341 (1990).
7. Gingerich, P. D., Smith, B. H. & Simons, E. L., *Science* **249,** 154–157 (1990).
8. Court, N. & Jaeger, J. J. *C. r. hebd. Séanc. Acad. Sci., Paris.* **3129,** 559–565 (1991).
9. Gregory, W. K. *Bull. Am. Mus. nat. Hist.* **27,** 1–542 (1910).
10. Simpson, G. G. *Bull. Am. Mus. nat. Hist.* **85,** 1–350 (1945).
11. Simpson, G. G. *Tempo and Mode in Evolution* (Columbia Univ. Press, New York, 1944).
12. Carroll, R. L. *Vertebrate Paleontology and Evolution* (Freeman, New York, 1988).

13. Avery, O. T., MacLeod, C. M. & McCarthy, M. *J. exp. Med.* **79,** 137–158 (1944).
14. Zuckerkandl, E. & Pauling, L. in *Horizons in Biochemistry* (eds. Kash, M. & Pullman, B.) 189–225 (Academic, New York, 1962).
15. Sarich, V. M. *Syst. Zool.* **18,** 286–295 (1969).
16. Goodman, M. & Moore, G. W. *Syst. Zool.* **20,** 19–62 (1971).
17. Goodman, M., Moore, G. W. & Matsuda, G. *Nature* **253,** 603–608 (1975).
18. Miyamoto, M. M. & Goodman, M. *Syst. Zool.* **35,** 230–240 (1986).
19. Irwin, D., Kocher, T. D. & Wilson, A. C. *J. molec. Evol.* **32,** 128–144 (1991).
20. Baker, R. J., Honeycutt, R. L. & Van Den Bussche, R. A. *Bull. Am. Mus. nat. Hist.* **206,** 42–53 (1991).
21. White, T. J., Arnheim, N. & Erlich, H. A. *Trends Genet.* **5,** 185–189 (1989).
22. McKenna, M. C. in *Phylogeny of the Primates* (eds. Luckett, W. P. & Szalay, F. S.) 21–46 (Plenum, New York and London, 1975).
23. Novacek, M. J. in *Macromolecular Sequences in Systematic and Evolutionary Biology* (ed. Goodman, M.) 3–41 (Plenum, New York, 1982).
24. Swofford, D. L. *PAUP: Phylogenetic Analysis Using Parsimony,* Version 3.0 (Illinois Natural History Survey, Champaign, IL, 1989).
25. Farris, J. S. *HENNIG 86,* Version 1.5 (Distributed by the author: 41 Admiral Street, Port Jefferson Station, New York, 1988).
26. Waterman, M. S. *Bull. math. Biol.* **46,** 473 (1984).
27. Wheeler, W. C. & Honeycutt, R. L. *Molec. Biol. Evol.* **5,** 90–96 (1988).
28. Faith, D. P. & Cranston, P. S. *Cladistics* **7,** 1–28 (1991).
29. Novacek, M. J., Wyss, A. R. & McKenna, M. C. in *The Phylogeny and Classification of the Tetrapods* Vol. 2 (ed. Benton, M. J.) 31–71 (Clarendon, Oxford, 1988).
30. McKenna, M. C. in *Molecules and Morphology in Evolution: Conflict or Compromise?* (ed. Patterson, C.) 55–93 (Cambridge Univ. Press, Cambridge, 1987).
31. Novacek, M. J. in *Current Mammalogy* Vol. 2 (ed. Genoways, H. H.) 507–543 (Plenum, New York, 1990).
32. Prothero, D. R., Manning, E. M. & Fischer, M. in *The Phylogeny and Classification of the Tetrapods* Vol. 2 (ed. Benton, M. J.) 201–234 (Clarendon, Oxford, 1988).
33. Owen, R. *J. geol. Soc. Lond.* **4,** 103–141 (1848).
34. Sarich, V. M. in *Evolutionary Relationships among Rodents* (eds. Luckett, W. P. & Hartenberger, J.-L.) 423–452 (Plenum, New York and London, 1985).
35. Li, W.-H., Guoy, M., Sharp, P. M., O'Huigin, C. & Yang, Y.-Y. *Proc. natn. Acad. Sci. USA* **87,** 6703–6707 (1990).
36. Janis, C. M. & Scott, K. M. *Am. Mus. Novit.* **2893,** 1–85 (1987).
37. Kraus, F. & Miyamoto, M. M. *Syst. Zool.* **40,** 117–130 (1991).
38. Pettigrew, J. D. *Science* **231,** 1304–1306 (1986).
39. Pettigrew, J. D. *Syst. Zool.* **40,** 199–216 (1991).
40. Baker, R. J., Novacek, M. J. & Simmons, N. B. *Syst. Zool.* **40,** 216–231 (1991).
41. Graur, D., Hide, W. A. & Li, W.-H. *Nature* **351,** 649–652 (1991).
42. Luckett, W. P. & Hartenberger, J.-L. in *Evolutionary Relationships among Rodents* (eds. Luckett, W. P. & Hartenberger, J.-L.) 685–712 (Plenum, New York and London, 1985).
43. Allard, M. W., Miyamoto, M. M. & Honeycutt, R. L. *Nature* **353,** 610–611 (1991).
44. Adkins, R. M. & Honeycutt, R. L. *Proc. natn. Acad. Sci. USA* **88,** 10317–10321 (1991).
45. Wyss, A. R. *Nature* **347,** 428–429 (1990).

46. Goodman, M., Czelusniak, J. & Beeber, J. E. *Cladistics* **1,** 171–185 (1985).
47. Gauthier, J., Kluge, A. G. & Rowe, T. *Cladistics* **4,** 105–209 (1988).
48. Novacek, M. J. *Syst. Biol.* (in the press).
49. Woodall, P. F. *J. submicrosc. Cytol. Pathol.* **23,** 47–58 (1991).
50. Simmons, E. L., Holroyd, P. A. & Bown, T. M. *Proc. natn. Acad. Sci. USA* **88,** 9734–9737 (1991).
51. McKenna, M. C. *Acta zool. fenn.* **191** (in the press).
52. Butler, P. M. in *Studies in Vertebrate Evolution* (eds. Joysey, K. A. & Kemp, T. S.) 253–265 (Winchester, New York, 1972).
53. McDowell, S. B. *Bull. Am. Mus. nat. Hist.* **115,** 113–214 (1958).
54. Butler, P. M. *Proc. zool. Soc. Lond.* **118,** 453–481 (1956).
55. Kielan-Jaworowska, Z. *Palaeont, pol.* **33,** 5–13 (1975).
56. Novacek, M. J. *Bull. Am. Mus. nat. Hist.* **183,** 1–111 (1986).
57. Flynn, J. J., Neff, N. A. & Tedford, R. H. in *The Phylogeny and Classification of the Tetrapods* Vol. 2 (ed. Benton, M. J.) 73–115 (Clarendon, Oxford, 1988).
58. Czelusniak, J., Goodman, M., Moncrief, N. D. & Kehoe, S. M. *Meth. Enzym.* **183,** 601–615 (1990).
59. Flower, W. H. *Proc. zool. Soc. Lond.* **1869,** 4–37 (1869).
60. Mivart, St. G. *Proc. zool. Soc. Lond.* **1885,** 484–500 (1885).
61. McLaren, I. A. *Syst. Zool.* **9,** 18–28 (1960).
62. Tedford, R. H. *Syst. Zool.* **25,** 363–374 (1976).
63. Wyss, A. R. *Am. Mus. nat. Hist. Novit.* **2871,** 1–13 (1987).
64. Arnason, U. *Hereditas* **76,** 179–226 (1974).
65. Goodman, M., Romero-Herrera, A. E., Dene, H., Czelusniak, J. & Tashian, R. E. in *Macromolecular Sequences in Systematics and Evolution* (ed. Goodman, M.) 115–191 (Plenum, New York, 1982).
66. Romero-Herrera, A. E., Lehmann, H., Joysey, K. A. & Friday, A. E. *Phil. Trans. R. Soc.* B **283,** 61–163 (1978).
67. Marshall, L. G., Case, J. A. & Woodburne, M. O. in *Current Mammalogy* Vol. 2 (ed. Genoways, H. H.) 433–505 (Plenum, New York, 1990).
68. Aplin, K. & Archer, M. in *Possums and Opossums: Studies in Evolution* (ed. Archer, M.) xv–lxxii (R. zool. Soc. NSW, Sydney, 1987).
69. Clemens, W. A., Richardson, B. J. & Baverstock, P. R. in *Fauna of Australia* Vol. 1B (eds. Walton, D. W. & Richardson, B. J.) 527–548 (Australian Government Publishing Service, Canberra, 1989).
70. Woodburne, M. O. & Zinsmeister, W. J. *J. Paleont.* **58,** 913–948 (1984).
71. Kirsch, J. A. W. *Austr. J. Zool.* **52** (suppl.), 1–152 (1977).
72. Baverstock, P. R., Birrell, J. & Krieg, M. in *Possums and Opossums: Studies in Evolution* (ed. Archer, M.) 229–234 (R. zool. Soc. NSW, Sydney, 1987).
73. Kirsch, J. A. W., Krajewski, C., Springer, M. S. & Archer, M. *Aust. J. Zool.* **38,** 673–696 (1990).
74. Springer, M. S., Kirsch, J. A. W., Aplin, K. & Flannery, T. *J. molec. Evol.* **30,** 298–311 (1990).
75. Westerman, M. & Edwards, D. *Aust. J. Zool.* **39,** 123–130 (1991).
76. Thomas, R. H., Schaffner, W., Wilson, A. C., Pääbo, S. *Nature* **340,** 465–467 (1989).
77. Archer, M., Hand, S. & Godthelp, H. *Uncovering Australia's Dreamtime* (Surrey Beatty, Chipping Norton, New South Wales, 1986).

78. Archer, M., Flannery, T. F., Ritchie, A. & Molnar, R. *Nature* **318,** 363–366 (1985).
79. Wyss, A. R., Novacek, M. J. & McKenna, M. C. *Molec. Biol. Evol.* **4,** 99–116 (1987).
80. Holmes, E. C. *J. molec. Evol.* **33,** 209–215 (1991).
81. Kessler, M. & Gruss, P. *Cell* **67,** 89–104 (1991).
82. Goodman, M. in *The Hierarchy of Life* (eds. Fernholm, B., Bremer, K. & Jörnvall, H.) 43–61 (Elsevier, Amsterdam, 1989).
83. Luckett, W. P. in *Evolutionary Relationships among Rodents* (eds. Luckett, W. P. & Hartenberger, J.-L.) 227–276 (Plenum, New York and London, 1985).
84. Simpson, G. G. *Proc. Am. Phil. Soc.* **122,** 318–328 (1978).
85. Simpson, G. G. in *Phylogeny of the Primates* (eds. Luckett, W. P. & Szalay, F. S.) 3–19 (Plenum, New York and London, 1975).
86. Czelusniak, J. *et al.* in *Current Mammalogy* Vol. 2 (ed. Genoways, H. H.) 545–572 (Plenum, New York, 1990).
87. Mindell, D. P., Dick, C. W. & Baker, R. J. *Proc. natn. Acad. Sci. USA* **88,** 10322–10326 (1991).
88. Ammerman, L. & Hillis, D. M. *Am. Zool.* **30,** 50 (1990).
89. Thewisson, J. G. M. & Babcock, S. K. *Science* **251,** 934–936 (1991).
90. Szalay, F. S. in *Major Patterns in Vertebrate Evolution* (eds. Hecht, M. K., Goody, P. C. & Hecht, B. M.) 315–374 (Plenum, New York, 1977).
91. Thiele, A., Vogelsang, M., Hoffmann, K.-P. *J. comp. Neur.* **314,** 671–683 (1991).

Acknowledgments

I thank E. Heck for preparation of figures and N. Simmons, M. McKenna, M. Norell and P. Vrana for discussion.

PART 5

CLIMBING DOWN: THE HISTORY OF PRIMATES

In 1990, R. D. Martin published a weighty volume on the evolution of the primates.[1] No sooner had he done this than a number of remarkable fossil discoveries came to light, findings that shook the base of the primate tree as traditionally reconstructed. Specifically, the status of extinct creatures called plesiadapiforms as primates was questioned—it was suggested that they were more closely related to the so-called flying lemur or colugo (order Dermoptera) instead.[2–6] Of particular recent significance has been the discovery of very early anthropoids (the clade that includes monkeys, apes and humans) from China and elsewhere in Asia.[7–11]

Following these discoveries I asked Martin if he'd write an article for *Nature* that would serve as a revised introduction to his book. This he did, and it is included as the first of the three contributions to this, the last part of this book. But he also raised a larger question, about the patchiness of the fossil record, which repeatedly leads us to underestimate times of origin of groups within the primate tree.[12,13] This lesson, of course, is applicable to the fossil record generally, not just to primates.[14]

Several episodes in the history of primates have attracted controversy—and new insight—in recent years. Not surprisingly, these questions have centered around the origins of humanity. Cladistics has helped to frame the questions precisely, if not always to solve them irrefutably. Then again, cladistic solutions are never anything except hypotheses, so this is perhaps to be expected.

In the second contributon to this section, Peter Andrews takes us back to the Miocene epoch, between around 10 and 20 million years ago, when a variety of apes lived in Africa and Eurasia. The question of which of these primates lies closest to the ancestries of humans and the (other) extant apes remains open: the absence of a fossil record for hominids between 10 and 5 million years ago remains, as Andrews notes, a "frustrating barrier"—and one that has not lifted since Andrews's review was published. References 15–17 deal with these questions in greater depth.

Light has been shed on the period between 4.5 and 3 million years ago with discoveries in East Africa of species assigned to *Ardipithecus ramidus, Australopithecus anamensis* and *Australopithecus afarensis* (see ref. 18 for a comprehensive, up-to-date and lavishly illustrated overview of hominid evolution). These creatures acquired the unique form of hominid bipedal locomotion but were otherwise rather apelike, retaining many adaptations that suggest tree-climbing ability. Notwithstanding the mystery of their history before 4.4 million years ago, the evolutionary picture seems relatively straightforward (so far).

In contrast, the interval between around 3 and 2 million years ago reveals a puzzling phylogenetic tangle, in which several species of hominid appeared: several kinds of the "robust australopithecine," *Paranthropus,* coexisted with *Australopithecus africanus* and several species usually assigned to *Homo.* Two of these, *Homo erectus* and *H. ergaster,* seem closely related to one another, and are recognizably human in their body shapes (see ref. 19 for a technical account of *H. erectus,* and ref. 20 for a highly readable account of the discovery and interpretation of a near-complete skeleton of *H. ergaster*).

Other species assigned to *Homo* seem less clear-cut. The arguments concerning the status of *H. habilis* and *H. rudolfensis* as members of our own genus cut to the quick of our understanding of generic limits in *Homo,* and more generally. It has been increasingly hard to accommodate these two species within *Homo,* and to differentiate them from apelike creatures such as *Australopithecus* and *Paranthropus.* This question is discussed by Bernard Wood in the final review in this volume—but is not resolved.

The debate about the origin of so-called anatomically modern humans *(Homo sapiens),* and the related question of our affinity with Neandertal man, is perhaps the most acrimonious of all, perhaps because it lies closest to home. It is also one area in which we at *Nature* have signally failed to commission a review article and have it pass through the peer-review process without mortal wounding. It seems to be virtually impossible to discuss the subject without raising irreconcilable passions. Given that no review of the subject appears here, it is perhaps inappropriate to go into details. Happily, the reader intent on discovering more can be referred to a plethora of excellent popular and semi-popular accounts[21-26] and some terser, professional texts.[27-28]

References

1. R. D. Martin. *Primate origins and evolution: A phylogenetic reconstruction.* London: Chapman and Hall, 1990.
2. K. C. Beard. Gliding behaviour and palaeoecology of the alleged primate family Paromomyidae (Mammalia, Dermoptera). *Nature* 345 (1990): 340–341.
3. R. F. Kay, R. W. Thorington and P. Houde. Eocene plesiadapiform shows affinities with flying lemurs not primates. *Nature* 345 (1990): 342–344.
4. R. D. E. MacPhee, ed. *Primates and their relatives in phylogenetic perspective.* New York: Plenum, 1993.
5. L. M. Van Valen. The origin of the plesiadapid primates and the nature of *Purgatorius. Evolutionary Monographs* 15 (1994): 1–79.
6. K. C. Beard and J.-W. Wang. The first Asian plesiadapoids (Mammalia: Primatomorpha). *Annals of the Carnegie Museum* 64 (1995): 1–33.
7. K. C. Beard, T. Qi, M. R. Dawson, B. Wang and C. Li. A diverse new primate fauna from middle Eocene fissure-fillings in southeastern China. *Nature* 368 (1994): 604–609.
8. K. C. Beard, Y. Tong, M. R. Dawson, J. Wang and X. Huang. Earliest complete dentition of an anthropoid primate from the late middle Eocene of Shanxi Province, China. *Science* 272 (1996): 82–85.
9. R. D. E. MacPhee, K. C. Beard and T. Qi. Significance of primate petrosal from Middle Eocene fissure-fillings at Shanghuang, Jiangsu Province, People's Republic of China. *Journal of Human Evolution* 29 (1995): 501–514.
10. S. Ducrocq and J.-J. Jaeger. Anthropoïdes: La piste asiatique. *La Recherche* 306 (1998): 45–47.
11. J.-J. Jaeger, Y. Chaimanee and S. Ducrocq. Origin and evolution of Asian hominoid primates: Paleontological data versus molecular data. *Comptes Rendus de l'Académie des Sciences* 321 (1998): 73–78.
12. P. D. Gingerich and M. D. Uhen. Time of origin of primates. *Journal of Human Evolution* 27 (1994): 443–445.
13. U. Arnason, A. Gullberg, A. Janke and X.-F. Xu. Pattern and timing of evolutionary divergences between hominoids based on analyses of complete mtDNAs. *Journal of Molecular Evolution* 43 (1996): 650–661.
14. M. A. Steel, A. C. Cooper and D. Penny. Confidence intervals for the divergence time of two clades. *Systematic Biology* 45 (1996): 127–134.
15. D. Pilbeam. Genetic and morphological records of the Hominoidea and hominid origins: A synthesis. *Molecular Phylogenetics and Evolution* 5 (1996): 155–168.
16. D. Begun. Relations among the great apes and humans: New interpretations based on the fossil great ape *Dryopithecus. Yearbook of Physical Anthropology* 37 (1994): 11–63.
17. P. Andrews, T. Harrison, E. Delson, R. Bernor and L. Martin. Distribution and

biochronology of European and Southwest Asian Miocene catarrhines. In R. Bernor, V. Fahlbusch and H. W. Mittmann, eds., *The evolution of western Eurasian Neogene mammal faunas,* 168–207. New York: Columbia University Press, 1996.

18. D. Johanson and B. Edgar. *From Lucy to language.* New York: Simon and Schuster, 1996.
19. G. P. Rightmire. *The evolution of* Homo erectus: *Comparative anatomical studies of an extinct human species.* Cambridge: Cambridge University Press, 1990.
20. A. Walker and P. Shipman. *The wisdom of bones: In search of human origins.* New York: Alfred A. Knopf, 1996.
21. R. Foley. *Humans before humanity.* Oxford: Blackwell, 1995.
22. J. Shreeve. *The Neandertal enigma: Solving the mystery of modern human origins.* New York: William Morrow, 1995.
23. C. Stringer and C. Gamble. *In search of the Neanderthals: Solving the puzzle of human origins.* London: Thames and Hudson, 1993.
24. E. Trinkaus and P. Shipman. *The Neandertals: Changing the image of mankind.* London: Jonathan Cape, 1993.
25. R. Lewin. *Bones of contention.* London: Penguin, 1987.
26. S. Stringer and R. McKie. *African exodus: The origins of modern humanity.* London: Jonathan Cape, 1996.
27. M. J. Aitken, C. B. Stringer and P. A. Mellars, eds. *The origin of modern humans and the impact of chronometric dating.* Princeton, NJ: Princeton University Press, 1993.
28. G. Bräuer and F. H. Smith, eds. *Continuity or replacement: controversies in* Homo sapiens *evolution.* Rotterdam: Balkema, 1992.

Robert D. Martin

PRIMATE ORIGINS: PLUGGING THE GAPS

Recent discoveries of fossil primate specimens have produced several surprises and challenged prevailing views of early primate evolution. Plesiadapiforms, long regarded as "archaic primates," may perhaps be linked to the peculiar colugos instead. Inferred relationships of the earliest known undoubted primates (adapids and omomyids) are in turmoil. Both groups have been proposed as sources for the simian primates. Although the origin of the simian primates is obscure, new fossil evidence could push it further back by at least 10 million years. Such uncertainties reflect the low sampling level of the primate fossil record, which can potentially also lead to underestimation of times of origin within the primate tree.

A picture of the earliest phases of primate evolution that seemed broadly acceptable just a decade ago has been increasingly challenged. Several recent discoveries of early fossil primates have opened up new possibilities for the pattern and timing of emergence of the main groups of primates between the late Cretaceous and the end of the Eocene. As part of the upheaval, the cohesion of various fossil groups as monophyletic units has also been questioned. Widespread acceptance of the essentially Paleocene Plesiadapiformes ("archaic primates") as a well-defined group of exclusive early relatives of primates has been shaken by new fossil evidence and analyses. Certain features have been interpreted as indications that some genera, at least, may instead be directly linked to the Dermoptera (gliding colugos, also called "flying lemurs"). New fossil specimens, especially postcranial parts, of early Tertiary Adapidae ("lemuroids") have increasingly confirmed that some European genera (*Adapis, Leptadapis* and relatives) belong to a distinct sideline relative to other European genera (for example *Europolemur, Pronycticebus*), which are morphologically closer to North American adapids (for example *Notharctus, Smilodectes*). A revived suggestion that simian primates (monkeys, apes and humans) may be derived from the Adapidae directly conflicts with an alternative interpretation that the early Tertiary Omomyidae ("tarsioids") are relatives of tarsiers and simians whereas adapids are specific allies of modern lemurs and lorises (strepsirhine primates). Yet the discovery of several skulls of the North American omomyid *Shoshonius* led to the opposing claim that this genus is in fact even more directly linked to tarsiers, at the same time indicating a much earlier date for tarsier origins. Finally, dental evidence of the new genus *Algeripithecus* from a North African site dating back to the middle or even early Eocene would increase the age of the earliest yet identified simians by a striking margin of at least 10 million years (Myr). Given all this new fossil material and the sometimes conflicting interpretations that have been made, can the outlines of an emerging revised view of early primate evolution be traced? Consideration of the new evidence in the light of exploratory estimates of the current sampling level of the primate fossil record (Box 1) suggests that we may still have much to learn about the early relationships of primates and that inferred times of origin are likely to be pushed back much further into the past.

In 1979, when the last comprehensive review of fossil primates[1] was published, it seemed that broad agreement might be emerging with respect to the early evolutionary history of the primates (Fig. 1). It was almost universally accepted that the primates originated toward the close of the Cretaceous, just over 65 Myr ago, and an overwhelming majority of primate paleontologists regarded the Plesiadapiformes as members of the first major adaptive radiation from ancestral primates. Although these "archaic primates" increasingly came to be seen as a side branch, with no

BOX 1 How significant are gaps in the primate fossil record?

The fossil record of primate evolution is obviously incomplete, despite major continuing achievements by field paleontologists. But how large are the gaps? Do we merely need to bridge over a few spaces here and there, or are there in fact yawning chasms? Our effectiveness in sampling past primate species has major implications for interpretations of primate evolution based on the known fossil record.

A crude estimate of our sampling level of the primate fossil record can be obtained relatively simply.[14,105] About 200 primate species exist today and it is widely accepted that the primates originated at least 65 Myr ago. To make a very rough calculation of the number of fossil primate species that have ever existed, it is necessary to make some assumption about the pattern of expansion of species numbers over time and to have an estimate of the average survival time for individual species. The simplest assumption for expansion of species numbers over time is a uniform rate of increase. If anything, this is likely to underestimate the numbers of extinct species, as it is commonly believed that the rate of speciation is initially high within a given group and then decelerates as available ecological niches become occupied. For fossil mammal species, several studies indicate that the average survival time is about 1 Myr.[105] Taking all of these elements together, it is possible to infer that about 6,500 extinct primate species preceded the modern fauna.

This estimate can be used to assess our sample of the primate fossil record. Just over a decade ago, one review recognized 64 species of archaic primates and 186 fossil primates of modern aspect.[1] These 250 species represent 3.8% of the estimated total of past primate species. If the calculation is restricted to primates of modern aspect, commonly thought to have emerged about 55 Myr ago, the sampling level (186 out of an estimated total of

5,500 species) declines to only 3.4%. The potential implications of such low sampling levels for interpretation of the primate fossil record can be illustrated with a simple tree (Fig. 2). Although random distribution of discovery of fossil species within the tree is in fact unlikely, this does not affect the two main points that can be recognized. First, times of origin for the group as a whole and for subgroups within the tree are likely to be considerably earlier than indicated by first known fossil representatives. It is common practice to date the time of origin of any group from just before the first known fossil representative, but this will obviously be seriously misleading at very low sampling levels. Indeed, some lineages within the tree may remain completely unrepresented in the known fossil record. This leads on to the second point that a very low sampling density may lead us to assign inappropriate ancestral positions to known fragmentary fossils.

Repeated simulations of the effects of different sampling levels for the tree shown in Fig. 2 yield average figures for the earliest fossil species that will be discovered within such a tree by chance and the correction factors required to infer the true time of origin of the group (Fig. 3). Preliminary calculations conducted by S. Tavaré (University of Southern California) have shown that the values derived by such simple simulation (Figs. 2, 3) are quite close to those that would be predicted using branching process models, although some minor correction may be necessary. Such a simulation approach is admittedly approximate and any real tree is likely to show significant departures from the simple model. In many cases, however, sampling of the fossil record has been even poorer than assumed in the model. For example, primates of modern aspect are particularly well known from the Eocene: 83 of the 186 species mentioned above are derived from that epoch. During this exceptionally warm period, primates, characteristically inhabitants of tropical or subtropical forests, temporarily expanded from the south into the higher latitudes of the northern continents, where the probabilities of fossilization and subsequent fossil discovery were markedly higher. Because of significant geographic expansion, there may well have been a major bulge in species numbers within the primate tree during the Eocene, so a uniform model would underestimate the total number of extinct primate species.

For illustration, the derived correction factors can be applied to the known primate fossil record. With a sampling level of 3.8%, it is necessary to add a correction factor of about 30% to the time of origin indicated by the earliest known species. Hence, if archaic primates are accepted as genuine primates, an inferred time of origin of 65 Myr ago based on the first known representative would be revised to 85 Myr, well back in the Cretaceous. If the calculation is restricted to primates of modern aspect, the sampling level falls to 3.4%, requiring a correction factor of 33%, and a time of origin of 55 Myr ago based on the first known representative would be revised to 72 Myr, also a Cretaceous date.

There is, however, a further reason why the calculations illustrated in Figs. 2 and 3 can be expected to underestimate the effects of a low sampling level for fossil primate species. An initial correction of the time of origin of primates of modern aspect based on the data in Fig. 3 yields a figure of 72 Myr. If, however, primates of modern aspect actually originated 72 Myr ago, rather than 55 Myr, the model in Fig. 2 yields a revised total of 7,200 extinct primate species. The sampling level for 186 known species correspondingly drops to 2.5%, which requires a correction factor of 40% instead of 33%, indicating an even earlier date of 77 Myr for the origin of primates. Serial recalculations of this kind eventually lead to a date of about 80 Myr for the origin of primates of modern aspect. A similar series of calculations can be applied to the simian primates, the earliest known fossil representative of which, until very recently, was dated at about 36 Myr, on the Eocene-Oligocene boundary. Application of a similar correction factor to this date would indicate an origin of simians at about 52 Myr, close to the Paleocene-Eocene boundary.

It should also be noted that such recalculation of times of origin within the primate tree would have secondary implications for calibration of phylogenetic trees based on biomolecular data. Calibration has often been based on the suspect procedure of taking the date of the first known representative of a given group as close to the time of origin. Quite aside from other problems associated with biomolecular trees, it might be advisable to add 40% to all dates of divergence based on calibration derived from properly identified fossil evidence.

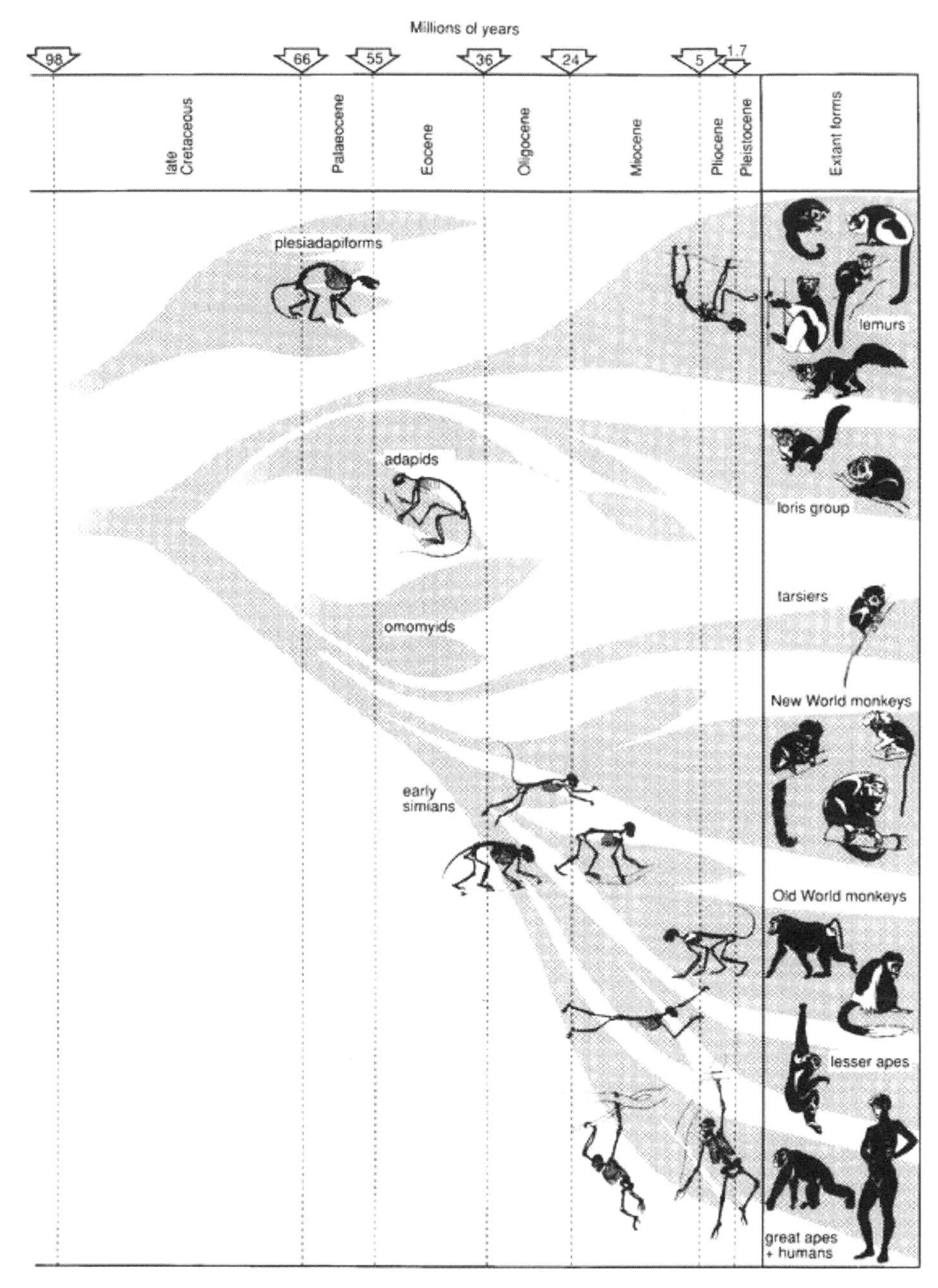

Fig. 1 Outline tree showing the main groups of living primates and one possible interpretation of their relationships. The tree includes fossil forms known from relatively complete skeletons and emphasizes remaining gaps (for example, no fossil lemurs other than recent subfossils; no fossil tarsiers apart from a single lower molar from the Miocene[117]). Four key groups are discussed with respect to the early evolution of primates: plesiadapiforms, adapids, omomyids, early simians. Because of relatively low fossil sampling levels (Box 1), the time of origin of primates of modern aspect has been set at about 80 Myr ago, rather than at the value of 65 Myr often cited. The time of origin of simian primates has similarly been increased to about 55 Myr. (Adapted from ref. 14; illustration by Lukrezia BielerBeerli.)

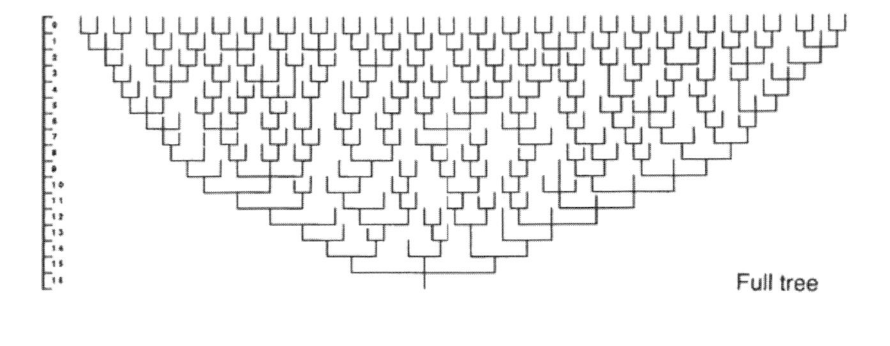

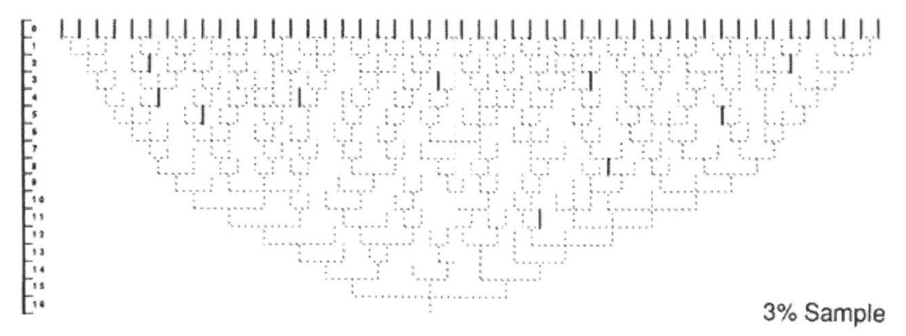

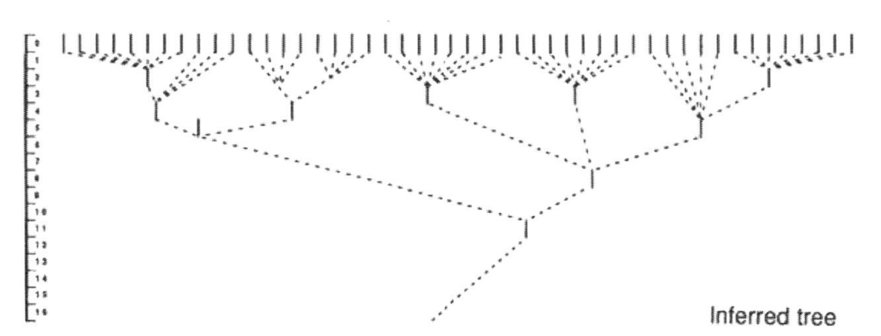

Fig. 2 *Full tree:* Model branching tree, with progressive expansion of number of species from 1 to 48 over a period of 16 Myr. Each species has been given a standard survival time of 1 Myr. *3% sample:* A typical example in which a 3% sample of fossil species (n = 10 species) has been randomly distributed throughout the tree. In this case, the earliest known species is dated at 11 Myr and underestimates the true time of origin of the group by 5 Myr (corresponding to a required correction factor of 45%). *Inferred tree:* An extreme illustration of the kind of tree that might be reconstructed if major gaps in the fossil record are not acknowledged. The time of origin of the group (based on the first known species) is likely to be seriously underestimated and fossil species may be forced into unrealistic ancestral positions.

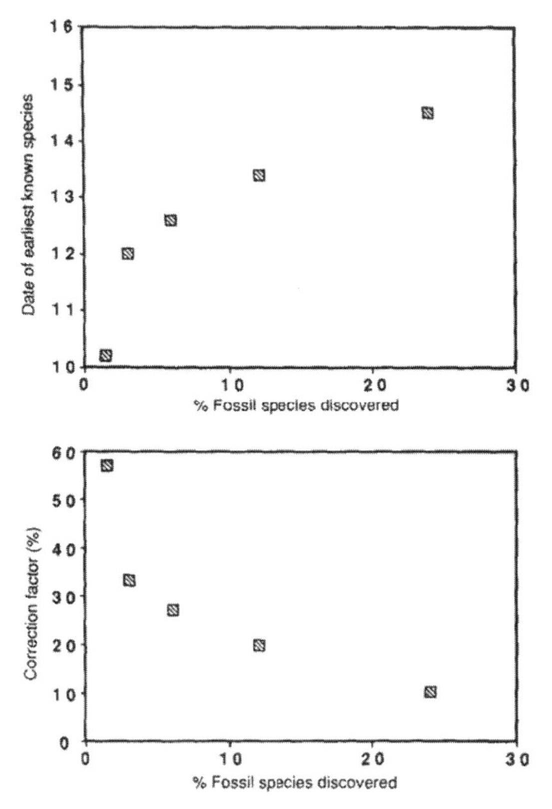

Fig. 3 *Upper curve:* Plots of average values obtained from 25 simple simulations of various sampling densities in the fossil record (1.5%, 3%, 6%, 12%, 24%), taking the tree in Fig. 2. Relative to a true origin at 16 Myr ago, the date of the earliest known species is close to 10 Myr at a sampling level of 1.5% and is below 15 Myr even with a sampling level of 24%. *Lower curve:* Correction factors (% value to be added) for such simple trees can be calculated for inferring the actual time of origin from the date of the first known fossil species at different sampling densities of the fossil record.

more than an early connection to the main radiation that led to modern primates, few doubted that the Plesiadapiformes were the exclusive sister group of primates of modern aspect ("euprimates"). There was also widespread, if not universal, agreement that the earliest known fossil primates of modern aspect from the Eocene fall into two distinct groups corresponding to the divergence of two main groups of living primates. The Adapidae were often regarded as relatives of the modern strepsirhine primates, whereas the Omomyidae were believed by many authors to be specifically allied to modern tarsiers. For the growing number of authors who accepted the existence of a haplorhine group containing tarsiers and simians, the omomyids almost automatically came to be seen as a key group close to the ancestry of all modern haplorhines.

Challenges to this picture of primate evolution have arisen in part because of new fossil material. But revised interpretations based on comparative studies and theoretical considerations have also contributed. In fact, exploration of the possible effects of a low sampling level on interpretations of the primate fossil record (Box 1) indicates that radical revision of prevailing views of primate evolution may be inevitable. Assessment of the possible effects of a low sampling level lead to certain predictions: (1) there is still enormous scope for the discovery of new fossil primate species; (2) with increased sampling of the primate fossil record, inferred times of origin are likely to increase, sometimes dramatically; (3) given the difficulties of interpreting fragmentary fossils, newly discovered specimens are likely to lead to repeated radical revision of inferred phylogenetic relationships; (4) the geographical context of primate evolution will also require reappraisal, as earlier times of origin necessarily imply different patterns of proximity between drifting continents.

Relationships of Plesiadapiformes

The revolutionary proposal that some or all plesiadapiforms may be directly related to the modern colugo *Cynocephalus* rather than to primates of modern aspect[2–4] rests on two main lines of evidence. The first stems from a study of postcranial features by Beard.[5] The most striking inference, stimulated by new skeletal material, is that the hand of early Eocene *Phenacolemur* and *Ignacius* (both members of the family Paromomyidae) resembled that of *Cynocephalus.* For both *Phenacolemur* and *Ignacius,* the intermediate digital bones (phalanges) of each finger were reported to be longer than the basal phalanges, as in *Cynocephalus,* possibly indicating adaptation to support webbing between the fingers as part of a patagium (gliding membrane).[2,6,7] Further, a cladistic analysis of the fossil evidence (based mainly on postcranial features) led to the conclusion that shared derived features link various plesiadapiforms more closely to colugos than to primates. Skeletons of *Plesiadapis,* however, do not seem to show any gliding adaptation. The second main source of evidence is a new, almost complete skull of *Ignacius.* Unlike primates of modern aspect, in which the auditory bulla grows out from the petrosal, the bulla of *Ignacius* can be confidently interpreted as the product of a separate entotympanic element (Fig. 4). As the presence of a petrosal bulla is generally regarded as a defining feature of primates, often stated to apply to

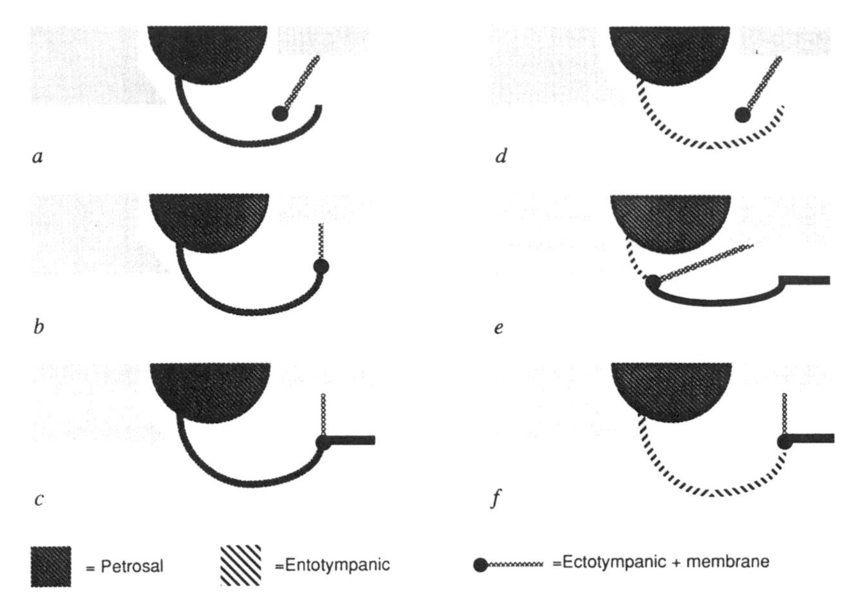

Fig. 4 Schematic transverse sections of the auditory bulla in primates *(left)* and non-primates *(right)*. In each case, the skull midline is on the left and the outer skull margin is on the right. In all living primates, the bulla wall is formed from the petrosal, but the relationship between the ectotympanic ring and the bulla varies between groups: enclosed in lemurs *(a)*; fused to the outer margin in the loris group and in New World monkeys *(b)*; fused to the outer margin and extended to form a tube in tarsiers and Old World simians *(c)*. In tree-shrews *(d)*, the bulla wall is formed from an entotympanic and the ectotympanic ring is enclosed. In colugos *(e)*, the bulla wall is formed mainly from the ectotympanic, with a small entotympanic contribution. Note the almost horizontal orientation of the tympanic membrane (eardrum). In the plesiadapiform *Ignacius* *(f)*, the bulla wall is formed mainly from an entotympanic, with the ectotympanic fused to the outer margin and extended to form a tube. There is convergence in general form (but not in constitution) between tree-shrews and lemurs *(d* versus *a)* and between plesiadapiforms and tarsiers/Old World simians *(f* versus *c)*.

plesiadapiforms as well, the entotympanic bulla of *Ignacius* sets paromomyids (at least) apart from primates. It has further been claimed that *Ignacius* specifically resembles colugos because the medial part of the entotympanic bulla contacts the basioccipital on the underside of the skull, that the bulla in *Plesiadapis* may well be formed from an entotympanic as well, and that *Ignacius, Plesiadapis* and colugos also share an unusual secondary regression of the internal carotid system.[3] The inferred relationship between *Ignacius, Plesiadapis* and colugos was later reinforced by a cladistic analysis of 33 cranial features.[8]

Reassessment of the relationships of the Plesiadapiformes has been increasingly hampered by mounting evidence that the group may not be monophyletic. A recent major review, focused on dentitions, divided the Plesiadapiformes into two superfamilies: Microsyopoidea (families Microsyopidae and Palaechthonidae) and Plesiadapoidea (families Plesiadapidae, Paromomyidae, Carpolestidae, Saxonellidae and Picrodontidae).[9] The family Microsyopidae has had a checkered history, being sometimes excluded from the plesiadapiform group[1,10] and sometimes included.[11,12] Beard[13] omitted Microsyopidae, Palaechthonidae and Picrodontidae from his analysis because of the lack of fossil evidence. For the remaining families, he proposed a new three-level arrangement for the order Dermoptera with colugos linked first to Paromomyidae, then to a group containing Plesiadapidae, Saxonellidae and Carpolestidae, and finally to Micromomyidae (Fig. 5). The plesiadapiform families included are hence not monophyletic. Beard combined his thus expanded order Dermoptera with primates of modern aspect in the new mirorder "Primatomorpha." By contrast, Kay *et al.*[8] conclude from their cladistic analysis of craniodental features that plesiadapiforms (excluding Microsyopidae) are a monophyletic sister group of colugos and propose that all plesiadapiforms should now be transferred to the Dermoptera.[3,8] Their analysis indicates no connection between plesiadapiforms and primates, confirming the conclusion from a previous re-examination of evidence.[14]

The evidence for a specific link between plesiadapiforms and colugos is open to question. In particular, Krause[15] challenged the inference of gliding adaptations in *Phenacolemur* and *Ignacius* on a number of grounds. His primary criticism is that there are doubts about identification of isolated postcranial bones attributed to *Phenacolemur* and *Ignacius,* particularly the assignment of phalanges to individual digits of hand or foot. As the interpretation of isolated postcranial bones is always difficult, the inference of gliding adaptations in *Phenacolemur* and *Ignacius* is provisional until associated skeletal remains are found. More complete skeletal evidence might, for instance, show whether the atlas vertebra was butterfly shaped for the attachment of the patagium, or whether the ulna was attenuated and united at its lower end with the radius, as in modern *Cynocephalus.* Evidence from the bulla of *Ignacius,* although eliminating one of the main arguments for linking plesiadapiforms to primates, does not necessarily indicate a link with colugos (Fig. 4). Numerous mammals incorporate one or more entotympanics in the bulla, and contact between the entotympanic and the basioccipital, presumably reflecting medial expansion of the bulla, could easily arise by convergence. The proposal that a lateral collar-shaped ectotympanic is a shared derived feature of plesiadapiforms and colugos[3] is unconvincing as the bulla of colugos is in fact formed almost entirely from the ectotympanic. In contrast to the condition in *Ignacius,* a small rostral entotympanic and an even smaller caudal entotympanic make only minor contributions to the bulla in *Cyno-*

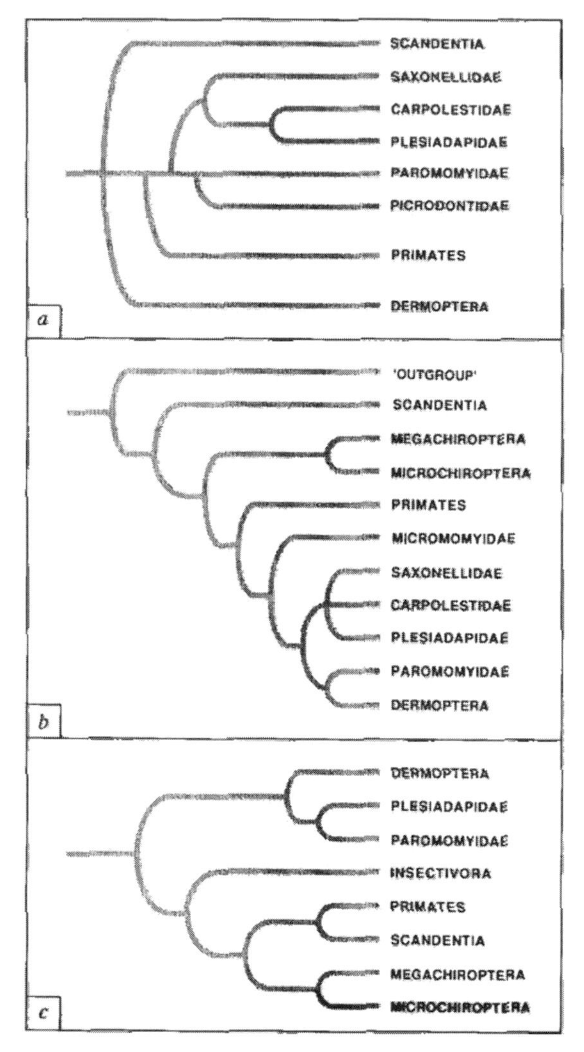

Fig. 5 Alternative interpretations of the relationships between dermopterans, plesiadapiforms and primates, following: *a,* Szalay and Delson;[1] *b,* Beard;[13] and *c,* Kay *et al.*[8]

cephalus.[16] This feature is linked to a peculiarity of *Cynocephalus:* the eardrum is almost horizontal. This orientation of the eardrum was primitive for mammals and there has been a general trend toward vertical re-alignment.[14] Among modern mammals, the quasi-horizontal condition has been retained only by monotremes and certain lipotyphlan insectivores. As a retained primitive feature of *Cynocephalus,* this would suggest a very early separation from all other eutherian mammals. *Cynocephalus* does show secondary reduction in the internal carotid system within the bulla, associated with increased emphasis on the vertebral arteries for supply of blood to the brain,[16] and a similar condition can be inferred for some Paromomyidae and Plesiadapidae, although not for Micromomyidae.[9,13] But such ontogenetic suppression of a pre-existing system could easily have arisen convergently.

Several other points are pertinent. First, the Paleocene/Eocene Plagiomenidae have long been regarded as fossil relatives of colugos because of dental resemblances.[17] But new basicranial evidence revealed that *Plagiomene* greatly differs from *Cynocephalus,* indicating that the dental similarity is convergent.[18] This provides yet another example of the danger of basing inferred relationships exclusively on dental evidence. More recently, *Dermotherium* from the late Eocene of Thailand has been identified as a direct relative of modern colugos.[19] Although the evidence is restricted to a lower jaw fragment containing only two molars, the resemblance to *Cynocephalus* is impressive and may well indicate that the colugos existed as a distinct lineage at least by the late Eocene. *Dermotherium* seems to rule out the Plagiomenidae[17] or the early Eocene Placentidentidae of Europe[20] as relatives of *Cynocephalus.*

Curiously, little has been said about possible dental evidence linking plesiadapiforms to colugos. For example, the cladistic analysis of Kay *et al.* included no dental evidence.[8] Supposed resemblances in the molar teeth between plesiadapiforms and primates of modern aspect provided the main basis for inferring a phylogenetic link, although this evidence has been questioned.[14] Surely, if plesiadapiforms are linked to colugos instead of primates, we should expect plesiadapiforms to show some dental resemblance to *Cynocephalus.* Modern colugos in fact have a highly unusual dentition, with comb-like crowns on the procumbent lower incisors and serrated upper incisors and upper and lower canines. The upper molar teeth are essentially triangular and bear three main cusps, presumably a primitive feature, and all the cheek teeth possess tall, sharp cusps adapted for shearing. *Cynocephalus* reportedly feeds primarily on leaves, but the dietary function of the comb-like lower incisors has yet to be explained. Comparison between *Cynocephalus* and Paromomyidae merely shows that they are dentally very different.[19] It is, of course, possible that the dental features of *Cynocephalus* have become so specialized that its relationships are no longer evident.[6] For instance, Beard proposes a *Nannopithex*-fold on the upper molars as a shared derived feature characterizing the "Primatomorpha" and then infers that it disappeared without trace from colugos through evolutionary reversal.[13] This dental feature is, in fact, lacking from the earliest known fossil primates of modern aspect, so it is more likely to have arisen by convergence even within primates.[8,14]

In sum, plesiadapiforms, formerly "dental primates," are no more than "postcranial colugos."

If colugos and plesiadapiforms together constitute the sister group of primates, all should possess shared derived features. Beard's cladistic analysis,[13] based mainly on postcranial traits, seems to demonstrate the existence of a special set of adaptations for vertical postures and climbing.[6] By contrast, the analysis of cranial features by Kay *et al.*[8] leads to the incompatible conclusion that primates are linked not to plesiadapiforms and colugos but to tree-shrews and bats. Moreover, in addition to the horizontal orientation of its eardrum (Fig. 4), *Cynocephalus* shows several features ignored in both analyses that would be highly unusual in any direct relative of primates. The brain, for instance, is remarkably small, with dominant olfactory components and diminutive cerebral hemispheres (Fig. 6). Unfortunately, a report on relative brain size in *Cyno-*

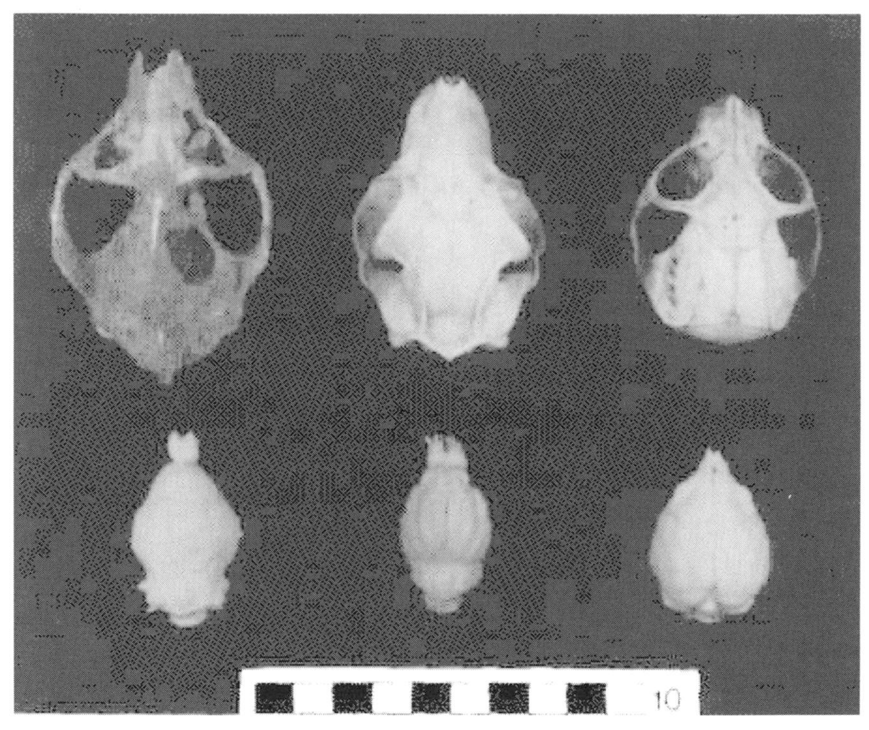

Fig. 6 Skulls and endocasts (casts of the brain cavity) of *Adapis parisiensis (left), Cynocephalus variegatus (center)* and *Perodicticus potto (right).* In both the Eocene *Adapis* and the modern *Cynocephalus* the endocast is very small relative to body size and the neocortex is diminutive, as is indicated by the dorsal exposure of the olfactory bulbs and the cerebellum. Note also the laterally facing orbits, the lack of a postorbital bar and the wide separation of the orbits in *Cynocephalus,* contrasting starkly with the typical primate pattern in *Adapis* and *Perodicticus.*

cephalus[21] cited an immature body weight of only 810 g. Taking a revised adult body weight of 1.4 kg and a cranial capacity of 6.9 ml, the index of cranial capacity (ICC)[14] is only 1.5, on the boundary between "basal" and "advanced" insectivores. This value is well below the lower limit for modern primates (2.4) and is instead close to the value of 1.4 for the Eocene primate *Adapis,* which lived some 40 Myr ago (Fig. 6). As a further strikingly primitive feature, although *Cynocephalus* has only a single neonate, it is apparently extremely poorly developed (altricial) and has been described as "marsupial-like."[22] Modern primates are uniformly characterized by neonates with well-defined precocial features.[14]

Inference of a link between primates, plesiadapiforms and colugos is connected with a recent revival of the concept of the superorder Archonta, originally grouping primates, colugos, bats, tree-shrews and elephant-shrews.[23] Although the terrestrial elephant-shrews are now generally excluded, there have been repeated proposals of relationships among some or all of the remaining forms, which typically have some connection with arboreal habits. Proposed inclusion of tree-shrews in the order Primates has a long history, but there is no convincing evidence of shared derived features and it is preferable to allocate them to their own order Scandentia.[14,24] MacKenna's proposal[25] that bats, colugos, tree-shrews and primates together form a phylogenetic unit was later supported on the basis of post-cranial evidence,[26] although the status of the bats remained uncertain, whereas immunological evidence apparently linked colugos, tree-shrews and primates to the exclusion of bats.[27] On morphological grounds, Shoshani identified a phylogenetic association between bats, colugos, tree-shrews and primates[28] and Novacek cited limited support for a concept of Archonta in which tree-shrews are allied with primates and colugos with bats.[29–31] By contrast, an analysis of amino-acid sequences in a tandem alignment of up to seven proteins[32,33] linked bats and tree-shrews with insectivores and carnivores and associated primates with lagomorphs and rodents.

In a new variation on the archontan theme, Pettigrew[34] reported evidence from the visual system (pattern of retinotectal projection) indicating that the fruit bats (Megachiroptera), but not other bats (Microchiroptera), are directly related to primates. Following a cladistic analysis of 24 neural characters, this proposal was then expanded into the "flying primate hypothesis" linking primates, fruit bats, colugos and tree-shrews (in that order),[35] with paromomyids being included as purported relatives of colugos in due course.[36] This hypothesis has been opposed because it conflicts with extensive morphological evidence indicating a monophyletic origin for bats.[31,37,38] It has also emerged that the pattern of retinotectal projection in the fruit bat *Rousettus aegyptaicus* is not primate-like but follows the general mammalian scheme,[39] suggesting that a primate-like pattern may have evolved convergently in some fruit bats (such as *Pteropus*). Further, a sequence analysis of ϵ-globin genes for 17 mammal species[40] produced very convincing evidence for a monophyletic origin of bats and linked tree-shrews to lagomorphs rather than to primates. Monophyly of bats was independently confirmed by an analysis of the mitochondrial cytochrome oxidase subunit II gene.[41] This analysis also groups *Tupaia* and *Cynocephalus* (but not bats) with primates, apparently providing support for a curtailed archontan assemblage. In fact, *Cynocephalus* consistently emerged as closer to primates than *Tupaia* did. As the comparisons include only nine placental mammal species, however, these results must be regarded as tentative.

Overall, the superordinal relationships of primates remain unclear. Even authors who accept the existence of a group "Archonta" generally do not agree on its composition or internal arrangement and no consistent picture is provided by molecular comparisons.[42] All such interpretations are based on undoubted morphological similarities between bats, colugos, tree-shrews and primates, but their interpretation is questionable. Most authors assume that the ancestral placental mammals were terrestrial and accordingly interpret any arboreal adaptations of "archontans" as shared derived features indicating a specific common ancestry. If, however, the ancestral placental mammals were at least partially arboreal in habits, the similarities shared by colugos, bats, tree-shrews and primates can be interpreted as a mixture of primitive retentions and convergent adaptations.[14] In this respect, an important new development has been the description of the postcranial skeleton of the Jurassic eupantothere *Henkelotherium,*[43] which is surprisingly advanced and shows several adaptations for arboreal life. Therian mammals (marsupials and placentals) are probably derived from a eupantothere stock, so this new evidence strengthens the inference that marsupials and placentals both had an arboreal origin.

Affinities of the Adapidae

New discoveries of fossil adapids ("lemuroids") have included important finds both in North America and in Europe. In North America, where all or most forms can be allocated to the subfamily Notharctinae, the main advance has been a significant increase in our knowledge of the early genera *Cantius* and *Pelycodus.* The later forms, *Notharctus* and *Smilodectes,* have long been documented by some of the best material available for fossil primates,[1,44–46] but the earlier North American adapids were known essentially from isolated teeth and jaw fragments. Indeed, well-preserved skulls of these earlier representatives have yet to be discovered. A study of more than 100 new foot elements from seven species of *Cantius* and (?)*Pelycodus*[47] has now partly improved our knowledge. Not surprisingly, *Notharctus* and *Smilodectes* are apparently derived in comparison with *Cantius,* although only in subtle details. Overall, these new finds fit expectation quite closely. It is interesting, however, that the tarsal bones changed relatively little during the early radiation of *Cantius* and (?)*Pelycodus,* although the dentition changed quite markedly over the 5-Myr period concerned. This is a particularly clear example of mosaic evolution. Comparison of foot bones of *Cantius, Notharctus* and *Smilodectes* with those of modern lemurs and lorises provides little evidence of a specific link between the Eocene forms and modern strepsirhine primates. The only features that may indicate a specific link between all adapids and modern strepsirhines are morphological features of the ankle region (sloping talo-fibular facet; lateral groove for the flexor hallucis longus; ventral extension of the navicular facet for the cuboid). Such features have been cited[48] in support of the interpretation that the adapids are the sister group of modern lemurs and lorises, and this inferred link has been reinforced with the suggestion that adapids and strepsirhines show shared derived features of the hand and wrist associated with an emphasis on hand postures involving ulnar deviation.[49] A link between adapids and modern strepsirhines has also been proposed on the basis of resemblances in the incisor-canine complex.[50] Taken together, these various lines of evidence provide suggestive but by no means conclusive indicators of a connection between adapids and strepsirhines.

The main surprises with respect to the adaptive radiation of the adapids have come from recent studies of European forms. The best-known genera with respect to craniodental remains continue to be *Adapis* and *Leptadapis,* still mainly represented by numerous well-preserved skulls discovered long ago in late Eocene deposits of France. Although several species were originally identified,[51] most recent authors have recognized just the two species *Adapis parisiensis* and *Leptadapis magnus.* A comprehensive re-examination of these skulls has now convincingly shown that quantitative variation among the skulls attributed to either species greatly exceeds that found within any modern primate species, even exceeding the level of variation across closely related genera for some features.[52,53] Excessive variation in dental dimensions of specimens allocated to *"Adapis parisiensis"* had already been noted by Gingerich[54,55] and it now seems highly likely that this taxon and *"Leptadapis magnus"* in fact group a number of species together.[56] An earlier suggestion[57,58] that quantitative differences among skulls of *Adapis* or *Leptadapis* reflect sexual dimorphism must now be re-examined. The new findings do not necessarily show that sexual dimorphism was lacking in *Adapis* and *Leptadapis,* but such sexual differences must now be sought within the individual species identifiable on the basis of Lanèque's analyses. For the time being, the only remaining evidence for sexual dimorphism in Eocene primates is that reported for the upper canine teeth of *Notharctus venticolus.*[59] In that case, the specimens concerned came from the same stratigraphic horizon and showed two different canine sizes despite the fact the molar teeth were all very similar in size and shape, so it is less likely that interspecific differences were confused with differences between the sexes.

In contrast to information on the North American representatives, our understanding of European adapids has long been severely constrained by the lack of associated postcranial remains. Isolated postcranial bones have been attributed to *Adapis* and *Leptadapis* on the grounds of size and primate-like features, but calcanei (heel bones) attributed to these two genera lack the relative elongation of the anterior segment that is otherwise universally found among fossil and living prosimian primates.[14,60,61] It is possible, as concluded from the detailed study conducted by Dagosto,[62] that *Adapis* and *Leptadapis* exhibited extensive secondary specialization of the postcranial skeleton, accounting for departures from the typical primate pattern. Additional evidence of the puzzling nature of postcranial elements attributed to *Adapis* and *Leptadapis* is provided by a recent study[63] in which body weight of fossil primates was estimated from dimensions of the tarsal bones. Overall, there is relatively good agreement between these postcranial estimates and those derived from skulls and/or teeth. There is one surprise in that *Adapis* skull measurements consistently indicate a body weight of at least 2 kg (ref. 14), whereas postcranial elements attributed to this genus indicate a body weight of only 1 kg; for *Leptadapis,* both skulls and postcranial elements indicate the same body weight of about 8.5 kg. It is possible that this disparity will disappear when the different species currently subsumed in *"Adapis parisiensis"* have been clearly separated, but only the discovery of associated cranial and postcranial remains would settle the issue clearly.

In recent years, postcranial remains of a number of other European adapids have been reported. Most finds have come from the Messel deposits of southern Germany, which have yielded five specimens.[64–68] Unfortunately, the specimens are all incomplete and three of them cannot be identified even to genus level, because the skull and teeth have not been preserved, and they have hence been recognized simply as adapids of some kind. One specimen, however, consists of the anterior part of the

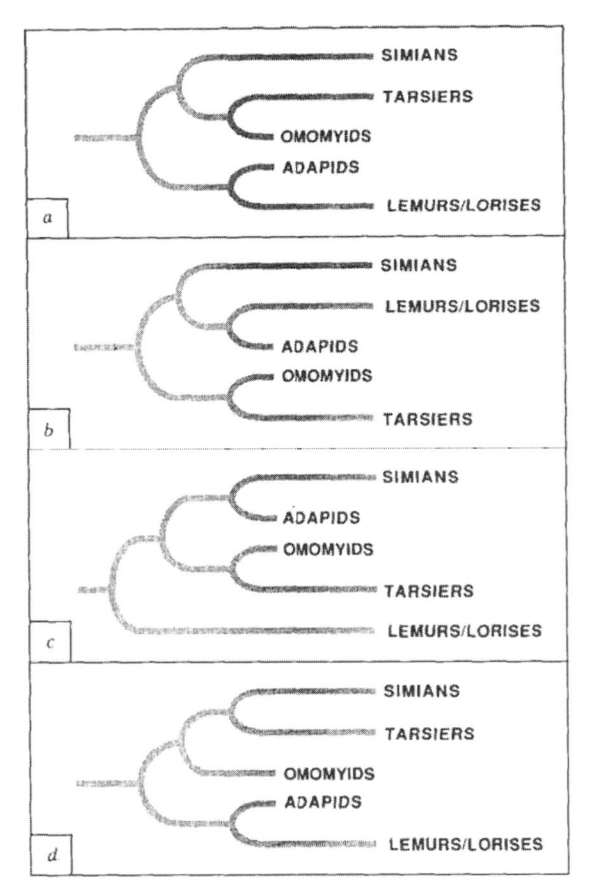

Fig. 7 Four alternative interpretations of the relationships between adapids, omomyids and living primates. *a,* The omomyophile hypothesis,[78] with simians deriving from an omomyid-like stock; *b,* one version of the adapid hypothesis (the lemurphile hypothesis[78]), with both simians and strepsirhine primates originating from an adapid-like stock; *c,* a second version of the adapid hypothesis (the lemurphobe hypothesis), with simians originating from an adapid-like stock, whereas strepsirhine primates had a distinct origin; *d,* the tarsiphile hypothesis,[78] with simians deriving from a tarsier-like ancestor.

body including a crushed skull and has been identified as *Europolemur koenigswaldi,* while a second specimen has been identified as the rear end of the same species. These specimens indicate that *Europolemur* was a relatively small-bodied arboreal climber and leaper with extremities adapted for grasping.[66] Another crucial new specimen from a different site is the crushed skull and skeleton of *Pronycticebus neglectus* from the Geiseltal deposits of eastern Germany.[69] The genus *Pronycticebus* was previously documented only by a single skull from France and study of the German skeleton indicates that this primate, which probably had a body weight close to 750 g, was an arboreal climber and leaper with an insectivorous/frugivorous diet. Finally, recent excavations in northern Spain have yielded more than 500 specimens representing most of the postcranial skeleton attributed to the hitherto poorly known genus *Anchomomys.*[70] This genus is of particular significance because its body size (about 120 g) is much smaller than that of other adapids. Interestingly, the postcranial elements of all of these new adapid finds are closely similar to those of the North American notharctines and unlike those attributed to *Adapis* and *Leptadapis.*

One possible interpretation of the new finds is that at least some postcranial elements attributed to *Adapis* and *Leptadapis* do not in fact belong to those genera. By contrast, if the allocations are correct, it follows that *Adapis* and *Leptadapis* are quite distinct from most other known adapids. Gingerich[71] originally distinguished between an "*Adapis* group" and a "*Protoadapis* group" in Europe, although he did not identify a major separation between them. Later, citing dental characteristics and postcranial features, Franzen[66] proposed that the European adapids should be split into two distinct groups: (1) *Adapis, Leptadapis, Microadapis* and (?)*Caenopithecus;* (2) *Cantius, Protoadapis* and *Europolemur.* The second group, characterized (for example) by lower-crowned molars with weakly defined crests, by a transverse oblong shape of the upper molars and by more limited molarization of the premolars, roughly corresponds to the tribe Protoadapini recognized by Szalay and Delson.[1] This tribe also includes *Pronycticebus* and the distinctive North American genus *Mahgarita.* Although the Protoadapini are morphologically closer to the North American forms and are represented in Europe throughout the Eocene, *Adapis, Leptadapis, Microadapis* and *Caenopithecus* appear abruptly toward the end of the middle Eocene.[72] For now, it is probably best to regard the *Adapis* group as an aberrant sideline that may or may not be a part of a monophyletic radiation of adapids. By contrast, the main group is of particular interest because of suggestions that the origins of simians may be traced specifically to the Protoadapini (Box 2).

It is also significant that a single lower molar tooth (M_2) from a presumed middle Eocene locality in Southeast Asia has been identified as an adapid.[73] Confirmation of this identification would greatly expand the known geographical range of Eocene adapids and could lead to new interpretations of early primate evolution.

Affinities of the Omomyidae

The status of the early Tertiary Omomyidae ("tarsioids") is arguably the most controversial issue concerning the evolutionary tree of primates and

BOX 2 What was the origin of the simian primates?

Four main hypotheses have been proposed to explain simian origins: (1) the tarsier hypothesis, deriving simians from a tarsier-like ancestor;[106] (2) the omomyid hypothesis, deriving simians from an omomyid stock;[107] (3) the adapid hypothesis, deriving simians from an adapid stock;[108] (4) the pre-simian hypothesis, deriving simians from hypothetical ancestors in Africa during the early Tertiary[109] or in Gondwanaland during the Upper Cretaceous.[110] A widely favored interpretation has been that omomyids gave rise to both tarsiers and simians (Fig. 7a). That interpretation has been increasingly challenged, however, and several authors now prefer some form of the adapid hypothesis instead. One version is that adapids gave rise to both simians and strepsirhine primates (Fig. 7b),[46,71,79] but this is in direct conflict with abundant evidence from living forms that tarsiers and simians form a monophyletic group. An alternative version is that adapids exclusively gave rise to simians (Fig. 7c).[81,96,111–114] Another possible interpretation,[78] according to which the omomyids branched away before the separation between tarsiers and simians (Fig. 7d), in effect amounts to a revival of the tarsier hypothesis.

With respect to the known fossil record, the discussion boils down to the question whether omomyids or adapids are most likely to have given rise to simians. There is little if any direct evidence for a link between omomyids and simians (Fig. 7a or d); instead there has generally been an indirect chain of argument that tarsiers are related to omomyids, that tarsiers are also related to simians and that omomyids must thus be related to simians. If tarsiers are not, in fact, related to omomyids, the argument for an omomyid origin of simians virtually collapses.

The alternative proposal that simians are derived from adapids has been backed on the basis of a seemingly impressive list of similarities, including increased body size, vertically implanted spatulate incisors, large interlocking canines with a honing relationship between the upper canine and the anterior lower premolar, sexual dimorphism in body size and canine size, a quadratic form of the molars, fusion of the anterior junction (symphysis) between the two halves of the lower jaw, an annular form of the ectotympanic, unfused tibia and fibula and lack of elongation of the calcaneum and navicular.[46,79,111] Some of these characters can be discounted because they are likely to be primitive for primates generally (for example, large canines; annular ectotympanic; unfused tibia and fibula; non-elongated calcaneum and navicular), whereas others have developed so often during mammalian evolution that they carry little weight (for example, increased body size; quadratic molars). Further, the evidence for sexual dimorphism in adapids now needs careful re-examination and such dimorphism is in any case neither unique to nor ubiquitously present in simian primates.

Spatulate incisors and fusion of the lower jaw symphysis have all arisen repeatedly in the evolution of primates and other mammals. Indeed, within the primates these features were developed independently in various large-bodied subfossil lemurs and similarities in jaws and teeth between *Archaeolemur* and Old World monkeys provide one of the most striking cases of convergent evolution known for primates. Further, fusion of the lower jaw is not found in all adapids. It was lacking in the earliest adapids *(Cantius, Pelycodus)* and was absent even in some later forms, such as *Smilodectes*[115,116] and *Pronycticebus,*[69] so this feature must have arisen several times independently within the Adapidae.[14,46,66] Symphyseal fusion occurred late in development in *Notharctus;* only in *Adapis, Leptadapis* and (presumably) *Mahgarita* did fusion occur early in life as in simians. Hence, if simians are to be derived from adapids on grounds of symphyseal fusion, they must presumably be derived from one of these genera. First, however, *Adapis, Leptadapis* and *Mahgarita* are all late Eocene forms, most probably too recent for simian ancestry. Second, it now seems that in the earliest yet known simians, such as late Eocene *Catopithecus*[96] and *Arsinoea,*[97] the symphysis may not have been fused, in which case symphyseal fusion must have developed after the origin of these simian primates.

Rasmussen and Simons[113] narrowed the possible source of simians among adapids to the tribe Protoadapini, and Rasmussen[114] subsequently noted additional specific resemblances between the North American genus *Mahgarita* and simians. These included: large promontory canal; stapedial canal reduced or absent; pneumatized petromastoid region; lateral transverse intrabullar septum; probably no free ectotympanic ring; mandible firmly fused and robust; mandibular symphysis with a transverse torus; short, deep maxilla; detailed similarities in occlusal features of upper molars and other teeth. Although these similarities are intriguing, they can only be of phylogenetic significance if simians are derived directly from *Mahgarita* or some similar ancestral form. One must bear in mind the possibility that sharing of a number of similarities between certain adapids and simians may be attributable to convergence.

Derivation of simians from either adapids or omomyids remains problematic because the known fossil skulls uniformly lack any indication of postorbital closure by a bony partition and the frontal bones remain unfused into adult life. Similarly, known adapid skulls show no sign of the early marked increase in relative brain size that distinguishes simians from prosimians overall and is already established in *Aegyptopithecus*, although such an increase is in fact identifiable in Eocene omomyids.[14,46] Finally, no known fossil skulls before the Fayum simians show the loss of the median gap between the upper incisors that characterizes tarsiers and simians and would indicate loss of the rhinarium (haplorhine condition). In short, even if simians can be derived from adapids or omomyids, vital transitional evidence is still lacking. In other words, the pre-simian hypothesis, tracing simians to an as yet undocumented ancestral stock in the southern continents, is still a viable option.

seems set to remain so for the foreseeable future. Not long ago, it was widely accepted that omomyids are directly linked to the ancestry of tarsiers. Similarities between tarsiers and at least some omomyids exist in the bell-shaped upper dental arcade, unusual foreshortening of the lower jaw associated with a reduction in number of teeth relative to the upper jaw, a tubular meatus in the bulla and adaptation of the postcranial skeleton for leaping. In addition, omomyids do seem to resemble both tarsiers and simians in possessing a short facial skull associated with a relatively small nasal fossa,[74] and there is some evidence of limited reduction in the size of the olfactory bulbs.[14] Although it was suggested that omomyids resemble tarsiers and simians in possessing an enlarged promontory artery that enters the bulla posteromedially,[75] *Necrolemur* apparently possessed a relatively large stapedial artery and the medial entry into the bulla may in fact be primitive for mammals.[76–78] Rosenberger[74] has noted further similarities in the basicranium between some omomyids and tarsiers, including a flexed basicranial axis, encroachment of the bulla on the pterygoid fossa associated with extensive laminar confluence between pterygoid and bulla, and a deeply guttered, slot-like shape of the glenoid fossa. But there has always been an undercurrent of doubt about the inference of a link between omomyids and tarsiers, and interpretation of the fossil record is hampered by the fact that modern tarsiers have molar teeth that seem extremely primitive. The upper molars have a simple three-cusped pattern, and the paraconid cusp (commonly lost in fossil and living primates of modern aspect) is still present in the lower molars. Further, while it is true that omomyids resemble tarsiers in possessing paraconids on their lower molars, this is no more than retention of a primitive feature. Hence, it is difficult to link omomyids to either tarsiers or simians through dental morphology.

One alternative view is that the omomyids are, indeed, related to tarsiers and that both are appropriately allocated to the infraorder Tarsiiformes, but that the latter has no connection with simians. An extreme version of this interpretation was advocated by Gingerich,[11] who allied Tarsiiformes with Plesiadapiformes in the suborder Plesitarsiiformes because of apparent similarities in the structure of the bulla and in the anterior lower dentition. All other primates of modern aspect (lemurs, lorises and simians) were combined in a second suborder Simiolemuriformes. This arrangement, however, requires convergent evolution of many features present in all living and fossil primates of modern aspect (including tarsiiforms) but lacking in plesiadapiforms. Gingerich subsequently accepted a model of primate evolution incorporating three adaptive grades: Plesiadapiformes, Prosimii (including lemurs, lorises, tarsiers, Adapidae and Omomyidae) and Anthropoidea[12] and later retracted his suggestion of a link between plesiadapiforms and tarsiiforms.[79] His revised model, however, incorporated no direct connection between tarsiiforms and simians. This left the way open for the less extreme proposal that the origin of simians within the prosimian grade is to be traced not to omomyids but to adapids (Box 2). In fact, both of these alternative proposals are in direct conflict with many similarities, several of which are unique, linking modern tarsiers to simians.[14,80] It is, of course, conceivable that some similarities have arisen convergently. For instance, developmental evidence seems to indicate that the almost complete bony eye socket of tarsiers and simians (a feature found in no other mammals) may have arisen independently.[81] Nevertheless the sheer numbers of apparent derived features in independent systems shared by tarsiers and simians render convergence in all of them extremely unlikely.

A different challenge to the consensus view linking omomyids, tarsiers and simians comes from the view that there is no specific link between omomyids and tarsiers.[77,82] One feature linking modern tarsiers and simians (haplorhine primates) is the complete loss of the rhinarium (an area of naked skin surrounding the nostrils). This derived feature, which is highly unusual among mammals, is correlated with closure of a median gap between the upper incisors that occurs in association with the rhinarium in lemurs and lorises (strepsirhine primates). It is now known that the European omomyids *Necrolemur* and *Pseudoloris* still had a gap between the upper incisors.[77,80] It is therefore possible that a rhinarium was still present in omomyids, although its loss could have preceded closure of the gap between the incisors. In any event, broad reviews of the evidence[77,78,80,83] have indicated fairly clearly that the relationship between tarsiers and simians is probably closer than that between tarsiers and omomyids. Indeed, Schmid[82] has suggested that omomyids should be classified together with lemurs, lorises and adapids in the suborder Strepsirhini.

By contrast, there have been a number of important developments concerning omomyids in recent years, and the authors involved have tended to support a link between omomyids and tarsiers. One interesting development has been identification of a reasonably complete skull of *Microchoerus* from the Eocene of France.[84] This skull had long been allocated to the closely related genus *Necrolemur,* which is documented by several skulls. The orbits of *Microchoerus* are somewhat larger than in *Necrolemur* and the upper dental arcade does not clearly show the bell shape typically found in other European omomyids. Postcranially, the European omomyids *Necrolemur* and *Microchoerus* differ from the North American forms in having far greater elongation of the anterior segment of the calcaneus,[85] representing a development toward the condition found in modern bushbabies and tarsiers. It has also been claimed that the European forms *Necrolemur* and *Nannopithex* specifically resemble modern tarsiers in showing distal fusion between the tibia and fibula, although the evidence for *Nannopithex* has not stood up to closer examination and that for *Necrolemur* is questionable (P. Schmid, personal communication). Rosenberger and Dagosto,[84] however, emphasize apparent derived features of the jaw joint and skull base shared by tarsiers and the European omomyids (Microchoerinae), concluding that there is a specific link between them and that the simians arose from less specialized North American omomyids (anaptomorphines or omomyines). This follows the general trend of inferring a specific link between tarsiers and microchoerines, which were in fact included in the family Tarsiidae by Simons.[86]

More dramatic was the discovery of four new skulls of the North American omomyid *Shoshonius,* previously documented only by fragmentary jaws and teeth.[87,88] Previously, the skull of Eocene omomyids from North America was known only from an incomplete cranial specimen of *Tetonius,* so the new skulls are in any case of great importance for our understanding of early omomyids. According to Beard *et al.,*[87] however, the skulls of *Shoshonius* are of special significance because some features, not found in other omomyids, indicate a direct relationship to modern *Tarsius.* This would simultaneously confirm the derivation of tarsiers from omomyid stock, although from omomyines rather than microchoerines, and would push the time of origin of omomyids well back into the Paleocene. *Shoshonius* resembles tarsiers in having relatively very large orbits, in having a very short snout and in a number of specific features of the ear region. These resemblances are all rather limited, however, and could have arisen through convergent evolution. The possibility of a specific relationship between *Shoshonius* and *Tarsius* could be tested with postcranial evidence. If, for instance, *Shoshonius* were found to show fusion of the tibia and fibula and the same peculiar adaptation of the ankle region as in *Tarsius,*[89] this would provide more convincing evidence of a specific link between these two genera. Postcranial elements and additional skull material of *Shoshonius* have been found, so more information should be forthcoming. At present, the interpretation of omomyid relationships remains confused, with several possibilities still open. On the basis of isolated cheek teeth, the late Paleocene *Altiatlasius* from Morocco has been interpreted as an early omomyid,[90] perhaps indicating an African connection.

Early Simian Primates

Until recently, it was often inferred that simian primates (Anthropoidea) first emerged close to the boundary between the Eocene and the Oligocene, just over 35 Myr ago.[1] The earliest known undoubted simians had been recovered from the Fayum deposits in Egypt, the best-documented genus being *Aegyptopithecus,* which exhibits numerous simian features in the skull and postcranial skeletons. A Middle Oligocene age was originally accepted for the Fayum deposits, but subsequently the date was revised to Lower Oligocene age.[91] Although it has now been suggested that the entire sequence of the Fayum deposits should be dated as late Eocene,[92] paleomagnetic reversal stratigraphy indicates that they straddle the Eocene-Oligocene boundary.[93] Even this, however, would indicate an earlier date for the common ancestry of the Fayum simians. Identification of five new fossil simian genera very low down in the deposits (largely based on fragmentary specimens, but including the relatively well preserved skull of *Catopithecus*[94–97]) have reinforced the effects of this redating to indicate that simians must have emerged by the late Eocene, if not earlier. The new discoveries have raised the total number of simian genera reported from the Fayum to 11 *(Aegyptopithecus, Apidium, Arsinoea, Catopithecus, Oligopithecus, Parapithecus, Plesiopithecus, Propliopithecus, Proteopithecus, Qatrania, Serapia),* indicating that considerable diversification had taken place at an early stage. One of the new early forms *(Plesiopithecus)* may have lacked incisors in the lower jaw. If the identification of *Plesiopithecus* as a simian on the basis of a single lower jaw fragment is correct, considerable diversity had apparently already developed following divergence from ancestral simians. On the basis of isolated teeth, simians have also now been reported from earliest Oligocene deposits in Oman (including *Oligopithecus*)[98] and from late Eocene deposits in Algeria *(Biretia).*[99]

Indications of an earlier date for the Fayum simians may have been superseded by the recovery of dental specimens representing an apparent simian, *Algeripithecus minutus,* from the Algerian site of Glib Zegdou, which is of middle or even early Eocene age.[100] The find is unfortunately confined to three cheek teeth, but the upper and lower molars exhibit a number of simian-like features. In particular, certain upper molar features are shared with *Aegyptopithecus* from the Fayum deposits. If these features are indicative of a sister-group relationship between *Algeripithecus* and *Aegyptopithecus,* as proposed,[11] then the origin of known simians must be pushed back to the Lower Eocene or even into the Paleocene. Against this background, the late Eocene *Amphipithecus*[101] and *Pondaungia*[102] from Burma clearly now deserve serious re-examination. Both of these fossil genera are of uncertain status but could well be linked to simian origins.[79] Overall, a date of at least 55 Myr instead of just over 35 Myr for the origin of simians now seems possible. It is, in fact, predictable that the simians originated far earlier than hitherto accepted by most authors (see Box 1) and an even earlier date for simian ancestry cannot be ruled out. Nevertheless, although it now seems quite probable that the simian primates emerged much earlier than previously believed, their origins remain obscure (Box 2).

Future Developments

It is clear that there are still many uncertainties regarding the course and timing of early primate evolution. There are, however, several indications that conclusions drawn from an exploratory assessment of the sampling density of the primate fossil record (Box 1) are correct in principle. Our current sample of the primate fossil record is very limited and it follows from this that times of divergence within the primate tree may have been commonly underestimated by a considerable margin. Against this background, it would not be surprising if the time of origin of simian primates were to be pushed back into the Paleocene, in which case direct migration of simians between Africa and South America[103] becomes far more likely. Limited sampling of the fossil record, combined with the fragmentary nature of most of the fossils concerned, also explains why interpretations of primate evolution have been subject to repeated, often extensive revision. In the face of major gaps in the fossil record, far-reaching interpretation of fragmentary fossil remains can easily lead to misinterpretation of phylogenetic relationships. As a general rule, there has been a tendency to overlook the significance of convergent evolution in the interpretation of fossil primates. If only a limited number of characters can be investigated, it is extremely difficult to test for possible convergence. Greater caution in the interpretation of fragmentary fossil specimens combined with a healthy respect for convergent evolution might help to reduce the level of misinterpretation in the future.

Recognition of the implications of the low sampling level of the primate fossil record in fact raises a new possibility with respect to the evolutionary relationships of the early Tertiary adapids and omomyids. At present, modern strepsirhines, tarsiers and simians have been traced back to adapids or omomyids because these are the only known fossil forms. Rather than seeking to establish direct links with modern primate groups, we should perhaps consider the possibility that the adapids and omomyids together are the outcome of an entirely separate radiation of early primates of modern aspect in the Northern Hemisphere. The divergence between adapids and omomyids could have evolved in parallel to a divergence between strepsirhine and haplorhine primates in the Southern Hemisphere, such that adapids and omomyids are in fact more closely related to one another than to any modern primates. Such an interpretation, which (like all of the potential solutions reviewed above) requires acceptance of a certain amount of convergent evolution, could explain certain features that clash with reconstructions linking adapids to modern strepsirhines and omomyids to modern haplorhines or with alternative hypotheses. It is, for instance, unusual that the adapids were typically larger than omomyids, whereas the reverse is the case for the comparison of modern strepsirhines with haplorhines.[104] Adapids were presumably predominantly diurnal and yet supposedly gave rise to the primary nocturnal strepsirhines, whereas the omomyids were presumably predominantly nocturnal and yet supposedly gave rise to the primarily diurnal haplorhines.[14] As with many other aspects of early primate evolution, more precise answers depend on a continuing, vital effort by field paleontologists, who will hopefully uncover many more unknown extinct primate species.

ROBERT D. MARTIN is at the Anthropologisches Institut und Museum, Universität Zürich-Irchel, Winterthurerstrasse 190, CH-8057 Zürich, Switzerland.

References

1. Szalay, F. S. & Delson, E. *Evolutionary History of the Primates* (Academic, New York, 1979).
2. Beard, K. C. *Nature* **345,** 340–341 (1990).
3. Kay, R. F., Thorington, R. W. & Houde, P. *Nature* **345,** 342–344 (1990).
4. Martin, R. D. *Nature* **345,** 291–292 (1990).

5. Beard, K. C. *Postcranial Anatomy, Locomotor Adaptations, and Paleoecology of Early Cenozoic Plesiadapidae, Paromomyidae and Micromomyidae (Eutheria, Dermoptera)* (Johns Hopkins University, Baltimore, 1989).
6. Beard, K. C. in *Origine(s) de la Bipédie chez les Hominidés* (eds. Coppens, Y. & Senut, B.) 79–87 (Éditions du CNRS, Paris, 1991).
7. Beard, K. C. *J. vert. Paleont.* **9,** 13A (1989).
8. Kay, R. F., Thewissen, J. G. M. & Yoder, A. D. *Am. J. phys. Anthrop.* **89,** 477–498 (1992).
9. Gunnell, G. F. *Univ. Mich. Pap. Paleont.* **27,** 1–157 (1989).
10. Hoffstetter, R. *C. r. Acad. Sci. Paris ser. II* **302,** 43–45 (1986).
11. Gingerich, P. D. *Univ. Mich. Pap. Paleont.* **15,** 1–140 (1976).
12. MacPhee, R. D. E., Cartmill, M. & Gingerich, P. D. *Nature* **301,** 509–511 (1983).
13. Beard, K. C. in *Mammal Phylogeny:* Vol. 2 *Placentals* (eds. Szalay, F. S., Novacek, M. J. & Mackenna, M. C.) 129–150 (Springer, New York, in the press).
14. Martin, R. D. *Primate Origins and Evolution: A Phylogenetic Reconstruction* (Chapman & Hall, London and Princeton Univ. Press, Princeton, NJ, 1990).
15. Krause, D. W. *J. hum. Evol.* **21,** 177–188 (1991).
16. Hunt, R. B. & Korth, W. W. *J. Morph.* **164,** 167–211 (1980).
17. Rose, K. D. & Simons, E. L. *Contrib. Mus. Paleont. Univ. Michigan* **24,** 221–236 (1977).
18. MacPhee, R. D. E., Cartmill, M. & Rose, K. D. *J. vert. Paleont.* **9,** 329–349 (1989).
19. Ducrocq, S. *et al. Palaeontology* **35,** 373–380 (1982).
20. Russell, D. E. *et al. Paleovertebrata Mémoire Extraordinaire* 1–77 (1982).
21. Pirlot, P. & Kamiya, T. *Can. J. Zool.* **60,** 565–572 (1982).
22. Lekagul, B. & McNeely, J. A. *Mammals of Thailand* 1–758 (Sahakarnbhat, Bangkok, 1977).
23. Gregory, W. K. *Bull. Am. Mus. nat. Hist.* **27,** 1–524 (1910).
24. Butler, P. M. in *Studies in Vertebrate Evolution* (eds. Joysey, K. A. & Kemp, T. S.) 253–265 (Oliver & Boyd, Edinburgh, 1972).
25. McKenna, M. C. in *Phylogeny of the Primates* (eds. Luckett, W. P. & Szalay, F. S.) 21–46 (Plenum, New York, 1975).
26. Szalay, F. S. & Drawhorn, G. in *Comparative Biology and Evolutionary Relationships of Tree Shrews* (ed. Luckett, W. P.) 133–169 (Plenum, New York, 1980).
27. Cronin, J. E. & Sarich, V. M. in *Comparative Biology and Evolutionary Relationships of the Tree Shrews* (ed. Luckett, W. P.) 293–312 (Plenum, New York, 1980).
28. Shoshani, J. *Molec. Biol. Evol.* **3,** 222–242 (1986).
29. Novacek, M. J. & Wyss, A. R. *Cladistics* **2,** 257–287 (1988).
30. Novacek, M. J., Wyss, A. R. & McKenna, M. C. in *The Phylogeny and Classification of the Tetrapods* (ed. Benton, M. J.) 99–116 (Clarendon, Oxford, 1988).
31. Novacek, M. J. *Nature* **356,** 121–125 (1992).
32. Goodman, M., Czelusniak, J. & Beeber, J. E. *Cladistics* **1,** 171–185 (1985).
33. Miyamoto, M. M. & Goodman, M. *Syst. Zool.* **35,** 230–240 (1986).
34. Pettigrew, J. D. *Science* 231, 1304–1306 (1986).
35. Pettigrew, J. D. *et al. Phil. Trans. R. Soc.* B **325,** 489–559 (1989).
36. Pettigrew, J. D. *Am. J. phys. Anthrop.* **18** (suppl.), 3–54 (1991).
37. Wible, J. R. & Novacek, M. J. *Am. Mus. Novitates* **2911,** 1–19 (1988).
38. Baker, R. J., Novacek, M. J. & Simmons, N. B. *Syst. Zool.* **40,** 216–231 (1991).
39. Thiele, A., Vogelsang, M. & Hoffmann, K.-P. *J. comp. Neurol.* **314,** 671–683 (1991).

40. Bailey, W. J., Slightom, J. L. & Goodman, M. *Science* **256,** 86–89 (1992).
41. Adkins, R. M. & Honeycutt, R. L. *Proc. natn. Acad. Sci. USA* **88,** 10317–10321 (1991).
42. Wyss, A. R., Novacek, M. J. & McKenna, M. C. *Molec. Biol. Evol.* **4,** 99–116 (1987).
43. Krebs, B. *Berliner geowiss, Abh.* **133,** 1–121 (1991).
44. Gregory, W. K. *Mem. Am. Mus. nat. Hist.* **3,** 49–243 (1920).
45. Gazin, C. L. *Smithson. Misc. Coll.* **136,** 1–112 (1958).
46. Gingerich, P. D. in *Evolutionary Biology of the New World Monkeys and Continental Drift* (eds. Ciochon, R. L. & Chiarelli, A. B.) 123–138 (Plenum, New York, 1980).
47. Gebo, D. L., Dagosto, M. & Rose, K. D. *Am. J. phys. Anthrop.* **86,** 51–72 (1991).
48. Beard, K. C., Dagosto, M., Gebo, D. L. & Godinot, M. *Nature* **331,** 712–714 (1988).
49. Godinot, M. & Beard, K. C. *Hum. Evol.* **6,** 307–354 (1991).
50. Rosenberger, A. L., Strasser, E. & Delson, E. *Folia Primatol.* **44,** 15–39 (1985).
51. Stehlin, H. G. *Abh. schweiz. pal. Ges.* **38,** 1165–1298 (1912).
52. Lanèque, L. *C. r. Acad. Sci. Paris ser. II* **314,** 1387–1393 (1992).
53. Lanèque, L. *Hum. Evol.* **7,** 1–16 (1992).
54. Gingerich, P. D. *Folia Primatol.* **28,** 60–80 (1977).
55. Gingerich, P. D. *Gébios Méc. spéc.* **1,** 165–182 (1977).
56. Godinot, M. *C. r. Acad. Sci. Paris ser. II* **314,** 237–242 (1992).
57. Gingerich, P. D. *Am. J. phys. Anthrop.* **56,** 217–234 (1981).
58. Gingerich, P. D. & Martin, R. D. *Am. J. phys. Anthrop.* **56,** 235–257 (1981).
59. Krishtalka, L., Stucky, R. K. & Beard, K. C. *Proc. natn. Acad. Sci. USA* **87,** 5223–5226 (1990).
60. Martin, R. D. *Phil. Trans. R. Soc.* B **264,** 295–352 (1972).
61. Martin, R. D. in *The Study of Prosimian Behavior* (eds. Doyle, G. A. & Martin, R. D.) 45–77 (Academic, New York, 1979).
62. Dagosto, M. *Folia Primatol.* **41,** 49–101 (1983).
63. Dagosto, M. & Terranova, C. J. *Int. J. Primatol.* **13,** 307–344 (1992).
64. von Koenigswald, W. *Paläont, Z.* **53,** 63–76 (1979).
65. von Koenigswaid, W. *Carolinea* **42,** 145–148 (1985).
66. Franzen, J. L. *Cour. Forsch.-Inst. Senckenberg* **91,** 151–187 (1987).
67. Franzen, J. L. *Cour. Forsch.-Inst. Senckenberg* **107,** 263–273 (1988).
68. Franzen, J. L. & Frey, E. *Abstract in International Conference Monument Grube Messel* (Darmstadt, 1991).
69. Thalmann, U., Haubold, H. & Martin, R. D. *Palaeovertebrata* **19,** 115–130 (1989).
70. Moya Sola, S. & Köhler, M. *Folia Primatol.* (in the press).
71. Gingerich, P. D. & Schoeninger, M. J. *J. hum. Evol.* **6,** 483–505 (1977).
72. Franzen, J. L. & Haubold, H. *Mod. Geol.* **10,** 159–170 (1986).
73. Suteethorn, V., Buffetaut, E., Helmcke-Ingavat, R., Jaeger, J.-J. & Jongkanjasoontorn, Y. N. *Jb. Geol. Paläont. Mh.* **9,** 563–570 (1988).
74. Rosenberger, A. L. *Folia Primatol.* **45,** 179–194 (1985).
75. Rosenberger, A. L. & Szalay, F. S. in *Evolutionary Biology of the New World Monkeys and Continental Drift* (eds. Ciochon, R. C. & Chiarelli, A. B.) 139–157 (Plenum, New York, 1980).

76. Schmid, P. in *Primate Evolutionary Biology* (eds. Chiarelli, A. B. & Corruccini, R. S.) 6–13 (Springer, Berlin, 1981).
77. Schmid, P. *Die Systematische Revision der europäischen Microchoeridae Lydekker, 1887 (Omomyiformes, Primates)* (University of Zürich, 1982).
78. MacPhee, R. D. E. & Cartmill, M. in *Comparative Primate Biology* (eds. Swindler, D. R. & Erwin, J.) 219–275 (Liss, New York, 1986).
79. Gingerich, P. D. *Yb. phys. Anthrop.* **27,** 57–72 (1984).
80. Aiello, L. C. in *Major Topics in Primate and Human Evolution* (eds. Wood, B. A., Martin, L. B. & Andrews, P. J.) 47–65 (Cambridge Univ. Press, Cambridge, 1986).
81. Simons, E. L. & Rasmussen, D. T. *Am. J. phys. Anthrop.* **79,** 1–23 (1989).
82. Schmid, P. *Folia Primatol.* **40,** 1–10 (1983).
83. Cartmill, M. in *Evolutionary Biology of the New World Monkeys and Continental Drift* (eds. Ciochon, R. L. & Chiarelli, A. B.) 243–274 (Plenum, New York, 1980).
84. Rosenberger, A. L. & Dagosto, M. in *Topics in Primatology:* Vol. 3 *Evolutionary Biology, Reproductive Endocrinology, and Virology* (eds. Matano, S., Tuttle, R. H., Ishida, H. & Goodman, M.) 37–51 (Univ. of Tokyo Press, Tokyo, 1992).
85. Schmid, P. *Folia Primatol.* **31,** 301–311 (1979).
86. Simons, E. L. *Primate Evolution: An Introduction to Man's Place in Nature* (Macmillan, New York, 1972).
87. Beard, K. C., Krishtalka, L. & Stucky, R. K. *Nature* **349,** 64–67 (1991).
88. Martin, R. D. *Nature* **349,** 19–20 (1991).
89. Jouffroy, F.-K., Berge, C. & Niemitz, C. in *Biology of Tarsiers* (eds. Niemitz, C.) 167–190 (Gustav Fischer, Stuttgart, 1984).
90. Sigé, B., Jaeger, J.-J., Sudre, J. & Vianey-Liaud, M. *Palaeontographica Abt.* A **214,** 31–56 (1990).
91. Fleagle, J. G., Bown, T. M., Obradovitch, J. D. & Simons, E. L. *Science* **234,** 1247–1249 (1986).
92. Van Couvering, J. A. & Harris, J. A. *J. hum. Evol.* **21,** 241–260 (1991).
93. Kappelman, J. *J. hum. Evol.* **6,** 495–503 (1992).
94. Simons, E. L. *Proc. natn. Acad. Sci. USA* **86,** 9956–9960 (1989).
95. Simons, E. L. *Science* **247,** 1567–1569 (1990).
96. Rasmussen, D. T. & Simons, E. L. *Int. J. Primatol.* **13,** 477–508 (1992).
97. Simons, E. L. *Proc. natn. Acad. Sci. USA* **89,** 10743–10747 (1992).
98. Thomas, H., Roger, J., Sen, S. & Al-Sulaimani, Z. *C. r. Acad. Sci. Paris, sér. II* **306,** 823–829 (1988).
99. de Bonis, L., Jaeger, J.-J., Coiffait, B. & Coiffait, P.-E. *C. r. Acad. Sci. Paris sér. II* **306,** 929–934 (1988).
100. Godinot, M. & Mahboubi, M. *Nature* **357,** 324–326 (1992).
101. Ciochon, R. L., Savage, D. E., Thaw, T. & Maw, B. *Science* **229,** 756–759 (1985).
102. Maw, B., Ciochon, R. L. & Savage, D. E. *Nature* **282,** 65–67 (1979).
103. Hoffstetter, R. in *Evolutionary Biology of the New World Monkeys and Continental Drift* (eds. Ciochon, R. C. & Chiarelli, A. B.) 103–188 (Plenum Press, New York, 1980).
104. Fleagle, J. G. *Paleobiology* **4,** 67–76 (1978).
105. Martin, R. D. in *Major Topics in Primate and Human Evolution* (eds. Wood, B. A., Martin, L. B. & Andrews, P.) 1–31 (Cambridge Univ. Press, Cambridge, 1986).

106. Wood Jones, F. *The Ancestry of Man* 1–35 (Gillies, Brisbane, 1923).
107. Wortman, J. L. *Am. J. Sci. ser. 4* **17,** 203–214 (1904).
108. Gidley, J. W. *Proc. U. S. Nat. Mus.* **63,** 1–38 (1923).
109. Hoffstetter, R. *J. hum. Evol.* **3,** 327–350 (1974).
110. Hershkovitz, P. *Folia Primatol.* **21,** 1–35 (1974).
111. Rasmussen, D. T. *J. hum. Evol.* **15,** 1–12 (1986).
112. Rasmussen, D. T. *Am. J. Primatol.* **22,** 263–277 (1990).
113. Rasmussen, D. T. & Simons, E. L. *Folia Primatol.* **51,** 182–208 (1989).
114. Rasmussen, D. T. *Int. J. Primatol.* **11,** 439–469 (1990).
115. Beecher, R. M. *Am. J. phys. Anthrop.* **50,** 418 (1979).
116. Beecher, R. M. *Int. J. Primatol.* **4,** 99–112 (1983).
117. Ginsburg, L. & Mein, P. *C. r. Acad. Sci. Paris sér. II* **304,** 1213–1215 (1986).

Peter Andrews

EVOLUTION AND ENVIRONMENT IN THE HOMINOIDEA

Between 10 and 20 million years ago, a variety of hominoid primates lived in Africa, Europe and Asia. The question of which of these, if any, lie closest to the ancestries of humans and modern apes remains a lively source of debate. Recent fossil discoveries, though, shed light on the environments in which the various groups of hominoid emerged and, it is hoped, on their evolution. But the lack of a hominid fossil record before about 5 million years ago—and any fossil record for the African apes—is still a frustrating barrier.

The hominoid primates are the group that encompasses the apes and humans. The origin of the group and the evolution of the various hominoid species are of relevance to the human condition, but of more immediate concern are the issues of when and where the human species originated and the context in which it evolved: the effects of climate, environment and behavior, for example, and the ecological relationships between communities. I shall review the evidence relating to these issues, particularly the systematic relationships of fossil hominoids and their paleoecology, insofar as it is currently available from the fossil record.

The hominoid apes include the lesser apes, the gibbons and siamangs of eastern Asia, and the great apes, chimpanzees and gorillas of Africa and the orang-utan of Southeast Asia. Molecular evidence has shown that the old concept of great apes (family Pongidae), as distinct from humans (family Hominidae), is no longer valid[1] because the African apes and humans are more closely related to each other than any of them are to the orang-utan. This is recognized taxonomically by the division between the Ponginae, which contains only the orang-utan and its putative fossil relatives, and the Homininae, for the African apes and humans.

This distinction is supported both by molecular evidence and by morphology. Neither source of evidence, however, offers a definitive solution on the relationships within the African ape and human group, the Homininae. The majority of molecular studies support a closer relationship between chimpanzees and humans, with gorillas more distantly related (see Box 1), but some molecular studies support the alternative of closer rela-

BOX 1 The relationships of the extant hominoid primates

The characters distinguishing the three branching points are listed below.

Node 1: Characters of the extant Hominoidea

Large numbers of characters defining the ancestral hominoid morphotype have been listed,[3,6,75,76] but the most useful way of summarizing them is to extend Harrison's[6] list of discrete functional complexes for the postcrania:

1. Differential usage of the forelimb, including increased potential for raising arms above the head, for extending the forelimb at the elbow joint, and for rotation of the forelimb.
2. Greater flexibility of the wrist and opposable thumb.
3. More erect posture during locomotion and feeding with broadening of thorax and loss of tail.
4. Greater mobility at the hip and ankle joints.
5. To this may be added the characters of the dentition related to diet, such as broad spatulate central incisors, low crowned premolars with loss of honing on P_3, and relatively broad molars with low rounded cusps.

Node 2: Characters of the Hominidae linking Ponginae and Homininae[3,6,76]

1. Skull with enlarged maxillary sinuses, orbits higher than broad, increased alveolar prognathism with elongated premaxilla, facial lengthening with elongation of nasal bones and narrow incisive foramen.
2. Great increase in size of incisors relative to molar size.
3. Robust and enlarged premolars relative to molars, upper premolar heteromorphy reduced.
4. Reduced molar cingula.
5. Mandible robust with large inferior transverse torus.
6. Distal humerus with deep sulci either side of lateral trochlear keel.

Node 3: Characters of the Homininae linking the African apes and humans

The characters distinguishing this node are both morphological (described in more detail in Box 3) and molecular,[76,77] for example:

1. Myoglobin chain, positions 23 and 110.
2. Two shared substitutions in the fibrinopeptide A and B chains.
3. Twenty-seven shared substitutions in the DNA and mitochondrial DNA chains and three deletions.
4. Fusion of os centrale in the wrist.
5. Enlarged supra-orbital torus.

Evidence resolving the trichotomy at node 3 is provided for a human-chimpanzee grouping[78–87] and for a chimpanzee-gorilla grouping.[88–96] Some of these workers recognize that the trichotomy is not yet resolved,[97–101] but others insist that it is resolved in favor of the chimpanzee-human grouping.[81,87,102]

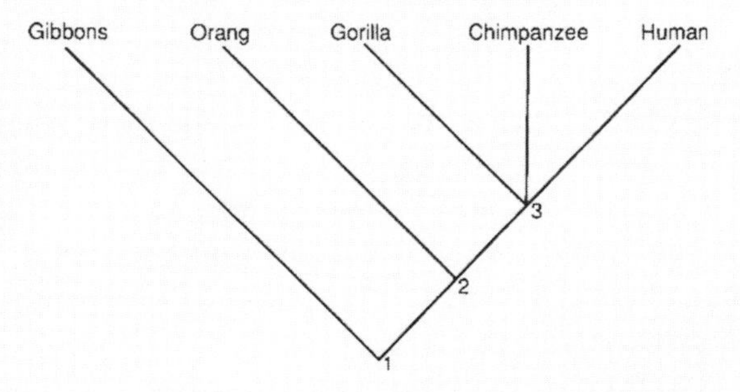

tionship between chimpanzees and gorillas, with humans the sister group to the combined African apes. Morphological evidence is also ambiguous, so in my opinion the trichotomy shown in Box 1 is not yet resolved.

Hominoid Origins

The earliest fossil record for the Hominoidea is in early Miocene deposits in East Africa. The genus *Proconsul* is widely distributed in Uganda and

Kenya in sediments ranging in age from 22 to 17 million years ago (Myr).[2] *Proconsul* is recognized as having mainly ancestral catarrhine characters of the skull and dentition, but to be linked with the Hominoidea on the basis of a few postcranial characters.[3–5] Some of these have been appropriately criticized,[6] but I consider that the characters listed in Box 2 are well established as hominoid synapomorphies shared by *Proconsul*. It has also been claimed recently that *Proconsul* lacked a tail,[7] as do extant hominoids, and that it had a relatively larger brain than comparably sized monkeys[8] (Box 2), and there is now little doubt that *Proconsul* is phylogenetically linked with the Hominoidea.

Species diversity in the early Miocene East African hominoids is high, with two to three species of *Proconsul*-related forms and two to three species of small-bodied apes present at many sites.[9] Differences are partly based on size: the two species of *Proconsul* from Rusinga Island are estimated at 9 kg and 26–38 kg based on good postcranial evidence.[10] The small species, formerly *P. africanus,* is now being referred to a new species,[11] with *P. africanus* being restricted to Koru and Songhor, together with the similarly sized but morphologically distinct *Rangwapithecus gordoni.* Also present at these sites is a much smaller form related to the latter and a larger species of *Proconsul, P. major.*

Associated with *Proconsul* are a number of less well known fossil primate taxa, many of which are inferred as being hominoid but which lack the necessary body parts to be sure. The "small-bodied ape" *Limnopithecus legetet* has been grouped with *Proconsul* in the Hominoidea on the basis of dental evidence.[12] An older discovery that has become significant is the taxon *"Proconsul (Xenopithecus) hamiltoni"* described for a maxilla fragment from Lothidok Hill in northern Kenya,[13] because the Eragaleit beds from which this fossil came (together with some additional undescribed specimens) have recently been dated at between 24.3 and 27.5 Myr.[14] This is several million years older than the previously oldest known fossil ape from Meswa Bridge in the Koru area,[15] and if this specimen is indeed a fossil hominoid it puts the date for hominoid origins at least to ~25 Myr.

Proconsul has been shown to have been mainly associated with forested paleoenvironments in East Africa. The faunas and floras from Songhor, Koru and Mfwangano in Kenya, and Napak in Uganda all have clear affinities with present-day rainforest biotas.[16] Some of the Rusinga Island floras are closer to more seasonal tropical woodlands, although there are wet forest elements as well,[17] and the faunas have both forest elements and non-forest at different stratigraphic horizons,[16] indicating a mixture of habitats during the Miocene. The community structure of the mammalian

BOX 2 Relationships of the fossil hominoids, with characters distinguishing the branching points

Two fossil clades are recognized here. The first consists of the Proconsulidae at node 0 and includes *Proconsul,* for which the best evidence is available. This clade also includes *Rangwapithecus, Nyanzapithecus,* and the Lothidok hominoid (see text). Some of the characters distinguishing node 0 are the same as for node 1, thereby providing evidence for the hominoid status of *Proconsul,* as follows:

1. Low crowned P_3 with length greater than height, cusp heteromorphy on upper premolars reduced, reduction in breadth of upper molars (breadth/length 115–121%), and I^1 nearly as broad as high (breath/length index 80–90%).[12]

2. Relative increase in brain size: encephalization quotient 48.8% compared with a range of values of 22 to 41% for 11 species of monkey.[8]

3. Medial torsion of the humeral head and elongated vertebral border of the scapula, leading to increased potential for raising arms above the head.[5]

4. Strong medial and lateral keels of the trochlea.[24]

5. Increased mobility at the ulnocarpal joint.[103]

6. Phalanges of hallux broad and robust, and complete opposability of the thumb.[23]

7. Absence of tail.[7]

The second node, at position 1a on the cladogram, is probably a heterogeneous association of taxa, grouped here because we do not at present understand their phylogenetic relationships. It is referred here to the Dryopithecinae and it includes three groups: (1) *Dryopithecus* itself in tribe Dryopithecini; (2) a newly proposed tribe, Afropithecini with *Afropithecus,*[26] *Heliopithecus*[32] and *Otavipithecus,*[33,34] together with material from Maboko Island and Nachola that has been referred in the past to *Kenyapithecus;* and (3) the Kenyapithecini, which includes *Kenyapithecus wickeri* from Fort Ternan,[38] *Griphopithecus alpani* from Pasalar, Turkey,[37] and the postcrania from Klein Hadersdorf.[36] The characters distinguishing node 1a are as follows:

1. Premolar enlargement combined with retention of varying degrees of cusp heteromorphy.[32]

2. Molars with low cusps, flattened occlusal surfaces,[47] reduction in cingulum and squarish shape.[34]

3. Reduction in size of superior transverse torus of the mandibular symphysis.[3]

4. Robust but low crowned canines.

At node 1a, the three groups have the following distinctive characters:

Afropithecini. Increase in enamel thickness of the molars to intermediate thick, with relative enamel thickness[48] of 17.35 (ref. 47); and further enlargement of the premolars. The postcrania (from Maboko[42]) retain primitive hominoid characters similar to the morphology seen in *Proconsul.*

Kenyapithecini. Increase in enamel thickness, but the enamel is even thicker than in afropithecins, as measured for the Pasalar sample, with relative enamel thickness of 19.71 (ref. 47), although it has yet to be measured for any of the African *Kenyapithecus;* this character may be diagnostic of node 1a if it can be shown to be ancestral for both pongines and hominines.[48]

Dryopithecini. Limb bones with rounded shafts; distal humerus with rounded capitulum, deep olecranon fossa and prominent trochlear keels.[49] These characters may be homologous with those of the orang-utan, or with those of the African apes and humans, but not with both as there is now evidence of independent derivation of orang postcranial characters from the other great apes.[54]

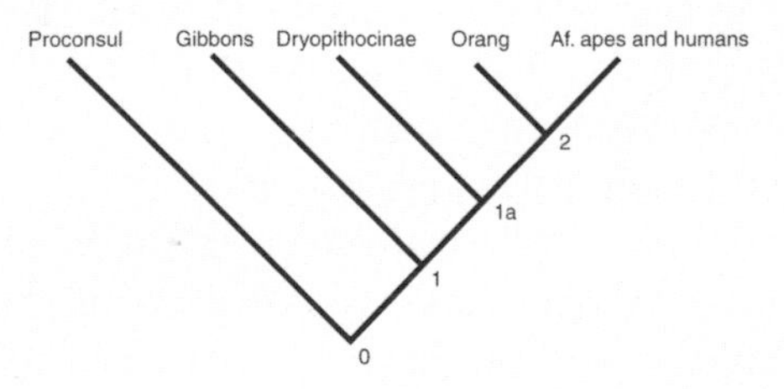

faunas is similar to modern forest communities,[18] again with the exception of some of the Rusinga faunas. It may be that the species difference between the Rusinga hominoids and those from all the other East African sites reflects these apparent ecological differences, for there are similar differences in several other groups of mammals.[18]

All of the *Proconsul* species appear to share a form of arboreal quadrupedalism[19] that was little changed from their catarrhine ancestors.[20] Evidence from the wear patterns of teeth shows that they also shared similar diets of soft fruit and young leaves,[21] and it seems likely that the various species operated within similar ecological constraints, although there is evidence of some ecological diversification in contemporaneous hominoids both in diet, for example the more folivorous *Rangwapithecus,*[21] and in locomotion, for example the long-limbed *Dendropithecus.*[22] The recent conclusion that the *Proconsul* hand had a fully opposable thumb, with rotation at the trapezium–first metacarpal joint at least as extensive as in other hominoids,[23] is of particular significance in combination with thumb length, for *Proconsul* lacks the relatively short thumb of the living great apes and so had hand proportions very similar to those of modern humans.[24] This is particularly interesting in view of the recent emphasis on tool use in chimpanzees,[25] and it would be interesting in this respect to ascertain if there is any evidence of tool use by natural living populations of gorillas and orang-utans.

The Dryopithecinae

In the period from 17 to 12 Myr there appears a second radiation of hominoid species. Their relationships with each other and with later hominoids is still uncertain, but as they represent different evolutionary trends I am going to distinguish them taxonomically at the level of tribe, where they have been distinguished at generic level before. Three groups of fossil hominoid may be recognized, manifesting two rather distinct evolutionary trends. The three groups, and their constituent fossils, are listed in Box 2.

The fossil record of the first group, the Afropithecini, is confined to the African continent. Three important fossil sites are known, the Turkana sites of Buluk and Kalodirr yielding *Afropithecus,* dated to older than 17 Myr,[26] the Arabian site of Ad Dabtiyah, yielding *Heliopithecus,* with a faunal age of about 17 Myr,[27] and the East African sites of Maboko Island and Nachola, yielding *"Kenyapithecus" africanus,* and both dated about 15 Myr.[28,29] These three genera seem to form a natural grouping (Box 2) and in the past they have been confused with one another: in commenting on the original description of the Buluk specimens, Delson[30] suggested that they should be attributed to *Kenyapithecus* (rather than to *Sivapithecus,* as they then were[31]), and in an addendum to their original description of *Heliopithecus,* Andrews and Martin[32] recognized the affinities of that genus to *Afropithecus,* which was published just before *Heliopithecus.* The recently described taxon *Otavipithecus namibiensis* from 13 Myr deposits in Namibia[33] has been compared with this group[34] on the basis of shared dental characters, but cranial and postcranial material is needed to clarify this point. This combination of specimens could be recognized either by grouping them into a single genus or into a tribe, and on present evidence I favor the latter alternative and suggest the Afropithecini as a suitable name. This follows from the probable distinction between the Fort Ternan and Maboko[35,36]/Nachola[29] material: because the type species of *Kenyapithecus* is *K. wickeri* from Fort Ternan, the generic name *Kenyapithecus* has to remain with the Fort Ternan material and forms the root for the second tribe Kenyapithecini.

Later in the Miocene, the hominoid fossil record becomes sparse in Africa. *Kenyapithecus* is known from middle Miocene sites such as Fort Ternan[35] (see above), about 14 Myr, but the rest of the fossil record is restricted to a small number of indeterminate isolated teeth and fragmentary hominine jaws (see below). Similar-aged sites in Eurasia are also rare, but hominoids very similar to *Kenyapithecus wickeri* are known from 15 to 14 Myr sites in Turkey, Pasalar and Candir,[37] and Czechoslovakia and Austria (Neudorf Sandberg and Klein Hadersdorf). These fossils constitute the second group recognized here, the Kenyapithecini.[38] The Pasalar hominoids have been assigned in the past to *Sivapithecus,* but new material shows that at least one of the species lacks the derived subnasal morphology of that genus. For this reason, we have provisionally allocated the Pasalar species to *Griphopithecus.*[39] The third group comprises the European fossils that are mostly attributed to the genus *Dryopithecus.* The Dryopithecini are known from 12 to 8 Myr deposits in France, Austria, Germany, Spain and Hungary, and similar fossils from Georgia through to China have also been attributed to this genus.

The Afropithecini is characterized by thickening of the enamel on the molars (only actually measured on *Heliopithecus*[32]), by broadening of the molars, reduction in molar cusps and ridges so that the occlusal surfaces become more flattened, hyperrobusticity of the canine, the breadth of which is almost as great as length, and enlargement of the premolars, particularly the uppers, which are 92–97% the size of the first molar. The premolar enlargement is greater than is seen in any other hominoid, living or fossil, except for the Moroto palate,[40] which on this basis is also included in the Afropithecini.[32,36] At least two premolar morphologies are included in this combination, *Afropithecus* and the Moroto palate having strong cusp heteromorphy on the third premolar and the Maboko *"Kenyapithecus"* having the cusps more nearly equal in size.

Enamel thickening of the molars and enlargement of the premolars seem likely to be linked to a dietary change in the afropithecin fossils. This may be an increased component of hard fruit in the diet, although the evidence for this is ambiguous.[41] The postcrania are only known for *"Kenyapithecus"* from Maboko Island, and they indicate little change from the generalized arboreal quadrupedalism present in the early Miocene hominoids like *Proconsul.*[42] Little is known of the ecological context of the afropithecin fossils, but while the best record is again from Maboko Island, there is some doubt about the ecological interpretation of this site. On the basis of the mammalian fauna, the hominoids appear to be associated with tropical forest, albeit a dry seasonal and probably deciduous forest.[43] The land gastropods appear to indicate more arid conditions, although still with tree cover,[44] indicating a high degree of climatic seasonality. It is hard to reconcile this interpretation with the apparent frugivorous adaptations of *Kenyapithecus* and the presence in the fauna of at least four other primates (one other hominoid, one monkey and two small-bodied apes[45,46]), but the interpretation of seasonal forest is consistent with the little that is known about the other afropithecin sites as well.

The Kenyapithecini appear to have thicker enamel than the afropithecins,[47,48] although enamel thickness has yet to be measured on any of the East African specimens. They occur in deposits slightly later than most afropithecins. Their postcranial remains are little different from those of the afropithecins (as determined by the Maboko postcrania[42] for the latter). The kenyapithecins are represented by the humerus from Fort Ternan,[36] undescribed remains from Pasalar and perhaps the Klein Hadersdorf specimens.[49] The paleoenvironments also appear to have been little different from those of the afropithecins, with closed woodland–forest indicated for Fort Ternan,[18,38,50] although some authors infer open country or even grassland for this site.[51] At Pasalar, the evidence is strongly in favor of subtropical forest, probably evergreen but with a pronounced dry season,[52] and current work on the microwear of the Pasalar specimens, which are the oldest ones known to have thick enamel, indicates a diet of small hard objects, probably fruits.[53]

The Dryopithecini retain the primitively thin enamel of early hominoids,[47] and in some aspects of the skull and dentition it is little advanced over proconsulids. The subnasal morphology is slightly different from that of *Proconsul,* with some reduction in size of the incisive foramen, but there is no premaxillary lengthening similar to that of extant great apes.

On the other hand, the morphology of the glabella region and the greater development of the brow ridges may indicate affinities with the African apes and humans, but the polarity of these characters is uncertain[3] (see Box 4). Postcranially, the limb bone shafts are rounded, the humerus shaft is straight with a convex deltoid region, and the elbow joint has increased ranges of extension and supination-pronation combined with stability, as seen in the distal humerus, which has a deep olecranon fossa, rounded capitulum and well-developed medial and lateral ridges of the trochlea.[49] These adaptations indicate upper arm and elbow function similar to that of the living great apes, with below-branch arboreal adaptations and elbow joints modified for stability;[42] this might be evidence of a relationship, although the absence of some of these characters in *Sivapithecus*[54] may invalidate this conclusion.

The environment of the dryopithecins appears to have been forested. The vegetation suggested by floral evidence is of mesophytic forests, subtropical to warm temperate in nature, but probably partly deciduous and strongly seasonal.[55] It is probable that the dryopithecin apes were mainly arboreal, but it is not clear at this stage what their diet was in such an environment. Their abundance in such sites as Rudabanya, Hungary, and Can Llobateres, Spain, and their wide distribution across the whole of southern Europe, demonstrates a successful adaptation spanning 3–4 Myr, but it is not possible with our present knowledge to say what was the basis of this success.

Ponginae

Over the past decade it has become accepted that some of the fossils formerly thought to be human ancestors are in fact on the line leading to the orang-utan. It has been shown earlier that this line branched off before the separation of humans and the African apes, so that these fossils are now considered remote from human ancestry. *"Ramapithecus"* was once considered to be a human ancestor, but it is now viewed as a synonym of *Sivapithecus.*[56] This has led some authors to make *Sivapithecus* a human ancestor,[57] and others to recognize the affinities of these fossil genera with the orang-utan but provide evidence linking the orang clade with humans.[58] The third option, and the one that I prefer,[59] is that *Sivapithecus,* including *"Ramapithecus,"* is related to the orang-utan and that they branch at node 2 in Box 1. The evidence for these options is summarized in Box 3.

The evidence supporting a *Sivapithecus*–orang-utan relationship has been questioned recently on the basis of new postcranial remains of *Siva-*

pithecus. Two new humeri from Pakistan are said to retain[54] ancestral characters similar to those seen in *Proconsul* and *Kenyapithecus,* with the proximal shaft curving laterally and anteriorly (that is, convex laterally as in Old World monkeys), and a flat deltoid plane,[54] whereas the distal articular surface has several derived great ape features. This combination of characters in *Sivapithecus* indicates that, if it is indeed related to the orang-utan, characters like the straight shaft and convex deltoid plane that are present in the extant great apes must have arisen independently in the orang-utan and the African apes. Alternatively, it may be that *Sivapithecus* precedes the splitting of the great apes, so that the combination of cranial characters shared by it and the orang-utan must have arisen independently. In my view, the evidence of the face (Boxes 3, 4) is of greater significance, for it involves several functional complexes of the nose, orbits and

BOX 3 Relationship of *Sivapithecus* with humans or the orang-utan

Two options have been put forward in the literature: (1) *Sivapithecus* is related to humans,[57] which has little support; and (2) *Sivapithecus* is related to the orang-utan.[57] Both of these may be combined with the third possibility, that the orang-utan is the nearest living relative to humans,[58] although there is much evidence against this proposition (Box 1). The characters proposed in support of these propositions are listed below.

Evidence for *Sivapithecus*-human relationship

1. Low cusped molars with thick enamel (although this is more likely to be an ancestral character for the pongine and hominine clades (Box 2).
2. Robust mandibles with shallow, broad mandibular bodies.
3. Reduced canines, upper canine being reduced mesiodistally, and with reduced canine sexual dimorphism.
4. Tendency toward enlarged P_3 metaconid.

Evidence for *Sivapithecus*–orang-utan relationship

1. Narrow interorbital distance.
2. Broad high zygomatic region.
3. Zygomaxillary foramen above lower rim of orbit.
4. Smooth unstepped nasal floor, with elongated premaxilla, rotated antero-superiorly, with extremely narrow incisive canal.
5. Facial profile concave with extreme airorhynchy (Box 4).
6. Great size discrepancy between the upper incisors.

A number of other characters shared by the orang-utan and *Sivapithecus* may also support this relationship, such as the lack of glabellar thickening and the shapes of the orbits and nose, but their significance is uncertain.

BOX 4 The characters of the homininae and the relationship of *Graecopithecus*

The following characters have been claimed to distinguish the hominine branching point, node 3, Box 1, and their condition in fossil apes is described.

1. Prominent supra-orbital torus, continuous and bar-like in African apes and discontinuous in australopithecines; moderately prominent and discontinuous in *Dryopithecus*[104] and *Graecopithecus.*[62]
2. Prominent glabella: prominent in *Afropithecus,*[26] *Dryopithecus*[104] and *Graecopithecus,*[62] and probably an ancestral hominid character.
3. Frontal sinus present: present in *Proconsul, Afropithecus* and *Dryopithecus,* and so it is almost certainly an ancestral hominoid character. Not known if present in *Graecopithecus,* but may be so if it is associated with increased brow-ridge development.[105]

4. Elongated nasoalveolar clivus of the premaxilla with narrowing of the incisive foramen. Least developed in the gorilla but greater in chimpanzees and australopithecines, with formation of incisive canal.[77] The hominine character state could either be gorilla-like lengthening of the premaxilla, which is not present in *Dryopithecus* but is in *Graecopithecus,* or it could be unlengthened as in *Dryopithecus,* but in both cases the elongated premaxilla and narrow incisive canal in *Graecopithecus* could constitute a hominine synapomorphy. The extreme reduction in diameter of the incisive canal and lengthening and rotation of the alveolar clivus of the premaxilla in the orang-utan are probably derived independently of the hominine condition.

5. Increased klinorhynchy, which has been linked with most of the above characters in a single functional complex relating to the angle of hafting of the face to the cranium.[105] It is difficult to identify this character in unsectioned skulls, but it is likely that some degree of airorhynchy is primitive for hominoids, greatly exaggerated in the pongine lineage and modified toward klinorhynchy in hominines, but as my observations lead me to believe that the bonobo is airorhynchous to some degree, it is difficult to be sure of the ancestral hominine character state. It is also highly doubtful that this character state can be identified in the broken and distorted partial skulls of *Dryopithecus* and *Graecopithecus,* although it is claimed that the former is klinorhynchous because of the brow-ridge development.[104] It is also claimed[106] that the molar cusp arrangement is related to cranial flexion, with the protocone being posteriorly placed relative to paracone in klinorhynchous skulls and transversely placed in airorhynchous forms, but this is the reverse of the usually accepted polarity for cusp orientation, with the primitive trigon having the protocone posteriorly placed.

6. Straight humeral shaft and convex deltoid plane: not known for *Graecopithecus* but present in *Dryopithecus.* Lateral and anterior curvature (that is, laterally and anteriorly convex) and flat deltoid plane represent the ancestral hominoid condition, as present in *Proconsul,* and these characters are present also in *Sivapithecus.*[54] The degree of curvature, however, far exceeds that of other fossil hominoids, including *Proconsul,* and it may be a uniquely derived and functionally related characteristic of the sivapithecines.

7. Characters such as broader incisors and enlarged maxillary sinuses[104] are the only characters apart from the premaxillary/incisive complex (see character 4) that Begun[104] designates for the hominine clade, but I would interpret these as basal hominid characters (node 2 on the cladogram Box 1), not diagnostic of the hominine clade. Thre further characters[106] all appear to be ancestral, non-pongine characters not diagnostic of the hominine clade. They are: canine facets in line with molars (ancestral hominoid character), lingual cusps of the molars posteriorly placed (ancestral catarrhine character; see character 5), and vertical orientation of incisors (non-pongine character).

Classification at family, subfamily and tribe level of the Hominoidea

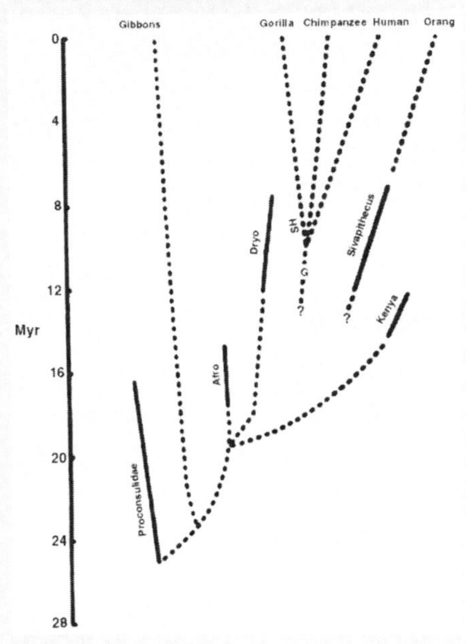

- Hominoidea
- Proconsulidae
- Hylobatidae
- Hominidae
  - Dryopithecinae
    - Afropithecini
    - Dryopithecini
    - Kenyapithecini
  - Ponginae
    - Sivapithecini
    - Pongini
  - Homininae
    - Gorillini
    - Hominini
  - Homininae indet
    - G = *Graecopithecus*
    - SH = Samburu Hills
- Hominidae indet
  - *Lufengpithecus*
  - *Oreopithecus*

palate,[59] and this supports the relationship of *Sivapithecus* with the orang-utan.

The presently accepted age range for *Sivapithecus* is 12.5 to 7 Myr,[60] and it is this date that provides the earliest evidence for the splitting of the orang-utan from the other hominoids. Geographic distribution covers from Indo-Pakistan to Turkey. Paleoenvironments were probably subtropical forests, strongly seasonal but most likely evergreen. On the evidence of postcrania, *Sivapithecus* was similar to other Miocene apes in being an arboreal quadruped[54] with perhaps a terrestrial component as well. Its diet appears to have been soft fruit,[41] which appeared to contradict the suggestion[61] that thick enamel was adaptive for hard-fruit diets; but the more recent conclusion that the microwear of the much earlier thick-enamelled hominoid from Pasalar[53] is indicative of a hard-fruit diet suggests that there was a dietary shift after the development of thick enamel,[47] with dietary change independent of morphology.

Homininae

There are two groups of fossils that can be attributed to the Homininae, that is the African ape and human clade. These are the sample from Greece that has been called *Ouranopithecus macedoniensis* by some,[62] or synonymized with *Graecopithecus freybergi* by others,[63] and the single maxilla from Samburu Hills[29] which has not yet been named. Both lack pongine characters, retaining the primitive condition for the characters that distinguish the pongine clade. The claim that on this basis the Greek fossils fit better with *Australopithecus afarensis*[62] is certainly correct, but based as it is on primitive characters this conclusion has no phylogenetic significance,[64] On the other hand, there are several characters linking *Graecopithecus* to the African ape and human clade.[64] The supra-orbital torus is moderately well developed, with broad lateral orbital rims, but the torus is discontinuous, unlike the continuous bar-like structure in the African apes. The region of glabella is also well developed, although the condition in *Graecopithecus* is similar to that seen in dryopithecins and afropithecins, so that the significance of this character is uncertain.[65] The premaxillary lengthening in *Graecopithecus* is a highly derived character shared with the African apes, and it is combined with the development of an incisive canal and an extended nasoalveolar clivus. These characters and some others are summarized in Box 4.

Hominidae Indet

There are two groups of middle Miocene hominoid that have unknown relationships with the three subfamilies recognized here. These are

Lufengpithecus lufengensis from the late Miocene of China and *Oreopithecus* from the late Miocene of Italy. The Chinese hominoid has been claimed to belong to the pongine clade,[66] but the characters are of doubtful significance, for example the enlarged maxillary sinus, which may be an ancestral hominid character, or the size discrepancy between central and lateral incisors, which, although correct, is linked with an extremely low crowned and heavily buttressed morphology. Other supposed pongine characters of the zygomatic and orbital regions appear more likely to be primitive for the Hominidae.

Oreopithecus is one of the best-known Miocene primates, and yet agreement has yet to be reached as to its taxonomic status. It is considered a separate family of the Cercopithecoidea by some[67] on the basis of its teeth, while others attribute it to a separate family of the Hominoidea[68] on the basis of its postcrania. All agree on the highly derived nature of the teeth, and it is the postcranial adaptations that are of particular interest, for *Oreopithecus* shares with living hominoids a number of apparent postcranial synapomorphies. Some of these characters are also present in *Dryopithecus,* and they have been used to support a relationship of this genus with the living great apes,[49] but their presence in *Oreopithecus,* which is so highly specialized cranially,[68] must cast some doubt on this interpretation.

Hominoid Evolutionary Trends and the Homininae

The hominoid primates appear to have emerged during the Oligocene at least 25 Myr ago as a tropical African group inhabiting tropical forests in an equable climatic regime. The majority were arboreal frugivores occupying much the same niche as equivalent-sized monkeys today.[9] Two trends are apparent during the Miocene period. The first led to increased enamel thickness on the molar teeth combined with little change in postcranial anatomy, and this is seen in the progression from *Proconsul* to *Afropithecus* to *Kenyapithecus.* This progression could be interpreted as an ancestor-descendent lineage, but the relationships of the later thick-enamelled hominoids to this putative lineage, *Graecopithecus* on the one hand and the pongine *Sivapithecus* on the other, are uncertain. These morphological changes may be interpreted functionally in dietary terms as a change to a diet with hard fruit as the main constituent, and this may be related in turn to the drier and more seasonal environments they lived in, but as there was little change in their postcranial skeletons it is likely that they retained their ancestral arboreal locomotor pattern.

The second trend produced change in the postcranial anatomy, leading to functional complexes similar to those of living great apes, but there was less change in the skull and particularly the teeth. There may have been some dietary change, despite the absence of change in enamel thickness, but this has yet to be tested. Postcranial change is seen in the progression from *Proconsul* and/or *Afropithecus* to *Dryopithecus* on the one hand, and to *Sivapithecus* on the other. *Dryopithecus* appears to share many of the great ape synapomorphies of the humerus, but these are present also in *Oreopithecus,* which makes their interpretation difficult. *Sivapithecus* combines some of these same characters of the distal humerus with primitive characters of the humeral shaft, and this confuses the issue still further if it is grouped in the orang-utan clade, for this indicates that some of the postcranial characters shared by the orang-utan and the African apes must have evolved independently. It is likely that locomotor and positional behaviors were related to environmental change, in this case the move from tropical forest habitats in Africa to the subtropical and warm temperate forest habitats of Europe, which would have been both more seasonal and more open, with less complex canopies.

The orang-utan lineage appears to have originated from within the first trend, with further modifications of skull and postcrania, but with little change in environments. There is no good evidence available at present to identify the origin of the African ape and human clade, the Homininae, but two alternatives can be mentioned. First, there may have been a continuation of the thick-enamelled trend independently of the pongine radiation,[48] such as is seen in *Graecopithecus* for instance. The morphology of the alveolar clivus supports this alternative, in which case *Graecopithecus* can be interpreted as a fossil hominine, but there is a possibility that this morphology represents an ancestral great ape character which was primitively retained in *Graecopithecus* and the hominines (and was independently further modified to produce the more derived orang-utan condition).

The second alternative is that there may have been a continuation of the dryopithecin radiation. This is supported by the morphology of the postcrania, which is more similar in *Dryopithecus fontani* to the living great apes and humans than that seen in any other fossil hominoid.[49] Two problems arise here, however, for *Oreopithecus* shares some of the same characters, and the pongine *Sivapithecus* is less advanced in this respect. *Oreopithecus* is clearly not a dryopithecin, which indicates these postcranial characters may be primitive for the great ape and human clade, but if *Sivapithecus* belongs in the orang-utan clade, as I have argued, the shared morphology of the orang-utan and the African apes must have arisen independently. This weakens the phylogenetic potential for these postcranial characters in *Dryopithecus,* and although it could be suggested that *Dryo-*

pithecus could belong in either of the two derivative clades (Ponginae or Homininae), additional supporting evidence would be needed before such a claim could be substantiated.

The earliest fossil evidence for the Homininae is based on fragments of jaw from latest Miocene to Pliocene deposits in Africa. An upper jaw from Samburu Hills[29] appears to have gorilla-like morphology of its teeth combined with thick enamel, which is very much what would be expected if thick enamel is identified as an ancestral hominine character. Its date is uncertain but is thought to have been about 9 Myr,[29] and it is placed at this date on the phylogeny in Box 4 (identified as SH). Two lower jaws, one from Lothagam in northern Kenya now dated at older than 5.6 Myr,[69] and a slightly younger lower jaw from Tabarin[70] from deposits just over 5 Myr, share very similar morphology with each other and with *Australopithecus afarensis.*[71] It may be asked, however, to what extent this morphology would differ from that of thick-enamelled hominoid ancestors from the Miocene, and on what basis, therefore, should these fossils from Lothagam and Tabarin be identified as human ancestors. Early australopithecines are linked with living humans on the basis of shared characters related to bipedalism, but it has yet to be shown that their jaws and teeth differ from putative hominine ancestors. As the Tabarin and Lothagam jaws share only primitive characters with the earliest australopithecines, this is not enough to establish them definitely as human ancestors. The presence of some arboreal retentions in early australopithecines,[72] and their association with paleoenvironments that may have been more wooded[73] than has formerly been thought, should also give rise to caution in identifying human origins.[74]

PETER ANDREWS is in the Department of Palaeontology, Natural History Museum, London SW7 5BD, UK.

References

1. Sarich, V. & Wilson, A. C. *Science* **158,** 1200–1203 (1967).
2. Pickford, M. & Andrews, P. *J. hum. Evol.* **10,** 11–33 (1981).
3. Andrews, P. in *Ancestors: The Hard Evidence* (ed. Delson, E.) 14–33 (Liss, New York, 1985).
4. Walker, A. C. & Pickford, M. in *New Interpretations of Ape and Human Ancestry* (eds. Ciochon, R. L. & Corruccini, R. S.) 325–351 (Plenum, New York, 1983).
5. Walker, A. C. & Teaford, M. *Sci. Am.* **260,** 76–82 (1989).
6. Harrison, T. *J. hum. Evol.* **16,** 41–80 (1987).
7. Ward, C. V., Walker, A. C. & Teaford, M. *J. hum. Evol.* **21,** 215–220 (1991).
8. Walker, A. C., Falk, D., Smith, R. & Pickford, M. *Nature* **305,** 525–527 (1983).

9. Andrews, P. in *Aspects of Human Evolution* (ed. Stringer, C. B.) 25–61 (Taylor & Francis, London, 1981).
10. Ruff, C. B., Walker, A. C. & Teaford, M. F. *J. hum. Evol.* **18,** 515–536 (1989).
11. Walker, A. C., Teaford, M. F., Martin, L. & Andrews, P. *J. hum. Evol.* (in the press).
12. Andrews, P. *Bull. Br. Mus. nat. Hist. (Geol.)* **30,** 85–224 (1978).
13. Madden, C. T. *Primates* **21,** 241–252 (1980).
14. Boschetto, H. B., Brown, F. H. & McDougall, I. *J. hum. Evol.* **22,** 47–71 (1992).
15. Bishop, W. W., Miller, J. A. & Fitch, F. J. *Am. J. Sci.* **267,** 669–699 (1969).
16. Andrews, P. & Van Couvering, J. H. in *Approaches to Primate Paleobiology* (ed. Szalay, F. S.) 62–103 (Karger, Basel, 1975).
17. Collinson, M. *IAAP Newslett.* **8,** 4–12 (1985).
18. Andrews, P., Lord, J. & Evans, E. *Biol. J. Linn. Soc.* **11,** 177–205 (1979).
19. Aiello, L. in *Aspects of Human Evolution* (ed. Stringer, C. B.) 63–97 (Taylor & Francis, London, 1981).
20. Harrison, T. thesis, Univ. London (1982).
21. Kay, R. F. *Nature* **268,** 628–630 (1977).
22. Le Gros Clark, W. E. & Thomas, D. P. *Fossil Mammals Afr.* **3,** 1–27 (1951).
23. Rose, M. D. *J. hum. Evol.* **22,** 255–266 (1992).
24. Napier, J. R. & Davis, P. R. *Fossil Mammals Afr.* **16,** 1–69 (1959).
25. Wynn, T. & McGrew, W. C. *Man* **24,** 383–398 (1989).
26. Leakey, R. E. & Leakey, M. G. *Nature* **324,** 143–146 (1986).
27. Whybrow, P. J., Mcclure, H. A. & Elliott, G. F. *Bull. Br. Mus. nat. Hist (Geol.)* **41,** 371–382 (1987).
28. Feibel, C. S. & Brown, F. H. *J. hum. Evol.* **21,** 221–225 (1991).
29. Ishida, H., Pickford, M., Nakaya, Y. & Nakano, Y. *Af. Study Monog.* **1,** 73–86 (1984).
30. Delson, E. *Nature* **318,** 107–108 (1985).
31. Leakey, R. E. & Walker, A. C. *Nature* **318,** 173 (1985).
32. Andrews, P. & Martin, L. *Bull. Br. Mus. nat. Hist. (Geol.)* **41,** 383–393 (1987).
33. Conroy, G., Pickford, M., Senut, B., Van Couvering, J. & Mein, P. *Nature* **356,** 144–148 (1992).
34. Andrews, P. *Nature* **356,** 106 (1992).
35. Pickford, M. *J. hum. Evol.* **14,** 113–143 (1985).
36. Harrison, T. *Primates* (in the press).
37. Alpagut, B., Andrews, P. & Martin, L. *J. hum. Evol.* **19,** 397–422 (1990).
38. Pickford, M. *J. hum. Evol.* **16,** 305–309 (1987).
39. Martin, L. & Andrews, P. *Am. J. phys. Anthrop.* **12** (suppl.), 126 (1991).
40. Pilbeam, D. *Bull. Peabody Mus. nat. Hist. New Haven* **31,** 1–185 (1969).
41. Teaford, M. & Walker, A. *Am. J. phys. Anthrop.* **64,** 191–200 (1984).
42. Rose, M. D. *J. hum. Evol.* **18,** 131–162 (1989).
43. Evans, E., Van Couvering, J. H. & Andrews, P. *J. hum. Evol.* **10,** 35–48 (1981).
44. Pickford, M. in *New Interpretations of Ape and Human Ancestry* (eds. Ciochon, R. L. & Corruccini, R. S.) 421–439 (Plenum, New York, 1983).
45. Benefit, B. R. & McCrossin, M. L. *J. hum. Evol.* **18,** 493–497 (1989).
46. Harrison, T. *J. hum. Evol.* **18,** 537–557 (1989).
47. Andrews, P. & Martin, L. *Phil. Trans. R. Soc.* **334,** 199–209 (1991).
48. Martin, L. *Nature* **314,** 260–263 (1985).

49. Begun, D. *Am. J. phys. Anthrop.* **87,** 311–347 (1992).
50. Kappelman, J. *J. hum. Evol.* **20,** 95–129 (1991).
51. Retallack, G. *Am. J. phys. Anthrop.* **75,** 260 (1988).
52. Andrews, P. *J. hum. Evol.* **19,** 569–582 (1990).
53. King, T. thesis Univ. London (1992).
54. Pilbeam, D. R., Rose, M. D., Barry, J. C. & Shah, I. *Nature* **348,** 237–239 (1990).
55. Kretzoi, M., Krolopp, E., Lorincz, H. & Palfalvy, I. *Foldt Int. Evi. Jel.* 365–394 (1974).
56. Kelley, J. & Pilbeam, D. in *Comparative Primate Biology* (ed. Swindler, D. R.) 361–411 (Liss, New York, 1986).
57. Kay, R. F. & Simons, E. L. in *New Interpretations of Ape and Human Ancestry* (eds. Ciochon, R. L. & Corruccini, R. S.) 577–624 (Plenum, New York, 1983).
58. Schwartz, J. H. *Nature* **308,** 501–515 (1984).
59. Andrews, P. & Cronin, J. E. *Nature* **297,** 541–546 (1982).
60. Kappelman, J. *et al. J. hum. Evol.* **21,** 61–73 (1991).
61. Kay, R. F. *Am. J. phys. Anthrop.* **55,** 141–151 (1981).
62. Bonis, L. de, Bouvrain, G., Geraads, D. & Koufos, G. *Nature* **345,** 712–714 (1990).
63. Martin, L. & Andrews, P. *Cour. Forsch. Senckenberg* **G9,** 25–40 (1984).
64. Andrews, P. *Nature* **345,** 664–665 (1990).
65. Martin, L. in *Major Topics in Ape and Human Evolution* (eds. Wood, B., Martin, L. & Andrews, P.) 161–187 (Cambridge Univ. Press, Cambridge, 1986).
66. Schwartz, J. *J. hum. Evol.* **19,** 591–605 (1990).
67. Szalay, F. & Delson, E. *Evolutionary History of the Primates* (Academic, New York, 1979).
68. Harrison, T. *J. hum. Evol.* **15,** 541–581 (1986).
69. Hill, A., Ward, S. & Brown, F. *J. hum. Evol.* **22,** 439–451 (1992).
70. Ward, S. & Hill, A. *Am. J. phys. Anthrop.* **72,** 21–37 (1987).
71. White, T. *Anthropos. Brno* **23,** 79–90 (1986).
72. Stern, J. & Susman, R. *Am. J. phys. Anthrop.* **60,** 279–317 (1983).
73. Andrews, P. *J. hum. Evol.* **18,** 173–181 (1990).
74. Wood, B. *Nature* **355,** 783–790 (1992).
75. Delson, E. & Andrews, P. in *Phylogeny of the Primates* (eds. Luckett, W. P. & Szalay, F. S.) 405–446 (Plenum, New York, 1975).
76. Andrews, P. & Martin, L. *J. hum. Evol.* **16,** 101–118 (1987).
77. Ward, S. C. & Pilbeam, D. R. in *New Interpretations of Ape and Human Ancestry* (eds. Ciochon, R. L. & Corruccini, R. S.) 211–238 (Plenum, New York, 1983).
78. Yunis, J. J. & Prakash, O. *Science* **215,** 1525–1530 (1982).
79. Goodman, M., Braunitzer, G., Stangl, A. & Shrank, B. *Nature* **303,** 546–548 (1983).
80. Slightom, J. L., Chang, L., Koop, B. & Goodman, M. *Molec. Biol. Evol.* **2,** 370–389 (1985).
81. Sibley, C. G. & Ahlquist, J. E. *J. molec. Evol.* **26,** 99–121 (1987).
82. Miyamoto, M. M., Slightom, J. L. & Goodman, M. *Science* **238,** 369–373 (1987).
83. Ruvolo, M. & Smith, T. F. *Molec. Biol. Evol.* **3,** 285–289 (1986).
84. Koop, B., Goodman, M., Xu, P., Chan, K. & Slightom, J. L. *Nature* **319,** 234–238 (1986).
85. Hasegawa, M., Kishino, H. & Yano, T. in *Statistical Theory and Data Analysis II* (ed. Matusita, K.) 1–113 (Elsevier, Amsterdam, 1988).

86. Holmes, E. C., Pesole, G. & Saccone, C. *J. hum. Evol.* **18,** 775–794 (1989).
87. Sibley, C. G., Comstock, J. A. & Ahlquist, J. E. *J. molec. Evol.* **30,** 202–236 (1990).
88. Dene, H. T., Goodman, M. & Prychodko, W. in *Molecular Anthropology* (eds. Goodman, M. & Tashian, R. E.) 171–196 (Plenum, New York, 1976).
89. Marks, J. *Cytogenet. Cell Genet.* **34,** 261–264 (1982).
90. Bianchi, N. O., Bianchi, M. S., Cleaver, J. E. & Wolff, S. *J. molec. Evol.* **22,** 323–333 (1985).
91. Dutrillaux, B. *J. Reprod. Fertil.* **28,** 105–111 (1980).
92. Brown, W. M., Prager, E. M., Wang, A. & Wilson, A. C. *J. molec. Evol.* **18,** 225–239 (1982).
93. Hixson, J. E. & Brown, W. M. *Molec. Biol. Evol.* **3,** 1–18 (1986).
94. Templeton, A. R. *Evolution* **37,** 221–244 (1983).
95. Templeton, A. R. *Molec. Biol.* **4,** 315–319 (1987).
96. Krajewski, C. & Dickerman, A. W. *Syst. Zool.* **39,** 383–390 (1990).
97. Hasegawa, M. & Kishino, H. in *New Aspects of the Genetics of Molecular Evolution* (eds. Kimura, M. & Takahata, N.) 303–317 (Springer, Berlin, 1991).
98. Djlan, P. & Green, H. *Proc. natn. Acad. Sci. USA* **86,** 8447–8451 (1989).
99. Smouse, P. E. & Li, W.-H. *Evolution* **41,** 1162–1176 (1987).
100. Holmquist, R., Miyamoto, M. M. & Goodman, M. *Molec. Biol. Evol.* **5,** 201–216 (1988).
101. Saitou, N. *Am. J. Phys. Anthrop.* **84,** 75–85 (1991).
102. Ruvolo, M., Disotell, T. R., Allard, M. W., Brown, W. M. & Honeycutt, R. L. *Proc. natn. Acad. Sci. USA* **88,** 1570–1574 (1991).
103. Gebo, D. L. *et al. J. hum. Evol.* **17,** 393–401 (1988).
104. Begun, D. *Science* **257,** 1929–1933 (1992).
105. Shea, B. T. in *Orang-utan Biology* (ed. Schwartz, J.) 233–245 (Oxford Univ. Press, Oxford, 1988).
106. Dean, D. & Delson, E. *Nature* **359,** 676–677 (1992).

Acknowledgments

I thank L. Aiello, E. Andrews, R. Bernor, D. Cameron, E. Delson, T. Harrison, F. Clark Howell, L. Martin, C. Stringer and M. Wolpoff for critically reading the manuscript.

Bernard Wood

ORIGIN AND EVOLUTION OF THE GENUS *HOMO*

It is remarkable that the taxonomy and phylogenetic relationships of the earliest known representatives of our own genus, *Homo,* remain obscure. Advances in techniques for absolute dating and reassessments of the fossils themselves have rendered untenable a simple unilineal model of human evolution, in which *Homo habilis* succeeded the australopithecines and then evolved via *H. erectus* into *H. sapiens*—but no clear alternative consensus has yet emerged.

Traditionally, *Homo* has been associated with brain enlargement, the acquisition of culture, a reduction in emphasis on mastication as a means of food preparation and breakdown, and a bipedal gait. The australopithecines, on the other hand, are judged not to have advanced much beyond extant apes in relative brain size, to have been dependent on large post-canine teeth for processing food and to have mixed climbing with bipedalism. Any proposal claiming to have identified the earliest evidence for *Homo* needs to demonstrate a shift from the australopithecine to the hominine grade, but the closer to the cusp between the grades the greater the difficulty in distinguishing them. This review draws upon recent developments in taxonomy and systematics and focuses on the evidence for, and the controversies about, the origin and subsequent evolution of *Homo.*

H. habilis: The Case for an Early *Homo* Taxon

A little over 25 years ago, Leakey, Tobias and Napier[1] claimed to have identified a new, and probably the earliest, species of our own genus, *Homo.* They proposed that the new taxon, *Homo habilis* from Olduvai Gorge, Tanzania, was distinct from contemporary australopithecines but acknowledged that with respect to brain and tooth size it was significantly more primitive than *H. erectus,* hitherto the oldest recognized *Homo* species. The proposal attracted strong but conflicting criticism. Some thought that *H. habilis* showed too few advanced features to separate it from *Australopithecus,*[2,3] but others complained that some of the fossils were indistinguishable from *H. erectus.*[4,5] Although both Leakey and Tobias subsequently modified their detailed perceptions of *H. habilis,*[6,7] Tobias still supports its morphological integrity and distinctiveness.[8] Additional fossil evidence[9–13] (Fig. 1) has confirmed the existence of a fossil hominid distinct from both *Australopithecus* and *Homo erectus,*[14,15] and yet this acceptance[16] persists in the absence of a consensus about the nature and relationships of this species. Many are content for the hypodigm (the list of fossils allocated to the species) to comprise most or all the material listed in Table 1 (refs. 17, 18); others opt for more restricted membership.[19–23] Doubts have also been expressed about the very large range of morphological variation in the postcranial material attributed to *H. habilis.*[24–27] Does the hypodigm subsume more than one species? If it is judged to be heterogeneous, which taxa are represented? Do they belong to *Homo,* or is one, or more, an australopithecine? What are the implications for the debate about the lack of "taxonomic space" between *Australopithecus* and *H. erectus?*[4,28] How have attempts to define the genus *Homo* fared when matched with more recent evidence about *H. habilis?* Does the definition of *Homo* need modification, and if so, how? Finally, have phylogenetic analyses clarified the relationships between *H. habilis* and other early hominid taxa?

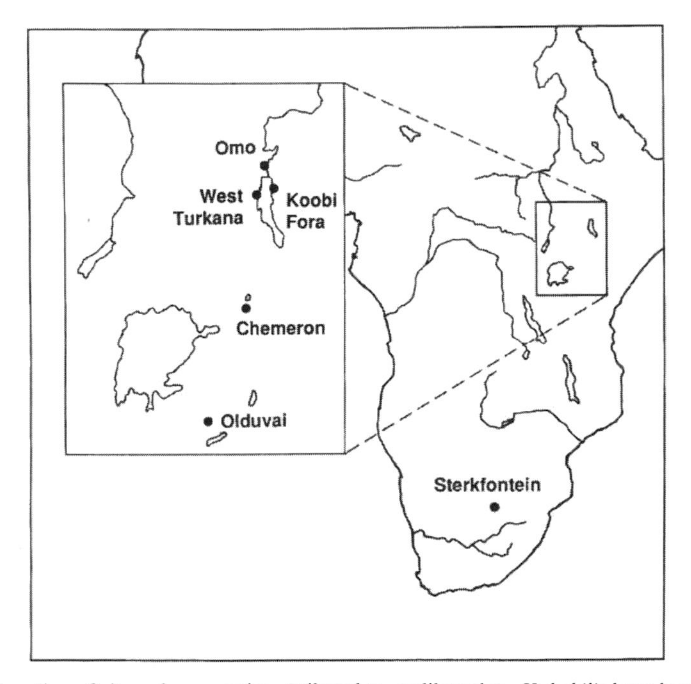

Fig. 1 Location of sites where remains attributed to, or likened to, *H. habilis* have been found.

Table 1 Fossil hominid remains (by site) that are either formally or informally allocated to or declared to have affiliations with *H. habilis* (better-preserved specimens in bold)

Sites	Specimens			
	Skulls and Crania	Mandibles	Teeth	Postcranial
Olduvai (OH)	6, **7, 13,** 14, **16, 24,** 52, 62	**7, 13,** 37, 62	4, 6, 15, **16,** 17, 21, 27, 31, 32, **39,** 40, 41, 42, 44, 45, 46, 47, 55, 56	**7, 8,** 10, **35,** 43, 48, 49, 50, **62**
Koobi Fora (KNM-ER)	807, **1470,** 1478, **1590, 1805, 1813, 3732,** 3735, 3891	819, **1482,** 1483, **1501,** 1502, 1506, 1801, **1802, 1805,** 3734	808, 809, 1462, 1480, 1508, 1814	**813, 1472, 1481, 3228,** 3735
Omo	**L894-1**	**Omo 75-14; Omo 222-2744**	L26-1g; L28-30, 31; L398-573, 1699; Omo 29-43; Omo 33-740, 3282, 5496; Omo 47-47; Omo 74-18; Omo 75s-15, 16; Omo 123-5495; Omo 166-781; Omo 177-4525; Omo 195-1630; Omo K7-19; Omo SH1-17; P933-1	—
West Turkana	Kangaki I site (no specimen no. given)	—	—	—
Chemeron	KNM-BC 1	—	—	—
Sterkfontein	**Stw 53,** SE 255, 1508, 1579, 1937, 2396; Sts 19	—	—	—

Fossil Evidence

The initial description of *H. habilis* referred to seven cranial and postcranial specimens (see Box 1). Of the Olduvai fossils subsequently added to the hypodigm (Table 1), the cranium OH 24 (OH, Olduvai hominid) and the partial skeleton OH 62 are the most important. Although OH 24 was found on the surface it was assigned to Lower Bed I, beneath Tuff 1B.[29] Its early date (Table 2) thus countered criticisms that *H. habilis* was a mistaken amalgamation of earlier, more "*Australopithecus*-like" and later, dentally more advanced, "*erectus*-like" remains from the middle strata of Bed II.[4] The partial skeleton OH 62 (ref. 10) is a potentially rich source of information about the proportions and detailed morphology of the limbs (Box 2). A fragment of a temporal bone recovered from Chemeron, a locality in the Gregory Rift Valley to the south of Lake Turkana, was not

BOX 1 First evidence of *Homo habilis*

In addition to the type specimen, six specimens were assigned to *H. habilis* in the original description.[1]

Type. The juvenile skeleton (OH 7) was found in 1960 at site FLKNN in Bed I. It comprises parts of both parietals, much of the alveolar process and dentition, but little else, of a mandible and at least 13 hand bones.

Paratypes. These are remains resembling the type specimen and include skull fragments and teeth (OH 4 and 6), part of an adult foot (OH 8) and an incomplete skull of an adolescent (OH 13).

Referred specimens. A collection of juvenile cranial pieces (OH 14) and the fragmented cranial vault and dentition (OH 16) of a young adult.

Howell[17] proposed that OH 6, 8 and 14 be excluded from *H. habilis;* Tobias includes OH 6 and 14 in his latest list,[8] but questions the allocation of OH 8.

Features. The authors of the original report listed the following features of *H. habilis* but they are not all distinctive:

Cranial and mandibular

1. Maxilla and mandible smaller than in *Australopithecus,* but equivalent in size to *H. erectus* and *H. sapiens.*
2. Brain size greater than *Australopithecus,* but smaller than *H. erectus.*
3. Slight to strong muscular markings.
4. Parietal bone curvature in the sagittal plane varying from slight (i.e. hominine) to moderate (i.e. australopithecine).
5. Relatively open-angled external sagittal curvature to occipital.
6. Retreating chin, with a slight or absent mental trigone.

Dental

1. Incisors large with respect to those of *Australopithecus* and *H. erectus.*
2. Molar size overlaps the ranges for *Australopithecus* and *H. erectus.*
3. Canines large relative to premolars.
4. Premolars narrower than in *Australopithecus* and within the range of *H. erectus.*
5. All teeth relatively narrow buccolingually and elongated mesiodistally, especially the mandibular molars and premolars.

Postcranial

1. Clavicle resembles *H. sapiens.*
2. Hand bones have broad terminal phalanges, capitate and MCP articulations resembling *H. sapiens,* but differ in respect of the scaphoid and trapezium, attachments of the superficial flexor tendons and the robusticity and curvature of the phalanges.
3. Foot bones resemble *H. sapiens* in the stout and adducted big toe, and well-marked foot arches, but differ in the shape of the trochlea surface of the talus and the relatively robust third metatarsal.

Table 2 Best estimates of the geological age of some of the better-preserved East African fossil evidence for *H. habilis*

Sites				
Olduvai (OH)	1.6	1.8	24	2.0
Koobi Fora (KNM-ER)	13	3891	1.9	
Omo	16	7, 8, 62	1470, 1802, 1813, 3732, 3735	
Myr	1.7	1805	L894-1	

BOX 2 OH 62: A skeleton belonging to *Homo habilis*

The partial skeleton, OH 62, was found in 1986[10]—more than 300 fragments were eventually recovered from the site. Fragmented skeletal material had previously been recovered from Olduvai (Box 1), but none of the earlier specimens provided such clear evidence about the overall size and proportions of the postcranial skeleton of *H. habilis.*
Skull. The palate is the best-preserved part of the skull; details of the floor of the nose, the shape of the tooth row and the size of the teeth point to it belonging to *Homo* and not *Paranthropus.*
Limbs. Sufficient of the humerus and forearm bones and of the femur and leg bones are preserved to enable their lengths to be estimated and to allow the relative lengths of the limbs to be determined.

Compared with modern humans, extant African apes have relatively shorter legs and longer forelimbs. The estimated lengths and robusticity of the humerus and forearm bones of OH 62 suggest that its proportions are remarkably ape-like.[68] Weight/stature relationships are also distinct between the African apes and modern humans; on these criteria, OH 62 is ape-like.[65]
Body weight. Limb bone dimensions predict a body weight of at least 30 kg.[86]

initially referred to *H. habilis,* but a substantial number of its traits were listed as compatible with it.[30] A new assessment of this fragment cites two features of the jaw joint that are apparently unique to *Homo.*[31]

The largest contribution to the *H. habilis* hypodigm in terms of numbers (Table 1) and quality comes from Koobi Fora, another site associated with the Gregory Rift, which lies on the northeast shore of Lake Turkana; recent examinations of the homogeneity of the *H. habilis* hypodigm have focused on the crania KNM-ER 1470 and 1813 which just antedate the evidence from Olduvai (Table 2). Hominid remains recovered from the northern group of Turkana basin sediments (specifically from members G and H of the Omo Shungura Formation) have also been likened to *H. habilis.* These include a fragmented cranium,[12] two mandibles and about 20 isolated teeth[32,33] (Table 1). A cranial fragment recovered from the Nachukui Formation, on the western shores of Lake Turkana, has also been referred to *H. habilis.*[34] The first assessment of the hominid remains recovered from Hadar in Ethiopia and Laetoli in Tanzania suggested that an early *Homo* species may be represented,[35,36] but these specimens have now all been accommodated within a single species, *Australopithecus afarensis.*[37]

A fragmentary skull, Stw 53, from the cave site of Sterkfontein in southern Africa, together with isolated teeth also recovered from Member 5 at Sterkfontein, have been said to resemble *H. habilis.*[38] Several authors have proposed that hominine material[39,40] can also be identified within Member 4 at Sterkfontein, a collection which is dominated by remains attributed to *A. africanus* (Table 1). Attribution to *H. habilis* is also among the options considered for both cranial (SK 847 and SK 27) and dental (SK 2635) remains from Member 1 at Swartkrans.[17,41,42] Although several authors have linked SK 847 with *H. erectus,* recent studies have emphasized its affinities with Stw 53[43,44] and thus indirectly with *H. habilis.*

Proposals that *H. habilis* has been identified at sites beyond Africa, in the Near East and Asia, have not been sustained. The hominid fragments from Ubeidiyah in Israel were listed[1] as possible members of the *H. habilis* hypodigm, and it was tentatively suggested that remains from Indonesia assigned to *Meganthropus palaeojavanicus* may be synonymous with *H. habilis;*[45] both proposals have since been abandoned.[8]

H. habilis: One Species or Two?

Doubts have been expressed about the taxonomic homogeneity of the Olduvai hypodigm of *H. habilis* from the outset (Table 3). The initial criticisms, that there was a temporal basis for the taxonomic heterogeneity,[4] were effectively countered by the discovery of the cranium OH 24 (see above), but Leakey still invoked time to explain at least some of the range of morphology subsumed within *H. habilis.* He cast the mandibles OH 7 and OH 13 as, respectively, the early and late representatives of a "*sapiens*-like" lineage, whose morphology he contrasted with the "protopithecanthropine" features of specimens such as the cranium OH 16 (ref. 7). A subsequent assessment sorted the specimens differently, concluding that morphological differences between OH 13 and 24, on the one hand, and OH 16 and OH 7, on the other, hinted at taxonomic variation.[9]

The additions to the *H. habilis* hypodigm provided by the Koobi Fora remains failed to clarify the situation. Several observers[17,18] stress the taxonomic unity of the augmented hypodigm, while others have re-emphasized the australopithecine affinities of part of the enlarged hypodigm.[46–48] The theme of the morphological dichotomy between OH 13 and 24, and OH 7 and 16 (ref. 9) has been taken up by others,[46,48] who suggest

Table 3 Multiple taxon solutions for crania and mandibles attributed to *H. habilis*

Author(s)	Specimens		Taxon Names	Comments
	OH	KNM-ER		
Robinson (1965)[4]	7	—	*A.* aff. *A. africanus*	C/P_3 ratio; $\bar{C}$ morphology
	13	—	*H.* aff. *H. erectus*	U-shaped mandibular contour; gracile mandibular corpus
Leakey *et al.* (1971)[9]	7, 16	—	*H. habilis*	*H. erectus*–type, evenly curved lambdoid suture
	13, 24	—	*H. habilis/Homo* sp.	*H. sapiens*–type, V-shaped lambdoid suture
Groves (1989)[89]	7, 13, 16, 24	—	*H. habilis*	Narrow premolars; upper face > midface
	—	1470, 1590, 1802	*H. rudolfensis*	Midface > upper face; P_4 large relative to canine size
	—	730, 820, 992, 1805, 1813	*H. ergaster*	Broad premolars; $P_4 < P_3$
Leakey *et al.* (1978)[90]	7	1470, 1590, 1802	*H. habilis*	Enlarged anterior and cheek teeth; robust mandibles; cranial capacity >750 cm^3
	13	992, 1813	*A.* aff. *A. africanus*	Small molars and premolars; cranial capacity approx. 600 cm^3
Stringer (1986)[21]	7, 24	1470, 1590, 1802, 3732	*H. habilis* (group 1)	Large cranial capacity approx. 750 cm^3; flat lower face; large teeth and jaws
	13, 16	992, 1805, 1813	*H. habilis* (group 2) or *H. ergaster*	Small cranial capacity approx. 510 cm^3; projecting lower face; small teeth and jaws
Chamberlain (1989)[23]	7, 13, 16, 24, 37	—	*H. habilis sensu stricto*	Premolars and molars mesiodistally elongated; reduced jaw size relative to neurocranium
	—	992, 1470, 1483, 1802, 1805, 1813, 3734	*Homo* sp.	Mandibular premolar roots complex; shares traits with "robust" australopithecines
Wood (1991)[50]	7, 13, 16, 24, 37, and so on	1478, 1501, 1805, 1813, 3735, and so on	*H. habilis*	Small neurocranium approx. 500 cm^3; upper face > midface; small buccolingually narrow P_3–M_1; M_3 reduction
	—	1470, 1482, 1590, 1802, 3732, and so on	*H. rudolfensis*	Large neurocranium approx. 750 cm^3; orthognathic with midface > upper face; large postcanine teeth; $M_3 > M_2$

that *H. habilis* should be restricted to OH 7 and 16, that is, the larger-toothed, bigger-brained, and presumably larger-bodied, component of the main hypodigm. In this scheme, specimens without these attributes, such as OH 13 and 24 and the crania KNM-ER 1805 and 1813 from Koobi Fora, were judged to be "late-surviving small *Australopithecus* individuals that were contemporaneous first with *H. habilis,* then with *H. erectus.*"[48]

These latter proposals involve removing a substantial part of the *H. habilis* hypodigm from *Homo,* but others have been prepared to accommodate both subsets within the same genus. Although some support a split of the hypodigm based on brain and tooth size,[20,21,49] among other considerations, others have suggested schemes that partition the remains geographically, with *H. habilis* confined to Olduvai, and one or more new species of *Homo* at Koobi Fora.[22,23] Groves and Mazak designated one of these new species *H. ergaster,* but others consider this as most probably synonymous with *H. erectus,*[23] or a subset thereof[50] (see below).

Recent research has compared variation within *H. habilis* with that within living primates in order to assess the probability that *H. habilis* samples a single species, but the results have been equivocal. A comparison of two crania from Koobi Fora (formerly East Rudolf), KNM-ER 1470 and 1813 (Fig. 2), cast doubt on a single-species solution;[51] another, focusing on brain size,[52] supported it. The former study concluded that the contrasts between KNM-ER 1470 and 1813 were greater in degree than equivalent differences between male and female gorilla crania, whereas the latter concluded that the endocranial volumes, 752 and 510 cm^3, of the two fossil crania did not imply a degree of intraspecific variation exceeding that in sexually dimorphic higher primate species. Such considerations have become more sophisticated to the extent that separate attention has been paid to the pattern as well as to the degree of morphological variation.[53] Variables are assessed as "good" or "poor" taxonomic discriminators, with greater emphasis placed on those variables that vary more between than within species closely related to the early hominids.

The two most recent studies of *H. habilis*[8,50] differ in emphasis but are partly consistent in their taxonomic conclusions. In his magisterial review of Olduvai remains attributed to *H. habilis,*[8] Tobias concludes that the features originally said to characterize *H. habilis*[1] (Box 1) have stood the test of time, and has no hesitation in retaining all of the Olduvai hypodigm within *H. habilis* and including the material from Koobi Fora that has

Fig. 2 Photograph of the crania KNM-ER 1470 *(left)* and 1813 *(right)*. These two crania, both dating from around 1.9 Myr, would need to be subsumed within a single early *Homo* species. Although they share a similarly shaped neurocranium, differences in their facial anatomy are less easy to accommodate in a single-species model.

been referred to *H. habilis;* his is a single-species interpretation of *H. habilis*. My own approach was different: I submitted the cranial remains attributed to *H. habilis* to examinations designed to test the null hypothesis of a single early *Homo* species. The degree of variation in the sample was compared with that in another synchronic hominid species, *Paranthropus boisei,* as well as with the degree of variation observed in *H. erectus* and the gorilla. Patterns of variation in early *Homo* were compared with those in *H. sapiens* and *Pan troglodytes,* the two species that are most closely related to early hominids and which share patterns of cranial variation.[53] In the event there were few cases in which the degree of variation in the main *H. habilis* hypodigm exceeded that in the comparative material, but there were several examples (facial shape and premolar tooth morphology, for example) in which the pattern of cranial, mandibular and dental variation differed. However, in each case the evidence for taxonomic heterogeneity came not from the Olduvai part of the hypodigm but from remains referred to *H. habilis* from the Koobi Fora collection. This led me to agree with Tobias and accept the Olduvai hypodigm of *H. habilis* as evidence for that species, but to suggest that the Koobi Fora hypodigm sampled two species of early *Homo,* one conspecific

BOX 3 Evidence for and features of *Homo habilis sensu stricto* and *Homo rudolfensis*

It has been suggested that *Homo habilis sensu stricto* and *H. rudolfensis* are species components of the larger *H. habilis* hypodigm.[50] Specimens allocated to the two taxa are set out below.

H. habilis sensu stricto

Olduvai: OH 4, 6, 8, 10, 13–16, 21, 24, 27, 35, 37, 39–45, 48–50, 52, 62

Koobi Fora: KNM-ER 1478, 1501, 1502, 1805, 1813, 3735

H. rudolfensis

Koobi Fora: KNM-ER 813, 819, 1470, 1472, 1481–1483, 1590, 1801, 1802, 3732, 3891

Skull and Teeth	*Homo habilis s.s.*	*Homo rudolfensis*
Absolute brain size (cm³)	$\bar{X} = 610$	$\bar{X} = 751$
Overall cranial vault morphology	Enlarged occipital contribution to the sagittal arc (but see ref. 8 for a contrary view)	Primitive condition
Endocranial morphology	Primitive sulcal pattern[91] (but see Holloway[92])	Frontal lobe asymmetry[91,93]
Suture pattern	Complex	Simple
Frontal	Incipient supraorbital torus	Torus absent
Parietal	Coronal > sagittal chord	Primitive condition
Face—overall	Upper-face > midface breadth	Midface > upper-face breadth; markedly orthognathic
Nose	Margins sharp and everted; evident nasal sill	Less everted margins; no nasal sill
Malar surface	Vertical, or near vertical	Anteriorly inclined
Palate	Foreshortened	Large
Upper teeth	Probably two-rooted premolars	Premolars three-rooted; absolutely and relatively large anterior teeth
Mandibular fossa	Relatively deep	Shallow
Foramen magnum	Orientation variable	Anteriorly inclined
Mandibular corpus	Moderate relief on external surface	Marked relief on external surface
	Rounded base	Everted base
Lower teeth	Buccolingually narrowed postcanine crowns	Broad postcanine crowns
	Reduced talonid on P_4	Relatively large P_4 talonid
	M_3 reduction	No M_3 reduction
	Mostly single-rooted mandibular premolars	Twin, plate-like P_4 roots, and bifid, or even twin, plate-like P_3 roots
Postcranium		
Limb proportions	Ape-like	?
Forelimb robusticity	Ape-like	?
Hand	Mosaic of ape-like and modern human–like features	?
Hindfoot	Retains climbing adaptations	Later *Homo*-like
Femur	Australopithecine-like	Later *Homo*-like

with *H. habilis* from Olduvai and the other a new species of early *Homo,* already designated *H. rudolfensis* Alexeev, 1986 (Box 3). *H. ergaster* Groves and Mazak, 1975 is not available as the name of the second species of early *Homo* because the type specimen KNM-ER 992 resembles, though is not necessarily conspecific with, *H. erectus.* Thus I have put forward *H. ergaster*[50] as the proper name for the probable African precursor of *H. erectus,* a taxon which would also include the crania KNM-ER 3733 and 3883 and probably also the skeleton KNM-WT 15000, although formal assignment of that specimen must await its detailed description and analysis.

Evidence from Limbs

Morphological characterization of the postcranial anatomy of *H. habilis,* together with its functional interpretation, has proved to be a topic every bit as controversial as those involving the cranial evidence.

The first integrated account of fossil evidence of the foot and leg of *H. habilis,* like that of the cranium, was based on the evidence from Olduvai, and emphasized the ways in which these remains resembled those of *H. sapiens.*[54] But functional conclusions based on more detailed descriptions of foot (OH 8, 10) and leg (OH 35) fossils, which are probably from the same individual,[55] were generally more cautious. They stressed that the unique striding gait of *H. sapiens* has not yet been achieved[56] and suggested that the knee joint may have been imperfectly adapted to bipedalism.[57] Despite these caveats, the inferred functional affinities between the fossils of *H. habilis* and *H. sapiens* were still emphasized, and a bipedal gait of the modern human type was widely claimed for the former species.

Subsequent reassessments of the anatomy of the OH 8 foot stressed its potential for climbing, and its retention of anatomical features seen in living non-human anthropoid apes.[55,58] Although OH 8 apparently possessed the mechanism that transforms the foot into a rigid, close-packed organ during the support phase of bipedal walking,[58] it lacks at least some of the refinements (such as lateral deviation of the heel) present in *H. sapiens.* The same author was also struck by the lack of evidence for a propulsive role for the big toe, an important factor in the bipedal gait of modern humans.[59]

A different and more complex interpretation of the hindlimb anatomy of early *Homo* has emerged from the Koobi Fora evidence, although Tobias[8] correctly cautions that the specimens providing much of the conflicting evidence (in particular the femora KNM-ER 1472 and 1481A and the talus KNM-ER 813) may not belong to *H. habilis*, but to *H. erectus* or *H. ergaster*. But the features that prompted the assignment of KNM-ER 1472 and 1481A to *H. erectus*[60] may not be diagnostic of that species[61] and are likely to be derived features of all *Homo* species, including *H. habilis*. Others have demonstrated that the two Koobi Fora specimens are distinct from australopithecine femora[62] and that a talus, KNM-ER 813, from a similar-aged horizon as the femora, resembles modern human tali much more closely than do australopithecine tali.[63] Thus, there are at Koobi Fora leg fossils whose later *Homo*-like morphology contrasts with that of the more australopithecine-like morphology of the Olduvai remains. These relatively derived remains from Koobi Fora are found alongside a specimen such as KNM-ER 3735, which is judged to resemble the more primitive OH 62 skeleton.[64]

An analysis of estimated stature/body weight relationships has also shown that, whereas predictions based on the two Koobi Fora femora are in line with modern human and archaic *H. sapiens* relationships, they are substantially different from predictions based on the australopithecine-like Olduvai *H. habilis* remains, which instead conform to predictions based on the living African apes.[65] This wide range of morphology has no apparent allometric basis, at least as far as the proximal femur is concerned,[66] and is thus additional evidence for taxonomic heterogeneity in early *Homo*.[26,50]

Evidence of the forelimb skeleton of Olduvai *H. habilis* is meager, with the OH 7 hand providing the most pertinent information. Its fragmentary nature precludes comprehensive morphological and functional analysis, but it is generally interpreted as demonstrating a mosaic of an ape-like carpus with a thumb which could have both been rotated within the hand to allow pulp-to-pulp opposition, as well as used to provide the kind of firm support essential for effective tool manufacture and manipulation.[55,58,67]

Considered with the evidence of the relatively primitive limb proportions of OH 62 (refs. 10, 68), the architecture of the limbs of *H. habilis sensu stricto* is closer to that of the australopithecines than to later *Homo*.

Defining the Genus *Homo*

The task would be greatly eased were agreement to be reached on definitive criteria for the genus *Homo*. Mayr has insisted that the genus is impossible to define and delimit on a purely morphological basis,[69] except in that *Homo* was characterized by "upright posture, with its shift to a terrestrial mode of living and the freeing of the anterior extremity for new functions which, in turn, have stimulated brain evolution." Thus defined, Mayr's genus *Homo* embraced the then known australopithecines, *H. erectus* and *H. sapiens.* Simpson[70] subsequently warned, in the context of what he described as "the chaos of anthropological nomenclature," that genera are "necessarily more arbitrary and less precise in definition than the species," but suggested that a genus can be defined as "a group of species believed to be more closely related among themselves than to any other species placed in other genera."

The dictum adumbrated by Mayr and Simpson, that it is the species that make the genus and not *vice versa,* is borne out by critical inspection of the definitions of *Homo* provided by Le Gros Clark[71] and Robinson.[72] The former is a concatenation of the features of what was then known of *H. sapiens, H. neanderthalensis* and *H. erectus.* This is reflected in some of the proposed components (such as brain size with a range of 900–2,000 cm^3), but the definition also includes references to morphological trends—for example a centrally situated foramen magnum, reduced lingual cusp on the third lower premolar, relative size reduction of the third molar—and follows Mayr to the extent that it specifies a "limb skeleton adapted for a fully erect posture and gait."[71] When announcing *H. habilis,* Leakey *et al.*[1] made only modest amendments to Le Gros Clark's definition of *Homo,* and Robinson later emphasized only three additional features: a nasal sill, "harmoniously proportioned" anterior and postcanine teeth and incompletely molarized crowns in deciduous first lower molars.[72]

Few explicit definitions of *Homo* have been offered since, but the increasing use of cladistic methods within hominid paleontology means that it is possible to deduce the character-state changes that define the *Homo* clade. Cladistic analyses of early hominids are at present confined to cranial evidence, and surprisingly few of these analyses combine data from early hominids, later *Homo* and extant *H. sapiens.* A. T. Chamberlain and I investigated the distribution of 90 cranial, mandibular and dental characters across a range of hominid species and outgroups[15] and found that the resulting *Homo* clade could be defined by eight character-state changes (Box 4). Others[20] have also emphasized reduced prognathism, dental reduction and brain enlargement as factors defining the *Homo* clade.

BOX 4 *Homo* defined cladistically

Cladistic, or phylogenetic, analysis is a method of investigating the relationships between taxa. It makes the assumption that morphological evolution is parsimonious, that is, similar morphological features, or characters, are more likely to be derived from a shared common ancestor than to have evolved independently. Each character has a hypothetical morphocline of states, from primitive to derived. The primitive condition is assumed to be that which is most widely distributed in related taxa and/or the condition which has the least complex developmental history. Character states uniquely defining a taxon are known as autapomorphies; their distribution does not help in establishing relationships. Character states distributed within part of the cladogram (defining a "node") are called synapomorphies, and states common to all the taxa in a cladogram are referred to as symplesiomorphies. Each character and the distribution of its states among the taxa to be assessed allows the generation of a cladogram. The cladogram, or pattern of relationships, supported by the majority of characters is assumed to be the most probable hypothesis of relationships.

The cladogram presented above is the most parsimonious generated from a set of 90 cranial, mandibular and dental characters.[15,50] The resulting *Homo* clade is defined by the following character-state changes at node A.

1. Increased cranial vault thickness.
2. Reduced postorbital constriction.
3. Increased contribution of the occipital bone to cranial sagittal arc length.
4. Increased cranial vault height.
5. More anteriorly situated foramen magnum.
6. Reduced lower facial prognathism.
7. Narrower tooth crowns, particularly mandibular premolars.
8. Reduction in length of the molar tooth row.

The two species comprising the original *H. habilis* hypodigm, *H. habilis sensu stricto* and *H. rudolfensis,*[50] share a hypothetical ancestor with each other which neither shares with any other taxon. That sister group is defined by the five character-state changes at node B.

1. Elongated anterior basicranium.
2. Higher cranial vault.

3, 4. Mesiodistally elongated M_1 and M_2.

5. Narrow mandibular fossa.

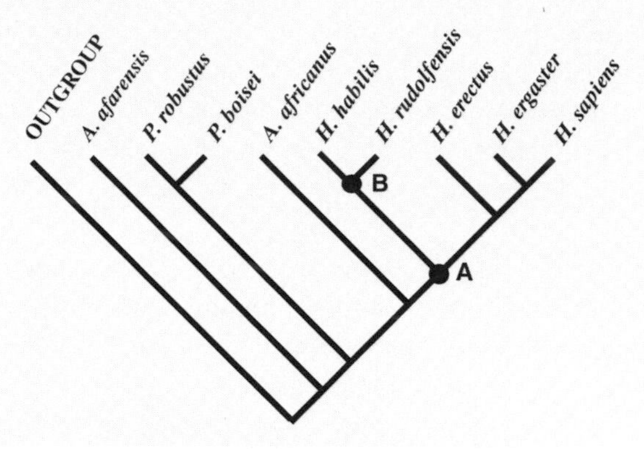

Diagnostic Features of *H. habilis*

H. habilis was characterized[1] by the features listed in Box 1. Of the eleven that relate to the cranium, mandible and dentition, four—the absolute size of the maxilla and mandible, brain size, the degree of parietal sagittal curvature and molar crown size—were said to be intermediate between the condition in the australopithecines and in the then oldest hominine taxon, *H. erectus.* In two characters the hypodigm was specified as resembling existing taxa, namely occipital sagittal curvature *(H. sapiens)* and premolar width *(H. erectus).* The reference to a retreating chin, with no mental trigone, does not suggest a significant departure from the australopithecine condition, and another character, the degree of cranial muscular marking, is too nonspecific to be helpful. The diagnostic dental features are relative incisor and canine size and the shape of the tooth crowns, with the buccolingual narrowing of the mandibular premolars and molars a particularly important feature. But others dispute the uniqueness of narrow premolars among Plio-Pleistocene hominids.[4,73,74] As for the implications of the postcranial skeleton on gait, all the remains that have been entertained as belonging to the hypodigm of *H. habilis* have been interpreted as belonging to a bipedal hominid, but the extent to which *H. habilis* was also a climber is still debated.[75,76]

The 1964 definition was later expanded[77] and emphasized the peculiarities of the morphology of the mandibular canines and premolars of *H. habilis* and differences in the trend of mandibular premolar crown size between *H. habilis* and australopithecines. Tobias[16] has cited additional "critical features of the morphology of *H. habilis.*" These include the relative crown size of the second and third lower molar, vertically thin brow ridges, minimal pneumatization of the cranium, a prognathous face, an anteriorly situated foramen magnum, a short basicranium, coronally orientated petrous bones and a "very robust" mandible with a slight chin. However, he does not indicate whether any of these features are unique to *H. habilis.* In an encyclopedic summary of the features of *H. habilis,* he refers to 344 cranial, mandibular and dental traits, but although their expression is meticulously recorded for *Australopithecus africanus, A. robustus, A. boisei* and *H. erectus,* their wider comparative context is not explored and so their taxonomic and phylogenetic value has yet to be demonstrated.

Cladistic analyses have made little contribution to the search for distinctive features, or autapomorphies, of *H. habilis.* In an early study,[78] an endocranial volume of more than 600 cm^3 was cited as a character distinguishing *A. habilis* (that is, *H. habilis*) from *A. africanus,* but in a wider taxonomic context this cannot be regarded as anything other than a part of what has been called a "combination" definition.[79] Others were unable to identify any autapomorphic features.[80] This was implicitly conceded in another analysis[47] but the value of this latter is reduced because no distinction was made between "gracile" australopithecines from southern Africa and the Olduvai hypodigm of *H. habilis.* A cladistic analysis based on quantitative cranial data[81] deduced only one *H. habilis* autapomorphy, a relatively narrow mid-face, but Chamberlain and I (more recently) identified two further *H. habilis* autapomorphies using an augmented data set[15]—an elongated anterior basicranium and a mesiodistally elongated crown in the first upper molar. When the two proposed species subdivisions of the early *Homo* hypodigm, *H. habilis sensu stricto* and *H. rudolfensis,* are included separately in a cladistic analysis, they are linked as sister taxa within a clade defined by five character states (Box 4).

Phylogenetic Relationships

Phylogenetic analyses that have included *H. habilis* as a separate and unified species have almost always concluded that it is part of a clade, or monophyletic group, corresponding to the genus *Homo,* along with *H. erectus* and *H. sapiens.* In this clade, *H. habilis* is usually, but not universally, nominated as the sister taxon of *H. erectus* and *H. sapiens,*[14,15,20,73,81–83] species which are themselves clearly defined cladistically.[15,20] There is, however, substantially less consensus about the relationships of *H. habilis,* and the *Homo* clade, with australopithecine taxa (Fig. 3). The position of *A. africanus* is pivotal. Some have interpreted it as exclusively ancestral to either the *Homo* (Fig. 3d, e) or the "robust" australopithecine clades (Fig. 3a), but others see it as the common ancestor of both (Fig. 3b, c). The results of two formal cladistic analyses[14,15] concur in that *A. africanus* is strongly associated with neither the *Homo* nor the "robust" australopithecine clades (Fig. 3f). This confusion is caused not for the lack of features shared by *A. africanus* and either clade, but because this species represents a mosaic of relatively primitive characters and traits that are derived either in the direction of *Homo* (such as the cranial vault), or of the "robust" australopithecines (face and masticatory apparatus). This suggests that *A. africanus* is too specialized to be ancestral to either the *Homo* or the "robust" australopithecine clades, or even for it to be the common ancestor of all later hominids (Fig. 3f).

Discussion

There is little doubt that the enlarged hypodigm of *H. habilis* subsumes remains that have traits more derived than those of *Australopithecus* and are yet distinct from approximately contemporary fossils usually attributed to African *H. erectus* or, by some, to *H. ergaster.* But if its taxonomic integrity is retained, it is a species that manifestly embraces an unusually large amount of variation.[21,26,49–51] Those who believe this range to be unacceptably wide, and thus for whom *H. habilis* represents more than one species, disagree about how the hypodigm should be apportioned.

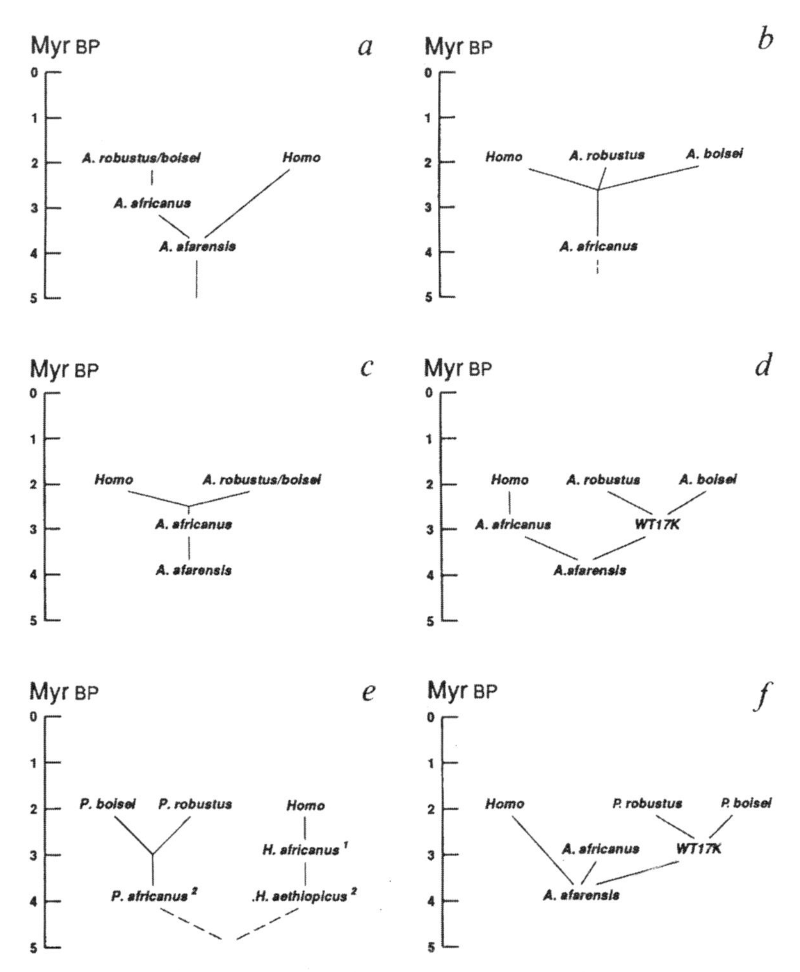

Fig. 3 Comparison of the relationships of the *Homo* lineage/clade in published early hominid phylogenetic schemes: *a,* ref. 83; *b,* ref. 87; c, ref. 14; *d,* ref. 88; *e,* ref. 47 *(1, A. africanus; 2, A. afarensis); f,* ref. 50.

Most (but not all[14,19,21]) regard the Olduvai hypodigm as taxonomically homogeneous (Table 1). The Koobi Fora evidence is either regarded as representative of a single species distinct from *H. habilis sensu stricto,*[43] or as a taxonomically heterogeneous hypodigm that samples both *H. habilis sensu stricto* and at least one other *Homo* species,[50] for which *H. rudolfensis* has apparent priority as a species name.[84] These two early *Homo* species are sufficiently distinct for them to have substantially different grounds for their inclusion in *Homo* (Box 3). Whereas *H. habilis sensu stricto* is hominine with respect to its masticatory complex, it retains an essentially australopithecine postcranial skeleton. *H. rudolfensis,* on the other hand, probably combines a later *Homo*-like postcranial skeleton with a face and dentition adaptively analogous to those of the "robust" australopithecines.

The earliest sound evidence for the larger *H. habilis sensu lato* hypodigm is dated to ~1.9 Myr BP (Table 2), but a series of isolated teeth from members E, F and G of the Omo Shungura Formation and the evidence of the Chemeron temporal fragment may extend its known time range to well beyond 2.0 Myr BP.[31,50,85] The former date, 1.9 Myr BP, is little different from the age of the earliest known evidence for *H. ergaster,* the early African equivalent of *H. erectus.* If early *Homo* does subsume more than one species, the period before 2.0 Myr BP may have witnessed not merely the emergence of one hominid of the *Homo* grade, but a substantial radiation of early hominids, namely *H. habilis sensu stricto, H. rudolfensis* and *H. ergaster* (Fig. 4), each demonstrating a significant and distinctive shift away from the australopithecine adaptive plateau.

This multiple-species solution must now be tested. This will involve establishing sound criteria for confirming hypotheses of taxonomic heterogeneity, including refining the distinctions between the degree and pat-

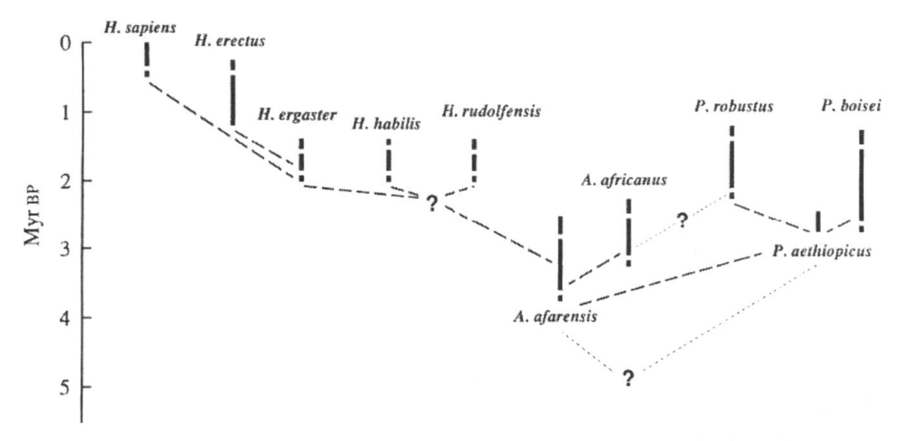

Fig. 4 A phylogenetic scheme for hominid evolution based on a recent analysis of hominid taxonomy and relationships.[50] The horizontal axis corresponds roughly to relative and absolute postcanine tooth size, so that forms with substantial tooth rows are to the right, and species demonstrating premolar and molar reduction and simplification are to the left. The phylogeny represented by the *bold broken lines* assumes that the two best-known "robust" australopithecine species, *Paranthropus robustus* and *P. boisei,* shared a common ancestor which was not unlike *P. aethiopicus* (that is, like KNM-WT 17000). But this monophyletic origin for *Paranthropus* is only marginally more parsimonious than deriving *P. robustus* from *Australopithecus africanus.* The similarities in facial form between *Homo rudolfensis* and a probable *Paranthropus* clade are most parsimoniously interpreted as convergent features.

tern of variability, as well as confirming the suitability of comparative analogues for the assessment of variation both within and between species. Other tests might also be explored, for instance, how realistic are the ranges of body weight estimated for each taxon? Do the proposed taxa imply either unacceptably large, or unreasonably small, levels of sexual dimorphism? Is there any evidence that they occupied the same, or different, parts of the paleolandscape?

If the multiple-species solution survives rigorous examination, other aspects of the hominid record, particularly the archeological evidence, must be integrated with this new interpretation of the paleontological data. Only when morphological studies, embracing both function and life history, are integrated with the contextual, and particularly the behavioral, evidence will we substantially increase our knowledge and understanding of the emergence and early evolution of our own genus. A sound taxonomic interpretation of the fossil record is the foundation upon which more complex hypotheses can be constructed.

BERNARD WOOD is in the Hominid Palaeontology Research Group, Department of Human Anatomy and Cell Biology, University of Liverpool, PO 247, Liverpool L69 3BX, UK.

References

1. Leakey, L. S. B., Tobias, P. V. & Napier, J. R. *Nature* **202,** 7–9 (1964).
2. Le Gros Clark, W. *Discovery* **25,** 49 (1964).
3. Holloway, R. L. *Nature* **208,** 205–206 (1965).
4. Robinson, J. T. *Nature* **205,** 121–124 (1965).
5. Brace, C. L., Mahler, P. E. & Rosen, R. B. *Yrbk. Phys. Anthrop.* **16,** 50–68 (1972).
6. Tobias, P. V. *Curr. Anthrop.* **6,** 391–411 (1965).
7. Leakey, L. S. B. *Nature* **209,** 1279–1281 (1966).
8. Tobias, P. V. *Olduvai Gorge IV: The Skulls, Endocasts and Teeth of Homo Habilis* (Cambridge Univ. Press, Cambridge, 1991).
9. Leakey, M. D., Clarke, R. J. & Leakey, L. S. B. *Nature* **232,** 308–312 (1971).
10. Johanson, D. C. *et al. Nature* **327,** 205–209 (1987).
11. Leakey, R. E. F. *Nature* **242,** 447–450 (1973).
12. Boaz, N. T. & Howell, F. C. *Am. J. Phys. Anthrop.* **46,** 93–108 (1977).
13. Coppens, Y. in *Current Argument on Early Man* (ed. Konigsson, L.-K.), 207–225 (Pergamon, Oxford, 1980).
14. Skelton, R. R., McHenry, H. M. & Drawhorn, G. M. *Curr. Anthrop.* **27,** 21–43 (1986).
15. Chamberlain, A. T. & Wood, B. A. *J. hum. Evol.* **16,** 119–133 (1987).
16. Tobias, P. V. in *Hominidae: Proc. 2nd Int. Congr. Human Paleontology* (ed. Giacobini, G.) 141–149 (Jaca, Milan, 1989).

17. Howell, F. C. in *Evolution of African Mammals* (eds. Maglio, V. J. & Cooke, H. B. S.) 154–248 (Harvard Univ. Press, Cambridge, MA, 1978).
18. Tobias, P. V. in *Ancestors: The Hard Evidence* (ed. Delson, E.) 94–101 (Liss, New York, 1985).
19. Walker, A. & Leakey, R. E. F. *Sci. Am.* **239,** 44–56 (1978).
20. Stringer, C. B. *J. hum. Evol.* **16,** 135–146 (1987).
21. Stringer, C. B. in *Major Topics in Primate and Human Evolution* (eds. Wood, B., Martin, L. & Andrews, P.) 266–294 (Cambridge Univ. Press, Cambridge, 1986).
22. Groves, C. P. & Mazak, V. *Cas. Miner. Geol.* **20,** 225–247 (1975).
23. Chamberlain, A. T. in *Hominidae* (ed. Giacobini, G.) 175–181 (Jaca, Milan, 1989).
24. Wood, B. A. *J. hum. Evol.* **3,** 373–378 (1974).
25. Kennedy, G. E. *J. hum. Evol.* **12,** 587–616 (1983).
26. Wood, B. A. *Nature* **327,** 187–188 (1987).
27. Lewis, O. J. *Functional Morphology of the Evolving Hand and Foot* (Clarendon, Oxford, 1989).
28. Robinson, J. T. *Nature* **240,** 239–240 (1972).
29. Hay. R. L. *Geology of the Olduvai Gorge* (Univ. California Press, Berkeley, 1976).
30. Tobias, P. V. *Nature* **215,** 479–480 (1967).
31. Hill, A., Ward, S., Deino, A., Curtis, G. & Drake, R. *Nature* **355,** 719–722 (1992).
32. Coppens. Y. *Bull. Soc. Geol. Fr.* **21,** 313–330 (1979).
33. Howell, F. C., Haesaerts, P. & de Heinzelin, J. *J. hum. Evol.* **16,** 665–700 (1987).
34. Harris, J. M., Brown, F. H., Leakey, M. G., Walker, A. C. & Leakey, R. E. *Science* **239,** 27–33 (1988).
35. Leakey. M. D. *et al. Nature* **262,** 460–466 (1976).
36. Johanson, D. C. & Taieb, M. *Nature* **260,** 293–297 (1976).
37. Johanson, D. C., White, T. D. & Coppens, Y. *Kirtlandia* **28,** 1–14 (1978).
38. Hughes, A. R. & Tobias, P. V. *Nature* **265,** 310–312 (1977).
39. Broom, R. & Robinson, J. T. *Trans. Mus. Mem.* **4,** 11–85 (1950).
40. Clarke, R. J. thesis, Univ. Witwatersrand (1977).
41. Clarke, R. J. & Howell, F. C. *Am. J. Phys. Anthrop.* **37,** 319–336 (1972).
42. Clarke, R. J. *S. Afr. J. Sci.* **73,** 46–49 (1977).
43. Chamberlain, A. T. thesis, Univ. Liverpool (1987).
44. Grine, F. E., Demes, B., Jungers, W. L. & Cole, T. M. (in the press).
45. Tobias, P. V. & von Koenigswald, G. H. R. *Nature* **204,** 515–518 (1964).
46. Leakey, R. E. F. *Nature* **248,** 653–656 (1974).
47. Olson, T. R. *J. hum. Evol.* **7,** 159–178 (1978).
48. Leakey, R. E. F. & Walker, A. *Science* **207,** 1103 (1980).
49. Wood, B. A. in *Ancestors: The Hard Evidence* (ed. Delson, E.) 206–214 (Liss, New York, 1985).
50. Wood, B. A. *Koobi Fora Research Project IV: Hominid Cranial Remains from Koobi Fora* (Clarendon, Oxford, 1991).
51. Lieberman, D. E., Pilbeam, D. R. & Wood, B. A. *J. hum. Evol.* **17,** 503–511 (1988).
52. Miller, J. A. *Am. J. Phys. Anthrop.* **84,** 385–398 (1991).
53. Wood, B. A., Li, Y. & Willoughby, C. *J. Anat.* **174,** 185–205 (1991).
54. Napier, J. R. *Arch. Biol., Liege* **75** (suppl.), 673–708 (1964).

55. Susman, R. L. & Stern, J. T. *Science* **217,** 931–934 (1982).
56. Day, M. H. & Napier, J. R. *Nature* **201,** 969–970 (1964).
57. Davis, P. R. *Nature* **201,** 967 (1964).
58. Lewis, O. J. *Functional Morphology of the Evolving Hand and Foot* (Clarendon, Oxford, 1989).
59. Lewis, O. J. *Am. J. Phys. Anthrop.* **37,** 13–34 (1972).
60. Kennedy, G. E. *Am. J. Phys. Anthrop.* **61,** 429–434 (1983).
61. Trinkaus, E. *Am. J. Phys. Anthrop.* **64,** 137–139 (1984).
62. McHenry, H. M. & Corruccini, R. S. *Am. J. Phys. Anthrop.* **49,** 473–488 (1978).
63. Wood, B. A. *Nature* **251,** 135–136 (1974).
64. Leakey, R. E., Walker, A., Ward, C. V. & Grausz, H. M. *Proc. 2nd. Int. Congr. Human Palaeontology* (ed. Giacobini, G.) 167–173 (Jaca, Milan, 1989).
65. Aiello, L. *J. hum. Evol.* (in the press).
66. Wood, B. A. & Wilson, G. B. in *Variation, Culture and Evolution in African Populations* (eds. Singer, R. & Lundy, J.) 101–108 (Witwatersrand Univ. Press, Johannesburg, 1986).
67. Trinkaus, E. *Am. J. Phys. Anthrop.* **80,** 411–416 (1989).
68. Hartwig-Schrerer, S. & Martin, R. D. *J. hum. Evol.* **21,** 439–449 (1991).
69. Mayr, E. *Cold Spr. Harb. Symp. quant. Biol.* **15,** 109–118 (1950).
70. Simpson, G. G. *Principles of Animal Taxonomy* (Columbia Univ. Press, New York, 1961).
71. Le Gros Clark, W. E. *The Fossil Evidence for Human Evolution* 1st edn. (Univ. Chicago Press, Chicago, 1955).
72. Robinson, J. T. in *Evolution and Hominisation* 2nd edn. (ed. Kurth, G.) 150–175 (Fischer, Stuttgart, 1968).
73. White, T. D., Johanson, D. C. & Kimbel, W. H. *S. Afr. J. Sci.* **77,** 445–470 (1981).
74. Robinson, J. T. *Nature* **209,** 957–960 (1966).
75. Prost, J. H. *Am. J. Phys. Anthrop.* **52,** 175–189 (1990).
76. Rose, M. D. in *Food Acquisition and Processing in Primates* (eds. Chivers, D. J., Wood. B. A. & Bilsborough, A.) 509–524 (Plenum, New York, 1984).
77. Vandebroek, G. in *Evolution des Vertébrés de leur Origine à l'Homme* 450–518 (Masson, Paris, 1969).
78. Eldredge, N. & Tattersall, I. in *Approaches to Primate Paleobiology* (ed. Szalay, F. S.) 218–242 (Karger, Basel, 1975).
79. Wood, B. A. *Cour. Forsch. Inst. Senckenberg* **69,** 99–111 (1984).
80. Delson, E., Eldredge, N. & Tattersall, I. *J. hum. Evol.* **6,** 263–278 (1977).
81. Wood, B. A. & Chamberlain, A. T. in *Major Topics in Primate and Human Evolution* (eds. Wood, B., Martin, L. & Andrews, P.) 220–248 (Cambridge Univ. Press, Cambridge, 1986).
82. Bonde, N. in *Major Patterns in Vertebrate Evolution* (eds. Hecht, M. K., Goody, P. C. & Hecht, B. M.) 741–804 (Plenum, New York, 1977).
83. Johanson, D. C. & White, T. D. *Science* **202,** 321–330 (1979).
84. Alexeev, V. P. *The Origin of the Human Race* (Progress, Moscow, 1986).
85. Feibel, C. S., Brown, F. H. & McDougall, I. *Am. J. Phys. Anthrop.* **78,** 595–622 (1989).
86. Hartwig-Schrerer, S. *Cour. Forsch. Inst. Senckenberg* (in the press).
87. Tobias, P. V. *Palaeont. afr.* **23,** 1–17 (1980).
88. Delson, E. *Nature* **322,** 496–497 (1986).

89. Groves, C. P. A. *Theory of Human and Primate Evolution* (Clarendon, Oxford, 1989).
90. Leakey, R. E., Leakey, M. G. & Behrensmeyer, A. K. in *Koobi Fora Research Project I: The Fossil Hominids and an Introduction to Their Context, 1968–1974* (eds. Leakey, M. G. & Leakey, R. E.) 86–182 (Clarendon, Oxford, 1978).
91. Falk, D. *Science* **221,** 1072–1074 (1983).
92. Holloway, R. L. in *Hominid Evolution* (ed. Tobias, P. V.) 47–62 (Liss, New York, 1985).
93. Tobias, P. V. *J. hum. Evol.* **16,** 741–761 (1987).

Acknowledgments

My researches on the evidence for early *Homo* from Koobi Fora were at the invitation of R. Leakey, and I thank the Director and Trustees of the National Museums of Kenya for their cooperation. I also thank L. Aiello, A. Chamberlain and C. B. Stringer for comments and discussion, and C. Engleman, P. Guest and R. Read for help with the manuscript. This work was supported by the Leverhulme Trust and the Royal Society.

Henry Gee

AFTERWORD

I hope that this book will give you a flavor of some of the most contentious, exciting and active research areas in modern comparative biology. I also hope that the selection of reviews in this volume have supported my claim, at the beginning, that molecular biology and cladistics have revitalized comparative biology—perhaps placing it, once again, at the heart of biological inquiry.

Those who came to this end of the book in search of easy answers will encounter only this note—they will, therefore, be disappointed. Such is the frustration of science, expressed in that old cliché: that research often raises many more questions than it answers. Managers and policymakers seeking to put a price tag on research fail to understand this phenomenon. Neither do they understand that in this greatest of scientific frustration lies its greatest joy.

I would like to thank all the authors for allowing us to reprint their articles, and for supplying additional information, references and comments on earlier drafts. Two anonymous referees provided additional helpful comments (for even *Nature* editors get to appreciate the necessary forbearance of harassed authors.) Thanks are also due to Ross Sturley and Elizabeth Allen of *Nature,* and Christie Henry of the University of Chicago Press, who steered my pet project to fruition, and to my wife, Penny, who put up with it as it happened.

The Cranley is gone but not forgotten.

INDEX